MODERN QUANTUM CHEMISTRY

Introduction to Advanced Electronic Structure Theory

ATTILA SZABO

Laboratory of Chemical Physics
National Institutes of Health
Bethesda, Maryland

NEIL S. OSTLUND

Hypercube, Inc.
Waterloo, Ontario

D0224007

DOVER PUBLICATIONS, INC.
Mineola, New York

Bibliographical Note

This Dover edition, first published in 1996, is an unabridged, unaltered republication of the "First Edition, Revised," originally published by the McGraw-Hill Publishing Company, New York, 1989. The original edition was published by the Macmillan Publishing Company, New York, 1982.

Library of Congress Cataloging-in-Publication Data

Szabo, Attila, 1947–
  Modern quantum chemistry : introduction to advanced electronic structure theory / Attila Szabo, Neil S. Ostlund.
       p.   cm.
  Previously published: 1st ed., rev. New York : McGraw-Hill, c1989.
  Includes bibliographical references and index.
  ISBN-13: 978-0-486-69186-2 (pbk.)
  ISBN-10: 0-486-69186-1 (pbk.)
  1. Quantum chemistry.   I. Ostlund, Neil S.   II. Title.
[QD462.S95   1996]
541.2'8—dc20   96-10775
                                                               CIP

Manufactured in the United States by Courier Corporation
69186112   2013
www.doverpublications.com

TABLE OF CONTENTS

Preface to Revised Edition   ix

Preface   xi

Chapter 1.   Mathematical Review   1

1.1 Linear Algebra   2
    1.1.1   Three-Dimensional Vector Algebra   *2*
    1.1.2   Matrices   *5*
    1.1.3   Determinants   *7*
    1.1.4   *N*-Dimensional Complex Vector Spaces   *9*
    1.1.5   Change of Basis   *13*
    1.1.6   The Eigenvalue Problem   *15*
    1.1.7   Functions of Matrices   *21*

1.2 Orthogonal Functions, Eigenfunctions, and Operators   24

1.3 The Variation Method   31
    1.3.1   The Variation Principle   *31*
    1.3.2   The Linear Variational Problem   *33*

Notes   38

Further Reading   38

Chapter 2.   Many Electron Wave Functions and Operators   39

2.1 The Electronic Problem   40
    2.1.1   Atomic Units   *41*
    2.1.2   The Born-Oppenheimer Approximation   *43*
    2.1.3   The Antisymmetry or Pauli Exclusion Principle   *45*

2.2 Orbitals, Slater Determinants, and Basis Functions   46
    2.2.1   Spin Orbitals and Spatial Orbitals   *46*
    2.2.2   Hartree Products   *47*

2.2.3   Slater Determinants   *49*
2.2.4   The Hartree-Fock Approximation   *53*
2.2.5   The Minimal Basis H_2 Model   *55*
2.2.6   Excited Determinants   *58*
2.2.7   Form of the Exact Wave Function and Configuration Interaction   *60*

2.3   Operators and Matrix Elements   64
2.3.1   Minimal Basis H_2 Matrix Elements   *64*
2.3.2   Notations for One- and Two-Electron Integrals   *67*
2.3.3   General Rules for Matrix Elements   *68*
2.3.4   Derivation of the Rules for Matrix Elements   *74*
2.3.5   Transition from Spin Orbitals to Spatial Orbitals   *81*
2.3.6   Coulomb and Exchange Integrals   *85*
2.3.7   Pseudo-Classical Interpretation of Determinantal Energies   *87*

2.4   Second Quantization   89
2.4.1   Creation and Annihilation Operators and Their
         Anticommutation Relations   *89*
2.4.2   Second-Quantized Operators and Their Matrix Elements   *95*

2.5   Spin-Adapted Configurations   97
2.5.1   Spin Operators   *97*
2.5.2   Restricted Determinants and Spin-Adapted Configurations   *100*
2.5.3   Unrestricted Determinants   *104*

Notes   107

Further Reading   107

Chapter 3.   The Hartree-Fock Approximation   108

3.1   The Hartree-Fock Equations   111
3.1.1   The Coulomb and Exchange Operators   *112*
3.1.2   The Fock Operator   *114*

3.2   Derivation of the Hartree-Fock Equations   115
3.2.1   Functional Variation   *115*
3.2.2   Minimization of the Energy of a Single Determinant   *117*
3.2.3   The Canonical Hartree-Fock Equations   *120*

3.3   Interpretation of Solutions to the Hartree-Fock Equations   123
3.3.1   Orbital Energies and Koopmans' Theorem   *123*
3.3.2   Brillouin's Theorem   *128*
3.3.3   The Hartree-Fock Hamiltonian   *130*

3.4   Restricted Closed-Shell Hartree-Fock: The Roothaan Equations   131
3.4.1   Closed-Shell Hartree-Fock: Restricted Spin Orbitals   *132*
3.4.2   Introduction of a Basis: The Roothaan Equations   *136*
3.4.3   The Charge Density   *138*
3.4.4   Expression for the Fock Matrix   *140*
3.4.5   Orthogonalization of the Basis   *142*

3.4.6   The SCF Procedure   *145*

3.4.7   Expectation Values and Population Analysis   *149*

3.5   Model Calculations on H_2 and HeH^+ 152

3.5.1   The $1s$ Minimal STO-3G Basis Set   *153*

3.5.2   STO-3G H_2 *159*

3.5.3   An SCF Calculation on STO-3G HeH^+ *168*

3.6   Polyatomic Basis Sets   180

3.6.1   Contracted Gaussian Functions   *180*

3.6.2   Minimal Basis Sets: STO-3G   *184*

3.6.3   Double Zeta Basis Sets: 4-31G   *186*

3.6.4   Polarized Basis Sets: 6-31G* and 6-31G**   *189*

3.7   Some Illustrative Closed-Shell Calculations   190

3.7.1   Total Energies   *191*

3.7.2   Ionization Potentials   *194*

3.7.3   Equilibrium Geometries   *200*

3.7.4   Population Analysis and Dipole Moments   *203*

3.8   Unrestricted Open-Shell Hartree-Fock:
The Pople-Nesbet Equations   205

3.8.1   Open-Shell Hartree Fock: Unrestricted Spin Orbitals   *206*

3.8.2   Introduction of a Basis: The Pople-Nesbet Equations   *210*

3.8.3   Unrestricted Density Matrices   *212*

3.8.4   Expression for the Fock Matrices   *214*

3.8.5   Solution of the Unrestricted SCF Equations   *215*

3.8.6   Illustrative Unrestricted Calculations   *216*

3.8.7   The Dissociation Problem and its Unrestricted Solution   *221*

Notes   229

Further Reading   229

Chapter 4.   Configuration Interaction   231

4.1   Multiconfigurational Wave Functions and the
Structure of the Full CI Matrix   233

4.1.1   Intermediate Normalization and an Expression for
the Correlation Energy   *237*

4.2   Doubly Excited CI   242

4.3   Some Illustrative Calculations   245

4.4   Natural Orbitals and the One-Particle Reduced Density Matrix   252

4.5   The Multiconfiguration Self-Consistent Field (MCSCF)
and Generalized Valence Bond (GVB) Methods   258

4.6   Truncated CI and the Size-Consistency Problem   261

Notes   269

Further Reading   269

Chapter 5.   Pair and Coupled-Pair Theories   271

5.1   The Independent Electron Pair Approximation (IEPA)   272
      5.1.1   Invariance under Unitary Transformations: An Example *277*
      5.1.2   Some Illustrative Calculations *284*

5.2   Coupled-Pair Theories   286
      5.2.1   The Coupled Cluster Approximation (CCA) *287*
      5.2.2   The Cluster Expansion of the Wave Function *290*
      5.2.3   Linear CCA and the Coupled Electron Pair
              Approximation (CEPA) *292*
      5.2.4   Some Illustrative Calculations *296*

5.3   Many-Electron Theories with Single Particle Hamiltonians   297
      5.3.1   The Relaxation Energy via CI, IEPA, CCA, and CEPA *303*
      5.3.2   The Resonance Energy of Polyenes in Hückel Theory *309*

Notes   318

Further Reading   319

Chapter 6.   Many-Body Perturbation Theory   320

6.1   Rayleigh-Schrödinger (RS) Perturbation Theory   322

*6.2 Diagrammatic Representation of RS Perturbation Theory   327
      6.2.1   Diagrammatic Perturbation Theory for 2 States *327*
      6.2.2   Diagrammatic Perturbation Theory for N States *335*
      6.2.3   Summation of Diagrams *336*

6.3   Orbital Perturbation Theory: One-Particle Perturbations   338

*6.4 Diagrammatic Representation of Orbital Perturbation Theory   348

6.5   Perturbation Expansion of the Correlation Energy   350

6.6   The N-Dependence of the RS Perturbation Expansion   354

*6.7 Diagrammatic Representation of the Perturbation
      Expansion of the Correlation Energy   356
      6.7.1   Hugenholtz Diagrams *356*
      6.7.2   Goldstone Diagrams *362*
      6.7.3   Summation of Diagrams *368*
      6.7.4   What Is the Linked Cluster Theorem? *369*

6.8   Some Illustrative Calculations   370

Notes   378

Further Reading   379

Chapter 7. The One-Particle Many-Body Green's Function   380

7.1   Green's Functions in Single Particle Systems   381

7.2   The One-Particle Many-Body Green's Function   387
    7.2.1   The Self-Energy   *389*
    7.2.2   The Solution of the Dyson Equation   *391*

7.3   Application of the Formalism to H_2 and HeH^+ 392

7.4   Perturbation Theory and the Green's Function Method   398

7.5   Some Illustrative Calculations   405

Notes   409

Further Reading   409

Appendix A.   Integral Evaluation with 1s Primitive Gaussians   410

Appendix B.   Two-Electron Self-Consistent-Field Program   417

Appendix C.   Analytic Derivative Methods and Geometry Optimization *by M.C. Zerner*   437

Appendix D.   Molecular Integrals for H_2 as a Function of Bond Length   459

Index   461

Szeretettel a szüleimnek

AS

*To my wonderful father and
the memory of my dear mother*

NSO

PREFACE TO REVISED EDITION

This revised edition differs from its predecessor essentially in three ways. First, we have included an appendix describing the important recent developments that have made the efficient generation of equilibrium geometries almost routine. We are fortunate that M. Zerner agreed to write this since our own recent interests have been channeled in other directions. Second, numerous minor but annoying errors have been corrected. For most of these we are indebted to K. Ohno, T. Sakai and Y. Mochizuki who detected them in the course of preparing the Japanese translation which has recently been published by Tokyo University Press. Finally, we have updated the Further Reading sections of all the chapters. We are extremely pleased by the many favorable comments we have received about our book and we hope that the next generation of readers will find this edition useful.

ATTILA SZABO
NEIL S. OSTLUND

PREFACE

The aim of this graduate textbook is to present and explain, at other than a superficial level, modern *ab initio* approaches to the calculation of the electronic structure and properties of molecules. The first three chapters contain introductory material culminating in a thorough discussion of the Hartree-Fock approximation. The remaining four chapters describe a variety of more sophisticated approaches, which improve upon this approximation.

We have strived to make advanced topics understandable to the beginning graduate student. Our goal was to teach more than cocktail party jargon; we wanted to give insight into the nature and validity of a variety of approximate formalisms and improve the reader's problem-solving skills. Needless to say, a book of this size cannot cover all of quantum chemistry; it would be easy to write down a long list of important topics which we have not covered. Nevertheless, we believe that anyone who has mastered the material in this book will have a solid foundation to build upon, and will be able to appreciate most research papers and seminars dealing with electronic structure theory and its applications.

The origins of this book go back to our student days when we were trying to understand a variety of quantum chemical formalisms that at first glance appeared to be extremely complicated. We found that by applying such formalisms to simple, analytically tractable, model systems we not only got a feeling as to what was involved in performing actual calculations but also gained insight into the interrelationships among different approximation schemes. The models we used then, and ones we devised or learned subsequently, play an important role in the pedagogical approach we adopt in this book. The writing began in 1976 when we jointly taught a special topics course at Indiana University. The manuscript gradually evolved in response to the needs and reactions of students enrolled in a second-semester quantum chemistry course we taught some five times at both Indiana University and the University of Arkansas.

An important feature of this book is that over 150 exercises are embedded in the body of the text. These problems were designed to help the reader acquire a working knowledge of the material. The level of difficulty has been kept reasonably constant by breaking up the more complicated ones into manageable parts. Much of the value of this book will be missed if the exercises are ignored.

In the following, we briefly describe some of the highlights of the seven chapters. Chapter 1 reviews the mathematics (mostly matrix algebra) required for the rest of the book. It is self-contained and suited for self-study. Its presence in this book is dictated by the deficiency in mathematics of most chemistry graduate students. The pedagogical strategy we use here, and in much of the book, is to begin with a simple example that illustrates most of the essential ideas and then gradually generalize the formalism to handle more complicated situations.

Chapter 2 introduces the basic techniques, ideas, and notations of quantum chemistry. A preview of Hartree-Fock theory and configuration interaction is used to motivate the study of Slater determinants and the evaluation of matrix elements between such determinants. A simple model system (minimal basis H_2) is introduced to illustrate the development. This model and its many-body generalization (N independent H_2 molecules) reappear in all subsequent chapters to illuminate the formalism. Although not essential for the comprehension of the rest of the book, we also present here a self-contained discussion of second quantization.

Chapter 3 contains a thorough discussion of the Hartree-Fock approximation. A unique feature of this chapter is a detailed illustration of the computational aspects of the self-consistent-field procedure for minimal basis HeH^+. Using the output of a simple computer program, listed in Appendix B, the reader is led iteration-by-iteration through an *ab initio* calculation. This chapter also describes the results of Hartree-Fock calculations on a standard set of simple molecules using basis sets of increasing sophistication. We performed most of these calculations ourselves, and in later chapters we use these same molecules and basis sets to show how the Hartree-Fock results are altered when more sophisticated approaches are used. Thus we illustrate the influence of both the quality of the one-electron basis set and the sophistication of the quantum chemical method on calculated results. In this way we hope to give the reader a feeling for the kind of accuracy that can be expected from a particular calculation.

Chapter 4 discusses configuration interaction (CI) and is the first of the four chapters that deal with approaches incorporating electron correlation. One-electron density matrices, natural orbitals, the multiconfiguration self-consistent-field approximation, and the generalized valence bond method are discussed from an elementary point of view. The size-consistency problem associated with truncated CI is illustrated using a model consisting of N independent hydrogen molecules. This example highlights the need for so-called many-body approaches, which do not suffer from this deficiency, that are discussed in subsequent chapters.

Chapter 5 describes the independent electron pair approximation and a variety of more sophisticated approaches that incorporate coupling between pairs. Since this chapter contains some of the most advanced material in the book, many illustrative examples are included. In the second half of the chapter, as a pedagogical device, we consider the application of many-electron approaches to an N-electron system described by a Hamiltonian containing only single particle interactions. This problem can be solved exactly in an elementary way. However, by seeing how "high-powered" approaches work in such a simple context, the student can gain insight into the nature of these approximations.

Chapter 6 considers the perturbative approach to the calculation of the correlation energy of many-electron systems. A novel pedagogical approach allows the reader to acquire a working knowledge of diagrammatic perturbation theory surprisingly quickly. Although the chapter is organized so that the sections on diagrams (which are starred) can be skipped without loss of continuity, we find that the diagrammatic approach is fun to teach and is extremely well received by students.

Chapter 7 contains a brief introduction to the use of the one-particle many-body Green's function in quantum chemistry. Our scope is restricted to discussing ionization potentials and electron affinities. The chapter is directed towards a reader having no knowledge of second quantization or Green's functions, even in a simple context.

This book is largely self-contained and, in principle, requires no prerequisite other than a solid undergraduate physical chemistry course. However, exposure to quantum chemistry at the level of the text by I. N. Levine (*Quantum Chemistry*, Allyn and Bacon) will definitely enhance the student's appreciation of the subject material. We would normally expect the present text to be used for the second semester of a two-semester sequence on quantum chemistry. It is also suitable for a special topics course. There is probably too much material in the book to be taught in-depth in a single semester. For students with average preparation, we suggest covering the first four chapters and then discussing any one of the last three, which are essentially independent. Our preferred choice is Chapter 6. For an exceptionally well-prepared class, the major fraction of the semester could be spent on the last four chapters. We have found that a course based on this text can be enriched in a number of ways. For example, it is extremely helpful for students to perform their own numerical calculations using, say, the Gaussian 80 system of programs. In addition, recent papers on the applications of electronic structure theory can be assigned at the beginning of the course and the students asked to give short in-class presentations on one or more of such papers at the end of the course.

We have placed special emphasis on using a consistent notation throughout the book. Since quantum chemists use a number of different notations, it is appropriate to define the notation we have adopted. Spatial molecular orbitals (with latin indices $i, j, k \ldots$) are denoted by ψ. These are usually expanded in a set of spatial (atomic) basis functions (with greek indices $\mu, \nu, \lambda, \ldots$) denoted by ϕ. Molecular spin orbitals are denoted by χ. Occupied molecular orbitals are specifically labeled by $a, b, c, \ldots$ and unoccupied (virtual) molecular orbitals are specifically labeled by $r, s, t, \ldots$ Many-electron operators are denoted by capital script letters (for example, the Hamiltonian is $\mathscr{H}$), and one-electron operators are denoted by lower case latin letters (for example, the Fock operator for electron-one is $f(1)$). The exact many-electron wave function is denoted by Φ, and we use Ψ to denote approximate many-electron wave functions (i.e., the Hartree-Fock ground state wave function is Ψ_0, while Ψ_{ab}^{rs} is a doubly excited wave function). Exact and approximate energies are denoted by $\mathscr{E}$ and E, respectively. All numerical quantities (energies, dipole moments, etc.) are given in atomic units.

It is a pleasure to thank those people who helped us with scientific aspects of this book. J. A. Pople has always been a source of inspiration. M. Karplus and W. Reinhardt introduced us to many-body theory. J.-P. Malrieu and K. Schulten independently showed us in the early 1970s how to illustrate the lack of size consistency of doubly excited configuration interaction using the model of N noninteracting H_2 molecules. This led to our extensive use of this model. M. Zerner constructively criticized the entire manuscript. The following colleagues gave advice, performed calculations just for the book, or assisted us in some other useful way: R. Bartlett, J. Binkley, M. Bowen, R. Counts, C. Dykstra, W. Eaton, D. Freeman, W. Goddard III, S. Hagstrom, A. Hernandez, D. Merrifield, J. Neece, I. Shavitt, and R. Whiteside. The following students helped us clarify our presentation and eliminate errors: B. Basinger, G. Caldwell, T. Croxton, R. Farren, R. Feeney, M. Flanagan, J. Freeze, V. Hess, K. Holtzclaw, J. Joens, J. Johnson, R. Jones, J. Kehayias, G. Lipari, D. Lupo, D. McMullen, J. Meek, S. Munchak, N. Peyghambarian, R. Russo, W. Schinzer, D. Shoup, B. Stone, E. Tsang and I. Waight. Of all the secretaries who typed various versions of the manuscript, K. Wagner, M. Mabie, and L. Ferro were the most helpful. J. Hawkins prepared the illustrations.

ATTILA SZABO
NEIL S. OSTLUND

CHAPTER ONE

MATHEMATICAL
REVIEW

This chapter provides the necessary mathematical background for the rest of the book. The most important mathematical tool used in quantum chemistry is matrix algebra. We have directed this chapter towards the reader who has some familiarity with matrices but who has not used them in some time and is anxious to acquire a working knowledge of linear algebra. Those with strong mathematical backgrounds can merely skim the material to acquaint themselves with the various notations we use. Our development is informal and rigour is sacrificed for the sake of simplicity. To help the reader develop those often neglected, but important, manipulative skills we have included carefully selected exercises within the body of the text. The material cannot be mastered without doing these simple problems.

In Section 1.1 we present the elements of linear algebra by gradually generalizing the ideas encountered in three-dimensional vector algebra. We consider matrices, determinants, linear operators and their matrix representations, and, most importantly, how to find the eigenvalues and eigenvectors of certain matrices. We introduce the very clever notation of Dirac, which expresses our results in a concise and elegant manner. This notation is extremely useful because it allows one to manipulate matrices and derive various theorems painlessly. Moreover, it highlights similarities between linear algebra and the theory of complete sets of orthonormal functions as will be seen in Section 1.2. Finally, in Section 1.3 we consider one of the cornerstones of quantum chemistry namely, the variation principle.

1.1 LINEAR ALGEBRA

We begin our discussion of linear algebra by reviewing three-dimensional vector algebra. The pedagogical strategy we use here, and in much of the book, is to start with the simplest example that illustrates the essential ideas and then gradually generalize the formalism to handle more complicated situations.

1.1.1 Three-Dimensional Vector Algebra

A three-dimensional vector can be represented by specifying its components a_i, $i = 1, 2, 3$ with respect to a set of three mutually perpendicular unit vectors $\{\vec{e}_i\}$ as

$$\vec{a} = \vec{e}_1 a_1 + \vec{e}_2 a_2 + \vec{e}_3 a_3 = \sum_i \vec{e}_i a_i \qquad (1.1)$$

The vectors $\vec{e}_i$ are said to form a *basis*, and are called *basis vectors*. The basis is complete in the sense that any three-dimensional vector can be written as a linear combination of the basis vectors. However, a basis is not unique; we could have chosen three different mutually perpendicular unit vectors, $\vec{\varepsilon}_i$, $i = 1, 2, 3$ and represented $\vec{a}$ as

$$\vec{a} = \vec{\varepsilon}_1 a'_1 + \vec{\varepsilon}_2 a'_2 + \vec{\varepsilon}_3 a'_3 = \sum_i \vec{\varepsilon}_i a'_i \qquad (1.2)$$

Given a basis, a vector is completely specified by its three components with respect to that basis. Thus we can represent the vector $\vec{a}$ by a *column matrix* as

$$\mathbf{a} = \begin{pmatrix} a_1 \\ a_2 \\ a_3 \end{pmatrix} \quad \text{in the basis } \{\vec{e}_i\} \qquad (1.3a)$$

or as

$$\mathbf{a}' = \begin{pmatrix} a'_1 \\ a'_2 \\ a'_3 \end{pmatrix} \quad \text{in the basis } \{\vec{\varepsilon}_i\} \qquad (1.3b)$$

The *scalar* or *dot product* of two vectors $\vec{a}$ and $\vec{b}$ is defined as

$$\vec{a} \cdot \vec{b} = a_1 b_1 + a_2 b_2 + a_3 b_3 = \sum_i a_i b_i \qquad (1.4)$$

Note that

$$\vec{a} \cdot \vec{a} = a_1^2 + a_2^2 + a_3^2 \equiv |\vec{a}|^2 \qquad (1.5)$$

is simply the square of the length ($|\vec{a}|$) of the vector $\vec{a}$. Let us evaluate the scalar product $\vec{a} \cdot \vec{b}$ using Eq. (1.1)

$$\vec{a} \cdot \vec{b} = \sum_i \sum_j \vec{e}_i \cdot \vec{e}_j a_i b_j \qquad (1.6)$$

For this to be identical to the definition of (1.4), we must have

$$\vec{e}_i \cdot \vec{e}_j = \delta_{ij} = \delta_{ji} = \begin{cases} 1 & \text{if } i = j \\ 0 & \text{otherwise} \end{cases} \tag{1.7}$$

where we have introduced the Kronecker delta symbol δ_{ij}. This relation is the mathematical way of saying that the basis vectors are mutually perpendicular (orthogonal) and have unit length (normalized); in other words they are *orthonormal*.

Given a vector $\vec{a}$, we can find its component along $\vec{e}_j$ by taking the scalar product of Eq. (1.1) with $\vec{e}_j$ and using the orthonormality relation (1.7)

$$\vec{e}_j \cdot \vec{a} = \sum_i \vec{e}_j \cdot \vec{e}_i a_i = \sum_i \delta_{ij} a_i = a_j \tag{1.8}$$

Hence we can rewrite Eq. (1.1) as

$$\vec{a} = \sum_i \vec{e}_i \vec{e}_i \cdot \vec{a} = \vec{1} \cdot \vec{a} \tag{1.9}$$

where

$$\vec{1} = \sum_i \vec{e}_i \vec{e}_i \tag{1.10}$$

is the unit *dyadic*. A dyadic is an entity which when dotted into a vector gives another vector. The unit dyadic gives the same vector back. Equation (1.10) is called the *completeness relation* for the basis $\{\vec{e}_i\}$ since it is an alternate form of Eq. (1.1), which states that any vector $\vec{a}$ can be written as a linear combination of the basis vectors $\{\vec{e}_i\}$.

We now define an *operator* $\mathcal{O}$ as an entity which when acting on a vector $\vec{a}$ converts it into a vector $\vec{b}$

$$\mathcal{O}\vec{a} = \vec{b} \tag{1.11}$$

The operator is said to be *linear* if for any numbers x and y

$$\mathcal{O}(x\vec{a} + y\vec{b}) = x\mathcal{O}\vec{a} + y\mathcal{O}\vec{b} \tag{1.12}$$

A linear operator is completely determined if its effect on every possible vector is known. Because any vector can be expressed in terms of the basis $\{\vec{e}_i\}$, it is sufficient to know what $\mathcal{O}$ does to the basis vectors. Now since $\mathcal{O}\vec{e}_i$ is a vector, it can be written as a linear combination of the basis vectors $\{\vec{e}_i\}$, i.e.,

$$\mathcal{O}\vec{e}_i = \sum_{j=1}^{3} \vec{e}_j O_{ji}, \quad i = 1, 2, 3 \tag{1.13}$$

The number O_{ji} is the component of the vector $\mathcal{O}\vec{e}_i$ along $\vec{e}_j$. The nine numbers O_{ji}, $i, j = 1, 2, 3$ can be arranged in a two-dimensional array called a *matrix* as

$$\mathbf{O} = \begin{pmatrix} O_{11} & O_{12} & O_{13} \\ O_{21} & O_{22} & O_{23} \\ O_{31} & O_{32} & O_{33} \end{pmatrix} \tag{1.14}$$

We say that **O** is the *matrix representation* of the operator $\mathscr{O}$ in the basis $\{\vec{e}_i\}$. The matrix **O** completely specifies how the operator $\mathscr{O}$ acts on an arbitrary vector since this vector can be expressed as a linear combination of the basis vectors $\{\vec{e}_i\}$ and we know what $\mathscr{O}$ does to each of these basis vectors.

Exercise 1.1 a) Show that $O_{ij} = \vec{e}_i \cdot \mathscr{O}\vec{e}_j$. b) If $\mathscr{O}\vec{a} = \vec{b}$ show that $b_i = \sum_j O_{ij} a_j$.

If **A** and **B** are the matrix representations of the operators $\mathscr{A}$ and $\mathscr{B}$, the matrix representation of the operator $\mathscr{C}$, which is the product of $\mathscr{A}$ and $\mathscr{B}$ ($\mathscr{C} = \mathscr{A}\mathscr{B}$), can be found as follows:

$$\mathscr{C}\vec{e}_j = \sum_i \vec{e}_i C_{ij}$$

$$= \mathscr{A}\mathscr{B}\,\vec{e}_j$$

$$= \mathscr{A}\sum_k \vec{e}_k B_{kj}$$

$$= \sum_{ik} \vec{e}_i A_{ik} B_{kj} \tag{1.15}$$

so that

$$C_{ij} = \sum_k A_{ik} B_{kj} \tag{1.16}$$

which is the definition of matrix multiplication and hence

$$\mathbf{C = AB} \tag{1.17}$$

Thus if we define matrix multiplication by (1.16), then the matrix representation of the product of two operators is just the product of their matrix representations.

The order in which two operators or two matrices are multiplied is crucial. In general $\mathscr{A}\mathscr{B} \neq \mathscr{B}\mathscr{A}$ or $\mathbf{AB} \neq \mathbf{BA}$. That is, two operators or matrices do not necessarily *commute*. For future reference, we define the *commutator* of two operators or matrices as

$$[\mathscr{A}, \mathscr{B}] = \mathscr{A}\mathscr{B} - \mathscr{B}\mathscr{A} \tag{1.18a}$$

$$[\mathbf{A}, \mathbf{B}] = \mathbf{AB} - \mathbf{BA} \tag{1.18b}$$

and their *anticommutator* as

$$\{\mathscr{A}, \mathscr{B}\} = \mathscr{A}\mathscr{B} + \mathscr{B}\mathscr{A} \tag{1.19a}$$

$$\{\mathbf{A}, \mathbf{B}\} = \mathbf{AB} + \mathbf{BA} \tag{1.19b}$$

Exercise 1.2 Calculate $[\mathbf{A}, \mathbf{B}]$ and $\{\mathbf{A}, \mathbf{B}\}$ when

$$\mathbf{A} = \begin{pmatrix} 1 & 1 & 0 \\ 1 & 2 & 2 \\ 0 & 2 & -1 \end{pmatrix} \qquad \mathbf{B} = \begin{pmatrix} 1 & -1 & 1 \\ -1 & 0 & 0 \\ 1 & 0 & 1 \end{pmatrix}$$

1.1.2 Matrices

Now that we have seen how 3×3 matrices naturally arise in three-dimensional vector algebra and how they are multiplied, we shall generalize these results. A set of numbers $\{A_{ij}\}$ that are in general complex and have ordered subscripts $i = 1, 2, \ldots, N$ and $j = 1, 2, \ldots, M$ can be considered elements of a rectangular $(N \times M)$ matrix $\mathbf{A}$ with N rows and M columns

$$\mathbf{A} = \begin{pmatrix} A_{11} & A_{12} & \cdots & A_{1M} \\ A_{21} & A_{22} & \cdots & A_{2M} \\ \vdots & \vdots & & \vdots \\ A_{N1} & A_{N2} & \cdots & A_{NM} \end{pmatrix} \tag{1.20}$$

If $N = M$ the matrix is square. When the number of columns in the $N \times M$ matrix $\mathbf{A}$ is the same as the number of rows in the $M \times P$ matrix $\mathbf{B}$, then $\mathbf{A}$ and $\mathbf{B}$ can be multiplied to give a $N \times P$ matrix $\mathbf{C}$

$$\mathbf{C} = \mathbf{AB} \tag{1.21}$$

where the elements of $\mathbf{C}$ are given by the matrix multiplication rule

$$C_{ij} = \sum_{k=1}^{M} A_{ik} B_{kj} \qquad \begin{array}{l} i = 1, \ldots, N \\ j = 1, \ldots, P \end{array} \tag{1.22}$$

The set of M numbers $\{a_i\}$ $i = 1, 2, \ldots, M$ can similarly be considered elements of a column matrix

$$\mathbf{a} = \begin{pmatrix} a_1 \\ a_2 \\ \vdots \\ a_M \end{pmatrix} \tag{1.23}$$

Note that for an $N \times M$ matrix $\mathbf{A}$

$$\mathbf{Aa} = \mathbf{b} \tag{1.24}$$

where $\mathbf{b}$ is a column matrix with N elements

$$b_i = \sum_{j=1}^{M} A_{ij} a_j \qquad i = 1, 2, \ldots, N \tag{1.25}$$

We now introduce some important definitions. The *adjoint* of an $N \times M$ matrix $\mathbf{A}$, denoted by $\mathbf{A}^\dagger$, is an $M \times N$ matrix with elements

$$(\mathbf{A}^\dagger)_{ij} = A_{ji}^* \tag{1.26}$$

i.e., we take the complex conjugate of each of the matrix elements of $\mathbf{A}$ and interchange rows and columns. If the elements of $\mathbf{A}$ are real, then the adjoint of $\mathbf{A}$ is called the *transpose* of $\mathbf{A}$.

Exercise 1.3 If $\mathbf{A}$ is an $N \times M$ matrix and $\mathbf{B}$ is a $M \times K$ matrix show that $(\mathbf{AB})^\dagger = \mathbf{B}^\dagger \mathbf{A}^\dagger$.

The adjoint of a column matrix is a *row matrix* containing the complex conjugates of the elements of the column matrix.

$$\mathbf{a}^\dagger = (a_1^* a_2^* \cdots a_M^*) \tag{1.27}$$

If $\mathbf{a}$ and $\mathbf{b}$ are two column matrices with M elements, then

$$\mathbf{a}^\dagger \mathbf{b} = (a_1^* a_2^* \cdots a_M^*) \begin{pmatrix} b_1 \\ b_2 \\ \vdots \\ b_M \end{pmatrix} = \sum_{i=1}^{M} a_i^* b_i \tag{1.28}$$

Note that if the elements of $\mathbf{a}$ and $\mathbf{b}$ are real and M is 3, this is simply the scalar product of two vectors, c.f. Eq. (1.4). Taking the adjoint of Eq. (1.24) and using the result of Exercise 1.3 we have

$$\mathbf{b}^\dagger = \mathbf{a}^\dagger \mathbf{A}^\dagger \tag{1.29}$$

where $\mathbf{b}^\dagger$ is a row matrix with the N elements

$$b_i^* = \sum_{j=1}^{M} a_j^* (\mathbf{A}^\dagger)_{ji} = \left(\sum_{j=1}^{M} a_j A_{ij} \right)^* \qquad i = 1, 2, \ldots, N \tag{1.30}$$

Note that Eq. (1.30) is simply the complex conjugate of Eq. (1.25).

We now give certain definitions and properties applicable only to *square* matrices:

1. A matrix $\mathbf{A}$ is *diagonal* if all its off-diagonal elements are zero

$$A_{ij} = A_{ii} \delta_{ij}. \tag{1.31}$$

2. The *trace* of the matrix $\mathbf{A}$ is the sum of its diagonal elements

$$\text{tr } \mathbf{A} = \sum_i A_{ii}. \tag{1.32}$$

3. The *unit* matrix is defined by

$$\mathbf{1A} = \mathbf{A1} = \mathbf{A} \tag{1.33}$$

for any matrix **A**, and has elements

$$(\mathbf{1})_{ij} = \delta_{ij}. \tag{1.34}$$

4. The *inverse* of the matrix **A**, denoted by $\mathbf{A}^{-1}$, is a matrix such that

$$\mathbf{A}^{-1}\mathbf{A} = \mathbf{A}\mathbf{A}^{-1} = \mathbf{1}. \tag{1.35}$$

5. A *unitary* matrix **A** is one whose inverse is its adjoint

$$\mathbf{A}^{-1} = \mathbf{A}^{\dagger}. \tag{1.36}$$

A real unitary matrix is called *orthogonal*.

6. A *Hermitian* matrix is self-adjoint, i.e.,

$$\mathbf{A}^{\dagger} = \mathbf{A} \tag{1.37a}$$

or

$$A_{ji}^{*} = A_{ij} \tag{1.37b}$$

A real Hermitian matrix is called *symmetric*.

Exercise 1.4 Show that

a. $\operatorname{tr} \mathbf{AB} = \operatorname{tr} \mathbf{BA}$.
b. $(\mathbf{AB})^{-1} = \mathbf{B}^{-1}\mathbf{A}^{-1}$.
c. If **U** is unitary and $\mathbf{B} = \mathbf{U}^{\dagger}\mathbf{AU}$, then $\mathbf{A} = \mathbf{UBU}^{\dagger}$.
d. If the product $\mathbf{C} = \mathbf{AB}$ of two Hermitian matrices is also Hermitian, then **A** and **B** commute.
e. If **A** is Hermitian then $\mathbf{A}^{-1}$, if it exists, is also Hermitian.
f. If $\mathbf{A} = \begin{pmatrix} A_{11} & A_{12} \\ A_{21} & A_{22} \end{pmatrix}$, then $\mathbf{A}^{-1} = \dfrac{1}{(A_{11}A_{22} - A_{12}A_{21})}\begin{pmatrix} A_{22} & -A_{12} \\ -A_{21} & A_{11} \end{pmatrix}$.

1.1.3 Determinants

We now define and give some properties of the determinant of a square matrix **A**. Recall that a *permutation* of the numbers $1, 2, 3, \ldots, N$ is simply a way of ordering these numbers, and there are $N!$ distinct permutations of N numbers. The *determinant* of an $N \times N$ matrix **A** is a *number* obtained by the prescription

$$\det(\mathbf{A}) = |\mathbf{A}| = \begin{vmatrix} A_{11} & \cdots & A_{1N} \\ \vdots & & \vdots \\ A_{N1} & \cdots & A_{NN} \end{vmatrix} = \sum_{i=1}^{N!} (-1)^{p_i}\mathscr{P}_i A_{11}A_{22}\cdots A_{NN} \tag{1.38}$$

where $\mathscr{P}_i$ is a permutation operator that permutes the column indices $1, 2, 3, \ldots, N$ and the sum runs over all $N!$ permutations of the indices; p_i

is the number of transpositions required to restore a given permutation $i_1, i_2, \ldots, i_N$ to natural order $1, 2, 3, \ldots, N$. Note that it is important only whether p_i is an even or odd number. As an illustration we evaluate the determinant of a 2×2 matrix $\mathbf{A}$.

$$\mathbf{A} = \begin{pmatrix} A_{11} & A_{12} \\ A_{21} & A_{22} \end{pmatrix}$$

There are two permutations of the column indices 1 and 2, i.e.,

$$1\ 2\ (p_1 = 0)$$
$$2\ 1\ (p_2 = 1)$$

hence

$$\begin{vmatrix} A_{11} & A_{12} \\ A_{21} & A_{22} \end{vmatrix} = (-1)^0 A_{11} A_{22} + (-1)^1 A_{12} A_{21} = A_{11} A_{22} - A_{12} A_{21} \quad (1.39)$$

Some important properties of determinants which follow from the above definition are:

1. If each element in a row or in a column is zero the value of the determinant is zero.
2. If $(\mathbf{A})_{ij} = A_{ii} \delta_{ij}$, then $|\mathbf{A}| = \prod_i A_{ii} = A_{11} A_{22} \ldots A_{NN}$.
3. A single interchange of any two rows (or columns) of a determinant changes its sign.
4. $|\mathbf{A}| = (|\mathbf{A}^\dagger|)^*$
5. $|\mathbf{AB}| = |\mathbf{A}|\,|\mathbf{B}|$.

Exercise 1.5 Verify the above properties for 2×2 determinants.

Exercise 1.6 Using properties (1)–(5) prove that in general

6. If any two rows (or columns) of a determinant are equal, the value of the determinant is zero.
7. $|\mathbf{A}^{-1}| = (|\mathbf{A}|)^{-1}$.
8. If $\mathbf{AA}^\dagger = \mathbf{1}$, then $|\mathbf{A}|(|\mathbf{A}|)^* = 1$.
9. If $\mathbf{U}^\dagger \mathbf{OU} = \mathbf{\Omega}$ and $\mathbf{U}^\dagger \mathbf{U} = \mathbf{UU}^\dagger = \mathbf{1}$, then $|\mathbf{O}| = |\mathbf{\Omega}|$.

Exercise 1.7 Using Eq. (1.39), note that the inverse of a 2×2 matrix $\mathbf{A}$ obtained in Exercise 1.4f can be written as

$$\mathbf{A}^{-1} = \frac{1}{|\mathbf{A}|} \begin{pmatrix} A_{22} & -A_{12} \\ -A_{21} & A_{11} \end{pmatrix}$$

and thus A^{-1} does not exist when $|A| = 0$. This result holds in general for $N \times N$ matrices. Show that the equation

$$Ac = 0$$

where A is an $N \times N$ matrix and c is a column matrix with elements c_i, $i = 1, 2, \ldots, N$ can have a nontrivial solution ($c \neq 0$) only when $|A| = 0$.

For a 2×2 determinant it is easy to verify by direct calculation that

$$\begin{vmatrix} c_1 B_{11} + c_2 B_{12} & A_{12} \\ c_1 B_{21} + c_2 B_{22} & A_{22} \end{vmatrix} = c_1 \begin{vmatrix} B_{11} & A_{12} \\ B_{21} & A_{22} \end{vmatrix} + c_2 \begin{vmatrix} B_{12} & A_{12} \\ B_{22} & A_{22} \end{vmatrix}$$

This result is a special case of the following property of determinants that we shall use several times in the book.

$$
\begin{vmatrix}
A_{11} & A_{12} & \cdots & \sum_{k=1}^{M} c_k B_{1k} & \cdots & A_{1N} \\
A_{21} & A_{22} & \cdots & \sum_{k=1}^{M} c_k B_{2k} & \cdots & A_{2N} \\
\vdots & \vdots & & \vdots & & \vdots \\
A_{N1} & A_{N2} & \cdots & \sum_{k=1}^{M} c_k B_{Nk} & \cdots & A_{NN}
\end{vmatrix}
$$

$$
= \sum_{k=1}^{M} c_k
\begin{vmatrix}
A_{11} & A_{12} & \cdots & B_{1k} & \cdots & A_{1N} \\
A_{21} & A_{22} & \cdots & B_{2k} & \cdots & A_{2N} \\
\vdots & \vdots & & \vdots & & \vdots \\
A_{N1} & A_{N2} & \cdots & B_{Nk} & \cdots & A_{NN}
\end{vmatrix}
\tag{1.40}
$$

A similar result holds for rows.

1.1.4 *N*-Dimensional Complex Vector Spaces

We need to generalize the ideas of three-dimensional vector algebra to an N-dimensional space in which the vectors can be complex. We will use the powerful notation introduced by Dirac, which expresses our results in an exceedingly concise and simple manner. In analogy to the basis $\{\tilde{e}_i\}$ in three dimensions, we consider N basis vectors denoted by the symbol $|i\rangle$, $i = 1, 2, \ldots, N$, which are called *ket vectors* or simply *kets*. We assume this basis is complete so that any ket vector $|a\rangle$ can be written as

$$|a\rangle = \sum_{i=1}^{N} |i\rangle a_i \tag{1.41}$$

This is a simple generalization of Eq. (1.1) rewritten in our new notation.

After we specify a basis, we can completely describe our vector $|a\rangle$ by giving its N components a_i, $i = 1, 2, \ldots, N$ with respect to the basis $\{|i\rangle\}$. Just as before, we arrange these numbers in a column matrix $\mathbf{a}$ as

$$\mathbf{a} = \begin{pmatrix} a_1 \\ a_2 \\ \vdots \\ a_N \end{pmatrix} \tag{1.42}$$

and we say that $\mathbf{a}$ is the matrix representation of the abstract vector $|a\rangle$ in the basis $\{|i\rangle\}$. Recall (Eq. (1.27)) that the adjoint of the column matrix $\mathbf{a}$ is the row matrix $\mathbf{a}^\dagger$

$$\mathbf{a}^\dagger = (a_1^* a_2^* \cdots a_N^*) \tag{1.43}$$

Now we introduce an abstract *bra vector* $\langle a|$ whose matrix representation is $\mathbf{a}^\dagger$. The scalar product between a bra $\langle a|$ and a ket $|b\rangle$ is defined as

$$\langle a||b\rangle \equiv \langle a|b\rangle = \mathbf{a}^\dagger \mathbf{b} = (a_1^* a_2^* \cdots a_N^*) \begin{pmatrix} b_1 \\ b_2 \\ \vdots \\ b_N \end{pmatrix} = \sum_{i=1}^{N} a_i^* b_i \tag{1.44}$$

which is the natural generalization of the scalar product defined in Eq. (1.4). The unusual names bra (for $\langle \ |$) and ket (for $| \ \rangle$) were chosen because the notation for the scalar product ($\langle \ | \ \rangle$) looks like a bra-c-ket. Note that

$$\langle a|a\rangle = \sum_{i=1}^{N} a_i^* a_i = \sum_{i=1}^{N} |a_i|^2 \tag{1.45}$$

is always real and positive and is just the generalization of the square of the length of a three-dimensional vector. In analogy to Eq. (1.41) it is natural to introduce a bra basis $\{\langle i|\}$ that is complete in the sense that any bra $\langle a|$ can be written as a linear combination of the bra basis vectors as

$$\langle a| = \sum_i a_i^* \langle i| \tag{1.46}$$

The scalar product between $\langle a|$ and $|b\rangle$ now becomes

$$\langle a|b\rangle = \sum_{ij} a_i^* \langle i|j\rangle b_j$$

For this to be identical to our definition (1.44) of the scalar product we must have that

$$\langle i|j\rangle = \delta_{ij} \tag{1.47}$$

which is a statement of the orthonormality of the basis and is a generalization of Eq. (1.7). In summary, a ket vector $|a\rangle$ is represented by a column matrix **a**, a bra vector $\langle b|$ is represented by a row matrix $\mathbf{b}^\dagger$, and their scalar product is just the matrix product of their representations.

We now ask, given a ket $|a\rangle$ or a bra $\langle a|$, how can we determine its components with respect to the basis $\{|i\rangle\}$ or $\{\langle i|\}$? We proceed in complete analogy to the three-dimensional case (c.f. Eq. (1.8)). We multiply Eq. (1.41) by $\langle j|$ on the left and Eq. (1.46) by $|j\rangle$ on the right and obtain

$$\langle j|a\rangle = \sum_i \langle j|i\rangle a_i = \sum_i \delta_{ji}a_i = a_j \tag{1.48a}$$

and

$$\langle a|j\rangle = \sum_i a_i^*\langle i|j\rangle = \sum_i a_i^*\delta_{ij} = a_j^* \tag{1.48b}$$

The expression "multiplying by $\langle j|$ on the left" is a shorthand way of saying "taking the scalar product with $\langle j|$." Note that

$$\langle j|a\rangle = (\langle a|j\rangle)^* = \langle a|j\rangle^* \tag{1.49}$$

Using these results we can rewrite Eqs. (1.41) and (1.46) as

$$|a\rangle = \sum_i |i\rangle a_i = \sum_i |i\rangle\langle i|a\rangle \tag{1.50a}$$

and

$$\langle a| = \sum_i a_i^*\langle i| = \sum_i \langle a|i\rangle\langle i| \tag{1.50b}$$

which suggests we write

$$1 = \sum_i |i\rangle\langle i| \tag{1.51}$$

which is the analogue of Eq. (1.10) and is a statement of the completeness of the basis. We will find that multiplying by unity and using Eq. (1.51) is an extremely powerful way of deriving many relations.

In analogy to Eq. (1.11), we define an operator $\mathcal{O}$ as an entity which when acting on a ket $|a\rangle$ converts it into a ket $|b\rangle$.

$$\mathcal{O}|a\rangle = |b\rangle \tag{1.52}$$

As before, the operator is completely determined if we know what it does to the basis $\{|i\rangle\}$:

$$\mathcal{O}|i\rangle = \sum_j |j\rangle O_{ji} = \sum_j |j\rangle(\mathbf{O})_{ji} \tag{1.53}$$

so that **O** is the matrix representation of the operator $\mathcal{O}$ in the basis $\{|i\rangle\}$. Multiplying (1.53) on the left by $\langle k|$, we have

$$\langle k|\mathcal{O}|i\rangle = \sum_j \langle k|j\rangle(\mathbf{O})_{ji} = \sum_j \delta_{kj}(\mathbf{O})_{ji} = (\mathbf{O})_{ki} \tag{1.54}$$

which provides a useful expression for the matrix elements of **O**. It should be noted that we can easily get the matrix representation of $\mathcal{O}$ by using the completeness relation (1.51) as follows

$$\mathcal{O}|i\rangle = 1\mathcal{O}|i\rangle = \sum_j |j\rangle\langle j|\mathcal{O}|i\rangle \tag{1.55}$$

which upon comparison to Eq. (1.53) yields

$$\langle j|\mathcal{O}|i\rangle = (\mathbf{O})_{ji} = O_{ji} \tag{1.56}$$

As another illustration of the use of the completeness relation and the built-in consistency and simplicity of Dirac notation, let us find the matrix representation of the operator $\mathcal{C} = \mathcal{A}\mathcal{B}$ in terms of the matrix representations of the operators $\mathcal{A}$ and $\mathcal{B}$ (c.f. Eq. (1.15))

$$\begin{aligned}\langle i|\mathcal{C}|j\rangle = (\mathbf{C})_{ij} = \langle i|\mathcal{A}\mathcal{B}|j\rangle &= \langle i|\mathcal{A}1\mathcal{B}|j\rangle \\ &= \sum_k \langle i|\mathcal{A}|k\rangle\langle k|\mathcal{B}|j\rangle \\ &= \sum_k (\mathbf{A})_{ik}(\mathbf{B})_{kj}\end{aligned}$$

We now introduce the *adjoint* of the operator $\mathcal{O}$, which we denote by $\mathcal{O}^\dagger$. If $\mathcal{O}$ changes a ket $|a\rangle$ into the ket $|b\rangle$ (c.f. Eq. (1.52)), then its adjoint changes the bra $\langle a|$ into the bra $\langle b|$, i.e.,

$$\langle a|\mathcal{O}^\dagger = \langle b| \tag{1.57}$$

This equation is said to be the adjoint of Eq. (1.52). Multiplying both sides of Eq. (1.52) by $\langle c|$ on the left and multiplying both sides of Eq. (1.57) by $|c\rangle$ on the right, we have

$$\langle c|\mathcal{O}|a\rangle = \langle c|b\rangle$$

and

$$\langle a|\mathcal{O}^\dagger|c\rangle = \langle b|c\rangle$$

Since $\langle b|c\rangle = \langle c|b\rangle^*$, it follows that

$$\langle a|\mathcal{O}^\dagger|c\rangle = \langle c|\mathcal{O}|a\rangle^* \tag{1.58}$$

Since the labels a, b, and c are arbitrary, we have shown that the matrix representation of $\mathcal{O}^\dagger$ is the adjoint of the matrix representation of $\mathcal{O}$ since

$$\langle i|\mathcal{O}^\dagger|j\rangle \equiv (\mathbf{O}^\dagger)_{ij} = \langle j|\mathcal{O}|i\rangle^* \equiv O_{ji}^* \tag{1.59}$$

Finally, we say that an operator is Hermitian when it is self-adjoint

$$\mathcal{O} = \mathcal{O}^\dagger \tag{1.60}$$

Thus the elements of the matrix representation of a Hermitian operator satisfy

$$\langle a|\mathcal{O}|b\rangle = \langle a|\mathcal{O}^\dagger|b\rangle = \langle b|\mathcal{O}|a\rangle^* \tag{1.61}$$

1.1.5 Change of Basis

In Subsection 1.1.1 we have seen that the choice of basis is not unique. Given two complete orthonormal bases $\{|i\rangle\}$ and $\{|\alpha\rangle\}$ we now wish to find the relationship between them. We use latin letters $i, j, k, \ldots$ to specify the bras and kets in the first basis and greek letters α, β, γ to specify the bras and kets of the second basis. Thus we have

$$\langle i|j\rangle = \delta_{ij}, \qquad \sum_i |i\rangle\langle i| = 1 \tag{1.62a}$$

and

$$\langle\alpha|\beta\rangle = \delta_{\alpha\beta}, \qquad \sum_\alpha |\alpha\rangle\langle\alpha| = 1 \tag{1.62b}$$

Since the basis $\{|i\rangle\}$ is complete, we can express any ket in the basis $\{|\alpha\rangle\}$ as a linear combination of kets in the basis $\{|i\rangle\}$ and vice versa. That is,

$$|\alpha\rangle = 1|\alpha\rangle = \sum_i |i\rangle\langle i|\alpha\rangle = \sum_i |i\rangle U_{i\alpha} = \sum_i |i\rangle(\mathbf{U})_{i\alpha} \tag{1.63}$$

where we have defined the elements of a transformation matrix $\mathbf{U}$ as

$$\langle i|\alpha\rangle = U_{i\alpha} = (\mathbf{U})_{i\alpha} \tag{1.64}$$

Transforming in the opposite direction, we have

$$|i\rangle = 1|i\rangle = \sum_\alpha |\alpha\rangle\langle\alpha|i\rangle = \sum_\alpha |\alpha\rangle U_{i\alpha}^* = \sum_\alpha |\alpha\rangle(\mathbf{U}^\dagger)_{\alpha i} \tag{1.65}$$

where we have used Eq. (1.49) and the definition of the adjoint matrix to show that

$$\langle\alpha|i\rangle = \langle i|\alpha\rangle^* = U_{i\alpha}^* = (\mathbf{U}^\dagger)_{\alpha i} \tag{1.66}$$

It is important to remember that since we have defined $\mathbf{U}$ via Eq. (1.64), $\langle\alpha|i\rangle \neq U_{\alpha i}$ but rather is given by Eq. (1.66). We now prove that the transformation matrix $\mathbf{U}$ is unitary. This is a consequence of the orthonormality of the bases:

$$\delta_{ij} = \langle i|j\rangle$$
$$= \sum_\alpha \langle i|\alpha\rangle\langle\alpha|j\rangle$$
$$= \sum_\alpha (\mathbf{U})_{i\alpha}(\mathbf{U}^\dagger)_{\alpha j}$$
$$= (\mathbf{U}\mathbf{U}^\dagger)_{ij}$$

which in matrix notation is just

$$1 = \mathbf{U}\mathbf{U}^\dagger \tag{1.67a}$$

In an analogous way, by starting with $\langle\alpha|\beta\rangle = \delta_{\alpha\beta}$, one can show that

$$1 = \mathbf{U}^\dagger\mathbf{U} \tag{1.67b}$$

and hence $\mathbf{U}$ is unitary. Thus we arrive at the important result that two orthonormal bases are related by a unitary matrix via Eq. (1.63) and its inverse, Eq. (1.65). As shown by Eq. (1.64), the elements of the transformation matrix $\mathbf{U}$ are scalar products between the two bases.

Let us now consider how the matrix representations of an operator $\mathcal{O}$ are related in two different complete orthonormal bases. The result we shall obtain plays a central role in the next subsection where we consider the eigenvalue problem. Suppose $\mathbf{O}$ is the matrix representation of $\mathcal{O}$ in the basis $\{|i\rangle\}$, while $\mathbf{\Omega}$ is its matrix representation in the basis $\{|\alpha\rangle\}$

$$\mathcal{O}|i\rangle = \sum_j |j\rangle\langle j|\mathcal{O}|i\rangle = \sum_j |j\rangle O_{ji} \tag{1.68a}$$

$$\mathcal{O}|\alpha\rangle = \sum_\beta |\beta\rangle\langle\beta|\mathcal{O}|\alpha\rangle = \sum_\beta |\beta\rangle\Omega_{\beta\alpha} \tag{1.68b}$$

To find the relationship between $\mathbf{O}$ and $\mathbf{\Omega}$ we use the, by now, familiar technique of introducing the unit operator

$$\begin{aligned}
\Omega_{\alpha\beta} = \langle\alpha|\mathcal{O}|\beta\rangle &= \langle\alpha|1\mathcal{O}1|\beta\rangle \\
&= \sum_{ij} \langle\alpha|i\rangle\langle i|\mathcal{O}|j\rangle\langle j|\beta\rangle \\
&= \sum_{ij} (\mathbf{U}^\dagger)_{\alpha i}(\mathbf{O})_{ij}(\mathbf{U})_{j\beta}
\end{aligned} \tag{1.69}$$

Thus

$$\mathbf{\Omega} = \mathbf{U}^\dagger\mathbf{O}\mathbf{U} \tag{1.70a}$$

or, multiplying on the left by $\mathbf{U}$ and on the right by $\mathbf{U}^\dagger$

$$\mathbf{O} = \mathbf{U}\mathbf{\Omega}\mathbf{U}^\dagger \tag{1.70b}$$

These equations show that the matrices $\mathbf{O}$ and $\mathbf{\Omega}$ are related by a *unitary transformation*. The importance of such transformations lies in the fact that for any Hermitian operator whose matrix representation in the basis $\{|i\rangle\}$ is not diagonal, it is always possible to find a basis $\{|\alpha\rangle\}$ in which the matrix representation of the operator is diagonal, i.e.,

$$\Omega_{\alpha\beta} = \omega_\alpha\delta_{\alpha\beta} \tag{1.71}$$

In the next subsection we consider the problem of diagonalizing Hermitian matrices by unitary transformations.

Exercise 1.8 Show that the trace of a matrix is invariant under a unitary transformation, i.e., if

$$\mathbf{\Omega} = \mathbf{U}^\dagger\mathbf{O}\mathbf{U}$$

then show that $\operatorname{tr}\mathbf{\Omega} = \operatorname{tr}\mathbf{O}$.

1.1.6 The Eigenvalue Problem

When an operator $\mathcal{O}$ acts on a vector $|\alpha\rangle$, the resulting vector is in general distinct from $|\alpha\rangle$. If $\mathcal{O}|\alpha\rangle$ is simply a constant times $|\alpha\rangle$, i.e.,

$$\mathcal{O}|\alpha\rangle = \omega_\alpha|\alpha\rangle \tag{1.72}$$

then we say that $|\alpha\rangle$ is an *eigenvector* of the operator $\mathcal{O}$ with an *eigenvalue* ω_α. Without loss of generality we can choose the eigenvectors to be normalized

$$\langle\alpha|\alpha\rangle = 1 \tag{1.73}$$

In this book we are interested in the eigenvectors and eigenvalues of Hermitian operators ($\mathcal{O}^\dagger = \mathcal{O}$). They have the following properties.

1. The eigenvalues of a Hermitian operator are real. This follows immediately from Eq. (1.61), which states that

$$\langle\alpha|\mathcal{O}|\alpha\rangle = \langle\alpha|\mathcal{O}^\dagger|\alpha\rangle = \langle\alpha|\mathcal{O}|\alpha\rangle^* \tag{1.74}$$

Multiplying the eigenvalue relation Eq. (1.72) by $\langle\alpha|$ and substituting into (1.74) we have

$$\omega_\alpha = \omega_\alpha^* \tag{1.75}$$

which is the required result.

2. The eigenvectors of a Hermitian operator are orthogonal. To prove this consider

$$\mathcal{O}|\beta\rangle = \omega_\beta|\beta\rangle$$

The adjoint of this equation is

$$\langle\beta|\mathcal{O}^\dagger = \langle\beta|\omega_\beta^*$$

where we have used (1.57) and the fact that the adjoint of a number is its complex conjugate. Since $\mathcal{O}$ is Hermitian and ω_β is real, we have

$$\langle\beta|\mathcal{O} = \langle\beta|\omega_\beta \tag{1.76}$$

Multiplying (1.72) by $\langle\beta|$ and (1.76) by $|\alpha\rangle$ and subtracting the resulting expressions, we obtain

$$(\omega_\beta - \omega_\alpha)\langle\beta|\alpha\rangle = 0 \tag{1.77}$$

so that $\langle\beta|\alpha\rangle = 0$ if $\omega_\alpha \neq \omega_\beta$. Thus orthogonality follows immediately if the two eigenvalues are not the same (nondegenerate). Two eigenvectors $|1\rangle$ and $|2\rangle$ are degenerate if they have the same eigenvalue

$$\mathcal{O}|1\rangle = \omega|1\rangle \qquad \mathcal{O}|2\rangle = \omega|2\rangle \tag{1.78}$$

We now show that degenerate eigenvectors can always be chosen to be orthogonal. We first note that any linear combination of degenerate eigenvectors is also an eigenvector with the same eigenvalue, i.e.,

$$\mathcal{O}(x|1\rangle + y|2\rangle) = x\mathcal{O}|1\rangle + y\mathcal{O}|2\rangle = \omega(x|1\rangle + y|2\rangle) \tag{1.79}$$

There are many ways we can find two linear combinations of $|1\rangle$ and $|2\rangle$, which are orthogonal. One such procedure is called *Schmidt* orthogonalization. We assume that $|1\rangle$ and $|2\rangle$ are normalized and let $\langle 1|2\rangle = S \neq 0$. We choose $|I\rangle = |1\rangle$ so that $\langle I|I\rangle = 1$. We set $|II'\rangle = |1\rangle + c|2\rangle$ and choose c so that $\langle I|II'\rangle = 0 = 1 + cS$. Finally we normalize $|II'\rangle$ to obtain

$$|II\rangle = (S^{-2} - 1)^{-1/2}(|1\rangle - S^{-1}|2\rangle) \tag{1.80}$$

Thus the eigenvectors $\{|\alpha\rangle\}$ of a Hermitian operator can be chosen to form an orthonormal set

$$\langle \alpha|\beta\rangle = \delta_{\alpha\beta} \tag{1.81}$$

The matrix representation of a Hermitian operator $\mathcal{O}$ in an arbitrary basis $\{|i\rangle\}$ is generally not diagonal. However, its matrix representation in the basis formed by its eigenvectors is diagonal. To show this we multiply the eigenvalue equation (1.72) by $\langle \beta|$ and use the orthonormality relation (1.81) to obtain

$$\langle \beta|\mathcal{O}|\alpha\rangle = \omega_\alpha \delta_{\alpha\beta} \tag{1.82}$$

The eigenvalue problem we wish to solve can be posed as follows. Given the matrix representation, $\mathbf{O}$, of a Hermitian operator $\mathcal{O}$ in the orthonormal basis $\{|i\rangle, i = 1, 2, \ldots, N\}$ we wish to find the orthonormal basis $\{|\alpha\rangle, \alpha = 1, 2, \ldots, N\}$ in which the matrix representation, $\mathbf{\Omega}$, of $\mathcal{O}$ is diagonal, i.e., $\Omega_{\alpha\beta} = \omega_\alpha \delta_{\alpha\beta}$. In other words, we wish to *diagonalize* the matrix $\mathbf{O}$. We have seen in the last subsection, that the two representations of the operator $\mathcal{O}$ are related by a unitary transformation (c.f. Eq. (1.70a))

$$\mathbf{\Omega} = \mathbf{U}^\dagger \mathbf{O} \mathbf{U}$$

Thus the problem of diagonalizing the Hermitian matrix $\mathbf{O}$ is equivalent to the problem of *finding* the unitary matrix $\mathbf{U}$ that converts $\mathbf{O}$ into a diagonal matrix

$$\mathbf{U}^\dagger \mathbf{O} \mathbf{U} = \omega = \begin{pmatrix} \omega_1 & & & \\ & \omega_2 & & 0 \\ & 0 & \ddots & \\ & & & \omega_N \end{pmatrix} \tag{1.83}$$

It is clear from this formulation that an $N \times N$ Hermitian matrix has N eigenvalues.

There exist numerous, efficient algorithms for diagonalizing Hermitian matrices.[1] For our purposes, computer programs based on such algorithms can be regarded as "black boxes," which when given O determine U and ω. In order to make contact with the discussion of the eigenvalue problem found in most elementary quantum chemistry texts, we now consider a computationally inefficient procedure that is based on finding the roots of the secular determinant.

The eigenvalue problem posed above can be reformulated as follows. Given an $N \times N$ Hermitian matrix O, we wish to find all distinct column vectors c (the eigenvectors of O) and corresponding numbers ω (the eigenvalues of O) such that

$$Oc = \omega c \tag{1.84a}$$

This equation can be rewritten as

$$(O - \omega 1)c = 0 \tag{1.84b}$$

As shown in Exercise 1.7, Eq. (1.84b) can have a nontrivial solution ($c \neq 0$) only when

$$|O - \omega 1| = 0 \tag{1.85}$$

This is called the *secular* determinant. This determinant is a polynomial of degree N in the unknown ω. A polynomial of degree N has N roots ω_α, $\alpha = 1, 2, \ldots, N$, which in this case are called the eigenvalues of the matrix O. Once we have found the eigenvalues, we can find the corresponding eigenvectors by substituting *each* ω_α into Eq. (1.84) and solving the resulting equations for c^α. In this way, c^α can be found to within a multiplicative constant, which is finally determined by requiring c^α to be normalized

$$\sum_i (c_i^\alpha)^* c_i^\alpha = 1 \tag{1.86}$$

In this way we can find N solutions to Eq. (1.84)

$$Oc^\alpha = \omega_\alpha c^\alpha \qquad \alpha = 1, 2, \ldots, N \tag{1.87}$$

Since O is Hermitian, the eigenvalues are real and the eigenvectors are orthogonal

$$\sum_i (c_i^\alpha)^* c_i^\beta = \delta_{\alpha\beta} \tag{1.88}$$

In order to establish the connection with our previous development, let us now construct a matrix U defined as $U_{i\alpha} = c_i^\alpha$, i.e.,

$$U = \begin{pmatrix} c_1^1 & c_1^2 & \cdots & c_1^N \\ c_2^1 & c_2^2 & \cdots & c_2^N \\ \vdots & \vdots & & \\ c_N^1 & c_N^2 & \cdots & c_N^N \end{pmatrix} = (c^1 c^2 \cdots c^N) \tag{1.89}$$

Thus the αth column of U is just the column matrix c^α. Then using (1.87) it can be shown that

$$OU = U \begin{pmatrix} \omega_1 & & & \\ & \omega_2 & & 0 \\ & 0 & \ddots & \\ & & & \omega_N \end{pmatrix} = U\omega \qquad (1.90)$$

Since $U_{i\alpha} = c_i^\alpha$, the orthonormality relation (1.88) is equivalent to

$$\sum_i U_{i\alpha}^* U_{i\beta} = \sum_i (U^\dagger)_{\alpha i}(U)_{i\beta} = \delta_{\alpha\beta} \qquad (1.91)$$

which in matrix notation is

$$U^\dagger U = 1 \qquad (1.92)$$

Finally, multiplying both sides of Eq. (1.90) by $U^\dagger$ and using Eq. (1.92) we have

$$U^\dagger OU = \omega \qquad (1.93)$$

which is identical to Eq. (1.83). Thus Eq. (1.89) gives the relationship between the unitary transformation (U), which diagonalizes the matrix O, and the eigenvectors (c^α) of O.

Exercise 1.9 Show that Eq. (1.90) contains Eq. (1.87) for all $\alpha = 1, 2, \ldots, N$.

As an illustration of the above formalism, we consider the problem of finding the eigenvalues and eigenvectors of the 2×2 symmetric matrix ($O_{12} = O_{21}$)

$$O = \begin{pmatrix} O_{11} & O_{12} \\ O_{21} & O_{22} \end{pmatrix}$$

or, equivalently, solving the eigenvalue problem

$$\begin{pmatrix} O_{11} & O_{12} \\ O_{21} & O_{22} \end{pmatrix} \begin{pmatrix} c_1 \\ c_2 \end{pmatrix} = \omega \begin{pmatrix} c_1 \\ c_2 \end{pmatrix} \qquad (1.94)$$

We shall solve this problem in two ways: first via the secular determinant (Eq. (1.85)) and second by directly finding the matrix U that diagonalizes O.

For Eq. (1.94) to have a nontrivial solution, the secular determinant must vanish

$$\begin{vmatrix} O_{11} - \omega & O_{12} \\ O_{21} & O_{22} - \omega \end{vmatrix} = \omega^2 - \omega(O_{22} + O_{11}) + O_{11}O_{22} - O_{12}O_{21} = 0 \qquad (1.95)$$

This quadratic equation has two solutions

$$\omega_1 = \tfrac{1}{2}[O_{11} + O_{22} - ((O_{22} - O_{11})^2 + 4O_{12}O_{21})^{1/2}] \qquad (1.96a)$$

$$\omega_2 = \tfrac{1}{2}[O_{11} + O_{22} + ((O_{22} - O_{11})^2 + 4O_{12}O_{21})^{1/2}] \qquad (1.96b)$$

which are the eigenvalues of the matrix **O**. To find the eigenvector corresponding to a given eigenvalue, say ω_2, we substitute ω_2 into Eq. (1.94) to obtain

$$O_{11}c_1^2 + O_{12}c_2^2 = \omega_2 c_1^2 \qquad (1.97a)$$

$$O_{21}c_1^2 + O_{22}c_2^2 = \omega_2 c_2^2 \qquad (1.97b)$$

where the superscripts "2" indicate that we are concerned with the second eigenvalue. We then use one of these two equivalent equations and the normalization condition

$$(c_1^2)^2 + (c_2^2)^2 = 1 \qquad (1.98)$$

to solve for c_1^2 and c_2^2. As a simple illustration, we consider the case when $O_{11} = O_{22} = a$ and $O_{12} = O_{21} = b$. From Eq. (1.96), the two eigenvalues are

$$\omega_1 = a - b \qquad (1.99a)$$

$$\omega_2 = a + b \qquad (1.99b)$$

To find the eigenvector corresponding to ω_2 we use Eq. (1.97a), which in this case gives

$$ac_1^2 + bc_2^2 = (a + b)c_1^2$$

from which we obtain

$$c_1^2 = c_2^2$$

Finally, the normalization condition (1.98) gives

$$c_1^2 = 2^{-1/2} \qquad c_2^2 = 2^{-1/2} \qquad (1.100a)$$

In an entirely analogous way, we find

$$c_1^1 = 2^{-1/2} \qquad c_2^1 = -2^{-1/2} \qquad (1.100b)$$

Exercise 1.10 Since the components of an eigenvector can be found from the eigenvalue equation only to within a multiplicative constant, which is later determined by the normalization, one can set $c_1 = 1$ and $c_2 = c$ in Eq. (1.94). If this is done, Eq. (1.94) becomes

$$O_{11} + O_{12}c = \omega$$

$$O_{21} + O_{22}c = \omega c$$

After eliminating c, find the two roots of the resulting equation and show that they are the same as those given in Eq. (1.96). This technique, which we shall use numerous times in the book for finding the lowest eigenvalue of a matrix, is basically the secular determinant approach *without* determinants. Thus one can use it to find the lowest eigenvalue of certain $N \times N$ matrices without having to evaluate an $N \times N$ determinant.

Now let us solve the 2×2 eigenvalue problem by directly finding the orthogonal matrix $\mathbf{U}$ that diagonalizes the symmetric matrix $\mathbf{O}$, i.e.,

$$\mathbf{U^\dagger O U} = \begin{pmatrix} U_{11} & U_{21} \\ U_{12} & U_{22} \end{pmatrix}\begin{pmatrix} O_{11} & O_{12} \\ O_{12} & O_{22} \end{pmatrix}\begin{pmatrix} U_{11} & U_{12} \\ U_{21} & U_{22} \end{pmatrix} = \omega = \begin{pmatrix} \omega_1 & 0 \\ 0 & \omega_2 \end{pmatrix} \quad (1.101)$$

The requirement that

$$\mathbf{U^\dagger U} = \begin{pmatrix} U_{11}U_{11} + U_{21}U_{21} & U_{11}U_{12} + U_{21}U_{22} \\ U_{12}U_{11} + U_{22}U_{21} & U_{12}U_{12} + U_{22}U_{22} \end{pmatrix} = 1 = \begin{pmatrix} 1 & 0 \\ 0 & 1 \end{pmatrix} \quad (1.102)$$

places three constraints (two diagonal and one off-diagonal) on the four elements of the matrix $\mathbf{U}$. Therefore $\mathbf{U}$ can be completely specified by only one parameter. Since

$$\begin{pmatrix} \cos\theta & \sin\theta \\ \sin\theta & -\cos\theta \end{pmatrix}\begin{pmatrix} \cos\theta & \sin\theta \\ \sin\theta & -\cos\theta \end{pmatrix} = \begin{pmatrix} \cos^2\theta + \sin^2\theta & 0 \\ 0 & \cos^2\theta + \sin^2\theta \end{pmatrix} = 1$$
$$(1.103)$$

for all values of the parameter θ, it is convenient to write

$$\mathbf{U} = \begin{pmatrix} \cos\theta & \sin\theta \\ \sin\theta & -\cos\theta \end{pmatrix} \quad (1.104)$$

This is the most general form of a 2×2 orthogonal matrix. Let us now choose θ such that

$$\mathbf{U^\dagger O U} = \begin{pmatrix} \cos\theta & \sin\theta \\ \sin\theta & -\cos\theta \end{pmatrix}\begin{pmatrix} O_{11} & O_{12} \\ O_{12} & O_{22} \end{pmatrix}\begin{pmatrix} \cos\theta & \sin\theta \\ \sin\theta & -\cos\theta \end{pmatrix}$$

$$= \begin{pmatrix} \begin{matrix} O_{11}\cos^2\theta + O_{22}\sin^2\theta \\ + O_{12}\sin 2\theta \end{matrix} & \frac{1}{2}(O_{11} - O_{22})\sin 2\theta - O_{12}\cos 2\theta \\ \frac{1}{2}(O_{11} - O_{22})\sin 2\theta - O_{12}\cos 2\theta & \begin{matrix} O_{11}\sin^2\theta + O_{22}\cos^2\theta \\ - O_{12}\sin 2\theta \end{matrix} \end{pmatrix}$$

is diagonal. This can be done if we choose θ such that

$$\tfrac{1}{2}(O_{11} - O_{22})\sin 2\theta - O_{12}\cos 2\theta = 0$$

This has the solution

$$\theta_0 = \frac{1}{2}\tan^{-1}\frac{2O_{12}}{O_{11} - O_{22}} \quad (1.105)$$

Thus the two eigenvalues of **O** are

$$\omega_1 = O_{11} \cos^2 \theta_0 + O_{22} \sin^2 \theta_0 + O_{12} \sin 2\theta_0 \qquad (1.106a)$$

and

$$\omega_2 = O_{11} \sin^2 \theta_0 + O_{22} \cos^2 \theta_0 - O_{12} \sin 2\theta_0 \qquad (1.106b)$$

Upon comparison of Eqs. (1.104) and (1.89), we find the two eigenvectors to be

$$\begin{pmatrix} c_1^1 \\ c_2^1 \end{pmatrix} = \begin{pmatrix} \cos \theta_0 \\ \sin \theta_0 \end{pmatrix} \qquad (1.107a)$$

and

$$\begin{pmatrix} c_1^2 \\ c_2^2 \end{pmatrix} = \begin{pmatrix} \sin \theta_0 \\ -\cos \theta_0 \end{pmatrix} \qquad (1.107b)$$

It should be mentioned that the Jacobi method for diagonalizing $N \times N$ matrices is a generalization of the above procedure. The basic idea of this method is to eliminate iteratively the off-diagonal elements of a matrix by repeated applications of orthogonal transformations, such as the ones we have considered here.

Exercise 1.11 Consider the matrices

$$\mathbf{A} = \begin{pmatrix} 3 & 1 \\ 1 & 3 \end{pmatrix}$$

$$\mathbf{B} = \begin{pmatrix} 3 & 1 \\ 1 & 2 \end{pmatrix}$$

Find numerical values for the eigenvalues and corresponding eigenvectors of these matrices by a) the secular determinant approach; b) the unitary transformation approach. You will see that approach (b) is much easier.

1.1.7 Functions of Matrices

Given a Hermitian matrix **A**, we can define a function of **A**, i.e., $f(\mathbf{A})$, in much the same way we define functions $f(x)$ of a simple variable x. For example, the square root of a matrix **A**, which we denote by $\mathbf{A}^{1/2}$, is simply that matrix which when multiplied by itself gives **A**, i.e.,

$$\mathbf{A}^{1/2}\mathbf{A}^{1/2} = \mathbf{A} \qquad (1.108)$$

The sine or the exponential of a matrix are defined by the Taylor series of the function, e.g.,

$$\exp(\mathbf{A}) = 1 + \frac{1}{1!}\mathbf{A} + \frac{1}{2!}\mathbf{A}^2 + \frac{1}{3!}\mathbf{A}^3 + \cdots$$

or in general

$$f(\mathbf{A}) = \sum_{n=0}^{\infty} c_n \mathbf{A}^n \qquad (1.109)$$

After these definitions, we are still faced with the problem of calculating $\mathbf{A}^{1/2}$ or exp $(\mathbf{A})$. If $\mathbf{A}$ is a diagonal matrix

$$(\mathbf{A})_{ij} = a_i \delta_{ij}$$

everything is simple, since

$$(\mathbf{A})^n = \begin{pmatrix} a_1^n & & & \\ & a_2^n & & \mathbf{0} \\ & \mathbf{0} & \ddots & \\ & & & a_N^n \end{pmatrix} \qquad (1.110)$$

so that

$$f(\mathbf{A}) = \sum_{n=0}^{\infty} c_n \mathbf{A}^n = \begin{pmatrix} \sum_n c_n a_1^n & & & \\ & \sum_n c_n a_2^n & & \mathbf{0} \\ & \mathbf{0} & \ddots & \\ & & & \sum_n c_n a_N^n \end{pmatrix}$$

$$= \begin{pmatrix} f(a_1) & & & \\ & f(a_2) & & \mathbf{0} \\ & \mathbf{0} & \ddots & \\ & & & f(a_N) \end{pmatrix} \qquad (1.111)$$

Similarly, the square root of a diagonal matrix is

$$\mathbf{A}^{1/2} = \begin{pmatrix} a_1^{1/2} & & & \\ & a_2^{1/2} & & \mathbf{0} \\ & \mathbf{0} & \ddots & \\ & & & a_N^{1/2} \end{pmatrix} \qquad (1.112)$$

What do we do if $\mathbf{A}$ is not diagonal? Since $\mathbf{A}$ is Hermitian, we can always find a unitary transformation that diagonalizes it, i.e.,

$$\mathbf{U}^\dagger \mathbf{A} \mathbf{U} = \mathbf{a} \qquad (1.113a)$$

The reverse transformation that "undiagonalizes" $\mathbf{a}$ is

$$\mathbf{A} = \mathbf{U} \mathbf{a} \mathbf{U}^\dagger \qquad (1.113b)$$

Now notice that

$$\mathbf{A}^2 = \mathbf{U} \mathbf{a} \mathbf{U}^\dagger \mathbf{U} \mathbf{a} \mathbf{U}^\dagger = \mathbf{U} \mathbf{a}^2 \mathbf{U}^\dagger$$

or in general

$$A^n = Ua^nU^\dagger \tag{1.114}$$

so that

$$f(A) = \sum_n c_n A^n = U\left(\sum_n c_n a^n\right)U^\dagger = Uf(a)U^\dagger$$

$$= U\begin{pmatrix} f(a_1) & & & \\ & f(a_2) & & 0 \\ & 0 & \ddots & \\ & & & f(a_N) \end{pmatrix}U^\dagger \tag{1.115}$$

Thus to calculate any function of a Hermitian matrix **A**, we first diagonalize **A** to obtain **a**, the diagonal matrix containing all the eigenvalues of **A**. We then calculate $f(a)$, which is easy because **a** is diagonal. Finally we "undiagonalize" $f(a)$ using (1.113b) to obtain (1.115). For example, we can find the square root of a matrix **A** as

$$A^{1/2} = Ua^{1/2}U^\dagger$$

since

$$A^{1/2}A^{1/2} = Ua^{1/2}U^\dagger Ua^{1/2}U^\dagger = Ua^{1/2}a^{1/2}U^\dagger = UaU^\dagger = A$$

If the above procedure were to yield a result for $f(A)$ that was infinite, then $f(A)$ does not exist. For example, if we try to calculate the inverse of a matrix **A** that has a zero eigenvalue (say $a_i = 0$), then $f(a_i) = 1/a_i = \infty$ and so A^{-1} does not exist. As Exercise 1.12(a) shows, the determinant of a matrix is just the product of its eigenvalues. Thus if one of the eigenvalues of **A** is zero, $\det(A)$ is zero and the above argument shows that A^{-1} does not exist. This same result was obtained in a different way in Exercise 1.7.

Exercise 1.12 Given that

$$U^\dagger AU = a = \begin{pmatrix} a_1 & & & \\ & a_2 & & 0 \\ & 0 & \ddots & \\ & & & a_N \end{pmatrix} \quad \text{or} \quad Ac^\alpha = a_\alpha c^\alpha \quad \alpha = 1, 2, \ldots, N$$

Show that

a. $\det(A^n) = a_1^n a_2^n \cdots a_N^n$.

b. $\operatorname{tr} A^n = \sum_{\alpha=1}^{N} a_\alpha^n$.

c. If $\mathbf{G}(\omega) = (\omega\mathbf{1} - \mathbf{A})^{-1}$, then

$$(\mathbf{G}(\omega))_{ij} = \sum_{\alpha=1}^{N} \frac{U_{i\alpha}U_{j\alpha}^*}{\omega - a_\alpha} = \sum_{\alpha=1}^{N} \frac{c_i^\alpha c_j^{\alpha*}}{\omega - a_\alpha}.$$

Show that using Dirac notation this can be rewritten as

$$(\mathbf{G}(\omega))_{ij} \equiv \langle i|\mathcal{G}(\omega)|j\rangle = \sum_{\alpha} \frac{\langle i|\alpha\rangle\langle\alpha|j\rangle}{\omega - a_\alpha}$$

As an interesting application of this relation consider the problem of solving the following set of inhomogeneous linear equations

$$(\omega\mathbf{1} - \mathbf{A})\mathbf{x} = \mathbf{c}$$

for $\mathbf{x}$. The most straightforward way to proceed is to invert $\omega\mathbf{1} - \mathbf{A}$, i.e.,

$$\mathbf{x} = (\omega\mathbf{1} - \mathbf{A})^{-1}\mathbf{c} = \mathbf{G}(\omega)\mathbf{c}$$

If we want $\mathbf{x}$ as a function of ω we need to invert the matrix for *each* value of ω. However, if we diagonalize $\mathbf{A}$, we can write

$$x_i = \sum_j (\mathbf{G}(\omega))_{ij}c_j = \sum_{j\alpha} \frac{U_{i\alpha}U_{j\alpha}^*c_j}{\omega - a_\alpha}$$

It is now computationally simple to evaluate $\mathbf{x}$ as a function of ω.

Exercise 1.13 If

$$\mathbf{A} = \begin{pmatrix} a & b \\ b & a \end{pmatrix}$$

show that

$$f(\mathbf{A}) = \begin{pmatrix} \frac{1}{2}[f(a+b) + f(a-b)] & \frac{1}{2}[f(a+b) - f(a-b)] \\ \frac{1}{2}[f(a+b) - f(a-b)] & \frac{1}{2}[f(a+b) + f(a-b)] \end{pmatrix}$$

1.2 ORTHOGONAL FUNCTIONS, EIGENFUNCTIONS, AND OPERATORS

We know from the theory of Fourier series that it is possible to represent a sufficiently well-behaved function $f(x)$ on some interval as an infinite linear combination of sines and cosines with coefficients that depend on the function. Thus any such function can be represented by specifying these coefficients. This seems very similar to the idea of expanding a vector in terms of a set of basis vectors. The purpose of this section is to explore this similarity. We consider an infinite set of functions $\{\psi_i(x), i = 1, 2, \ldots\}$ that satisfy the orthonormality condition

$$\int_{x_1}^{x_2} dx \; \psi_i^*(x)\psi_j(x) = \delta_{ij} \qquad (1.116)$$

on the interval $[x_1, x_2]$. From now on, we shall drop the integration limits x_1 and x_2.

Let us suppose that any function $a(x)$ can be expressed as a linear combination of the set of functions $\{\psi_i\}$

$$a(x) = \sum_i \psi_i(x)a_i \qquad (1.117)$$

In other words, the basis $\{\psi_i(x)\}$ is complete. Given a function $a(x)$ we can determine its components a_j with respect to the basis $\{\psi_i\}$ by multiplying Eq. (1.117) by $\psi_j^*(x)$ and integrating over x, since

$$\int dx \; \psi_j^*(x)a(x) = \sum_i \int dx \; \psi_j^*(x)\psi_i(x)a_i = \sum_i \delta_{ji}a_i = a_j \qquad (1.118)$$

Substituting this result for the coefficients into the original expansion (1.117), we have

$$a(x) = \int dx' \left[\sum_i \psi_i(x)\psi_i^*(x') \right] a(x') \qquad (1.119)$$

The quantity in square brackets is a function of x and x', and has the unusual property that when multiplied by $a(x')$ and integrated over all x' one obtains $a(x)$. An entity with this property is called the *Dirac delta function* $\delta(x - x')$

$$\sum_i \psi_i(x)\psi_i^*(x') = \delta(x - x') \qquad (1.120)$$

The Dirac delta function is the continuous generalization of the familiar δ_{ij}, i.e.,

$$a_i = \sum_j \delta_{ij}a_j \leftrightarrow a(x) = \int dx' \; \delta(x - x')a(x') \qquad (1.121)$$

Moreover, just as $\delta_{ij} = \delta_{ji}$,

$$\delta(x - x') = \delta(x' - x) \qquad (1.122)$$

By setting, $x = 0$ in (1.121) and using (1.122), we obtain

$$a(0) = \int dx' \; a(x') \, \delta(x') \qquad (1.123)$$

Finally, taking $a(x') = 1$ in (1.123), we have

$$1 = \int dx' \; \delta(x') \qquad (1.124)$$

as long as the integration interval includes $x = 0$. Thus $\delta(x)$ is a rather peculiar "function." Equation (1.124) shows that it has unit area. Moreover,

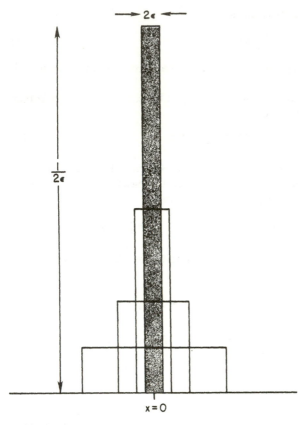

Figure 1.1 Successive approximations to the Dirac delta function $\delta(x)$.

when multiplied by a function $a(x)$ and integrated over any interval containing $x = 0$, it "plucks" out the value of the function at $x = 0$ (see Eq. (1.123)). The Dirac delta function $\delta(x)$ can be thought of as the limit of a sequence of functions that simultaneously become more and more peaked about $x = 0$ and narrower and narrower such that the area is always unity. For example, one representation of $\delta(x)$, which is shown in Fig. 1.1, is

$$\delta(x) = \lim_{\varepsilon \to 0} \delta_\varepsilon(x) \qquad (1.125)$$

where

$$\delta_\varepsilon(x) = \frac{1}{2\varepsilon} \qquad -\varepsilon \leq x \leq \varepsilon$$

$$= 0 \qquad \text{otherwise} \qquad (1.126)$$

Since the height of $\delta_\varepsilon(x)$ is $1/2\varepsilon$ and its width is 2ε, the area under $\delta_\varepsilon(x)$ is unity for all ε.

Exercise 1.14 Using the above representation of $\delta(x)$, show that

$$a(0) = \int_{-\infty}^{\infty} dx \, a(x)\delta(x)$$

Our development thus far is quite similar to the one we used in Subsection 1.1.4, where we discussed N-dimensional vector spaces. Indeed, the theory of complete orthonormal functions can be regarded as a generalization of ordinary linear algebra. To make the analogy explicit, it is convenient to introduce the shorthand notation

$$\psi_i(x) \equiv |i\rangle \qquad \psi_i^*(x) \equiv \langle i| \qquad (1.127a)$$

or more generally

$$a(x) \equiv |a\rangle \qquad a^*(x) \equiv \langle a| \qquad (1.127b)$$

and define the scalar product of two functions as

$$\int dx \, a^*(x)b(x) = \langle a|b\rangle \qquad (1.128)$$

Thus the orthonormality relation (1.116) becomes

$$\langle i|j\rangle = \delta_{ij} \qquad (1.129)$$

In this notation, Eq. (1.118) is

$$\langle j|a\rangle = a_j \qquad (1.130)$$

and thus Eq. (1.117) becomes

$$|a\rangle = \sum_i |i\rangle\langle i|a\rangle \qquad (1.131)$$

It is important to note that Eqs. (1.129), (1.130), and (1.131) are formally identical to Eqs. (1.47), (1.48a), and (1.50a), respectively of the previous section.

We define an operator $\mathcal{O}$ as an entity that acts on function $a(x)$ to yield the function $b(x)$

$$\mathcal{O}a(x) = b(x) \qquad (1.132a)$$

which can be rewritten using our shorthand notation as

$$\mathcal{O}|a\rangle = |b\rangle \qquad (1.132b)$$

which is identical to Eq. (1.52). An operator $\mathcal{O}$ is said to be *nonlocal* if we calculate $b(x)$ as

$$b(x) = \mathcal{O}a(x) = \int dx' \, O(x, x')a(x') \qquad (1.133)$$

Thus to find $b(x)$ at the point $x = x_0$ we need to know $a(x)$ for all x. Nonlocal operators are sometimes called integral operators. Note that (1.133) is the continuous generalization of

$$b_i = \sum_j O_{ij} a_j \qquad (1.134)$$

and hence we can regard $O(x, x')$ as a continuous matrix. An operator $\mathcal{O}$ is *local* when we can find $b(x)$ at the point x_0 by knowing $a(x)$ only in an infinitesimally small neighborhood of x_0. The derivative (d/dx) is an example of a local operator. If

$$\mathcal{O}\phi_\alpha(x) = \omega_\alpha \phi_\alpha(x) \qquad (1.135a)$$

or in our shorthand notation

$$\mathcal{O}|\alpha\rangle = \omega_\alpha|\alpha\rangle \qquad (1.135b)$$

then in analogy to matrix algebra, we say that $\phi_\alpha(x)$ is an *eigenfunction* of $\mathcal{O}$ with an eigenvalue ω_α. We can choose the eigenfunctions to be normalized

$$\int dx\ \phi_\alpha^*(x)\phi_\alpha(x) \equiv \langle\alpha|\alpha\rangle = 1 \qquad (1.136)$$

Multiplying (1.135a) by $\phi_\alpha^*(x)$, integrating over all x and using (1.136), we have

$$\omega_\alpha = \int dx\ \phi_\alpha^*(x)\mathcal{O}\phi_\alpha(x) \equiv \langle\alpha|\mathcal{O}|\alpha\rangle \qquad (1.137)$$

where we have introduced the notation

$$\int dx\ a^*(x)\mathcal{O}b(x) = \langle a|\mathcal{O}|b\rangle \qquad (1.138)$$

Note that we could have obtained (1.137) by multiplying (1.135b) on the left by $\langle\alpha|$ and using the normalization condition (1.136). In the rest of the book "multiplying by $\langle a|$" will be just a simple way of saying "multiplying by $a^*(x)$ and integrating over all x."

Just as before, we shall be interested in the eigenfunctions and eigenvalues of Hermitian operators. These are defined as operators for which

$$\int dx\ a^*(x)\mathcal{O}b(x) = \int dx\ b(x)(\mathcal{O}a(x))^* = \left(\int dx\ b^*(x)\mathcal{O}a(x)\right)^* \qquad (1.139)$$

Using the notation introduced in (1.138), Eq. (1.139) becomes

$$\langle a|\mathcal{O}|b\rangle = \langle b|\mathcal{O}|a\rangle^* \qquad (1.140)$$

which is identical to Eq. (1.61). The beauty of Dirac notation is that it allows us to manipulate vectors and functions, as well as the operators acting on them, in a formally identical way. For example, the proofs in Subsection 1.1.6 concerning the properties of the eigenvectors and eigenvalues of Hermitian operators acting on vectors apply immediately to Hermitian operators acting on functions. Thus the eigenfunctions of a Hermitian operator are ortho-normal and the corresponding eigenvalues are real.

Exercise 1.15 As a further illustration of the consistency of our notation, consider the matrix representation of an operator $\mathcal{O}$ in the basis $\{\psi_i(x)\}$. Starting with

$$\mathcal{O}\psi_i(x) = \sum_j \psi_j(x)O_{ji} \tag{1}$$

Show that

$$O_{ji} = \int dx\, \psi_j^*(x)\mathcal{O}\psi_i(x)$$

Then using Eqs. (1.127a) and (1.138) rewrite (1) in bra-ket notation and show that it is identical to Eq. (1.55).

Exercise 1.16 Consider the eigenvalue problem

$$\mathcal{O}\phi(x) = \omega\phi(x)$$

By expanding ϕ in the complete set $\{\psi_i(x), i = 1, 2, \ldots\}$ as

$$\phi(x) = \sum_{i=1}^{\infty} c_i\psi_i(x)$$

show that it becomes equivalent to the matrix eigenvalue problem

$$\mathbf{Oc} = \omega\mathbf{c}$$

where $(\mathbf{c})_i = c_i$ and $(\mathbf{O})_{ij} = \int dx\, \psi_i^*(x)\mathcal{O}\psi_j(x)$. Do this with and without using bra-ket notation. Note that $\mathbf{O}$ is an infinite matrix. In practice, we cannot handle infinite matrices. To keep things manageable, one uses only a finite subset of the set $\{\psi_i(x)\}$, i.e., $\{\psi_i(x), i = 1, 2, \ldots, N\}$. If the above analysis is repeated in this subspace, we obtain an $N \times N$ eigenvalue problem. As we shall see in Section 1.3, the corresponding N eigenvalues approximate the true eigenvalues. In particular, we shall prove that the lowest eigenvalue of the truncated eigenvalue problem is greater or equal to the exact lowest eigenvalue.

Exercise 1.17 In this subsection we have used a watered-down version of Dirac notation that is sufficient for our purposes but that oversimplifies the deep relationship between vectors and functions. The purpose of this exercise is to provide a glimpse at Dirac notation in its full glory. Consider a denumerably infinite set of complete orthonormal basis vectors, i.e.,

$$\sum_{i=1}^{\infty} |i\rangle\langle i| = 1 \tag{1a}$$

$$\langle i|j\rangle = \delta_{ij} \tag{1b}$$

Let us introduce a continuously infinite complete set of basis vectors denoted by $|x\rangle$. The analogue of (1a) is

$$\int dx\, |x\rangle\langle x| = 1 \tag{2a}$$

that is, we have replaced the summation in (1a) by an integral. If we multiply (2a) on the left by $\langle a|$ and on the right by $|b\rangle$ we have

$$\int dx\, \langle a|x\rangle\langle x|b\rangle = \langle a|b\rangle$$

Comparing this to Eq. (1.128) suggests that we identify $a^*(x)$ with $\langle a|x\rangle$ and $b(x)$ with $\langle x|b\rangle$. Recall that $\langle i|a\rangle$ is the component of $|a\rangle$ along the basis vector $|i\rangle$. Thus we can regard a function $b(x)$ as the x component of the abstract vector $|b\rangle$ in a coordinate system with a continuously infinite number of axes.

a. Multiply (2a) on the left by $\langle i|$ and on the right by $|j\rangle$. Using (1b) show that the resulting equation is identical to (1.116) if

$$\psi_i^*(x) = \langle i|x\rangle \qquad \psi_j(x) = \langle x|j\rangle.$$

b. Multiply (1a) by $\langle x|$ on the left and $|x'\rangle$ on the right. Show that the resulting equation is identical to (1.120) if

$$\langle x|x'\rangle = \delta(x - x') \tag{2b}$$

This is just the continuous analogue of (1b).

c. Multiply (2a) by $\langle x'|$ on the left and $|a\rangle$ on the right. Show that the resulting expression is identical to (1.121).

d. Consider an abstract operator $\mathcal{O}$. Its matrix elements in the continuous basis $|x\rangle$ are

$$\langle x|\mathcal{O}|x'\rangle = O(x, x')$$

Starting with the relation $\mathcal{O}|a\rangle = |b\rangle$ and inserting unity, we have

$$\mathcal{O}|a\rangle = \mathcal{O}1|a\rangle = \int dx\, \mathcal{O}|x\rangle\langle x|a\rangle = |b\rangle$$

Multiply this equation by $\langle x'|$ and show that the result is identical to Eq. (1.133).

e. If $O_{ij} = \langle i|\mathcal{O}|j\rangle$, show that

$$O(x, x') = \sum_{ij} \psi_i(x)O_{ij}\psi_j^*(x').$$

1.3 THE VARIATION METHOD

In this section we discuss an important approach to finding approximate solutions to the eigenvalue problem

$$\mathcal{O}\phi(x) = \omega\phi(x) \tag{1.141}$$

We are interested in eigenvalue problems because the time-independent Schrödinger equation is an eigenvalue equation:

$$\mathcal{H}|\Phi\rangle = \mathcal{E}|\Phi\rangle \tag{1.142}$$

where $\mathcal{H}$ is a Hermitian operator called the Hamiltonian, $|\Phi\rangle$ is the wave function, and $\mathcal{E}$ is the energy. We are interested in finding approximate solutions to eigenvalue equations because the Schrödinger equation cannot be solved exactly, except for the simplest cases. Although the subsequent development is applicable to any eigenvalue problem, we shall use the notation and terminology associated with the Schrödinger equation (1.142).

Given the operator $\mathcal{H}$, there exists a set of exact solutions to the Schrödinger equation, infinite in number, labeled by the index α

$$\mathcal{H}|\Phi_\alpha\rangle = \mathcal{E}_\alpha|\Phi_\alpha\rangle \qquad \alpha = 0, 1, \ldots \tag{1.143}$$

where

$$\mathcal{E}_0 \le \mathcal{E}_1 \le \mathcal{E}_2 \le \cdots \le \mathcal{E}_\alpha \le \cdots$$

We have assumed for the sake of simplicity that the set of eigenvalues $\{\mathcal{E}_\alpha\}$ is discrete. Since $\mathcal{H}$ is a Hermitian operator, the eigenvalues $\mathcal{E}_\alpha$ are real and the corresponding eigenfunctions are orthonormal

$$\langle\Phi_\alpha|\Phi_\beta\rangle = \delta_{\alpha\beta} \tag{1.144}$$

Thus by multiplying Eq. (1.143) by $\langle\Phi_\beta|$ on the left, we find

$$\langle\Phi_\beta|\mathcal{H}|\Phi_\alpha\rangle = \mathcal{E}_\alpha\delta_{\alpha\beta} \tag{1.145}$$

Furthermore, we assume that the eigenfunctions of $\mathcal{H}$ form a complete set and hence any function $|\tilde{\Phi}\rangle$ that satisfies the same boundary conditions as the set $\{|\Phi_\alpha\rangle\}$ can be written as a linear combination of the $|\Phi_\alpha\rangle$'s

$$|\tilde{\Phi}\rangle = \sum_\alpha |\Phi_\alpha\rangle c_\alpha = \sum_\alpha |\Phi_\alpha\rangle\langle\Phi_\alpha|\tilde{\Phi}\rangle \tag{1.146}$$

and

$$\langle\tilde{\Phi}| = \sum_\alpha c_\alpha^*\langle\Phi_\alpha| = \sum_\alpha \langle\tilde{\Phi}|\Phi_\alpha\rangle\langle\Phi_\alpha| \tag{1.147}$$

1.3.1 The Variation Principle

We are now in a position to state and prove an important theorem, called the *variation principle*: Given a normalized wave function $|\tilde{\Phi}\rangle$ that satisfies the appropriate boundary conditions (usually the requirement that the wave function vanishes at infinity), then the expectation value of the Hamiltonian is an upper bound to the exact ground state energy. That is, if

$$\langle \tilde{\Phi} | \tilde{\Phi} \rangle = 1 \tag{1.148}$$

then

$$\langle \tilde{\Phi} | \mathscr{H} | \tilde{\Phi} \rangle \geq \mathscr{E}_0 \tag{1.149}$$

The equality holds only when $|\tilde{\Phi}\rangle$ is identical to $|\Phi_0\rangle$.

The proof of this theorem is simple. First we consider

$$\langle \tilde{\Phi} | \tilde{\Phi} \rangle = 1 = \sum_{\alpha\beta} \langle \tilde{\Phi} | \Phi_\alpha \rangle \langle \Phi_\alpha | \Phi_\beta \rangle \langle \Phi_\beta | \tilde{\Phi} \rangle = \sum_{\alpha\beta} \langle \tilde{\Phi} | \Phi_\alpha \rangle \delta_{\alpha\beta} \langle \Phi_\beta | \tilde{\Phi} \rangle$$

$$= \sum_\alpha \langle \tilde{\Phi} | \Phi_\alpha \rangle \langle \Phi_\alpha | \tilde{\Phi} \rangle = \sum_\alpha |\langle \Phi_\alpha | \tilde{\Phi} \rangle|^2 \tag{1.150}$$

where we have used Eqs. (1.144), (1.146), and (1.147). Then

$$\langle \tilde{\Phi} | \mathscr{H} | \tilde{\Phi} \rangle = \sum_{\alpha\beta} \langle \tilde{\Phi} | \Phi_\alpha \rangle \langle \Phi_\alpha | \mathscr{H} | \Phi_\beta \rangle \langle \Phi_\beta | \tilde{\Phi} \rangle = \sum_\alpha \mathscr{E}_\alpha |\langle \Phi_\alpha | \tilde{\Phi} \rangle|^2 \tag{1.151}$$

where we have used Eq. (1.145). Finally, since $\mathscr{E}_\alpha \geq \mathscr{E}_0$ for all α, we have

$$\langle \tilde{\Phi} | \mathscr{H} | \tilde{\Phi} \rangle \geq \sum_\alpha \mathscr{E}_0 |\langle \Phi_\alpha | \tilde{\Phi} \rangle|^2 = \mathscr{E}_0 \sum_\alpha |\langle \Phi_\alpha | \tilde{\Phi} \rangle|^2 = \mathscr{E}_0 \tag{1.152}$$

where we have used the normalization condition (1.150).

The variation principle for the ground state tells us that the energy of an approximate wave function is always too high. Thus one measure of the quality of a wave function is its energy: The lower the energy, the better the wave function. This is the basis of the variation method in which we take a normalized trial function $|\tilde{\Phi}\rangle$, which depends on certain parameters, and vary these parameters until the expectation value $\langle \tilde{\Phi} | \mathscr{H} | \tilde{\Phi} \rangle$ reaches a minimum. This minimum value of $\langle \tilde{\Phi} | \mathscr{H} | \tilde{\Phi} \rangle$ is then our variational estimate of the exact ground state energy.

Exercise 1.18 The Schrödinger equation (in atomic units) of an electron moving in one dimension under the influence of the potential $-\delta(x)$ is

$$\left(-\frac{1}{2} \frac{d^2}{dx^2} - \delta(x) \right) |\Phi\rangle = \mathscr{E} |\Phi\rangle$$

Use the variation method with the trial function

$$|\tilde{\Phi}\rangle = N e^{-\alpha x^2}$$

to show that $-\pi^{-1}$ is an upper bound to the exact ground state energy (which is -0.5). You will need the integral

$$\int_{-\infty}^{\infty} dx \, x^{2m} e^{-\alpha x^2} = \frac{(2m)! \pi^{1/2}}{2^{2m} m! \alpha^{m+1/2}}$$

Exercise 1.19 The Schrödinger equation (in atomic units) for the hydrogen atom is

$$\left(-\frac{1}{2}\nabla^2 - \frac{1}{r}\right)|\Phi\rangle = \mathscr{E}|\Phi\rangle.$$

Use the variation method with the trial function

$$|\tilde{\Phi}\rangle = Ne^{-\alpha r^2}$$

to show that $-4/3\pi = -0.4244$ is an upper bound to the exact ground state energy (which is -0.5). You will need the relations

$$\nabla^2 f(r) = r^{-2}\frac{d}{dr}\left(r^2\frac{d}{dr}\right)f(r)$$

$$\int_0^\infty dr\, r^{2m}e^{-\alpha r^2} = \frac{(2m)!\pi^{1/2}}{2^{2m+1}m!\alpha^{m+1/2}}$$

$$\int_0^\infty dr\, r^{2m+1}e^{-\alpha r^2} = \frac{m!}{2\alpha^{m+1}}$$

Exercise 1.20 The variation principle as applied to matrix eigenvalue problems states that if $\mathbf{c}$ is a normalized ($\mathbf{c}^\dagger\mathbf{c} = 1$) column vector, then $\mathbf{c}^\dagger\mathbf{O}\mathbf{c}$ is greater or equal to the lowest eigenvalue of $\mathbf{O}$. For the 2×2 symmetric matrix ($O_{12} = O_{21}$)

$$\mathbf{O} = \begin{pmatrix} O_{11} & O_{12} \\ O_{12} & O_{22} \end{pmatrix}$$

consider the trial vector

$$\mathbf{c} = \begin{pmatrix} \cos\theta \\ \sin\theta \end{pmatrix}$$

which is normalized for any value of θ. Calculate

$$\omega(\theta) = \mathbf{c}^\dagger\mathbf{O}\mathbf{c}$$

and find the value of θ (i.e., θ_0) for which $\omega(\theta)$ is a minimum. Show that $\omega(\theta_0)$ is exactly equal to the lowest eigenvalue of $\mathbf{O}$ (see Eqs. (1.105) and (1.106a)). Why should you have anticipated this result?

1.3.2 The Linear Variational Problem

Given a trial function $|\tilde{\Phi}\rangle$, which depends on a set of parameters, then the expectation value $\langle\tilde{\Phi}|\mathscr{H}|\tilde{\Phi}\rangle$ will be a function of these parameters. In general, it will be such a complicated function that there is no simple way of determining the values of the parameters for which $\langle\tilde{\Phi}|\mathcal{H}|\tilde{\Phi}\rangle$ is a minimum. However, if only linear variations of the trial function are allowed, i.e., if

$$|\tilde{\Phi}\rangle = \sum_{i=1}^{N} c_i|\Psi_i\rangle \qquad (1.153)$$

where $\{|\Psi_i\rangle\}$ is a *fixed* set of N basis functions, then the problem of finding the optimum set of coefficients, $\{c_i\}$, can be reduced to a matrix diagonalization.

To show this we assume that the basis functions are real and orthonormal

$$\langle\Psi_i|\Psi_j\rangle = \langle\Psi_j|\Psi_i\rangle = \delta_{ij} \qquad (1.154)$$

The case where the basis functions are complex and not orthogonal will be considered in Chapter 3. The matrix representation of the Hamiltonian operator in the basis $\{|\Psi_i\rangle\}$ is an $N \times N$ matrix $\mathbf{H}$ with elements

$$(\mathbf{H})_{ij} = H_{ij} = \langle\Psi_i|\mathcal{H}|\Psi_j\rangle \qquad (1.155)$$

Since the Hamiltonian is Hermitian and the basis is real, $\mathbf{H}$ is symmetric, i.e., $H_{ij} = H_{ji}$. The trial function is normalized, so that

$$\langle\tilde{\Phi}|\tilde{\Phi}\rangle = \sum_{ij} c_i c_j \langle\Psi_i|\Psi_j\rangle = \sum_i c_i^2 = 1 \qquad (1.156)$$

The expectation value

$$\langle\tilde{\Phi}|\mathcal{H}|\tilde{\Phi}\rangle = \sum_{ij} c_i \langle\Psi_i|\mathcal{H}|\Psi_j\rangle c_j = \sum_{ij} c_i c_j H_{ij} \qquad (1.157)$$

is a function of the expansion coefficients.

Our problem is to find the set of parameters for which $\langle\tilde{\Phi}|\mathcal{H}|\tilde{\Phi}\rangle$ is a minimum. Unfortunately, we cannot simply solve the equations

$$\frac{\partial}{\partial c_k}\langle\tilde{\Phi}|\mathcal{H}|\tilde{\Phi}\rangle = 0 \qquad k = 1, 2, \ldots, N \qquad (1.158)$$

because our N parameters are not independent. Since our trial function is normalized, the expansion coefficients are related by Eq. (1.156), and thus only $N - 1$ of them are independent. The problem of minimizing a function (Eq. (1.157)) subject to a constraint (Eq. (1.156)) is elegantly solved by Lagrange's *method of undetermined multipliers*. Let us construct a function

$$\mathcal{L}(c_1, \ldots, c_N, E) = \langle\tilde{\Phi}|\mathcal{H}|\tilde{\Phi}\rangle - E(\langle\tilde{\Phi}|\tilde{\Phi}\rangle - 1)$$

$$= \sum_{ij} c_i c_j H_{ij} - E\left(\sum_i c_i^2 - 1\right) \qquad (1.159)$$

Since the trial function is normalized, we have merely added zero to Eq. (1.157), and so the minimum of both $\langle\tilde{\Phi}|\mathcal{H}|\tilde{\Phi}\rangle$ and $\mathcal{L}$ occurs at the same values of the coefficients. If we arbitrarily chose $c_1, c_2, \ldots, c_{N-1}$ as independent so that c_N is determined from the normalization condition (1.156), then we have

$$\frac{\partial \mathscr{L}}{\partial c_k} = 0 \qquad k = 1, 2, \ldots, N - 1 \tag{1.160}$$

but $\partial \mathscr{L}/\partial c_N$ is *not* necessarily zero. However, we still have the undetermined multiplier E at our disposal. We now *choose* this multiplier so that $\partial \mathscr{L}/\partial c_N$ does equal zero and thus

$$\frac{\partial \mathscr{L}}{\partial c_k} = 0 \qquad k = 1, 2, \ldots, N - 1, N \tag{1.161}$$

so that

$$\frac{\partial \mathscr{L}}{\partial c_k} = 0 = \sum_j c_j H_{kj} + \sum_i c_i H_{ik} - 2E c_k \tag{1.162}$$

but since $H_{ij} = H_{ji}$, we have

$$\sum_j H_{ij} c_j - E c_i = 0 \tag{1.163}$$

By introducing a column vector $\mathbf{c}$ with elements c_i, this set of equations can be written in matrix notation as

$$\mathbf{Hc} = E\mathbf{c} \tag{1.164}$$

which is the standard eigenvalue problem for the matrix $\mathbf{H}$.

Since $\mathbf{H}$ is symmetric, Eq. (1.164) can be solved to yield N orthonormal eigenvectors $\mathbf{c}^\alpha$ and corresponding eigenvalues E_α, which for convenience are arranged so that $E_0 \leq E_1 \leq \cdots \leq E_{N-1}$. Thus

$$\mathbf{Hc}^\alpha = E_\alpha \mathbf{c}^\alpha \qquad \alpha = 0, 1, \ldots, N - 1 \tag{1.165}$$

with

$$(\mathbf{c}^\alpha)^\dagger \mathbf{c}^\beta = \sum_i c_i^\alpha c_i^\beta = \delta_{\alpha\beta} \tag{1.166}$$

Introducing the diagonal matrix $\mathbf{E}$ containing the eigenvalues E_α (i.e., $(\mathbf{E})_{\alpha\beta} = E_\alpha \delta_{\alpha\beta}$) and the matrix of eigenvectors $\mathbf{C}$ defined as $C_{i\alpha} = c_i^\alpha$, the N relations contained in (1.165) can be written as

$$\mathbf{HC} = \mathbf{CE} \tag{1.167}$$

Thus instead of finding just one solution for $|\tilde{\Phi}\rangle$ and the expansion coefficients, we actually have found N solutions

$$|\tilde{\Phi}_\alpha\rangle = \sum_{i=1}^N c_i^\alpha |\Psi_i\rangle = \sum_{i=1}^N C_{i\alpha} |\Psi_i\rangle \qquad \alpha = 0, 1, \ldots, N - 1 \tag{1.168}$$

which are orthonormal, since

$$\langle \tilde{\Phi}_\alpha | \tilde{\Phi}_\beta \rangle = \sum_{ij} c_i^\alpha c_j^\beta \langle \Psi_i | \Psi_j \rangle = \sum_{ij} c_i^\alpha c_j^\beta \delta_{ij} = \sum_i c_i^\alpha c_i^\beta = \delta_{\alpha\beta} \tag{1.169}$$

where we have used Eqs. (1.154) and (1.166). To discover the significance of the E's, we consider

$$\langle \tilde{\Phi}_\beta | \mathscr{H} | \tilde{\Phi}_\alpha \rangle = \sum_{ij} c_i^\beta \langle \Psi_i | \mathscr{H} | \Psi_j \rangle c_j^\alpha$$

$$= \sum_{ij} c_i^\beta H_{ij} c_j^\alpha$$

$$= (\mathbf{c}^\beta)^\dagger \mathbf{H} \mathbf{c}^\alpha$$

$$= E_\alpha (\mathbf{c}^\beta)^\dagger \mathbf{c}^\alpha = E_\alpha \delta_{\alpha\beta} \qquad (1.170)$$

where we have used Eqs. (1.165) and (1.166). Thus the eigenvalue E_α is the expectation value of the Hamiltonian with respect to $|\tilde{\Phi}_\alpha\rangle$. In particular, the lowest eigenvalue E_0 is the best possible approximation to the ground state energy of $\mathscr{H}$ in the space spanned by the basis functions $\{|\Psi_i\rangle\}$. Moreover, the variation principle assures us that

$$E_0 = \langle \tilde{\Phi}_0 | \mathscr{H} | \tilde{\Phi}_0 \rangle \geq \mathscr{E}_0 \qquad (1.171)$$

What is the significance of the remaining E's? It can be shown (see Exercise 1.21) that $E_\alpha \geq \mathscr{E}_\alpha$, $\alpha = 1, 2, \ldots$. Thus E_1 is an upper bound to the energy of the first excited state of $\mathscr{H}$ and so on.

Exercise 1.21 Consider a normalized trial function $|\tilde{\Phi}'\rangle$ that is orthogonal to the exact ground state wave function, i.e., $\langle \tilde{\Phi}' | \Phi_0 \rangle = 0$.

a. Generalize the proof of the variation principle of Subsection 1.3.1 to show that

$$\langle \tilde{\Phi}' | \mathscr{H} | \tilde{\Phi}' \rangle \geq \mathscr{E}_1.$$

b. Consider the function

$$|\tilde{\Phi}'\rangle = x|\tilde{\Phi}_0\rangle + y|\tilde{\Phi}_1\rangle$$

where $|\tilde{\Phi}_\alpha\rangle$, $\alpha = 0, 1$ are given by Eq. (1.168). Show that if it is normalized, then

$$|x|^2 + |y|^2 = 1.$$

c. When x and y are chosen so that $|\tilde{\Phi}'\rangle$ is normalized and so that $\langle \tilde{\Phi}' | \Phi_0 \rangle = 0$, then from part (a) it follows that $\langle \tilde{\Phi}' | \mathscr{H} | \tilde{\Phi}' \rangle \geq \mathscr{E}_1$. Show that

$$\langle \tilde{\Phi}' | \mathscr{H} | \tilde{\Phi}' \rangle = E_1 - |x|^2 (E_1 - E_0)$$

Since $E_1 \geq E_0$, conclude that $E_1 \geq \mathscr{E}_1$. The above argument can be generalized to show that $E_\alpha \geq \mathscr{E}_\alpha$, $\alpha = 2, 3, \ldots$.

In summary, the linear variational method is a procedure for finding the best possible approximate solutions to the eigenvalue problem

$$\mathscr{H}|\Phi\rangle = \mathscr{E}|\Phi\rangle \qquad (1.172)$$

given a fixed set of orthonormal functions $\{|\Psi_i\rangle, i = 1, 2, \ldots, N\}$. The procedure entails forming the matrix representation of the operator $\mathscr{H}$ in the finite basis $\{|\Psi_i\rangle\}$, i.e., $(\mathbf{H})_{ij} = \langle \Psi_i|\mathscr{H}|\Psi_j\rangle$ and solving the matrix eigenvalue problem

$$\mathbf{Hc} = E\mathbf{c} \qquad (1.173)$$

that is, diagonalizing the $N \times N$ matrix $\mathbf{H}$.

We derived this result by explicitly minimizing the expectation value of the Hamiltonian. However, we can obtain Eq. (1.173) in an alternate way, which we will find useful. In an attempt to solve (1.172), let us approximate $|\Phi\rangle$ as

$$|\Phi\rangle = \sum_{j=1}^{N} c_j|\Psi_j\rangle \qquad (1.174)$$

and substitute this expansion into (1.172)

$$\sum_{j} c_j\mathscr{H}|\Psi_j\rangle = \mathscr{E}\sum_{j} c_j|\Psi_j\rangle \qquad (1.175)$$

Multiplying (1.175) by $\langle\Psi_i|$ on the left and replacing $\mathscr{E}$ by E as a reminder that the expansion (1.174) is approximate, we find

$$\sum_{j} c_j\langle\Psi_i|\mathscr{H}|\Psi_j\rangle = E\sum_{j} c_j\langle\Psi_i|\Psi_j\rangle = Ec_i$$

or

$$\sum_{j} H_{ij}c_j = Ec_i \qquad (1.176)$$

which in matrix notation is identical to Eq. (1.173). If we had used a *complete* orthonormal basis $\{|\Psi_i\rangle, i = 1, 2, \ldots, N, N + 1, \ldots\}$, we would have obtained an equation identical to (1.173) except, of course, $\mathbf{H}$ would have been an infinite matrix. The eigenvalues of this matrix are exactly equal to the eigenvalues of the operator $\mathscr{H}$. Thus the linear variational method is equivalent to solving the eigenvalue equation, (1.172) in a finite subspace spanned by $\{|\Psi_i\rangle, i = 1, 2, \ldots, N\}$.

Exercise 1.22 The Schrödinger equation (in atomic units) for a hydrogen atom in a uniform electric field F in the z direction is

$$\left(-\frac{1}{2}\nabla^2 - \frac{1}{r} + Fr\cos\theta\right)|\Phi\rangle = (\mathscr{H}_0 + Fr\cos\theta)|\Phi\rangle = \mathscr{E}(F)|\Phi\rangle$$

Use the trial function

$$|\tilde{\Phi}\rangle = c_1|1s\rangle + c_2|2p_z\rangle$$

where $|1s\rangle$ and $|2p_z\rangle$ are the normalized eigenfunctions of $\mathscr{H}_0$, i.e.,

$$|1s\rangle = \pi^{-1/2}e^{-r}$$

$$|2p_z\rangle = (32\pi)^{-1/2}re^{-r/2}\cos\theta$$

to find an upper bound to $\mathscr{E}(F)$. In constructing the matrix representation of $\mathscr{H}$, you can avoid a lot of work by noting that

$$\mathscr{H}_0|1s\rangle = -\tfrac{1}{2}|1s\rangle, \qquad \mathscr{H}_0|2p_z\rangle = -\tfrac{1}{8}|2p_z\rangle$$

Using $(1 + x)^{1/2} \simeq 1 + x/2$, expand your answer in a Taylor series in F, i.e.,

$$E(F) = E(0) - \tfrac{1}{2}\alpha F^2 + \cdots$$

Show that the coefficient α, which is the approximate dipole polarizability of the system, is equal to 2.96. The exact result is 4.5.

NOTES

1. J. H. Wilkinson, *The Algebraic Eigenvalue Problem*, Oxford University Press, New York, 1965. The classic reference work on the subject.

FURTHER READING

Acton, F. S., *Numerical Methods that Work*, Harper & Row, New York, 1970. Chapter 13 of this excellent book contains a nice discussion of a variety of efficient algorithms for finding the eigenvalues of matrices.

Cushing, J. T., *Applied Analytical Mathematics for Physical Scientists*, Wiley, New York, 1975. The first four chapters contain a more rigorous, yet reasonably accessible, treatment of the mathematics of this chapter.

Merzbacher, E., *Quantum Mechanics*, 2nd ed., Wiley, New York, 1970. Most graduate quantum mechanics texts discuss Dirac notation in its full glory, and Chapter 14 of this book contains a good introduction.

Press, W. H., Flannery, B. P., Teukolsky, S. A., and Vetterling, W. T., *Numerical Recipes*, Cambridge University Press, Cambridge, 1986. Chapter 11 of this bible of scientific computing discusses and provides FORTRAN codes for a variety of algorithms for finding the eigenvalues and eigenvectors of matrices.

CHAPTER
TWO

MANY-ELECTRON WAVE FUNCTIONS AND OPERATORS

This chapter introduces the basic concepts, techniques, and notations of quantum chemistry. We consider the structure of many-electron operators (e.g., the Hamiltonian) and discuss the form of many-electron wave functions (Slater determinants and linear combinations of these determinants). We describe the procedure for evaluating matrix elements of operators between Slater determinants. We introduce the basic ideas of the Hartree-Fock approximation. This allows us to develop the material of this chapter in a form most useful for subsequent chapters where the Hartree-Fock approximation and a variety of more sophisticated approaches, which use the Hartree-Fock method as a starting point, are considered in detail.

In Section 2.1, the electronic problem is formulated, i.e., the problem of describing the motion of electrons in the field of fixed nuclear point charges. This is one of the central problems of quantum chemistry and our sole concern in this book. We begin with the full nonrelativistic time-independent Schrödinger equation and introduce the Born-Oppenheimer approximation. We then discuss a general statement of the Pauli exclusion principle called the *antisymmetry principle*, which requires that many-electron wave functions must be antisymmetric with respect to the interchange of any two electrons.

In Section 2.2, we describe one-electron functions (spatial and spin orbitals) and then construct many-electron functions (Hartree products and

Slater determinants) in terms of these one-electron functions. We then consider the Hartree-Fock approximation in which the exact wave function of the system is approximated by a single Slater determinant and describe its qualitative features. At this point, we introduce a simple system, the minimal basis ($1s$ orbital on each atom) *ab initio* model of the hydrogen molecule. We shall use this model throughout the book as a pedagogical tool to illustrate and illuminate the essential features of a variety of formalisms that at first glance appear to be rather formidable. Finally, we discuss the multideterminantal expansion of the exact wave function of an N-electron system.

Section 2.3 is concerned with the form of the one- and two-electron operators of quantum chemistry and the rules for evaluating matrix elements of these operators between Slater determinants. The conversion of expressions for matrix elements involving spin orbitals to expressions involving spatial orbitals is discussed. Finally, we describe a mnemonic device for obtaining the expression for the energy of any single determinant.

Section 2.4 introduces creation and annihilation operators and the formalism of second quantization. Second quantization is an approach to dealing with many-electron systems, which incorporates the Pauli exclusion principle but avoids the explicit use of Slater determinants. This formalism is widely used in the literature of many-body theory. It is, however, not required for a comprehension of most of the rest of this book, and thus this section can be skipped without loss of continuity.

Section 2.5 discusses electron spin and spin operators in many-electron systems and contains a description of restricted and unrestricted spin orbitals and spin-adapted configurations. Spin-adapted configurations, unlike many single determinants derived from restricted spin orbitals, are correct eigenfunctions of the total electron spin operator. Singlet, doublet, and triplet spin-adapted configurations as well as unrestricted wave functions, which are not eigenfunctions of the total electron spin operator, are described.

2.1 THE ELECTRONIC PROBLEM

Our main interest in this book is finding approximate solutions of the non-relativistic time-independent Schrödinger equation

$$\mathcal{H}|\Phi\rangle = \mathcal{E}|\Phi\rangle \tag{2.1}$$

where $\mathcal{H}$ is the Hamiltonian operator for a system of nuclei and electrons described by position vectors $\mathbf{R}_A$ and $\mathbf{r}_i$, respectively. A molecular coordinate system is shown in Fig. 2.1. The distance between the ith electron and Ath nucleus is $r_{iA} = |\mathbf{r}_{iA}| = |\mathbf{r}_i - \mathbf{R}_A|$; the distance between the ith and jth electron is $r_{ij} = |\mathbf{r}_i - \mathbf{r}_j|$, and the distance between the Ath nucleus and the Bth nucleus is $R_{AB} = |\mathbf{R}_A - \mathbf{R}_B|$. In atomic units, the Hamiltonian for N electrons

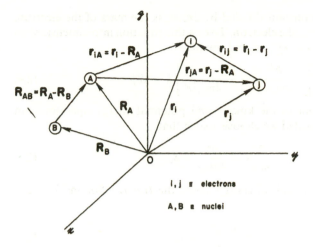

Figure 2.1 A molecular coordinate system: i, j = electrons; A, B = nuclei.

and M nuclei is

$$
\mathscr{H} = -\sum_{i=1}^{N} \frac{1}{2} \nabla_i^2 - \sum_{A=1}^{M} \frac{1}{2M_A} \nabla_A^2 - \sum_{i=1}^{N} \sum_{A=1}^{M} \frac{Z_A}{r_{iA}}
$$
$$
+ \sum_{i=1}^{N} \sum_{j>i}^{N} \frac{1}{r_{ij}} + \sum_{A=1}^{M} \sum_{B>A}^{M} \frac{Z_A Z_B}{R_{AB}} \tag{2.2}
$$

In the above equation, M_A is the ratio of the mass of nucleus A to the mass of an electron, and Z_A is the atomic number of nucleus A. The Laplacian operators ∇_i^2 and ∇_A^2 involve differentiation with respect to the coordinates of the ith electron and the Ath nucleus. The first term in Eq. (2.2) is the operator for the kinetic energy of the electrons; the second term is the operator for the kinetic energy of the nuclei; the third term represents the coulomb attraction between electrons and nuclei; the fourth and fifth terms represent the repulsion between electrons and between nuclei, respectively.

2.1.1 Atomic Units

The units we use throughout this book are called atomic units. To see how these units arise naturally let us consider the Schrödinger equation for the hydrogen atom. In SI units, we have

$$
\left[-\frac{\hbar^2}{2m_e} \nabla^2 - \frac{e^2}{4\pi\varepsilon_0 r} \right] \phi = \mathscr{E} \phi \tag{2.3}
$$

where $\hbar$ is Planck's constant divided by 2π, m_e is the mass of the electron, and $-e$ is the charge on the electron. To cast this equation into dimensionless form we let $x, y, z \rightarrow \lambda x', \lambda y', \lambda z'$ and obtain

$$\left[-\frac{\hbar^2}{2m_e\lambda^2} \nabla'^2 - \frac{e^2}{4\pi\varepsilon_0\lambda r'} \right] \phi' = \mathscr{E}\phi' \tag{2.4}$$

The constants in front of the kinetic and potential energy operators can then be factored, provided we choose λ such that

$$\frac{\hbar^2}{m_e\lambda^2} = \frac{e^2}{4\pi\varepsilon_0\lambda} = \mathscr{E}_a \tag{2.5}$$

where $\mathscr{E}_a$ is the atomic unit of energy called the *Hartree*. Solving Eq. (2.5) for λ we find

$$\lambda = \frac{4\pi\varepsilon_0\hbar^2}{m_e e^2} = a_0 \tag{2.6}$$

Thus λ is just the Bohr radius a_0 which is the atomic unit of length called a *Bohr*. Finally, since

$$\mathscr{E}_a\left[-\frac{1}{2} \nabla'^2 - \frac{1}{r'} \right] \phi' = \mathscr{E}\phi' \tag{2.7}$$

if we let $\mathscr{E}' = \mathscr{E}/\mathscr{E}_a$, we obtain the dimensionless equation

$$\left(-\frac{1}{2} \nabla'^2 - \frac{1}{r'} \right) \phi' = \mathscr{E}'\phi' \tag{2.8}$$

which is the Schrödinger equation in atomic units. The solution of this equation for the ground state of the hydrogen atom yields an energy $\mathscr{E}'$ equal to -0.5 atomic units $\equiv -0.5$ Hartrees. Table 2.1 gives the conversion factors X between atomic units and SI units, such that the SI value of any

Table 2.1 Conversion of atomic units to SI units

Physical quantity	Conversion factor X	Value of X (SI)
Length	a_0	5.2918×10^{-11} m
Mass	m_e	9.1095×10^{-31} kg
Charge	e	1.6022×10^{-19} C
Energy	$\mathscr{E}_a$	4.3598×10^{-18} J
Angular momentum	$\hbar$	1.0546×10^{-34} J s
Electric dipole moment	ea_0	8.4784×10^{-30} Cm
Electric polarizability	$e^2a_0^2\mathscr{E}_a^{-1}$	1.6488×10^{-41} C^{2}m^{2}J^{-1}
Electric field	$\mathscr{E}_a e^{-1}a_0^{-1}$	5.1423×10^{11} V m^{-1}
Wave function	$a_0^{-3/2}$	2.5978×10^{15} m$^{-3/2}$

quantity Q is related to its value in atomic units Q' by

$$Q = XQ' \tag{2.9}$$

Conversion factors for a few other units, which are not related to SI but which are necessary to read the existing literature, are as follows. One atomic unit of length equals 0.52918 Angströms (Å). One atomic unit of dipole moment (two unit charges separated by a_0) equals 2.5418 Debyes (D), and one atomic unit of energy equals 27.211 electron volts (eV) or 627.51 kcal/mole.

From now on we drop the primes and all our quantities will be in atomic units.

2.1.2 The Born-Oppenheimer Approximation

The Born-Oppenheimer approximation is central to quantum chemistry. Our discussion of this approximation is qualitative. The quantitative aspects of this approximation, including the problem of deriving corrections to it, are clearly discussed by Sutcliffe.[1] Since nuclei are much heavier than electrons, they move more slowly. Hence, to a good approximation, one can consider the electrons in a molecule to be moving in the field of fixed nuclei. Within this approximation, the second term of (2.2), the kinetic energy of the nuclei, can be neglected and the last term of (2.2), the repulsion between the nuclei, can be considered to be constant. Any constant added to an operator only adds to the operator eigenvalues and has no effect on the operator eigenfunctions. The remaining terms in (2.2) are called the electronic Hamiltonian or Hamiltonian describing the motion of N electrons in the field of M point charges,

$$\mathscr{H}_{elec} = -\sum_{i=1}^{N} \frac{1}{2} \nabla_i^2 - \sum_{i=1}^{N} \sum_{A=1}^{M} \frac{Z_A}{r_{iA}} + \sum_{i=1}^{N} \sum_{j>i}^{N} \frac{1}{r_{ij}} \tag{2.10}$$

The solution to a Schrödinger equation involving the electronic Hamiltonian,

$$\mathscr{H}_{elec}\Phi_{elec} = \mathscr{E}_{elec}\Phi_{elec} \tag{2.11}$$

is the electronic wave function,

$$\Phi_{elec} = \Phi_{elec}(\{\mathbf{r}_i\}; \{\mathbf{R}_A\}) \tag{2.12}$$

which describes the motion of the electrons and *explicitly* depends on the electronic coordinates but depends *parametrically* on the nuclear coordinates, as does the electronic energy,

$$\mathscr{E}_{elec} = \mathscr{E}_{elec}(\{\mathbf{R}_A\}) \tag{2.13}$$

By a parametric dependence we mean that, for different arrangements of the nuclei, Φ_{elec} is a different function of the electronic coordinates. The nuclear coordinates do not appear explicitly in Φ_{elec}. The total energy for fixed nuclei must also include the constant nuclear repulsion.

$$\mathscr{E}_{tot} = \mathscr{E}_{elec} + \sum_{A=1}^{M} \sum_{B>A}^{M} \frac{Z_A Z_B}{R_{AB}} \qquad (2.14)$$

Equations (2.10) to (2.14) constitute the electronic problem, which is our interest in this book.

If one has solved the electronic problem, it is subsequently possible to solve for the motion of the nuclei under the same assumptions as used to formulate the electronic problem. As the electrons move much faster than the nuclei, it is a reasonable approximation in (2.2) to replace the electronic coordinates by their average values, averaged over the electronic wave function. This then generates a nuclear Hamiltonian for the motion of the nuclei in the average field of the electrons,

$$\begin{aligned}
\mathscr{H}_{nucl} &= -\sum_{A=1}^{M} \frac{1}{2M_A} \nabla_A^2 + \left\langle -\sum_{i=1}^{N} \frac{1}{2} \nabla_i^2 - \sum_{i=1}^{N} \sum_{A=1}^{M} \frac{Z_A}{r_{iA}} + \sum_{i=1}^{N} \sum_{j>i}^{N} \frac{1}{r_{ij}} \right\rangle \\
&\quad + \sum_{A=1}^{M} \sum_{B>A}^{M} \frac{Z_A Z_B}{R_{AB}} \\
&= -\sum_{A=1}^{M} \frac{1}{2M_A} \nabla_A^2 + \mathscr{E}_{elec}(\{\mathbf{R}_A\}) + \sum_{A=1}^{M} \sum_{B>A}^{M} \frac{Z_A Z_B}{R_{AB}} \\
&= -\sum_{A=1}^{M} \frac{1}{2M_A} \nabla_A^2 + \mathscr{E}_{tot}(\{\mathbf{R}_A\}) \qquad (2.15)
\end{aligned}$$

The total energy $\mathscr{E}_{tot}(\{\mathbf{R}_A\})$ provides a potential for nuclear motion. This function constitutes a potential energy surface as shown schematically in Fig. 2.2. Thus the nuclei in the Born-Oppenheimer approximation move on a potential energy surface obtained by solving the electronic problem. Solu-

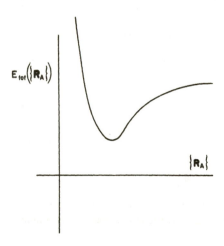

Figure 2.2 Schematic illustration of a potential surface.

tions to a nuclear Schrödinger equation,

$$\mathscr{H}_{\text{nucl}}\Phi_{\text{nucl}} = \mathscr{E}\Phi_{\text{nucl}} \tag{2.16}$$

describe the vibration, rotation, and translation of a molecule,

$$\Phi_{\text{nucl}} = \Phi_{\text{nucl}}(\{\mathbf{R}_A\}) \tag{2.17}$$

and $\mathscr{E}$, which is the Born-Oppenheimer approximation to the total energy of (2.1), includes electronic, vibrational, rotational, and translational energy. The corresponding approximation to the total wave function of (2.1) is,

$$\Phi(\{\mathbf{r}_i\}; \{\mathbf{R}_A\}) = \Phi_{\text{elec}}(\{\mathbf{r}_i\}; \{\mathbf{R}_A\})\Phi_{\text{nucl}}(\{\mathbf{R}_A\}) \tag{2.18}$$

From now on, we will not consider the vibrational-rotational problem but concentrate solely on the electronic problem of (2.11) to (2.14). We thus drop the subscript "elec" and only consider electronic Hamiltonians and electronic wave functions. Where it is convenient or necessary, we will distinguish between the electronic energy of (2.13) and the total energy of (2.14), which includes nuclear-nuclear repulsion.

2.1.3 The Antisymmetry or Pauli Exclusion Principle

The electronic Hamiltonian in Eq. (2.10) depends only on the spatial coordinates of the electrons. To completely describe an electron it is necessary, however, to specify its *spin*. We do this in the context of our nonrelativistic theory by introducing two spin functions $\alpha(\omega)$ and $\beta(\omega)$, corresponding to spin up and down, respectively. These are functions of an unspecified spin variable ω; from the operational point of view we need only specify that the two spin functions are complete and that they are orthonormal,

$$\int d\omega \, \alpha^*(\omega)\alpha(\omega) = \int d\omega \, \beta^*(\omega)\beta(\omega) = 1 \tag{2.19a}$$

$$\langle\alpha|\alpha\rangle = \quad \langle\beta|\beta\rangle \quad = 1 \tag{2.19b}$$

and

$$\int d\omega \, \alpha^*(\omega)\beta(\omega) = \int d\omega \, \beta^*(\omega)\alpha(\omega) = 0 \tag{2.20a}$$

$$\langle\alpha|\beta\rangle = \quad \langle\beta|\alpha\rangle \quad = 0 \tag{2.20b}$$

where the integration has been used in a formal way. In this formalism an electron is described not only by the three spatial coordinates $\mathbf{r}$ but also by one spin coordinate ω. We denote these four coordinates collectively by $\mathbf{x}$,

$$\mathbf{x} = \{\mathbf{r}, \omega\} \tag{2.21}$$

The wave function for an N-electron system is then a function of $\mathbf{x}_1$, $\mathbf{x}_2, \ldots, \mathbf{x}_N$. That is, we write $\Phi(\mathbf{x}_1, \mathbf{x}_2, \ldots, \mathbf{x}_N)$.

Because the Hamiltonian operator makes no reference to spin, simply making the wave function depend on spin (in the way just described) does not lead anywhere. A satisfactory theory can be obtained, however, if we make the following additional requirement on a wave function: *A many-electron wave function must be antisymmetric with respect to the interchange of the coordinate* $\mathbf{x}$ (*both space and spin*) *of any two electrons*,

$$\Phi(\mathbf{x}_1, \ldots, \mathbf{x}_i, \ldots, \mathbf{x}_j, \ldots, \mathbf{x}_N) = -\Phi(\mathbf{x}_1, \ldots, \mathbf{x}_j, \ldots, \mathbf{x}_i, \ldots, \mathbf{x}_N) \quad (2.22)$$

This requirement, sometimes called the *antisymmetry principle*, is a very general statement of the familiar Pauli exclusion principle. It is an independent postulate of quantum mechanics. Thus the exact wave function not only has to satisfy the Schrödinger equation, it also must be antisymmetric in the sense of Eq. (2.22). As we shall see, the requirement of antisymmetry is easily enforced by using Slater determinants.

2.2 ORBITALS, SLATER DETERMINANTS, AND BASIS FUNCTIONS

In this section we are concerned with the nomenclature, the conventions, and the procedure for writing down the wave functions that we use to describe many-electron systems. We will only consider many-electron wave functions that are either a single Slater determinant or a linear combination of Slater determinants. Sometimes, for very small systems, special functional forms are used for the wave function, but in most cases quantum chemists use Slater determinants. Before considering wave functions for many electrons, however, it is necessary to discuss wave functions for a single electron.

2.2.1 Spin Orbitals and Spatial Orbitals

We define an *orbital* as a wave function for a single particle, an electron. Because we are concerned with molecular electronic structure, we will be using *molecular orbitals* for the wave functions of the electrons in a molecule. A *spatial orbital* $\psi_i(\mathbf{r})$, is a function of the position vector $\mathbf{r}$ and describes the spatial distribution of an electron such that $|\psi_i(\mathbf{r})|^2 \, d\mathbf{r}$ is the probability of finding the electron in the small volume element $d\mathbf{r}$ surrounding $\mathbf{r}$. Spatial molecular orbitals will usually be assumed to form an orthonormal set

$$\int d\mathbf{r} \, \psi_i^*(\mathbf{r})\psi_j(\mathbf{r}) = \delta_{ij} \quad (2.23)$$

If the set of spatial orbitals $\{\psi_i\}$ were complete, then any arbitrary function $f(\mathbf{r})$ could be exactly expanded as

$$f(\mathbf{r}) = \sum_{i=1}^{\infty} a_i \psi_i(\mathbf{r}) \quad (2.24)$$

where the a_i are constant coefficients. In general, the set would have to be infinite to be complete; however, in practice we will never have available a complete set, but only a finite set $\{\psi_i | i = 1, 2, \ldots, K\}$ of K such orbitals. This finite set will only span a certain region of the complete space, but we can, however, describe results as being "exact" within the subspace spanned by the finite set of orbitals.

To completely describe an electron, it is necessary to specify its spin. A complete set for describing the spin of an electron consists of the two ortho-normal functions $\alpha(\omega)$ and $\beta(\omega)$, i.e., spin up (↑) and spin down (↓). The wave function for an electron that describes both its spatial distribution and its spin is a *spin orbital*, $\chi(\mathbf{x})$, where $\mathbf{x}$ indicates both space and spin coordinates (see Eq. (2.21)). From each spatial orbital, $\psi(\mathbf{r})$, one can form two different spin orbitals—one corresponding to spin up and the other to spin down—by multiplying the spatial orbital by the α or β spin function, respectively, i.e.,

$$\chi(\mathbf{x}) = \begin{cases} \psi(\mathbf{r})\alpha(\omega) \\ \quad \text{or} \\ \psi(\mathbf{r})\beta(\omega) \end{cases} \tag{2.25}$$

Given a set of K spatial orbitals $\{\psi_i | i = 1, 2, \ldots, K\}$, one can thus form a set of $2K$ spin orbitals $\{\chi_i | i = 1, 2, \ldots, 2K\}$ as

$$\left. \begin{array}{l} \chi_{2i-1}(\mathbf{x}) = \psi_i(\mathbf{r})\alpha(\omega) \\ \chi_{2i}(\mathbf{x}) = \psi_i(\mathbf{r})\beta(\omega) \end{array} \right\} i = 1, 2, \ldots, K \tag{2.26}$$

If the spatial orbitals are orthonormal, so are the spin orbitals

$$\int d\mathbf{x} \, \chi_i^*(\mathbf{x})\chi_j(\mathbf{x}) = \langle \chi_i | \chi_j \rangle = \delta_{ij} \tag{2.27}$$

Exercise 2.1 Given a set of K orthonormal spatial functions, $\{\psi_i^\alpha(\mathbf{r})\}$, and another set of K orthonormal functions, $\{\psi_i^\beta(\mathbf{r})\}$, such that the first set is not orthogonal to the second set, i.e.,

$$\int d\mathbf{r} \, \psi_i^{\alpha*}(\mathbf{r})\psi_j^\beta(\mathbf{r}) = S_{ij}$$

where S is an overlap matrix, show that the set $\{\chi_i\}$ of $2K$ spin orbitals, formed by multiplying $\psi_i^\alpha(\mathbf{r})$ by the α spin function and $\psi_i^\beta(\mathbf{r})$ by the β spin function, i.e.,

$$\left. \begin{array}{l} \chi_{2i-1}(\mathbf{x}) = \psi_i^\alpha(\mathbf{r})\alpha(\omega) \\ \chi_{2i}(\mathbf{x}) = \psi_i^\beta(\mathbf{r})\beta(\omega) \end{array} \right\} i = 1, 2, \ldots, k$$

is an orthonormal set.

2.2.2 Hartree Products

Having seen that the appropriate wave function describing a single electron is a spin orbital, we now consider wave functions for a collection of electrons, i.e., N-electron wave functions. Before considering the form of the exact wave function for a fully interacting system, let us first consider a simpler system containing noninteracting electrons having a Hamiltonian of the form

$$\mathcal{H} = \sum_{i=1}^{N} h(i) \tag{2.28}$$

where $h(i)$ is the operator describing the kinetic energy and potential energy of electron i. If we neglect electron-electron repulsion, then the full electronic Hamiltonian has this form. Alternatively, $h(i)$ might be an effective one-electron Hamiltonian that includes the effects of electron-electron repulsion in some average way.

Now, the operator $h(i)$ will have a set of eigenfunctions that we can take to be a set of spin orbitals $\{\chi_j\}$,

$$h(i)\chi_j(\mathbf{x}_i) = \varepsilon_j\chi_j(\mathbf{x}_i) \tag{2.29}$$

We now ask, "What are the corresponding eigenfunctions of $\mathcal{H}$?" Because $\mathcal{H}$ is a sum of one-electron Hamiltonians, a wave function which is a simple product of spin orbital wave functions for each electron,

$$\Psi^{\text{HP}}(\mathbf{x}_1, \mathbf{x}_2, \dots, \mathbf{x}_N) = \chi_i(\mathbf{x}_1)\chi_j(\mathbf{x}_2) \cdots \chi_k(\mathbf{x}_N) \tag{2.30}$$

is an eigenfunction of $\mathcal{H}$,

$$\mathcal{H}\Psi^{\text{HP}} = E\Psi^{\text{HP}} \tag{2.31}$$

with eigenvalue E, which is just the sum of the spin orbital energies of each of the spin orbitals appearing in Ψ^{HP},

$$E = \varepsilon_i + \varepsilon_j + \cdots + \varepsilon_k \tag{2.32}$$

Such a many-electron wave function is termed a *Hartree product*, with electron-one being described by the spin orbital χ_i, electron-two being described by the spin orbital χ_j, etc.

Exercise 2.2 Show that the Hartree product of (2.30) is an eigenfunction of $\mathcal{H} = \sum_{i=1}^{N} h(i)$ with an eigenvalue given by (2.32).

The Hartree product is an uncorrelated or independent-electron wave function because

$$|\Psi^{\text{HP}}(\mathbf{x}_1, \dots, \mathbf{x}_N)|^2 \, d\mathbf{x}_1 \cdots d\mathbf{x}_N$$

which is the simultaneous probability of finding electron-one in the volume element $d\mathbf{x}_1$, centered at $\mathbf{x}_1$, electron-two in $d\mathbf{x}_2$, etc., is just equal, from (2.30), to the product of probabilities

$$|\chi_i(\mathbf{x}_1)|^2 \, d\mathbf{x}_1 \, |\chi_j(\mathbf{x}_2)|^2 \, d\mathbf{x}_2 \cdots |\chi_k(\mathbf{x}_N)|^2 \, d\mathbf{x}_N$$

that electron-one is in $d\mathbf{x}_1$, times the probability that electron-two is in $d\mathbf{x}_2$, etc. The situation is analogous to the probability of drawing an ace of hearts (1/52) being equal to the probability of drawing an ace (1/13) times the probability of drawing a heart (1/4), since the probability of a particular card being an ace is independent or uncorrelated with the probability that the given card is a heart. The probability of finding electron-one at a given point in space is independent of the position of electron-two when a Hartree product wave function is used. In reality, electron-one and electron-two will be instantaneously repelled by the two-electron coulomb interaction, and electron-one will "avoid" regions of space occupied by electron-two so that the motion of the two electrons will be explicitly correlated. An example of correlated probabilities is provided by 2 hot potatoes and 2 cold apples in a bucket. The probability of obtaining a hot potato upon randomly withdrawing an object from the bucket (1/2) is not equal to the product of the probability of getting a hot object (1/2) times the probability of getting a potato (1/2), since whether the object is hot is perfectly correlated with whether the object is a potato.

Assuming independent electrons and a Hamiltonian of the form of Eq. (2.28), there is still a basic deficiency in the Hartree product; it takes no account of the indistinguishability of electrons, but specifically distinguishes electron-one as occupying spin orbital χ_i, electron-two as occupying χ_j, etc. The antisymmetry principle does not distinguish between identical electrons and requires that electronic wave functions be antisymmetric (change sign) with respect to the interchange of the space *and* spin coordinates of any two electrons.

2.2.3 Slater Determinants

The Hartree product does not satisfy the antisymmetry principle. However, we can obtain correctly antisymmetrized wave functions as follows. Consider a two-electron case in which we occupy the spin orbitals χ_i and χ_j. If we put electron-one in χ_i and electron-two in χ_j, we have

$$\Psi^{HP}_{12}(\mathbf{x}_1, \mathbf{x}_2) = \chi_i(\mathbf{x}_1)\chi_j(\mathbf{x}_2) \tag{2.33a}$$

On the other hand, if we put electron-one in χ_j and electron-two in χ_i, we have

$$\Psi^{HP}_{21}(\mathbf{x}_1, \mathbf{x}_2) = \chi_i(\mathbf{x}_2)\chi_j(\mathbf{x}_1) \tag{2.33b}$$

Each of these Hartree products clearly distinguishes between electrons; however, we can obtain a wave function which does not, and which satisfies the requirement of the antisymmetry principle by taking the appropriate linear combination of these two Hartree products,

$$\Psi(\mathbf{x}_1, \mathbf{x}_2) = 2^{-1/2}(\chi_i(\mathbf{x}_1)\chi_j(\mathbf{x}_2) - \chi_j(\mathbf{x}_1)\chi_i(\mathbf{x}_2)) \tag{2.34}$$

The factor $2^{-1/2}$ is a normalization factor. The minus sign insures that $\Psi(\mathbf{x}_1, \mathbf{x}_2)$ is antisymmetric with respect to the interchange of the coordinates of electrons one and two. Clearly,

$$\Psi(\mathbf{x}_1, \mathbf{x}_2) = -\Psi(\mathbf{x}_2, \mathbf{x}_1) \tag{2.35}$$

From the form of Eq. (2.34), it is evident that the wave function vanishes if both electrons occupy the same spin orbital (i.e., if $i = j$). Thus the anti-symmetry requirement immediately leads to the usual statement of the Pauli exclusion principle namely, that no more than one electron can occupy a spin orbital.

Exercise 2.3 Show that $\Psi(\mathbf{x}_1, \mathbf{x}_2)$ of Eq. (2.34) is normalized.

Exercise 2.4 Suppose the spin orbitals χ_i and χ_j are eigenfunctions of a one-electron operator h with eigenvalues ε_i and ε_j as in Eq. (2.29). Show that the Hartree products in Eqs. (2.33a, b) and the antisymmetrized wave function in Eq. (2.34) are eigenfunctions of the independent-particle Hamiltonian $\mathcal{H} = h(1) + h(2)$ (c.f. Eq. (2.28)) and have the same eigenvalue namely, $\varepsilon_i + \varepsilon_j$.

The antisymmetric wave function of Eq. (2.34) can be rewritten as a determinant (see Eq. (1.39))

$$\Psi(\mathbf{x}_1, \mathbf{x}_2) = 2^{-1/2} \begin{vmatrix} \chi_i(\mathbf{x}_1) & \chi_j(\mathbf{x}_1) \\ \chi_i(\mathbf{x}_2) & \chi_j(\mathbf{x}_2) \end{vmatrix} \tag{2.36}$$

and is called a *Slater determinant*. For an N-electron system the generalization of Eq. (2.36) is

$$\Psi(\mathbf{x}_1, \mathbf{x}_2, \ldots, \mathbf{x}_N) = (N!)^{-1/2} \begin{vmatrix} \chi_i(\mathbf{x}_1) & \chi_j(\mathbf{x}_1) & \cdots & \chi_k(\mathbf{x}_1) \\ \chi_i(\mathbf{x}_2) & \chi_j(\mathbf{x}_2) & \cdots & \chi_k(\mathbf{x}_2) \\ \vdots & \vdots & & \vdots \\ \chi_i(\mathbf{x}_N) & \chi_j(\mathbf{x}_N) & \cdots & \chi_k(\mathbf{x}_N) \end{vmatrix} \tag{2.37}$$

The factor $(N!)^{-1/2}$ is a normalization factor. This Slater determinant has N electrons occupying N spin orbitals $(\chi_i, \chi_j, \ldots, \chi_k)$ without specifying which electron is in which orbital. Note that the rows of an N-electron Slater determinant are labeled by electrons: first row $(\mathbf{x}_1)$, second row $(\mathbf{x}_2)$, etc., and the columns are labeled by spin orbitals: first column (χ_i), second column (χ_j), etc. Interchanging the coordinates of two electrons corresponds to inter-changing two rows of the Slater determinant, which changes the sign of the determinant. Thus Slater determinants meet the requirement of the anti-symmetry principle. Having two electrons occupying the same spin orbital corresponds to having two columns of the determinant equal, which makes the determinant zero. Thus no more than one electron can occupy a spin orbital (Pauli exclusion principle). It is convenient to introduce a short-hand notation for a normalized Slater determinant, which *includes the normaliza-*

tion constant and only shows the diagonal elements of the determinant,

$$\Psi(\mathbf{x}_1, \mathbf{x}_2, \ldots, \mathbf{x}_N) = |\chi_i(\mathbf{x}_1)\chi_j(\mathbf{x}_2) \cdots \chi_k(\mathbf{x}_N)\rangle \qquad (2.38)$$

If we always choose the electron labels to be in the order $\mathbf{x}_1, \mathbf{x}_2, \ldots, \mathbf{x}_N$, then Eq. (2.38) can be further shortened to

$$\Psi(\mathbf{x}_1, \mathbf{x}_2, \ldots, \mathbf{x}_N) = |\chi_i\chi_j \cdots \chi_k\rangle \qquad (2.39)$$

Because the interchange of any two columns changes the sign of a determinant, the ordering of spin orbital labels in Eq. (2.39) is important. In our short-hand notation, the antisymmetry property of Slater determinants is

$$|\cdots \chi_m \cdots \chi_n \cdots\rangle = -|\cdots \chi_n \cdots \chi_m \cdots\rangle \qquad (2.40)$$

To within a sign, a Slater determinant is completely specified by the spin orbitals from which it is formed (i.e., the spin orbitals that are occupied). Slater determinants formed from orthonormal spin orbitals are normalized. N-electron Slater determinants that have different orthonormal spin orbitals occupied are orthogonal.

Exercise 2.5 Consider the Slater determinants

$$|K\rangle = |\chi_i\chi_j\rangle$$

$$|L\rangle = |\chi_k\chi_l\rangle$$

Show that

$$\langle K|L\rangle = \delta_{ik}\delta_{jl} - \delta_{il}\delta_{jk}$$

Note that the overlap is zero unless: 1) $k = i$ and $l = j$, in which case $|L\rangle = |K\rangle$ and the overlap is unity and 2) $k = j$ and $l = i$ in which case $|L\rangle = |\chi_j\chi_i\rangle = -|K\rangle$ and the overlap is minus one.

We have seen that a Hartree product is truly an independent-electron wave function since the simultaneous probability of finding electron-one in $d\mathbf{x}_1$ at $\mathbf{x}_1$, electron-two in $d\mathbf{x}_2$ at $\mathbf{x}_2$, etc. is simply equal to the product of the probabilities that electron-one is in $d\mathbf{x}_1$, electron-two is in $d\mathbf{x}_2$, etc. Antisymmetrizing a Hartree product to obtain a Slater determinant introduces *exchange* effects, so-called because they arise from the requirement that $|\Psi|^2$ be invariant to the exchange of the space and spin coordinates of any two electrons. In particular, a Slater determinant incorporates *exchange correlation*, which means that the motion of two electrons with parallel spins is correlated. Since the motion of electrons with opposite spins remains uncorrelated, it is customary to refer to a single determinantal wave function as an uncorrelated wave function.

To see how exchange correlation arises, we now investigate the effect of antisymmetrizing a Hartree product on the electron density. Consider a two-electron Slater determinant in which spin orbitals χ_1 and χ_2 are occupied

$$\Psi(\mathbf{x}_1, \mathbf{x}_2) = |\chi_1(\mathbf{x}_1)\chi_2(\mathbf{x}_2)\rangle \tag{2.41}$$

If the two electrons have opposite spins and occupy different spatial orbitals,

$$\chi_1(\mathbf{x}_1) = \psi_1(\mathbf{r}_1)\alpha(\omega_1) \tag{2.42}$$

$$\chi_2(\mathbf{x}_2) = \psi_2(\mathbf{r}_2)\beta(\omega_2) \tag{2.43}$$

then by expanding the determinant, one obtains

$$|\Psi|^2 \, d\mathbf{x}_1 \, d\mathbf{x}_2 = \tfrac{1}{2}|\psi_1(\mathbf{r}_1)\alpha(\omega_1)\psi_2(\mathbf{r}_2)\beta(\omega_2) - \psi_1(\mathbf{r}_2)\alpha(\omega_2)\psi_2(\mathbf{r}_1)\beta(\omega_1)|^2 \, d\mathbf{x}_1 \, d\mathbf{x}_2 \tag{2.44}$$

for the simultaneous probability of electron-one being in $d\mathbf{x}_1$ and electron-two being in $d\mathbf{x}_2$. Let $P(\mathbf{r}_1, \mathbf{r}_2) \, d\mathbf{r}_1 d\mathbf{r}_2$ be the probability of finding electron-one in $d\mathbf{r}_1$ at $\mathbf{r}_1$ and simultaneously electron-two in $d\mathbf{r}_2$ at $\mathbf{r}_2$, as shown in Fig. 2.3. This probability is obtained by integrating (averaging) Eq. (2.44) over the spins of the two electrons,

$$P(\mathbf{r}_1, \mathbf{r}_2) \, d\mathbf{r}_1 \, d\mathbf{r}_2 = \int d\omega_1 \, d\omega_2 \, |\Psi|^2 \, d\mathbf{r}_1 \, d\mathbf{r}_2$$

$$= \tfrac{1}{2}[|\psi_1(\mathbf{r}_1)|^2|\psi_2(\mathbf{r}_2)|^2 + |\psi_1(\mathbf{r}_2)|^2|\psi_2(\mathbf{r}_1)|^2] \, d\mathbf{r}_1 \, d\mathbf{r}_2 \tag{2.45}$$

The first term in (2.45) is the product of the probability of finding electron-one in $d\mathbf{r}_1$ at $\mathbf{r}_1$ times the probability of finding electron-two in $d\mathbf{r}_2$ at $\mathbf{r}_2$, *if* electron-one occupies ψ_1 and electron-two occupies ψ_2. The second term has electron-one occupying ψ_2 and electron-two occupying ψ_1. Since electrons are indistinguishable, the correct probability is the average of the two

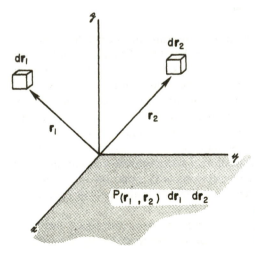

Figure 2.3 Probability of electron-one being in $d\mathbf{r}_1$ and electron-two being in $d\mathbf{r}_2$.

terms as shown. Thus the motion of the two electrons is uncorrelated. This is particularily obvious if $\psi_1 = \psi_2$, for in that case

$$P(\mathbf{r}_1, \mathbf{r}_2) = |\psi_1(\mathbf{r}_1)|^2|\psi_1(\mathbf{r}_2)|^2 \tag{2.46}$$

Note that $P(\mathbf{r}_1, \mathbf{r}_1) \neq 0$ so that there is a finite probability of finding two electrons with opposite spins at the same point in space.

If the two electrons have the same spin (say β), we have

$$\chi_1(\mathbf{x}_1) = \psi_1(\mathbf{r}_1)\beta(\omega_1) \tag{2.47}$$

$$\chi_2(\mathbf{x}_2) = \psi_2(\mathbf{r}_2)\beta(\omega_2) \tag{2.48}$$

then, by steps identical to the above, we obtain

$$P(\mathbf{r}_1, \mathbf{r}_2) = \tfrac{1}{2}\{|\psi_1(\mathbf{r}_1)|^2|\psi_2(\mathbf{r}_2)|^2 + |\psi_1(\mathbf{r}_2)|^2|\psi_2(\mathbf{r}_1)|^2$$
$$- [\psi_1^*(\mathbf{r}_1)\psi_2(\mathbf{r}_1)\psi_2^*(\mathbf{r}_2)\psi_1(\mathbf{r}_2) + \psi_1(\mathbf{r}_1)\psi_2^*(\mathbf{r}_1)\psi_2(\mathbf{r}_2)\psi_1^*(\mathbf{r}_2)]\} \tag{2.49}$$

where we now have an extra cross term, making the probabilities correlated. This is exchange correlation between electrons of parallel spin. Note that $P(\mathbf{r}_1, \mathbf{r}_1) = 0$, and thus the probability of finding two electrons with parallel spins at the same point in space is zero. A *Fermi hole* is said to exist around an electron. In summary, within the single Slater determinantal description, the motion of electrons with parallel spins is correlated but the motion of electrons with opposite spins is not.

2.2.4 The Hartree-Fock Approximation

Finding and describing approximate solutions to the electronic Schrödinger equation has been a major preoccupation of quantum chemists since the birth of quantum mechanics. Except for the very simplest cases like H_2^+, quantum chemists are faced with many-electron problems. Central to attempts at solving such problems, and central to this book, is the Hartree-Fock approximation. It has played an important role in elucidating modern chemistry. In addition, it usually constitutes the first step towards more accurate approximations. We are now in a position to consider some of the basic ideas which underlie this approximation. A detailed description of the Hartree-Fock method is given in Chapter 3.

The simplest antisymmetric wave function, which can be used to describe the ground state of an N-electron system, is a single Slater determinant,

$$|\Psi_0\rangle = |\chi_1\chi_2 \cdots \chi_N\rangle \tag{2.50}$$

The variation principle states that the best wave function of this functional form is the one which gives the lowest possible energy

$$E_0 = \langle\Psi_0|\mathscr{H}|\Psi_0\rangle \tag{2.51}$$

where $\mathscr{H}$ is the full electronic Hamiltonian. The variational flexibility in the wave function (2.50) is in the choice of spin orbitals. By minimizing E_0 with respect to the choice of spin orbitals, one can derive an equation, called the Hartree-Fock equation, which determines the optimal spin orbitals. We shall show in Chapter 3 that the Hartree-Fock equation is an eigenvalue equation of the form

$$f(i)\chi(\mathbf{x}_i) = \varepsilon\chi(\mathbf{x}_i) \qquad (2.52)$$

where $f(i)$ is an effective one-electron operator, called the *Fock* operator, of the form

$$f(i) = -\frac{1}{2}\nabla_i^2 - \sum_{A=1}^{M}\frac{Z_A}{r_{iA}} + v^{\mathrm{HF}}(i) \qquad (2.53)$$

where $v^{\mathrm{HF}}(i)$, which will be explicitly defined in Chapter 3, is the *average* potential experienced by the ith electron due to the presence of the other electrons. The essence of the Hartree-Fock approximation is to replace the complicated many-electron problem by a one-electron problem in which electron-electron repulsion is treated in an average way.

The Hartree-Fock potential $v^{\mathrm{HF}}(i)$, or equivalently the "field" seen by the ith electron, depends on the spin orbitals of the other electrons (i.e., the Fock operator depends on its eigenfunctions). Thus the Hartree-Fock equation (2.52) is nonlinear and must be solved iteratively. The procedure for solving the Hartree-Fock equation is called the self-consistent-field (SCF) method.

The basic idea of the SCF method is simple. By making an initial guess at the spin orbitals, one can calculate the average field (i.e., v^{HF}) seen by each electron and then solve the eigenvalue equation (2.52) for a new set of spin orbitals. Using these new spin orbitals, one can obtain new fields and repeat the procedure until self-consistency is reached (i.e., until the fields no longer change and the spin orbitals used to construct the Fock operator are the same as its eigenfunctions).

The solution of the Hartree-Fock eigenvalue problem (2.52) yields a set $\{\chi_k\}$ of orthonormal Hartree-Fock spin orbitals with orbital energies $\{\varepsilon_k\}$. The N spin orbitals with the lowest energies are called the *occupied* or *hole* spin orbitals. The Slater determinant formed from these orbitals is the Hartree-Fock ground state wave function and is the best variational approximation to the ground state of the system, of the single determinant form. We shall label occupied spin orbitals by the indices $a, b, c, \ldots$ (i.e., $\chi_a, \chi_b, \ldots$). The remaining members of the set $\{\chi_k\}$ are called *virtual, unoccupied,* or *particle* spin orbitals. We shall label virtual spin orbitals by the indices $r, s, t, \ldots$ (i.e., $\chi_r, \chi_s, \ldots$).

In principle, there are an infinite number of solutions to the Hartree-Fock equation (2.52) and an infinite number of virtual spin orbitals. In practice, the Hartree-Fock equation is solved by introducing a finite set of

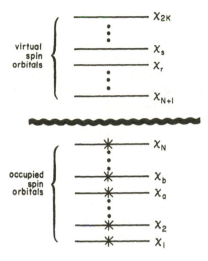

Figure 2.4 The Hartree-Fock ground state determinant, $|\chi_1 \chi_2 \cdots \chi_a \chi_b \cdots \chi_N\rangle$.

spatial basis functions $\{\phi_\mu(\mathbf{r}) | \mu = 1, 2, \ldots, K\}$. The spatial parts of the spin orbitals with the α spin function can then be expanded in terms of the known set of functions $\{\phi_\mu\}$. The spatial parts of the spin orbitals with the β spin can be expanded in the same way and both expansions substituted into the eigenvalue problem (2.52) to obtain matrix eigenvalue equations for the expansion coefficients. These matrix equations (e.g., the Roothaan equations) will be studied in some detail in Chapter 3. It is sufficient for this discussion to realize that using a basis set of K spatial functions $\{\phi_\mu\}$ leads to a set of $2K$ spin orbitals (K with α spin and K with β spin). This leads to a set of N occupied spin orbitals $\{\chi_a\}$ and a complementary set of $2K - N$ unoccupied or virtual spin orbitals $\{\chi_r\}$. A single Slater determinant formed from the set $\{\chi_a\}$ is the variational Hartree-Fock ground state, for which we will use the symbol Ψ_0 or $|\Psi_0\rangle$. A pictorial representation of $|\Psi_0\rangle$ is presented in Fig. 2.4. The $2K$ Hartree-Fock spin orbitals have been ordered according to their energy, and we have neglected possible degeneracies. The occupancy of the N lowest energy spin orbitals—one electron per spin orbital—is indicated by the asterisks.

The larger and more complete the set of basis functions $\{\phi_\mu\}$, the greater is the degree of flexibility in the expansion for the spin orbitals and the lower will be the expectation value $E_0 = \langle \Psi_0 | \mathcal{H} | \Psi_0 \rangle$. Larger and larger basis sets will keep lowering the Hartree-Fock energy E_0 until a limit is reached, called the *Hartree-Fock limit*. In practice, any finite value of K will lead to an energy somewhat above the Hartree-Fock limit.

2.2.5 The Minimal Basis H_2 Model

At this point we introduce a simple model system, which we will be using throughout this book to illustrate many of the methods and ideas of quantum chemistry. The model we use is the familiar minimal basis MO-LCAO description of H_2.

In this model, each hydrogen atom has a $1s$ atomic orbital and, as the two atoms approach, molecular orbitals (MOs) are formed as a linear combination of atomic orbitals (LCAO). The coordinate system is shown in Fig. 2.5. The first atomic orbital, ϕ_1, is centered on atom 1 at $\mathbf{R}_1$. The value of ϕ_1 at a point in space $\mathbf{r}$ is $\phi_1(\mathbf{r})$ or, since its value depends on the distance from its origin, we sometimes write $\phi_1 \equiv \phi_1(\mathbf{r} - \mathbf{R}_1)$. The second atomic orbital is centered on atom 2 at $\mathbf{R}_2$, i.e., $\phi_2 \equiv \phi_2(\mathbf{r} - \mathbf{R}_2)$. The exact $1s$ orbital of a hydrogen atom centered at $\mathbf{R}$ has the form

$$\phi(\mathbf{r} - \mathbf{R}) = (\zeta^3/\pi)^{1/2}e^{-\zeta|\mathbf{r} - \mathbf{R}|} \tag{2.54}$$

where ζ, the orbital exponent, has a value of 1.0. This is an example of a *Slater orbital*. In this book we will be concerned mostly with *Gaussian orbitals*, which lead to simpler integral evaluations than Slater orbitals. The $1s$ Gaussian orbital has the form

$$\phi(\mathbf{r} - \mathbf{R}) = (2\alpha/\pi)^{3/4}e^{-\alpha|\mathbf{r} - \mathbf{R}|^2} \tag{2.55}$$

where α is the Gaussian orbital exponent. For the present, we need not be concerned with the particular form of the $1s$ atomic orbitals. The two atomic orbitals ϕ_1 and ϕ_2 can be assumed to be normalized, but they will not be orthogonal. They will *overlap*, such that the overlap integral is

$$S_{12} = \int d\mathbf{r}\ \phi_1^*(\mathbf{r})\phi_2(\mathbf{r}) \tag{2.56}$$

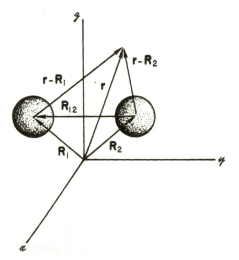

Figure 2.5 Coordinate system for minimal basis H_2.

The overlap will depend on the distance $R_{12} = |\mathbf{R}_1 - \mathbf{R}_2|$, such that $S_{12} = 1$ when $R_{12} = 0$ and $S_{12} = 0$ when $R_{12} = \infty$.

From the two localized atomic orbitals, ϕ_1 and ϕ_2, one can form, by linear combination, two delocalized molecular orbitals. The symmetric combination leads to a bonding molecular orbital of *gerade* symmetry (i.e., symmetric with respect to inversion about the point centered between the nuclei)

$$\psi_1 = [2(1 + S_{12})]^{-1/2}(\phi_1 + \phi_2) \qquad (2.57)$$

whereas, the antisymmetric combination leads to an antibonding molecular orbital of *ungerade* symmetry (i.e., antisymmetric with respect to inversion about the point centered between the nuclei)

$$\psi_2 = [2(1 - S_{12})]^{-1/2}(\phi_1 - \phi_2) \qquad (2.58)$$

Exercise 2.6 Show that ψ_1 and ψ_2 form an orthonormal set.

The above procedure is the simplest example of the general technique of expanding a set of spatial molecular orbitals in a set of known spatial basis functions

$$\psi_i(\mathbf{r}) = \sum_{\mu=1}^{K} C_{\mu i} \phi_\mu(\mathbf{r}) \qquad (2.59)$$

To obtain the exact molecular orbitals of H_2 one would need an infinite number of terms in such an expansion. Using only two basis functions for H_2 is an example of a *minimal* basis set and an obvious choice for the two functions ϕ_1 and ϕ_2 is the $1s$ atomic orbitals of the atoms. The correct linear combinations for this simple choice are determined by symmetry, and one need not solve the Hartree-Fock equations; ψ_1 and ψ_2 of Eqs. (2.57) and (2.58) are the Hartree-Fock spatial orbitals in the space spanned by ϕ_1 and ϕ_2.

Given the two spatial orbitals ψ_1 and ψ_2, we can form four spin orbitals

$$\chi_1(\mathbf{x}) = \psi_1(\mathbf{r})\alpha(\omega)$$
$$\chi_2(\mathbf{x}) = \psi_1(\mathbf{r})\beta(\omega)$$
$$\chi_3(\mathbf{x}) = \psi_2(\mathbf{r})\alpha(\omega) \qquad (2.60)$$
$$\chi_4(\mathbf{x}) = \psi_2(\mathbf{r})\beta(\omega)$$

The orbital energies associated with these spin orbitals can be obtained only by explicitly considering the Hartree-Fock operator. But, as might be expected, χ_1 and χ_2 are degenerate and have the lower energy corresponding to a bonding situation, while χ_3 and χ_4 are also degenerate having a higher energy corresponding to an antibonding situation. The Hartree-Fock ground state in this model is the single determinant

$$|\Psi_0\rangle = |\chi_1\chi_2\rangle \qquad (2.61)$$

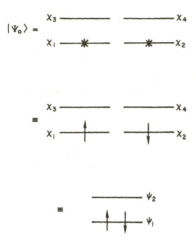

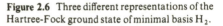

Figure 2.6 Three different representations of the Hartree-Fock ground state of minimal basis H_2.

shown pictorially in Fig. 2.6. Sometimes it is convenient to use a notation that indicates a spin orbital by its spatial part, using a bar or lack of a bar to denote whether it has the β or α spin function. Thus

$$\chi_1 \equiv \psi_1 \qquad \chi_2 \equiv \bar{\psi}_1$$
$$\chi_3 \equiv \psi_2 \qquad \chi_4 \equiv \bar{\psi}_2 \qquad (2.62)$$

In this notation the Hartree-Fock ground state is

$$|\Psi_0\rangle = |\psi_1\bar{\psi}_1\rangle = |1\bar{1}\rangle \qquad (2.63)$$

which indicates that both electrons occupy the same spatial orbital ψ_1, but one has an α spin and one has a β spin. It will be apparent from the context whether ψ_1 denotes a spatial orbital, or a spin orbital made up of the ψ_1 spatial orbital and the α spin function.

2.2.6 Excited Determinants

The Hartree-Fock procedure produces a set $\{\chi_i\}$ of $2K$ spin orbitals. The Hartree-Fock ground state,

$$|\Psi_0\rangle = |\chi_1\chi_2\cdots\chi_a\chi_b\cdots\chi_N\rangle \qquad (2.64)$$

is the best (in a variational sense) approximation to the ground state, of the single determinant form. However, it clearly is only one of many determinants that could be formed from the $2K > N$ spin orbitals. The number of combinations of $2K$ objects taken N at a time is the binomial coefficient

$$\binom{2K}{N} = \frac{(2K)!}{N!(2K - N)!}$$

This is the same as the number of different single determinants that one can form from N electrons and $2K$ spin orbitals; the Hartree-Fock ground state is just one of these. A convenient way of describing these other determinants is to consider the Hartree-Fock ground state (2.64) to be a reference state and to classify other possible determinants by how they differ from the reference state, i.e., by stating which occupied or hole spin orbitals of the set $\{\chi_a\}$ in (2.64), have been replaced by which virtual or particle spin orbitals of the set $\{\chi_r\}$. These other determinants can be taken to represent approximate excited states of the system or, as we shall see shortly, they can be used in linear combination with $|\Psi_0\rangle$ for a more accurate description of the ground state or any excited state of the system.

A singly excited determinant is one in which an electron, which occupied χ_a in the Hartree-Fock ground state (2.64), has been promoted to a virtual spin orbital χ_r, as shown in Fig. 2.7,

$$|\Psi_a^r\rangle = |\chi_1\chi_2 \cdots \chi_r\chi_b \cdots \chi_N\rangle \tag{2.65}$$

A doubly excited determinant, shown in Fig. 2.8, is one in which electrons have been excited from χ_a and χ_b to χ_r and χ_s,

$$|\Psi_{ab}^{rs}\rangle = |\chi_1\chi_2 \cdots \chi_r\chi_s \cdots \chi_N\rangle \tag{2.66}$$

All $\binom{2K}{N}$ determinants can thus be classified as either the Hartree-Fock ground state or singly, doubly, triply, quadruply, . . . , N-tuply excited states. The importance of these determinants as approximate representations of the true states of the system diminishes, in some sense, in the above order. While the excited determinants are not accurate representations of the

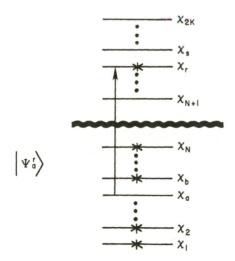

Figure 2.7 A singly excited determinant.

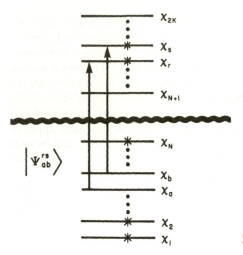

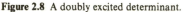

Figure 2.8 A doubly excited determinant.

excited states of the system, they are important as N-electron basis functions for an expansion of the exact N-electron states of the system.

2.2.7 Form of the Exact Wave Function and Configuration Interaction

We now consider the use of excited determinants as N-electron basis functions. Suppose we have a complete set of functions $\{\chi_i(x)\}$. Any function $\Phi(x_1)$ of a single variable can then be exactly expanded as

$$\Phi(x_1) = \sum_i a_i \chi_i(x_1) \tag{2.67}$$

where a_i is an expansion coefficient. How can we expand a function of two variables $\Phi(x_1, x_2)$ in an analogous way? If we think of x_2 as being held fixed, then we can expand $\Phi(x_1, x_2)$ as

$$\Phi(x_1, x_2) = \sum_i a_i(x_2) \chi_i(x_1) \tag{2.68}$$

where the expansion coefficients are now functions of x_2. Since $a_i(x_2)$ is a function of a single variable, it can be expanded in the complete set $\{\chi_i\}$ as

$$a_i(x_2) = \sum_j b_{ij} \chi_j(x_2) \tag{2.69}$$

Substituting this result in (2.68) gives

$$\Phi(x_1, x_2) = \sum_{ij} b_{ij} \chi_i(x_1) \chi_j(x_2) \tag{2.70}$$

If, however, we require Φ to be antisymmetric,

$$\Phi(x_1, x_2) = -\Phi(x_2, x_1) \tag{2.71}$$

then $b_{ij} = -b_{ji}$ and $b_{ii} = 0$, or

$$\Phi(x_1, x_2) = \sum_i \sum_{j>i} b_{ij}[\chi_i(x_1)\chi_j(x_2) - \chi_j(x_1)\chi_i(x_2)]$$

$$= \sum_{i<j} 2^{1/2} b_{ij} |\chi_i \chi_j\rangle \tag{2.72}$$

Thus an arbitrary antisymmetric function of the two variables can be exactly expanded in terms of all unique determinants formed from a complete set of one-variable functions $\{\chi_i(x)\}$. This argument is readily extended to more than two variables, so that the exact wave function for the ground and excited states of our N-electron problem can be written as a linear combination of all possible N-electron Slater determinants formed from a complete set of spin orbitals $\{\chi_i\}$.

Since all possible determinants can be described by reference to the Hartree-Fock determinant, we can write the exact wave function for any state of the system as

$$|\Phi\rangle = c_0|\Psi_0\rangle + \sum_{ra} c_a^r|\Psi_a^r\rangle + \sum_{\substack{a<b \\ r<s}} c_{ab}^{rs}|\Psi_{ab}^{rs}\rangle + \sum_{\substack{a<b<c \\ r<s<t}} c_{abc}^{rst}|\Psi_{abc}^{rst}\rangle + \cdots \tag{2.73}$$

By summing over $a < b$, we mean summing over all a *and* over all b greater than a (i.e., over all unique pairs of occupied spin orbitals). Similarly, summing over $r < s$ means summing over all unique pairs of virtual spin orbitals. Thus all unique doubly excited configurations are included in the expansion. The situation is analogous for triply and higher excited determinants. Thus the infinite set of N-electron determinants $\{|\Psi_i\rangle\} = \{|\Psi_0\rangle, |\Psi_a^r\rangle, |\Psi_{ab}^{rs}\rangle, \ldots\}$ is a complete set for the expansion of any N-electron wave function. The exact energies of the ground and excited states of the system are the eigenvalues of the Hamiltonian matrix (i.e., the matrix with elements $\langle\Psi_i|\mathscr{H}|\Psi_j\rangle$) formed from the complete set $\{|\Psi_i\rangle\}$. Since every $|\Psi_i\rangle$ can be defined by specifying a "configuration" of spin orbitals from which it is formed, this procedure is called *configuration interaction* (CI); CI will be considered in some detail in Chapter 4. The lowest eigenvalue of the Hamiltonian matrix, denoted by $\mathscr{E}_0$, is the exact nonrelativistic ground state energy of the system within the Born-Oppenheimer approximation. The difference between this exact energy, $\mathscr{E}_0$, and the Hartree-Fock-limit energy, E_0, is called the *correlation energy*

$$E_{\text{corr}} = \mathscr{E}_0 - E_0 \tag{2.74}$$

since the motion of electrons with opposite spins is not correlated within the Hartree-Fock approximation.

Unfortunately, the above procedure for the complete solution to the many-electron problem cannot be implemented in practice because one cannot handle infinite basis sets. If we work with a finite set of spin orbitals $\{\chi_i | i = 1, 2, \ldots, 2K\}$, then the $\binom{2K}{N}$ determinants formed from these spin orbitals do not form a complete N-electron basis. Nevertheless, diagonalizing the finite Hamiltonian matrix formed from this set of determinants leads to solutions that are exact within the one-electron subspace spanned by the $2K$ spin orbitals or, equivalently, within the N-electron subspace spanned by the $\binom{2K}{N}$ determinants. This procedure is called *full* CI. Even for relatively small systems and minimal basis sets, the number of determinants that must be included in a full CI calculation is extremely large. Thus in practice one must truncate the full CI expansion and use only a small fraction of the $\binom{2K}{N}$ possible determinants. Figure 2.9 schematically shows how the exact nonrelativistic Born-Oppenheimer wave function is approached as the size of the one-electron and N-electron basis sets increases.

Exercise 2.7 A minimal basis set for benzene consists of 72 spin orbitals. Calculate the size of the full CI matrix if it would be formed from determinants. How many singly excited determinants are there? How many doubly excited determinants are there?

Let us illustrate the above ideas with our minimal basis H_2 model. Recall (see Eq. (2.60)) that there are four ($2K = 4$) spin orbitals $\chi_1, \chi_2, \chi_3,$

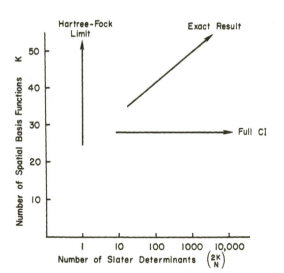

Figure 2.9 Dependence of calculations on size of one-electron and N-electron basis sets.

and χ_4 in this model. Since $N = 2$ we can form $\begin{pmatrix} 4 \\ 2 \end{pmatrix} = \dfrac{4!}{2!2!} = 6$ unique determinants. The Hartree-Fock ground state determinant is

$$|\Psi_0\rangle = |\chi_1\chi_2\rangle = |\psi_1\bar{\psi}_1\rangle = |1\bar{1}\rangle \qquad \begin{matrix} u & \text{———} & 2 \\ g & \text{—↿⇂—} & 1 \end{matrix} \qquad (2.75)$$

The singly excited determinants are

$$|\Psi_1^2\rangle = |2\bar{1}\rangle \qquad \begin{matrix} u & \text{—↑—} & 2 \\ g & \text{—↓—} & 1 \end{matrix} \qquad (2.76a)$$

$$|\Psi_{\bar{1}}^{\bar{2}}\rangle = |\bar{2}1\rangle \qquad \begin{matrix} u & \text{—↓—} & 2 \\ g & \text{—↓—} & 1 \end{matrix} \qquad (2.76b)$$

$$|\Psi_{\bar{1}}^2\rangle = |12\rangle \qquad \begin{matrix} u & \text{—↑—} & 2 \\ g & \text{—↑—} & 1 \end{matrix} \qquad (2.76c)$$

$$|\Psi_{\bar{1}}^{\bar{2}}\rangle = |1\bar{2}\rangle \qquad \begin{matrix} u & \text{—↓—} & 2 \\ g & \text{—↑—} & 1 \end{matrix} \qquad (2.76d)$$

There is only one doubly excited determinant,

$$|\Psi_{1\bar{1}}^{2\bar{2}}\rangle = |2\bar{2}\rangle = |\chi_3\chi_4\rangle = |\Psi_{12}^{34}\rangle \qquad \begin{matrix} u & \text{—↿⇂—} & 2 \\ g & \text{———} & 1 \end{matrix} \qquad (2.77)$$

Within the space spanned by the minimal basis set, the exact wave functions will be linear combinations of these six determinants. The Hartree-Fock ground state has two electrons in a gerade orbital and is of g symmetry (plus times plus equals plus). The doubly excited determinant has two electrons in an ungerade orbital and hence is also of g symmetry (minus times minus equals plus). The singly excited determinants, on the other hand, have one electron in a gerade orbital and one electron in an ungerade orbital and, therefore, are of u symmetry (plus times minus equals minus). The exact ground state wave function of minimal basis H_2, $|\Phi_0\rangle$, like the Hartree-Fock approximation to it, $|\Psi_0\rangle$, is of g symmetry. Therefore, only determinants of g symmetry can appear in the expansion of $|\Phi_0\rangle$ and thus we have

$$|\Phi_0\rangle = c_0|\Psi_0\rangle + c_{1\bar{1}}^{2\bar{2}}|\Psi_{1\bar{1}}^{2\bar{2}}\rangle = c_0|\Psi_0\rangle + c_{12}^{34}|\Psi_{12}^{34}\rangle \qquad (2.78)$$

The exact value of the coefficients in (2.78), which describe the wave function $|\Phi_0\rangle$ and the value of the exact energy $\langle\Phi_0|\mathcal{H}|\Phi_0\rangle$, can be found by diagonalizing the full CI matrix, i.e., the 2×2 Hamiltonian matrix in the basis $|\Psi_0\rangle$ and $|\Psi_{1\bar{1}}^{2\bar{2}}\rangle$,

$$\mathbf{H} = \begin{pmatrix} \langle\Psi_0|\mathcal{H}|\Psi_0\rangle & \langle\Psi_0|\mathcal{H}|\Psi_{1\bar{1}}^{2\bar{2}}\rangle \\ \langle\Psi_{1\bar{1}}^{2\bar{2}}|\mathcal{H}|\Psi_0\rangle & \langle\Psi_{1\bar{1}}^{2\bar{2}}|\mathcal{H}|\Psi_{1\bar{1}}^{2\bar{2}}\rangle \end{pmatrix} \qquad (2.79)$$

To proceed any further with this problem, or with most other formulations encountered in quantum chemistry, we need to be able to evaluate matrix elements of the Hamiltonian between determinants. The evaluation of such matrix elements is discussed in the next section.

2.3 OPERATORS AND MATRIX ELEMENTS

This section considers the problem of evaluating matrix elements of operators between Slater determinants formed from orthonormal orbitals. Given an operator $\mathcal{O}$ and two N-electron determinants $|K\rangle$ and $|L\rangle$, our problem is to evaluate $\langle K|\mathcal{O}|L\rangle$. By evaluating such matrix elements, we mean reducing them to integrals involving the individual spin orbitals χ_i occupied in $|K\rangle$ and $|L\rangle$, and ultimately to integrals involving spatial orbitals ψ_i. Before giving general rules for evaluating such matrix elements, we illustrate the procedure with our minimal basis H_2 model.

2.3.1 Minimal Basis H_2 Matrix Elements

Let us evaluate the matrix elements that appear in the full CI matrix of minimal basis H_2 (see Eq. (2.79)). The exact ground state of this model is a linear combination of the Hartree-Fock ground state $|\Psi_0\rangle = |\chi_1\chi_2\rangle = |1\bar{1}\rangle$ and the doubly excited state $|\Psi_{12}^{34}\rangle = |\chi_3\chi_4\rangle \equiv |\Psi_{1\bar{1}}^{2\bar{2}}\rangle = |2\bar{2}\rangle$. We need to evaluate the diagonal elements $\langle\Psi_0|\mathcal{H}|\Psi_0\rangle$ and $\langle\Psi_{12}^{34}|\mathcal{H}|\Psi_{12}^{34}\rangle$ (the Hartree-Fock ground state energy and the energy of the doubly excited state, respectively), as well as the off-diagonal elements $\langle\Psi_0|\mathcal{H}|\Psi_{12}^{34}\rangle$ and $\langle\Psi_{12}^{34}|\mathcal{H}|\Psi_0\rangle$.

The Hamiltonian for any two-electron system is

$$\mathcal{H} = \left(-\frac{1}{2}\nabla_1^2 - \sum_A \frac{Z_A}{r_{1A}}\right) + \left(-\frac{1}{2}\nabla_2^2 - \sum_A \frac{Z_A}{r_{2A}}\right) + \frac{1}{r_{12}}$$

$$= h(1) + h(2) + \frac{1}{r_{12}} \tag{2.80}$$

where $h(1)$ is a *core-Hamiltonian* for electron-one, describing its kinetic energy and potential energy in the field of the nuclei (the "core"). It will be convenient to separate the total Hamiltonian into its one- and two-electron parts

$$\mathcal{O}_1 = h(1) + h(2) \tag{2.81}$$

$$\mathcal{O}_2 = r_{12}^{-1} \tag{2.82}$$

Let us first consider the matrix element $\langle\Psi_0|\mathcal{O}_1|\Psi_0\rangle$ which, from (2.81), is a sum of two terms. The first term is

$$\langle\Psi_0|h(1)|\Psi_0\rangle = \int dx_1\, dx_2\, [2^{-1/2}(\chi_1(x_1)\chi_2(x_2) - \chi_2(x_1)\chi_1(x_2))]^*$$
$$\times h(r_1)[2^{-1/2}(\chi_1(x_1)\chi_2(x_2) - \chi_2(x_1)\chi_1(x_2))]$$

$$= \frac{1}{2}\int dx_1\, dx_2\, \{\chi_1^*(x_1)\chi_2^*(x_2)h(r_1)\chi_1(x_1)\chi_2(x_2) + \chi_2^*(x_1)\chi_1^*(x_2)h(r_1)\chi_2(x_1)\chi_1(x_2)$$

$$- \chi_1^*(x_1)\chi_2^*(x_2)h(r_1)\chi_2(x_1)\chi_1(x_2) - \chi_2^*(x_1)\chi_1^*(x_2)h(r_1)\chi_1(x_1)\chi_2(x_2)\}$$

$$(2.83)$$

In the above four terms the integration over x_2 produces either 1 (first two terms) or 0 (last two terms) from the orthonormality of the spin orbitals. Thus

$$\langle\Psi_0|h(1)|\Psi_0\rangle = \frac{1}{2}\int dx_1\, \chi_1^*(x_1)h(r_1)\chi_1(x_1) + \frac{1}{2}\int dx_1\, \chi_2^*(x_1)h(r_1)\chi_2(x_1)$$

$$(2.84)$$

By exactly the same procedure, one finds that $\langle\Psi_0|h(2)|\Psi_0\rangle = \langle\Psi_0|h(1)|\Psi_0\rangle$ and thus

$$\langle\Psi_0|\mathcal{O}_1|\Psi_0\rangle = \int dx_1\, \chi_1^*(x_1)h(r_1)\chi_1(x_1) + \int dx_1\, \chi_2^*(x_1)h(r_1)\chi_2(x_1) \quad (2.85)$$

The integrals in this expression are *one-electron integrals*, i.e., the integration is over the coordinates of a single electron. The dummy variables of integration are, by convention, chosen to be the coordinates of electron-one. Introducing the following notation for one-electron integrals involving spin orbitals,

$$\langle i|h|j\rangle = \langle\chi_i|h|\chi_j\rangle = \int dx_1\, \chi_i^*(x_1)h(r_1)\chi_j(x_1) \quad (2.86)$$

we have

$$\langle\Psi_0|\mathcal{O}_1|\Psi_0\rangle = \langle1|h|1\rangle + \langle2|h|2\rangle \quad (2.87)$$

Exercise 2.8 Show that

$$\langle\Psi_{12}^{34}|\mathcal{O}_1|\Psi_{12}^{34}\rangle = \langle3|h|3\rangle + \langle4|h|4\rangle$$

and

$$\langle\Psi_0|\mathcal{O}_1|\Psi_{12}^{34}\rangle = \langle\Psi_{12}^{34}|\mathcal{O}_1|\Psi_0\rangle = 0$$

Now, let us evaluate matrix elements of $\mathcal{O}_2$.

$$\langle\Psi_0|\mathcal{O}_2|\Psi_0\rangle = \int dx_1\,dx_2\,[2^{-1/2}(\chi_1(\mathbf{x}_1)\chi_2(\mathbf{x}_2) - \chi_2(\mathbf{x}_1)\chi_1(\mathbf{x}_2))]^*$$
$$\times\; r_{12}^{-1}[2^{-1/2}(\chi_1(\mathbf{x}_1)\chi_2(\mathbf{x}_2) - \chi_2(\mathbf{x}_1)\chi_1(\mathbf{x}_2))]$$
$$= \frac{1}{2}\int dx_1\,dx_2\,\{\chi_1^*(\mathbf{x}_1)\chi_2^*(\mathbf{x}_2)r_{12}^{-1}\chi_1(\mathbf{x}_1)\chi_2(\mathbf{x}_2) + \chi_2^*(\mathbf{x}_1)\chi_1^*(\mathbf{x}_2)r_{12}^{-1}\chi_2(\mathbf{x}_1)\chi_1(\mathbf{x}_2)$$
$$-\;\chi_1^*(\mathbf{x}_1)\chi_2^*(\mathbf{x}_2)r_{12}^{-1}\chi_2(\mathbf{x}_1)\chi_1(\mathbf{x}_2) - \chi_2^*(\mathbf{x}_1)\chi_1^*(\mathbf{x}_2)r_{12}^{-1}\chi_1(\mathbf{x}_1)\chi_2(\mathbf{x}_2)\}$$
$$(2.88)$$

Since $r_{12} = r_{21}$, we can interchange the dummy variables of integration in the second term of the above expression and show that it is equal to the first term. Similarly, the third and fourth terms are equal. Thus

$$\langle\Psi_0|\mathcal{O}_2|\Psi_0\rangle = \int dx_1\,dx_2\,\chi_1^*(\mathbf{x}_1)\chi_2^*(\mathbf{x}_2)r_{12}^{-1}\chi_1(\mathbf{x}_1)\chi_2(\mathbf{x}_2)$$
$$-\int dx_1\,dx_2\,\chi_1^*(\mathbf{x}_1)\chi_2^*(\mathbf{x}_2)r_{12}^{-1}\chi_2(\mathbf{x}_1)\chi_1(\mathbf{x}_2) \qquad (2.89)$$

The integrals in this expression are examples of *two-electron integrals*, i.e., the integration is over the eight space and spin coordinates of electron 1 and 2. It is conventional *always* to choose the dummy variables of integration in a two-electron integral to be the coordinates of electrons 1 and 2. Introducing the following notation for two-electron integrals involving spin orbitals,

$$\langle ij|kl\rangle = \langle\chi_i\chi_j|\chi_k\chi_l\rangle = \int dx_1\,dx_2\,\chi_i^*(\mathbf{x}_1)\chi_j^*(\mathbf{x}_2)r_{12}^{-1}\chi_k(\mathbf{x}_1)\chi_l(\mathbf{x}_2) \quad (2.90)$$

we have

$$\langle\Psi_0|\mathcal{O}_2|\Psi_0\rangle = \langle 12|12\rangle - \langle 12|21\rangle \qquad (2.91)$$

and the Hartree-Fock ground state energy is

$$\langle\Psi_0|\mathscr{H}|\Psi_0\rangle = \langle\Psi_0|\mathcal{O}_1 + \mathcal{O}_2|\Psi_0\rangle$$
$$= \langle 1|h|1\rangle + \langle 2|h|2\rangle + \langle 12|12\rangle - \langle 12|21\rangle \qquad (2.92)$$

Exercise 2.9 Using the above approach, show that the full CI matrix for minimal basis H_2 is

$$\mathscr{H} = \begin{pmatrix} \langle 1|h|1\rangle + \langle 2|h|2\rangle \\ +\langle 12|12\rangle - \langle 12|21\rangle & \langle 12|34\rangle - \langle 12|43\rangle \\ \langle 34|12\rangle - \langle 34|21\rangle & \langle 3|h|3\rangle + \langle 4|h|4\rangle \\ & +\langle 34|34\rangle - \langle 34|43\rangle \end{pmatrix}$$

and that it is Hermitian.

2.3.2 Notations for One- and Two-Electron Integrals

Before generalizing the above results and presenting general expressions for matrix elements involving N-electron determinants, it is appropriate to summarize the different notations we use in this book for one- and two-electron integrals. The notation for two-electron integrals over spin orbitals that we have introduced in Eq. (2.90), i.e.,

$$\langle ij|kl \rangle = \langle \chi_i \chi_j | \chi_k \chi_l \rangle = \int d\mathbf{x}_1 \, d\mathbf{x}_2 \, \chi_i^*(\mathbf{x}_1)\chi_j^*(\mathbf{x}_2)r_{12}^{-1}\chi_k(\mathbf{x}_1)\chi_l(\mathbf{x}_2) \quad (2.93)$$

is often referred to as the physicists' notation. Note that the complex conjugated spin orbitals appear side by side on the left and the space-spin coordinate of electron-one appears first. It is clear from this definition that

$$\langle ij|kl \rangle = \langle ji|lk \rangle \quad (2.94)$$

and that

$$\langle ij|kl \rangle = \langle kl|ij \rangle^* \quad (2.95)$$

Because two-electron integrals often appear in the following combination, we introduce a special symbol for an *antisymmetrized* two-electron integral

$$\langle ij||kl \rangle = \langle ij|kl \rangle - \langle ij|lk \rangle$$
$$= \int d\mathbf{x}_1 \, d\mathbf{x}_2 \, \chi_i^*(\mathbf{x}_1)\chi_j^*(\mathbf{x}_2)r_{12}^{-1}(1 - \mathscr{P}_{12})\chi_k(\mathbf{x}_1)\chi_l(\mathbf{x}_2) \quad (2.96)$$

where $\mathscr{P}_{12}$ is an operator which interchanges the coordinates of electron one and two. Note that

$$\langle ij||kk \rangle = 0 \quad (2.97)$$

It is an unfortunate fact of life that there is another notation for two-electron integrals over spin orbitals in common use, particularly in the literature of Hartree-Fock theory. This notation, often referred to as the chemists' notation, is

$$[ij|kl] = \int d\mathbf{x}_1 \, d\mathbf{x}_2 \, \chi_i^*(\mathbf{x}_1)\chi_j(\mathbf{x}_1)r_{12}^{-1}\chi_k^*(\mathbf{x}_2)\chi_l(\mathbf{x}_2) \quad (2.98)$$

Note that in this notation spin orbitals, which are functions of the coordinate of electron-one, appear side by side on the left and the complex conjugated spin orbital appears first. By interchanging the dummy variables of integration, one has

$$[ij|kl] = [kl|ij] \quad (2.99a)$$

In addition, if the spin orbitals are real, as is almost always the case in molecular Hartree-Fock calculations, one has

$$[ij|kl] = [ji|kl] = [ij|lk] = [ji|lk] \quad (2.99b)$$

Table 2.2 Notations for one- and two-electron integrals over spin (χ) and spatial (ψ) orbitals

SPIN ORBITALS

$$[i|h|j] = \langle i|h|j \rangle = \int dx_1 \, \chi_i^*(x_1)h(r_1)\chi_j(x_1)$$

$$\langle ij|kl \rangle = \langle \chi_i\chi_j|\chi_k\chi_l \rangle = \int dx_1 \, dx_2 \, \chi_i^*(x_1)\chi_j^*(x_2)r_{12}^{-1}\chi_k(x_1)\chi_l(x_2) = [ik|jl]$$

$$[ij|kl] = [\chi_i\chi_j|\chi_k\chi_l] = \int dx_1 \, dx_2 \, \chi_i^*(x_1)\chi_j(x_1)r_{12}^{-1}\chi_k^*(x_2)\chi_l(x_2) = \langle ik|jl \rangle$$

$$\langle ij||kl \rangle = \langle ij|kl \rangle - \langle ij|lk \rangle = \int dx_1 \, dx_2 \, \chi_i^*(x_1)\chi_j^*(x_2)r_{12}^{-1}(1 - \mathscr{P}_{12})\chi_k(x_1)\chi_l(x_2)$$

SPATIAL ORBITALS

$$(i|h|j) = h_{ij} = (\psi_i|h|\psi_j) = \int dr_1 \, \psi_i^*(r_1)h(r_1)\psi_j(r_1)$$

$$(ij|kl) = (\psi_i\psi_j|\psi_k\psi_l) = \int dr_1 \, dr_2 \, \psi_i^*(r_1)\psi_j(r_1)r_{12}^{-1}\psi_k^*(r_2)\psi_l(r_2)$$

$$J_{ij} = (ii|jj) \quad \text{Coulomb integrals}$$

$$K_{ij} = (ij|ji) \quad \text{Exchange integrals}$$

For one-electron integrals over spin orbitals, the chemists' and physicists' notations are essentially the same.

$$[i|h|j] = \langle i|h|j \rangle = \int dx_1 \, \chi_i^*(x_1)h(r_1)\chi_j(x_1) \tag{2.100}$$

Table 2.2 summarizes all the notations for one- and two-electron integrals used in this book. When we consider the reduction of integrals over spin orbitals to integrals over spatial orbitals later in this chapter, we will introduce a new notation for spatial integrals, which we have included in the table for the sake of completeness and ease of future reference.

2.3.3 General Rules for Matrix Elements

We have seen that it is fairly easy to evaluate matrix elements between two-electron Slater determinants. The N-electron case is more complicated, and here we simply present a set of rules that can be used to evaluate matrix elements and leave their derivation to the next subsection, which can be skipped, if desired.

There are two types of operators in quantum chemistry. The first type is a sum of one-electron operators

$$\mathcal{O}_1 = \sum_{i=1}^{N} h(i) \tag{2.101}$$

where $h(i)$ is any operator involving only the ith electron. These operators represent dynamic variables that depend only on the position or momentum of the electron in question, independent of the position or momentum of other electrons. Examples are operators for the kinetic energy, attraction of an electron to a nucleus, dipole moment, and most of the other operators that one encounters. The second type of operator is a sum of two-electron operators

$$\mathcal{O}_2 = \sum_{i=1}^{N} \sum_{j>i}^{N} v(i,j) \equiv \sum_{i<j} v(i,j) \qquad (2.102)$$

where $v(i,j)$ is an operator that depends on the position (or momentum) of both the ith and jth electron. The sum in (2.102) is over all unique pairs of electrons. The coulomb interaction between two electrons

$$v(i,j) = r_{ij}^{-1} \qquad (2.103)$$

is a two-electron operator.

The rules for evaluating the matrix element $\langle K|\mathcal{O}|L\rangle$ between the determinants $|K\rangle$ and $|L\rangle$ depend on whether the operator $\mathcal{O}$ is a sum of one-electron operators ($\mathcal{O}_1$) or a sum of two-electron operators ($\mathcal{O}_2$). In addition, the value of $\langle K|\mathcal{O}|L\rangle$ depends on the degree to which the two determinants $|K\rangle$ and $|L\rangle$ differ. We can distinguish three cases. The first, Case 1, is when the two determinants are identical, i.e., the matrix element is a diagonal matrix element $\langle K|\mathcal{O}|K\rangle$. For this case, we choose the determinant to be

$$|K\rangle = |\cdots \chi_m \chi_n \cdots\rangle \qquad (2.104)$$

The second, Case 2, is when the two determinants differ by one spin orbital, χ_m in $|K\rangle$ being replaced by χ_p in $|L\rangle$.

$$|L\rangle = |\cdots \chi_p \chi_n \cdots\rangle \qquad (2.105)$$

The third, Case 3, is when the two determinants differ by two spin orbitals, χ_m and χ_n in $|K\rangle$ being replaced by χ_p and χ_q, respectively, in $|L\rangle$,

$$|L\rangle = |\cdots \chi_p \chi_q \cdots\rangle \qquad (2.106)$$

When the two determinants differ by three or more spin orbitals the matrix element is always zero.

Tables 2.3 and 2.4 summarize the rules for the three cases. Note that the larger the difference in the two determinants, the simpler is the matrix element, i.e., the fewer number of terms it involves. The one-electron matrix elements are zero if the two determinants differ by *two* or more spin orbitals, in the same way that the two-electron matrix elements are zero if the two determinants differ by *three* or more spin orbitals. In the tables, m and n denote spin orbitals occupied in $|K\rangle$, so that sums over these indices include all N spin orbitals in that determinant.

To use the rules, the two determinants must first be in *maximum coincidence*. Consider, for example, a matrix element between $|\Psi_1\rangle$ and $|\Psi_2\rangle$

Table 2.3 Matrix elements between determinants for one-electron operators in terms of spin orbitals

$$\mathcal{O}_1 = \sum_{i=1}^{N} h(i)$$

Case 1: $|K\rangle = |\cdots mn \cdots\rangle$

$$\langle K|\mathcal{O}_1|K\rangle = \sum_{m}^{N} [m|h|m] = \sum_{m}^{N} \langle m|h|m\rangle$$

Case 2: $|K\rangle = |\cdots mn \cdots\rangle$
$|L\rangle = |\cdots pn \cdots\rangle$

$$\langle K|\mathcal{O}_1|L\rangle = [m|h|p] = \langle m|h|p\rangle$$

Case 3: $|K\rangle = |\cdots mn \cdots\rangle$
$|L\rangle = |\cdots pq \cdots\rangle$

$$\langle K|\mathcal{O}_1|L\rangle = 0$$

Table 2.4 Matrix elements between determinants for two-electron operators in terms of spin orbitals

$$\mathcal{O}_2 = \sum_{i=1}^{N} \sum_{j>i}^{N} r_{ij}^{-1}$$

Case 1: $|K\rangle = |\cdots mn \cdots\rangle$

$$\langle K|\mathcal{O}_2|K\rangle = \frac{1}{2}\sum_{m}^{N}\sum_{n}^{N} [mm|nn] - [mn|nm] = \frac{1}{2}\sum_{m}^{N}\sum_{n}^{N} \langle mn||mn\rangle$$

Case 2: $|K\rangle = |\cdots mn \cdots\rangle$
$|L\rangle = |\cdots pn \cdots\rangle$

$$\langle K|\mathcal{O}_2|L\rangle = \sum_{n}^{N} [mp|nn] - [mn|np] = \sum_{n}^{N} \langle mn||pn\rangle$$

Case 3: $|K\rangle = |\cdots mn \cdots\rangle$
$|L\rangle = |\cdots pq \cdots\rangle$

$$\langle K|\mathcal{O}_2|L\rangle = [mp|nq] - [mq|np] = \langle mn||pq\rangle$$

where

$$|\Psi_1\rangle = |abcd\rangle$$
$$|\Psi_2\rangle = |crds\rangle$$

At first glance, it might appear that the two determinants differ in all four columns; however, by interchanging columns of $|\Psi_2\rangle$ and keeping track of the sign, we have

$$|\Psi_2\rangle = |crds\rangle = -|crsd\rangle = |srcd\rangle$$

After being placed in maximum coincidence, they differ in two columns, and we can use the Case 3 rules. Using the following correspondence

$$|K\rangle \equiv |\Psi_1\rangle \qquad\qquad |L\rangle \equiv |\Psi_2\rangle$$

$$m \equiv a \qquad\qquad p \equiv s$$

$$n \equiv b \qquad\qquad q \equiv r$$

we thus have $\langle\Psi_1|\mathcal{O}_1|\Psi_2\rangle = 0$ and $\langle\Psi_1|\mathcal{O}_2|\Psi_2\rangle = \langle ab||sr\rangle$.

Using Tables 2.3 and 2.4, we can immediately write down the expression for the energy of a single determinant $|K\rangle$, i.e.,

$$\langle K|\mathscr{H}|K\rangle = \langle K|\mathcal{O}_1 + \mathcal{O}_2|K\rangle = \sum_m^N \langle m|h|m\rangle + \frac{1}{2}\sum_m^N\sum_n^N \langle mn||mn\rangle \quad (2.107)$$

where

$$h(i) = -\frac{1}{2}\nabla_i^2 - \sum_A \frac{Z_A}{r_{iA}} \qquad (2.108)$$

The sum in (2.107) is over the spin orbitals occupied in $|K\rangle$. Since (see Eq. (2.97))

$$\langle mm||mm\rangle = \langle nn||nn\rangle = 0 \qquad (2.109a)$$

and

$$\langle mn||mn\rangle = \langle nm||nm\rangle \qquad (2.109b)$$

the expression (2.107) can be rewritten as

$$\langle K|\mathscr{H}|K\rangle = \sum_m^N \langle m|h|m\rangle + \sum_m^N\sum_{n>m}^N \langle mn||mn\rangle$$

$$= \sum_m [m|h|m] + \sum_m^N\sum_{n>m}^N [mm|nn] - [mn|nm] \qquad (2.110)$$

The summation of antisymmetrized two-electron integrals is thus over *all unique pairs* of spin orbitals χ_m and χ_n occupied in $|K\rangle$. This observation suggests a simple mnemonic device for writing down the energy of any single determinant in terms of one- and two-electron integrals over spin orbitals. *Each occupied spin orbital χ_i contributes a term $\langle i|h|i\rangle$ to the energy, and every unique pair of occupied spin orbitals χ_i, χ_j contributes a term $\langle ij||ij\rangle$ to the energy.* Thus we can think of the total energy of an N-electron system, which is described by a Slater determinant, as the sum of "one-electron-energies" ($\langle i|h|i\rangle$ for an electron in spin orbital χ_i) plus the sum of unique pair-wise "interaction-energies" ($\langle ij||ij\rangle$ for a pair of electrons in spin orbitals χ_i and χ_j). In using this language, remember that it is only a mnemonic device. The physical interaction between two electrons is described by the coulomb repulsion term (r_{ij}^{-1}) in the Hamiltonian and not by an antisymmetrized two-electron integral.

Exercise 2.10 Derive Eq. (2.110) from Eq. (2.107).

Exercise 2.11 If $|K\rangle = |\chi_1\chi_2\chi_3\rangle$ show that

$$\langle K|\mathscr{H}|K\rangle = \langle 1|h|1\rangle + \langle 2|h|2\rangle + \langle 3|h|3\rangle + \langle 12||12\rangle$$
$$+ \langle 13||13\rangle + \langle 23||23\rangle$$

In this book we will often need matrix elements involving the Hartree-Fock ground state. For convenience, we have rewritten the rules in Tables 2.3 and 2.4 by identifying the labels m and n with a and b (occupied Hartree-Fock spin orbitals) and the labels p and q with r and s (unoccupied Hartree-Fock spin orbitals). Tables 2.5 and 2.6 contain matrix elements between the Hartree-Fock ground state and either itself (Case 1), a singly excited deter-

Table 2.5 Matrix elements with the Hartree-Fock ground state for one-electron operators

$$\mathscr{O}_1 = \sum_{i=1}^{N} h(i)$$

Case 1:	$\langle\Psi_0|\mathscr{O}_1|\Psi_0\rangle = \sum_a^N [a|h|a] = \sum_a^N \langle a|h|a\rangle$
Case 2:	$\langle\Psi_0|\mathscr{O}_1|\Psi_a^r\rangle = [a|h|r] = \langle a|h|r\rangle$
Case 3:	$\langle\Psi_0|\mathscr{O}_1|\Psi_{ab}^{rs}\rangle = 0$

Table 2.6 Matrix elements with the Hartree-Fock ground state for two-electron operators

$$\mathscr{O}_2 = \sum_{i=1}^{N}\sum_{j>i}^{N} r_{ij}^{-1}$$

Case 1:	$\langle\Psi_0|\mathscr{O}_2|\Psi_0\rangle = \frac{1}{2}\sum_a^N\sum_b^N [aa|bb] - [ab|ba] = \frac{1}{2}\sum_a^N\sum_b^N \langle ab||ab\rangle$
Case 2:	$\langle\Psi_0|\mathscr{O}_2|\Psi_a^r\rangle = \sum_b^N [ar|bb] - [ab|br] = \sum_b^N \langle ab||rb\rangle$
Case 3:	$\langle\Psi_0|\mathscr{O}_2|\Psi_{ab}^{rs}\rangle = [ar|bs] - [as|br] = \langle ab||rs\rangle$

minant (Case 2), or a doubly excited determinant (Case 3). Using these tables, we see that the energy of the Hartree-Fock ground state is

$$E_0 = \langle \Psi_0 | \mathscr{H} | \Psi_0 \rangle = \sum_a^N [a|h|a] + \frac{1}{2} \sum_a^N \sum_b^N [aa|bb] - [ab|ba] \quad (2.111)$$

using the chemists' notation, or equivalently

$$E_0 = \sum_a^N \langle a|h|a \rangle + \frac{1}{2} \sum_a^N \sum_b^N \langle ab||ab \rangle \quad (2.112)$$

using the physicists' notation. As shown above, expression (2.112) can be rewritten as

$$E_0 = \sum_a^N \langle a|h|a \rangle + \sum_a^N \sum_{b>a}^N \langle ab||ab \rangle \quad (2.113)$$

For minimal basis set H_2, $|\Psi_0\rangle = |\chi_1\chi_2\rangle$ so that from (2.113), we have

$$\begin{aligned} E_0 &= \langle 1|h|1 \rangle + \langle 2|h|2 \rangle + \langle 12||12 \rangle \\ &= \langle 1|h|1 \rangle + \langle 2|h|2 \rangle + \langle 12|12 \rangle - \langle 12|21 \rangle \quad (2.114) \end{aligned}$$

in agreement with our previous result in Eq. (2.92).

Exercise 2.12 Evaluate the matrix elements that occur in the minimal basis H_2 full CI matrix (Eq. (2.79)) using the rules. Compare with the result obtained in Exercise 2.9.

Exercise 2.13 Show that $\langle \Psi_a^r | \mathcal{O}_1 | \Psi_b^s \rangle$

$$\begin{aligned} &= 0 &&\text{if } a \neq b, r \neq s \\ &= \langle r|h|s \rangle &&\text{if } a = b, r \neq s \\ &= -\langle b|h|a \rangle &&\text{if } a \neq b, r = s \\ &= \sum_c^N \langle c|h|c \rangle - \langle a|h|a \rangle + \langle r|h|r \rangle &&\text{if } a = b, r = s \end{aligned}$$

Exercise 2.14 The Hartree-Fock ground state energy for an N-electron system is ${}^N E_0 = \langle {}^N\Psi_0 | \mathscr{H} | {}^N\Psi_0 \rangle$. Consider a state of the ionized system (in which an electron has been removed from spin orbital χ_a) with energy ${}^{N-1}E_a = \langle {}^{N-1}\Psi_a | \mathscr{H} | {}^{N-1}\Psi_a \rangle$, where $|{}^{N-1}\Psi_a\rangle$ is a single determinant with all spin orbitals but χ_a occupied,

$$|{}^{N-1}\Psi_a\rangle = |\chi_1\chi_2 \cdots \chi_{a-1}\chi_{a+1} \cdots \chi_N\rangle$$

Show, using the rules in the tables, that the energy required for this ionization process is

$${}^N E_0 - {}^{N-1}E_a = \langle a|h|a \rangle + \sum_b^N \langle ab||ab \rangle$$

To show the power and simplicity of the mnemonic device introduced in this subsection, let us derive the above result without doing any algebra. Consider the representation of $|^N\Psi_0\rangle$ in Fig. 2.4. If we remove an electron from χ_a, we lose the "one-electron energy" contribution $\langle a|h|a\rangle$ to $^NE_0$. Moreover, we lose the pair-wise contributions arising from the "interaction" of the electron in χ_a with the remaining electrons $\left(\text{i.e., } \sum_{b \neq a}^N \langle ab||ab\rangle\right)$ Because $\langle aa||aa\rangle = 0$, the above result follows immediately.

2.3.4 Derivation of the Rules for Matrix Elements

In this section we derive the rules in Tables 2.3 and 2.4 for matrix elements of one- and two-electron operators between N-electron determinants formed from orthonormal spin orbitals. The definition of an N-electron Slater determinants containing the spin orbitals $\chi_i(\mathbf{x}_1), \chi_j(\mathbf{x}_2), \ldots, \chi_k(\mathbf{x}_N)$ is (see Eq. (1.38))

$$|\chi_i\chi_j\cdots\chi_k\rangle = (N!)^{-1/2}\sum_{n=1}^{N!}(-1)^{p_n}\mathcal{P}_n\{\chi_i(1)\chi_j(2)\cdots\chi_k(N)\} \quad (2.115)$$

where we have let $\chi(\mathbf{x}_l) \equiv \chi(l)$. $\mathcal{P}_n$ is an operator that generates the nth permutation of the electron labels $1, 2, \ldots, N$ and p_n is the number of transpositions (simple interchanges) required to obtain this permutation.

Exercise 2.15 Generalize the result of Exercise 2.4 to N-electron Slater determinants. Show that the Slater determinant $|\chi_i\chi_j\cdots\chi_k\rangle$ formed from spin orbitals, which are eigenfunctions of the one-electron operator h as in Eq. (2.29), is an eigenfunction of the independent-electron Hamiltonian (2.28), $\mathcal{H} = \sum_{i=1}^N h(i)$, with an eigenvalue $\varepsilon_i + \varepsilon_j + \cdots + \varepsilon_k$. *Hint:* Since $\mathcal{H}$ is invariant to permutations of the electron labels, it commutes with the permutation operator $\mathcal{P}_n$.

We wish to evaluate matrix elements of the form $\langle K|\mathcal{O}|L\rangle$ where

$$|K\rangle = |\chi_m(1)\chi_n(2)\cdots\rangle \quad (2.116)$$

is a determinant, which occupies the spin orbitals $\chi_m, \chi_n, \ldots$. The determinant $|L\rangle$ differs from $|K\rangle$ in some known way. Prior to considering one- and two-electron operators and Cases 1, 2, and 3 let us set $\mathcal{O}$ equal to the unit operator and evaluate the overlap $\langle K|L\rangle$ between $|K\rangle$ and an arbitrary determinant $|L\rangle$ formed from the same set of spin orbitals,

$$|L\rangle = |\chi'_m(1)\chi'_n(2)\cdots\rangle \quad (2.117)$$

It is assumed that the two determinants have been placed in maximum coincidence. Using expression (2.115) for a determinant, we then have

$$\langle K|L\rangle = (N!)^{-1} \sum_{i}^{N!} \sum_{j}^{N!} (-1)^{p_i}(-1)^{p_j} \int dx_1 \, dx_2 \cdots dx_N$$

$$\times \mathscr{P}_i\{\chi_m^*(1)\chi_n^*(2)\cdots\}\mathscr{P}_j\{\chi_m'(1)\chi_n'(2)\cdots\} \tag{2.118}$$

The spin orbitals are assumed to form an orthonormal set. If the above overlap is to be nonzero, the primed spin orbitals must be identical with the unprimed spin orbitals. Otherwise a zero would always result from the orthogonality of some spin orbital χ_n' in $|L\rangle$ to the spin orbitals $\chi_m, \chi_n, \ldots$ in $|K\rangle$. Thus a determinant $|K\rangle$ is orthogonal to any other determinant that does not contain identical spin orbitals. If two determinants contain identical spin orbitals and are in perfect coincidence, i.e., if they are the same determinant, then

$$\langle K|K\rangle = (N!)^{-1} \sum_{i}^{N!} \sum_{j}^{N!} (-1)^{p_i}(-1)^{p_j} \int dx_1 \, dx_2 \cdots dx_N$$

$$\times \mathscr{P}_i\{\chi_m^*(1)\chi_n^*(2)\cdots\}\mathscr{P}_j\{\chi_m(1)\chi_n(2)\cdots\} \tag{2.119}$$

Now, in the above sum, integration will give zero unless each electron occupies the same spin orbital in both the ith permutation and the jth permutation. Thus the two permutations must be identical ($i = j$) and, since $(-1)^{2p_i} = 1$, we have

$$\langle K|K\rangle = (N!)^{-1} \sum_{i}^{N!} \int dx_1 \, dx_2 \cdots dx_N \, \mathscr{P}_i\{\chi_m^*(1)\chi_n^*(2)\cdots\}\mathscr{P}_i\{\chi_m(1)\chi_n(2)\cdots\}$$

$$\tag{2.120}$$

Each term in this sum is unity and, therefore,

$$\langle K|K\rangle = (N!)^{-1} \sum_{i}^{N!} 1 = 1 \tag{2.121}$$

showing that $|K\rangle$ is normalized. Thus we have

$$\langle K|K\rangle = 1 \qquad \text{Case 1}$$

$$\langle K|L\rangle = 0 \qquad \text{Case 2} \tag{2.122}$$

Next let us consider matrix elements of a sum of one-electron operators,

$$\langle K|\mathscr{O}_1|L\rangle = \langle K|h(1) + h(2) + \cdots + h(N)|L\rangle \tag{2.123}$$

Because the electrons in a determinant are indistinguishable, matrix elements of $h(1)$ will be identical to those of $h(2)$, $h(3)$, etc. Thus each term of the sum in (2.123) is identical, and we can write

$$\langle K|\mathcal{O}_1|L\rangle = N\langle K|h(1)|L\rangle \tag{2.124}$$

where, by convention, we choose to use the operator for electron 1. We begin with Case 1,

$$\langle K|\mathcal{O}_1|K\rangle = N\langle K|h(1)|K\rangle$$

$$= N(N!)^{-1} \sum_i^{N!} \sum_j^{N!} (-1)^{p_i}(-1)^{p_j} \int dx_1\, dx_2 \cdots dx_N$$

$$\times \mathcal{P}_i\{\chi_m^*(1)\chi_n^*(2)\cdots\}h(1)\mathcal{P}_j\{\chi_m(1)\chi_n(2)\cdots\} \tag{2.125}$$

Now in the integration over electrons 2, 3, ..., N, we will obtain zero unless these electrons occupy the same spin orbitals in the ith permutation as in the jth permutation, since the spin orbitals are orthonormal. If electrons 2, 3, ..., N occupy identical spin orbitals in both permutations, it must be that electron 1 also occupies the same spin orbital in both permutations. Thus only if the two permutations are identical ($i = j$) will we obtain a result different from zero.

$$\langle K|\mathcal{O}_1|K\rangle = [(N-1)!]^{-1} \sum_i^{N!} \int dx_1\, dx_2 \cdots dx_N$$

$$\times \mathcal{P}_i\{\chi_m^*(1)\chi_n^*(2)\cdots\}h(1)\mathcal{P}_i\{\chi_m(1)\chi_n(2)\cdots\} \tag{2.126}$$

In the sum over the $N!$ permutations, electron 1 will occupy each of the spin orbitals, $\{\chi_m | m = 1, 2, \ldots, N\}$, $(N-1)!$ times, i.e., if electron 1 is in a specific spin orbital χ_m, there will be $(N-1)!$ ways of arranging electrons 2, 3, ..., N amongst the other $N-1$ spin orbitals. Integration over electrons 2, 3, ..., N will always give a factor of 1 since the spin orbitals are normalized and thus,

$$\langle K|\mathcal{O}_1|K\rangle = (N-1)![(N-1)!]^{-1} \sum_m^N \int dx_1\, \chi_m^*(1)h(1)\chi_m(1)$$

$$= \sum_m^N \langle m|h|m\rangle \qquad \text{Case 1} \tag{2.127}$$

We now turn to Case 2, in which the two determinants differ by a single spin orbital, χ_p appearing in $|L\rangle$ where χ_m appears in $|K\rangle$,

$$|K\rangle = |\chi_m(1)\chi_n(2)\cdots\rangle \tag{2.128}$$

$$|L\rangle = |\chi_p(1)\chi_n(2)\cdots\rangle \tag{2.129}$$

By the same arguments we used for Case 1, to obtain (2.126) from (2.125), identical permutations must appear on either side of the operator to obtain a result different from zero

$$\langle K|\mathcal{O}_1|L\rangle = [(N-1)!]^{-1} \sum_i^{N!} \int dx_1 \, dx_2 \cdots dx_N$$

$$\times \mathscr{P}_i\{\chi_m^*(1)\chi_n^*(2)\cdots\}h(1)\mathscr{P}_i\{\chi_p(1)\chi_n(2)\cdots\} \qquad (2.130)$$

Because the spin orbital χ_m in the first permutation is orthogonal to any spin orbital in the second permutation, it must be occupied by electron 1, to "associate" it with $h(1)$ and yield a nonzero result. There are $(N-1)!$ ways of permuting the remaining electrons $2, 3, \ldots, N$ amongst the other $N-1$ spin orbitals $\chi_n, \ldots$. Integrating over these electrons always yields a factor of 1 from their normalization and, hence,

$$\langle K|\mathcal{O}_1|L\rangle = (N-1)![(N-1)!]^{-1} \int dx_1 \, \chi_m^*(1)h(1)\chi_p(1)$$

$$= \langle m|h|p\rangle \qquad \text{Case 2} \qquad (2.131)$$

Case 3 has the two determinants differing by two spin orbitals, χ_p and χ_q appearing in $|L\rangle$, where χ_m and χ_n appear in $|K\rangle$

$$|K\rangle = |\chi_m(1)\chi_n(2)\cdots\rangle \qquad (2.132)$$

$$|L\rangle = |\chi_p(1)\chi_q(2)\cdots\rangle \qquad (2.133)$$

Analogous to (2.125) we write

$$\langle K|\mathcal{O}_1|L\rangle = N(N!)^{-1} \sum_i^{N!} \sum_j^{N!} (-1)^{p_i}(-1)^{p_j} \int dx_1 \, dx_2 \cdots dx_N$$

$$\times \mathscr{P}_i\{\chi_m^*(1)\chi_n^*(2)\cdots\}h(1)\mathscr{P}_j\{\chi_p(1)\chi_q(2)\cdots\} \qquad (2.134)$$

Because χ_m and χ_n are orthogonal to any spin orbital in the second permutation, and because they both cannot be occupied by electron 1 to "associate" with $h(1)$, no combination of permutations is possible that does not result in zero by spin orbital orthogonality. Hence,

$$\langle K|\mathcal{O}_1|L\rangle = 0 \qquad \text{Case 3} \qquad (2.135)$$

We now turn to two-electron operators. The general matrix element is

$$\langle K|\mathcal{O}_2|L\rangle = \langle K|r_{12}^{-1} + r_{13}^{-1} + r_{14}^{-1} + \cdots + r_{23}^{-1} + r_{24}^{-1} + \cdots + r_{N-1,N}^{-1}|L\rangle$$

$$(2.136)$$

where the sum is over all pairs of electrons. Because determinants do not distinguish between identical electrons, each of the terms in this equation will give the same result, and we may replace $\mathcal{O}_2$ by a single operator r_{12}^{-1}, provided we multiply by the number of pairs of electrons,

$$\langle K|\mathcal{O}_2|L\rangle = \frac{N(N-1)}{2} \langle K|r_{12}^{-1}|L\rangle \qquad (2.137)$$

We begin again with Case 1,

$$\langle K|\mathcal{O}_2|K\rangle = \frac{N(N-1)}{2}(N!)^{-1}\sum_i^{N!}\sum_j^{N!}(-1)^{p_i}(-1)^{p_j}\int d\mathbf{x}_1\, d\mathbf{x}_2\cdots d\mathbf{x}_N$$
$$\times \mathcal{P}_i\{\chi_m^*(1)\chi_n^*(2)\cdots\}r_{12}^{-1}\mathcal{P}_j\{\chi_m(1)\chi_n(2)\cdots\} \tag{2.138}$$

Because the operator in (2.138) involves only electrons 1 and 2, it must be that electrons $3, 4, \ldots, N$ occupy the same spin orbitals in both the ith permutation and the jth permutation or we would obtain zero by orthogonality on integrating over the coordinates of these electrons. If electrons $3, 4, \ldots, N$ occupy the same spin orbitals in the two permutations and electrons 1 and 2 occupy two spin orbitals, say χ_k and χ_l in the permutation $\mathcal{P}_i$, then there are two possibilities for electrons 1 and 2 in the permutation $\mathcal{P}_j$: they could occupy the same spin orbitals as in the permutation $\mathcal{P}_i$ (i.e., $\mathcal{P}_j = \mathcal{P}_i$) or they could occupy the spin orbitals χ_l and χ_k (i.e., $\mathcal{P}_j$ differs from $\mathcal{P}_i$ by an interchange of the coordinates of electrons 1 and 2). Thus if

$$\mathcal{P}_i\{\chi_m(1)\chi_n(2)\cdots\} = [\chi_k(1)\chi_l(2)\cdots] \tag{2.139}$$

then

$$\mathcal{P}_j\{\chi_m(1)\chi_n(2)\cdots\} = [\chi_k(1)\chi_l(2)\cdots] \quad \text{or} \quad [\chi_k(2)\chi_l(1)\cdots] \tag{2.140}$$

If $\mathcal{P}_{12}$ is an operator that interchanges the coordinates of electrons 1 and 2, we can thus write our matrix element as

$$\langle K|\mathcal{O}_2|K\rangle = [2(N-2)!]^{-1}\sum_i^{N!}\int d\mathbf{x}_1\, d\mathbf{x}_2\cdots d\mathbf{x}_N\, \mathcal{P}_i\{\chi_m^*(1)\chi_n^*(2)\cdots\}$$
$$\times r_{12}^{-1}[\mathcal{P}_i\{\chi_m(1)\chi_n(2)\cdots\} - \mathcal{P}_{12}\mathcal{P}_i\{\chi_m(1)\chi_n(2)\cdots\}] \tag{2.141}$$

where there is a minus sign in front of $\mathcal{P}_{12}$ because the permutation $\mathcal{P}_{12}\mathcal{P}_i$ differs from the permutation $\mathcal{P}_i$ by the interchange of the coordinates of electrons 1 and 2, and hence will be an odd permutation if $\mathcal{P}_i$ is an even permutation, and vice versa. In the sum of $N!$ permutations $\mathcal{P}_i$, electrons 1 and 2 of (2.141) will occupy any two different spin orbitals χ_m and χ_n of the set of N spin orbitals contained in $|K\rangle$. For each choice of these two spin orbitals there are $(N-2)!$ ways of permuting the other $N-2$ electrons amongst the $N-2$ remaining spin orbitals, and hence

$$\langle K|\mathcal{O}_2|K\rangle = \frac{(N-2)!}{2(N-2)!}\sum_m^N\sum_{n\neq m}^N\int d\mathbf{x}_1\, d\mathbf{x}_2\, \chi_m^*(1)\chi_n^*(2)r_{12}^{-1}(1-\mathcal{P}_{12})\{\chi_m(1)\chi_n(2)\}$$
$$= \frac{1}{2}\sum_m^N\sum_{n\neq m}^N\int d\mathbf{x}_1\, d\mathbf{x}_2\, \chi_m^*(1)\chi_n^*(2)r_{12}^{-1}[\chi_m(1)\chi_n(2) - \chi_m(2)\chi_n(1)]$$
$$= \frac{1}{2}\sum_m^N\sum_{n\neq m}^N\langle mn|mn\rangle - \langle mn|nm\rangle \tag{2.142}$$

Since $\langle mn\|mn\rangle = \langle mn|mn\rangle - \langle mn|nm\rangle$ vanishes when $m = n$, we can eliminate the restriction on the summation above and write

$$\langle K|\mathcal{O}_2|K\rangle = \frac{1}{2}\sum_m^N\sum_n^N \langle mn\|mn\rangle \qquad \text{Case 1} \qquad (2.143)$$

For Case 2 we replace χ_m in $|K\rangle$ by χ_p in $|L\rangle$ and obtain

$$\langle K|\mathcal{O}_2|L\rangle = \frac{N(N-1)}{2}(N!)^{-1}\sum_i^{N!}\sum_j^{N!}(-1)^{p_i}(-1)^{p_j}\int dx_1\,dx_2\cdots dx_N$$
$$\times \mathcal{P}_i\{\chi_m^*(1)\chi_n^*(2)\cdots\}r_{12}^{-1}\mathcal{P}_j\{\chi_p(1)\chi_n(2)\cdots\} \qquad (2.144)$$

By the same argument that leads to Eq. (2.141) for Case 1, we can write for Case 2,

$$\langle K|\mathcal{O}_2|L\rangle = [2(N-2)!]^{-1}\sum_i^{N!}\int dx_1\,dx_2\cdots dx_N$$
$$\times \mathcal{P}_i\{\chi_m^*(1)\chi_n^*(2)\cdots\}r_{12}^{-1}(1-\mathcal{P}_{12})\mathcal{P}_i\{\chi_p(1)\chi_n(2)\cdots\} \qquad (2.145)$$

Now, since the spin orbital χ_m in the first permutation is orthogonal to any spin orbital in the second permutation, it must be occupied by either electron 1 or electron 2, to associate it with r_{12}^{-1}, and yield a nonzero result. If χ_m is occupied by electron 1, electron 2 can be in any of the remaining $N-1$ spin orbitals common to both $|K\rangle$ and $|L\rangle$. If χ_m is occupied by electron 2, then electron 1 can be in any of the remaining $N-1$ spin orbitals common to both $|K\rangle$ and $|L\rangle$. There are $(N-2)!$ ways of permuting electrons $3, 4, \ldots, N$ and integrating over these electrons gives

$$\langle K|\mathcal{O}_2|L\rangle = \frac{(N-2)!}{2(N-2)!}\sum_{n\neq m}^N\int dx_1\,dx_2\,[\chi_m^*(1)\chi_n^*(2)r_{12}^{-1}(1-\mathcal{P}_{12})\{\chi_p(1)\chi_n(2)\}$$
$$+ \chi_n^*(1)\chi_m^*(2)r_{12}^{-1}(1-\mathcal{P}_{12})\{\chi_n(1)\chi_p(2)\}] \qquad (2.146)$$

where the two terms arise from placing electron 1 in χ_m or electron 2 in χ_m. Since $r_{12}^{-1} = r_{21}^{-1}$ and $\mathcal{P}_{12} = \mathcal{P}_{21}$, we can interchange the definition of the two dummy variables of integration in the second term and show that it is equal to the first,

$$\int dx_1\,dx_2\,\chi_n^*(1)\chi_m^*(2)r_{12}^{-1}(1-\mathcal{P}_{12})\{\chi_n(1)\chi_p(2)\}$$
$$= \int dx_2\,dx_1\,\chi_n^*(2)\chi_m^*(1)r_{21}^{-1}(1-\mathcal{P}_{21})\{\chi_n(2)\chi_p(1)\}$$
$$= \int dx_1\,dx_2\,\chi_m^*(1)\chi_n^*(2)r_{12}^{-1}(1-\mathcal{P}_{12})\{\chi_p(1)\chi_n(2)\} \qquad (2.147)$$

We thus obtain

$$\langle K|\mathcal{O}_2|L\rangle = \sum_{n \neq m}^{N} \int dx_1 \, dx_2 \, \chi_m^*(1)\chi_n^*(2) r_{12}^{-1}(1 - \mathscr{P}_{12})\{\chi_p(1)\chi_n(2)\}$$

$$= \sum_{n \neq m}^{N} \int dx_1 \, dx_2 \, \chi_m^*(1)\chi_n^*(2) r_{12}^{-1}[\chi_p(1)\chi_n(2) - \chi_n(1)\chi_p(2)]$$

$$= \sum_{n \neq m}^{N} \langle mn|pn\rangle - \langle mn|np\rangle = \sum_{n}^{N} \langle mn||pn\rangle \qquad \text{Case 2}$$

$$(2.148)$$

where we have removed the restriction on the summation since $\langle mm||pm\rangle = 0$.

For Case 3, we replace χ_m and χ_n in $|K\rangle$ by χ_p and χ_q in $|L\rangle$ and use the same argument as in the previous two cases to begin with

$$\langle K|\mathcal{O}_2|L\rangle = [2(N-2)!]^{-1} \sum_{i}^{N!} \int dx_1 \, dx_2 \cdots dx_N$$

$$\times \mathscr{P}_i\{\chi_m^*(1)\chi_n^*(2)\cdots\} r_{12}^{-1}(1 - \mathscr{P}_{12})\mathscr{P}_i\{\chi_p(1)\chi_q(2)\cdots\} \quad (2.149)$$

Because χ_m and χ_n are orthogonal to any spin orbitals in the second permutation, they must be occupied by electrons 1 and 2 (or 2 and 1). There are $(N-2)!$ permutations of the remaining electrons $3, 4, \ldots, N$, and integrating over these electrons gives

$$\langle K|\mathcal{O}_2|L\rangle = \frac{1}{2} \int dx_1 \, dx_2 \, [\chi_m^*(1)\chi_n^*(2) r_{12}^{-1}(1 - \mathscr{P}_{12})\{\chi_p(1)\chi_q(2)\}$$

$$+ \chi_n^*(1)\chi_m^*(2) r_{12}^{-1}(1 - \mathscr{P}_{12})\{\chi_q(1)\chi_p(2)\}] \quad (2.150)$$

As in the last case, we can show that the two terms are identical by interchanging the dummy variables of integration, so that

$$\langle K|\mathcal{O}_2|L\rangle = \int dx_1 \, dx_2 \, \chi_m^*(1)\chi_n^*(2) r_{12}^{-1}(1 - \mathscr{P}_{12})\{\chi_p(1)\chi_q(2)\}$$

$$= \int dx_1 \, dx_2 \, \chi_m^*(1)\chi_n^*(2) r_{12}^{-1}[\chi_p(1)\chi_q(2) - \chi_q(1)\chi_p(2)]$$

$$= \langle mn|pq\rangle - \langle mn|qp\rangle = \langle mn||pq\rangle \qquad \text{Case 3}$$

$$(2.151)$$

In the same way that matrix elements of a sum of one-electron operators are zero if the determinants differ by two or more spin orbitals, matrix elements of a sum of two-electron operators are zero if the determinants differ by three or more spin orbitals,

$$\langle K|\mathcal{O}_2|L\rangle = 0 \qquad (2.152)$$

This completes the derivation of the rules for matrix elements between Slater determinants.

Exercise 2.16 A different procedure for deriving the above matrix elements uses the theorem that $\langle K|\mathcal{H}|L\rangle = (N!)^{1/2}\langle K^{HP}|\mathcal{H}|L\rangle$ where $|K^{HP}\rangle$ is the Hartree product corresponding to the determinant $|K\rangle$, i.e.,

$$|K\rangle = |\chi_m(\mathbf{x}_1)\chi_n(\mathbf{x}_2)\cdots\rangle$$

and

$$|K^{HP}\rangle = \chi_m(\mathbf{x}_1)\chi_n(\mathbf{x}_2)\cdots$$

Prove this theorem. Use it to derive the matrix elements of a sum of one-electron operators.

2.3.5 Transition from Spin Orbitals to Spatial Orbitals

All of our development so far has involved spin orbitals χ_i rather than spatial orbitals ψ_i. The use of spin orbitals simplifies the algebraic manipulations and notation associated with the general formulation of various theories encountered in quantum chemistry. For most computational purposes, however, the spin functions α and β must be integrated out, to reduce spin orbital formulations to ones which involve only spatial functions and spatial integrals that are amenable to numerical computation. We will show how this is done and introduce a notation for spatial integrals.

To illustrate the procedure in the simplest possible context, consider the Hartree-Fock energy of our minimal basis H_2 model (see Eq. (2.92))

$$E_0 = \langle \chi_1|h|\chi_1\rangle + \langle \chi_2|h|\chi_2\rangle + \langle \chi_1\chi_2|\chi_1\chi_2\rangle - \langle \chi_1\chi_2|\chi_2\chi_1\rangle \quad (2.153)$$

using the physicists' notation, or

$$E_0 = [\chi_1|h|\chi_1] + [\chi_2|h|\chi_2] + [\chi_1\chi_1|\chi_2\chi_2] - [\chi_1\chi_2|\chi_2\chi_1] \quad (2.154)$$

using the chemists' notation. Recall (see Eq. (2.60)) that

$$\chi_1(\mathbf{x}) \equiv \psi_1(\mathbf{x}) = \psi_1(\mathbf{r})\alpha(\omega) \quad (2.155)$$

$$\chi_2(\mathbf{x}) \equiv \bar{\psi}_1(\mathbf{x}) = \psi_1(\mathbf{r})\beta(\omega) \quad (2.156)$$

Substituting these expressions for the spin orbitals in Eq. (2.154), we have

$$E_0 = [\psi_1|h|\psi_1] + [\bar{\psi}_1|h|\bar{\psi}_1] + [\psi_1\psi_1|\bar{\psi}_1\bar{\psi}_1] - [\psi_1\bar{\psi}_1|\bar{\psi}_1\psi_1] \quad (2.157)$$

Consider the one-electron integral

$$[\bar{\psi}_1|h|\bar{\psi}_1] = \int d\mathbf{r}_1\, d\omega_1\, \psi_1^*(\mathbf{r}_1)\beta^*(\omega_1)h(\mathbf{r}_1)\psi_1(\mathbf{r}_1)\beta(\omega_1) \quad (2.158)$$

where we have assumed (as is the case for nonrelativistic Hamiltonians) that the one-electron operator does not depend on spin. Integrating over the spin variable ω_1 and using $\langle\beta|\beta\rangle = 1$, we have

$$[\bar{\psi}_1|h|\bar{\psi}_1] = \int d\mathbf{r}_1 \, \psi_1^*(\mathbf{r}_1)h(\mathbf{r}_1)\psi_1(\mathbf{r}_1) \equiv (\psi_1|h|\psi_1) \tag{2.159}$$

where we have introduced a new notation for a one-electron spatial integral (see Table 2.2). Since $\langle\alpha|\alpha\rangle = \langle\beta|\beta\rangle = 1$ and $\langle\alpha|\beta\rangle = \langle\beta|\alpha\rangle = 0$, the general reduction is

$$[\psi_i|h|\psi_j] = [\bar{\psi}_i|h|\bar{\psi}_j] = (\psi_i|h|\psi_j) \tag{2.160}$$

$$[\psi_i|h|\bar{\psi}_j] = [\bar{\psi}_i|h|\psi_j] = 0 \tag{2.161}$$

so that the one-electron contribution to E_0 is $2(\psi_1|h|\psi_1)$.

Consider, next, the first of the two-electron integrals in expression (2.157) for the ground state energy,

$$[\psi_1\psi_1|\bar{\psi}_1\bar{\psi}_1] = \int d\mathbf{r}_1 \, d\omega_1 \, d\mathbf{r}_2 \, d\omega_2 \, \psi_1^*(\mathbf{r}_1)\alpha^*(\omega_1)\psi_1(\mathbf{r}_1)\alpha(\omega_1)r_{12}^{-1}$$
$$\times \psi_1^*(\mathbf{r}_2)\beta^*(\omega_2)\psi_1(\mathbf{r}_2)\beta(\omega_2) \tag{2.162}$$

Integrating over the spin variables ω_1 and ω_2 and using $\langle\alpha|\alpha\rangle = \langle\beta|\beta\rangle = 1$, we have

$$[\psi_1\psi_1|\bar{\psi}_1\bar{\psi}_1] = \int d\mathbf{r}_1 \, d\mathbf{r}_2 \, \psi_1^*(\mathbf{r}_1)\psi_1(\mathbf{r}_1)r_{12}^{-1}\psi_1^*(\mathbf{r}_2)\psi_1(\mathbf{r}_2)$$
$$\equiv (\psi_1\psi_1|\psi_1\psi_1) \tag{2.163}$$

where we have introduced a new notation for spatial two-electron integrals (see Table 2.2). This notation for spatial integrals is just the chemists' notation with *round*, instead of square, brackets. We shall not introduce a comparable notation for spatial integrals written using the physicists' notation. Thus whether $\langle ij|kl\rangle$ refers to an integral over spin or over spatial orbitals can be determined only from the context. The last integral in (2.157),

$$[\psi_1\bar{\psi}_1|\bar{\psi}_1\psi_1] = \int d\mathbf{r}_1 \, d\omega_1 \, d\mathbf{r}_2 \, d\omega_2 \, \psi_1^*(\mathbf{r}_1)\alpha^*(\omega_1)\psi_1(\mathbf{r}_1)\beta(\omega_1)r_{12}^{-1}$$
$$\times \psi_1^*(\mathbf{r}_2)\beta(\omega_2)\psi_1(\mathbf{r}_2)\alpha(\omega_2) = 0 \tag{2.164}$$

since $\langle\alpha|\beta\rangle = \langle\beta|\alpha\rangle = 0$. In general, when only a single bar appears on either side of the two electron integral (e.g., $[\psi_i\bar{\psi}_j|\psi_k\psi_l]$), the integral vanishes by spin orthogonality. The general reduction is

$$[\psi_i\psi_j|\psi_k\psi_l] = [\psi_i\psi_j|\bar{\psi}_k\bar{\psi}_l] = [\bar{\psi}_i\bar{\psi}_j|\psi_k\psi_l] = [\bar{\psi}_i\bar{\psi}_j|\bar{\psi}_k\bar{\psi}_l] = (\psi_i\psi_j|\psi_k\psi_l) \tag{2.165}$$

with all other combinations of bars giving zero. Therefore, the Hartree-Fock energy of minimal basis H_2 is

$$E_0 = 2(\psi_1|h|\psi_1) + (\psi_1\psi_1|\psi_1\psi_1)$$
$$= 2(1|h|1) + (11|11) \tag{2.166}$$

Exercise 2.17 By integrating out spin, show that the full CI matrix for minimal basis H_2 (see Exercise 2.9) is

$$\mathbf{H} = \begin{pmatrix} 2(1|h|1) + (11|11) & (12|12) \\ (21|21) & 2(2|h|2) + (22|22) \end{pmatrix}$$

Let us generalize the above results to obtain an expression involving spatial integrals for the Hartree-Fock energy of an N-electron system containing an *even* number of electrons. The analogue of the minimal basis H_2 Hartree-Fock wave function,

$$|\Psi_0\rangle = |\chi_1\chi_2\rangle = |\psi_1\bar{\psi}_1\rangle \tag{2.167}$$

in an N-electron system is the *closed-shell restricted* Hartree-Fock wave function

$$\begin{aligned} |\Psi_0\rangle &= |\chi_1\chi_2\chi_3\chi_4 \cdots \chi_{N-1}\chi_N\rangle \\ &= |\psi_1\bar{\psi}_1\psi_2\bar{\psi}_2 \cdots \psi_{N/2}\bar{\psi}_{N/2}\rangle \end{aligned} \tag{2.168}$$

This wave function is represented in Fig. 2.10. Note that the spatial orbitals are restricted to be the same for α and β spins, and each spatial orbital is occupied by two electrons with different spin. The energy of this wave function, expressed in terms of the set of spin orbitals $\{\chi_a | a = 1, 2, \ldots, N\}$, is given by Eq. (2.111),

$$E_0 = \sum_a^N [a|h|a] + \frac{1}{2}\sum_a^N\sum_b^N [aa|bb] - [ab|ba] \tag{2.169}$$

Since the wave function (2.168) contains $N/2$ spin orbitals with α spin function and $N/2$ spin orbitals with β spin function, we can write a sum over all N spin oribtals χ_a as

$$\sum_a^N \chi_a = \sum_a^{N/2} \psi_a + \sum_a^{N/2} \bar{\psi}_a \tag{2.170}$$

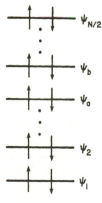

Figure 2.10 A closed-shell restricted Hartree-Fock ground state determinant, $|\psi_1\bar{\psi}_1\psi_2\bar{\psi}_2 \cdots \psi_a\bar{\psi}_a\psi_b\bar{\psi}_b \cdots \psi_{N/2}\bar{\psi}_{N/2}\rangle$.

where we have used the bar notation. Symbolically this becomes

$$\sum_a^N = \sum_a^{N/2} + \sum_{\bar{a}}^{N/2} \tag{2.171}$$

which means that the sum over all spin orbitals is equal to the sum of those with spin up and those with spin down. For double sums, we have

$$\sum_a^N \sum_b^N \chi_a \chi_b = \sum_a^N \chi_a \sum_b^N \chi_b$$

$$= \sum_a^{N/2} (\psi_a + \bar{\psi}_a) \sum_b^{N/2} (\psi_b + \bar{\psi}_b)$$

$$= \sum_a^{N/2} \sum_b^{N/2} \psi_a \psi_b + \psi_a \bar{\psi}_b + \bar{\psi}_a \psi_b + \bar{\psi}_a \bar{\psi}_b \tag{2.172}$$

or symbolically,

$$\sum_a^N \sum_b^N = \sum_a^{N/2} \sum_b^{N/2} + \sum_a^{N/2} \sum_{\bar{b}}^{N/2} + \sum_{\bar{a}}^{N/2} \sum_b^{N/2} + \sum_{\bar{a}}^{N/2} \sum_{\bar{b}}^{N/2} \tag{2.173}$$

Let us use these to reduce (2.169) to an equation involving spatial orbitals. We treat the one-electron integrals first,

$$\sum_a^N [a|h|a] = \sum_a^{N/2} [a|h|a] + \sum_a^{N/2} [\bar{a}|h|\bar{a}] = 2 \sum_a^{N/2} (\psi_a|h|\psi_a) \tag{2.174}$$

The two-electron integral term is

$$\frac{1}{2} \sum_a^N \sum_b^N [aa|bb] - [ab|ba]$$

$$= \frac{1}{2} \left\{ \sum_a^{N/2} \sum_b^{N/2} [aa|bb] - [ab|ba] + \sum_a^{N/2} \sum_b^{N/2} [aa|\bar{b}\bar{b}] - [a\bar{b}|\bar{b}a] \right.$$

$$\left. + \sum_a^{N/2} \sum_b^{N/2} [\bar{a}\bar{a}|bb] - [\bar{a}b|b\bar{a}] + \sum_a^{N/2} \sum_b^{N/2} [\bar{a}\bar{a}|\bar{b}\bar{b}] - [\bar{a}\bar{b}|\bar{b}\bar{a}] \right\}$$

$$= \sum_a^{N/2} \sum_b^{N/2} 2(\psi_a \psi_a|\psi_b \psi_b) - (\psi_a \psi_b|\psi_b \psi_a) \tag{2.175}$$

Thus the Hartree-Fock energy of a closed-shell ground state is

$$E_0 = 2 \sum_a^{N/2} (\psi_a|h|\psi_a) + \sum_a^{N/2} \sum_b^{N/2} 2(\psi_a \psi_a|\psi_b \psi_b) - (\psi_a \psi_b|\psi_b \psi_a) \tag{2.176}$$

The upper limits of summation, which indicate that we are summing over spatial orbitals, are redundant since we are using the round brackets. Thus

Eq. (2.176) can be rewritten as

$$E_0 = 2 \sum_a (a|h|a) + \sum_{ab} 2(aa|bb) - (ab|ba) \qquad (2.177)$$

When using the physicists' notation, it is necessary to show the upper limits of summation, since we have not introduced a notation analogous to round brackets. The convention we use is as follows. If no upper limit appears, the sum is over spin orbitals. If the upper limit is $N/2$, the sum is over spatial orbitals. Thus using the physicists' notation Eq. (2.177) is

$$E_0 = 2 \sum_a^{N/2} \langle a|h|a \rangle + \sum_{ab}^{N/2} 2\langle ab|ab \rangle - \langle ab|ba \rangle \qquad (2.178)$$

Exercise 2.18 In Chapter 6, where we consider perturbation theory, we show that the leading correction to the Hartree-Fock ground state energy is

$$E_0^{(2)} = \frac{1}{4} \sum_{abrs} \frac{|\langle ab||rs \rangle|^2}{\varepsilon_a + \varepsilon_b - \varepsilon_r - \varepsilon_s}$$

Show that for a closed-shell system (where $\varepsilon_i = \varepsilon_{\bar{i}}$) this becomes

$$E_0^{(2)} = \sum_{a,b=1}^{N/2} \sum_{r,s=(N/2+1)}^{K} \frac{\langle ab|rs \rangle (2\langle rs|ab \rangle - \langle rs|ba \rangle)}{\varepsilon_a + \varepsilon_b - \varepsilon_r - \varepsilon_s},$$

2.3.6 Coulomb and Exchange Integrals

Let us consider the physical interpretation of the result given in Eq. (2.177) for the Hartree-Fock energy of a closed-shell ground state, i.e.,

$$E_0 = 2 \sum_a (a|h|a) + \sum_{ab} 2(aa|bb) - (ab|ba) \qquad (2.179)$$

Consider the one-electron terms first,

$$(a|h|a) \equiv h_{aa} = \int d\mathbf{r}_1 \, \psi_a^*(\mathbf{r}_1) \left(-\frac{1}{2}\nabla_1^2 - \sum_A \frac{Z_A}{r_{1A}} \right) \psi_a(\mathbf{r}_1) \qquad (2.180)$$

Thus h_{aa} is the average kinetic and nuclear attraction energy of an electron described by the wave function $\psi_a(\mathbf{r}_1)$. Next consider the two-electron integral

$$(aa|bb) = \int d\mathbf{r}_1 \, d\mathbf{r}_2 \, |\psi_a(\mathbf{r}_1)|^2 r_{12}^{-1} |\psi_b(\mathbf{r}_2)|^2 \qquad (2.181)$$

which is the classical coulomb repulsion between the charge clouds $|\psi_a(\mathbf{r}_1)|^2$ and $|\psi_b(\mathbf{r}_2)|^2$. This integral is called a *coulomb* integral and is denoted by J_{ab}. In general,

$$J_{ij} = (ii|jj) = \langle ij|ij \rangle \qquad (2.182)$$

Finally, consider the two-electron integral

$$(ab|ba) = \int d\mathbf{r}_1 \, d\mathbf{r}_2 \, \psi_a^*(\mathbf{r}_1)\psi_b(\mathbf{r}_1)r_{12}^{-1}\psi_b^*(\mathbf{r}_2)\psi_a(\mathbf{r}_2) \qquad (2.183)$$

This integral does not have a simple classical interpretation. It is called an *exchange* integral and is denoted by K_{ab}. In general,

$$K_{ij} = (ij|ji) = \langle ij|ji \rangle \qquad (2.184)$$

Both exchange and coulomb integrals have positive values. We will now show that the appearance of exchange integrals in the expression for the energy of a determinant is the result of *exchange correlation* (i.e., the motion of electrons with parallel spins is correlated within the single determinantal approximation to the wave function). We have seen in Subsection 2.2.3 that antisymmetrizing a Hartree product to yield a Slater determinant introduces correlation. Before proceeding, let us rewrite the Hartree-Fock energy of a closed-shell system given in (2.179) in terms of coulomb and exchange integrals

$$E_0 = 2\sum_a h_{aa} + \sum_{ab} 2J_{ab} - K_{ab} \qquad (2.185)$$

Exercise 2.19 Prove the following properties of coulomb and exchange integrals

$$J_{ii} = K_{ii}$$

$$J_{ij}^* = J_{ij} \qquad\qquad K_{ij}^* = K_{ij}$$

$$J_{ij} = J_{ji} \qquad\qquad K_{ij} = K_{ji}$$

Exercise 2.20 Show that for *real* spatial orbitals

$$K_{ij} = (ij|ij) = (ji|ji)$$
$$= \langle ii|jj \rangle = \langle jj|ii \rangle$$

Exercise 2.21 Show that the full CI matrix for minimal basis H_2 (see Exercise 2.17) is

$$\mathbf{H} = \begin{pmatrix} 2h_{11} + J_{11} & K_{12} \\ K_{12} & 2h_{22} + J_{22} \end{pmatrix}$$

The spatial molecular orbitals of this model are real because they were constructed as linear combinations of real atomic orbitals (see Eqs. (2.54), (2.55), (2.57), and (2.58)).

A feeling for the occurance of exchange integrals can be gained by reconsidering the example discussed at the end of Subsection 2.2.3 from the energetic point of view. We have seen that in a system containing two electrons with parallel spin, described by the wave function $|\bar{\psi}_1\bar{\psi}_2\rangle$, the probability of finding two electrons at the same point in space is zero, whereas in a system containing two electrons with opposite spin, described by the wave function $|\psi_1\bar{\psi}_2\rangle$, it is not. Therefore, it is reasonable to expect that the energy of $|\bar{\psi}_1\bar{\psi}_2\rangle$ is lower than the energy of $|\psi_1\bar{\psi}_2\rangle$ when the coulomb repulsion between electrons is taken into account. Using Eq. (2.110), the energy of $|\psi_1\bar{\psi}_2\rangle$, denoted by $E(\uparrow\downarrow)$, is

$$
\begin{aligned}
E(\uparrow\downarrow) &= [\psi_1|h|\psi_1] + [\bar{\psi}_2|h|\bar{\psi}_2] + [\psi_1\psi_1|\bar{\psi}_2\bar{\psi}_2] - [\psi_1\bar{\psi}_2|\bar{\psi}_2\psi_1] \\
&= (1|h|1) + (2|h|2) + (11|22) \\
&= h_{11} + h_{22} + J_{12}
\end{aligned}
\tag{2.186}
$$

and the energy of $|\bar{\psi}_1\bar{\psi}_2\rangle$, denoted by $E(\downarrow\downarrow)$, is

$$
\begin{aligned}
E(\downarrow\downarrow) &= [\bar{\psi}_1|h|\bar{\psi}_1] + [\bar{\psi}_2|h|\bar{\psi}_2] + [\bar{\psi}_1\bar{\psi}_1|\bar{\psi}_2\bar{\psi}_2] - [\bar{\psi}_1\bar{\psi}_2|\bar{\psi}_2\bar{\psi}_1] \\
&= (1|h|1) + (2|h|2) + (11|22) - (12|21) \\
&= h_{11} + h_{22} + J_{12} - K_{12}
\end{aligned}
\tag{2.187}
$$

where we have used Eqs. (2.160), (2.161), and (2.165) to integrate out the spin. Because K_{12} is positive, $E(\downarrow\downarrow)$ is indeed lower than $E(\uparrow\downarrow)$. Thus the appearance of exchange integrals in the energy of a Slater determinant is a manifestation of the fact that, even within the single determinantal approximation to the wave function, the motion of electrons with parallel spins is correlated.

Exercise 2.22 Show that the energies of the Hartree products

$$
\Psi_{\uparrow\downarrow}^{HP} = \psi_1(\mathbf{r}_1)\alpha(\omega_1)\psi_2(\mathbf{r}_2)\beta(\omega_2)
$$

and

$$
\Psi_{\downarrow\downarrow}^{HP} = \psi_1(\mathbf{r}_1)\beta(\omega_1)\psi_2(\mathbf{r}_2)\beta(\omega_2)
$$

are the same and equal to $E(\uparrow\downarrow)$ as to be expected since the motion of electrons with parallel spin is not correlated within the Hartree product approximation to the wave function.

2.3.7 Pseudo-Classical Interpretation of Determinantal Energies

In Subsection 2.3.3, we introduced a simple mnemonic device for writing down the energy of a single determinant, constructed from a set of spin orbitals $\{\chi_i\}$, in terms of one-electron integrals over spin orbitals $(\langle i|h|i\rangle)$ and antisymmetrized two-electron integrals over spin orbitals $(\langle ij||ij\rangle)$. Here we will show how one can express, with equal ease, the energy of any restricted determinant, constructed from spin orbitals $\{\psi_i\alpha\}$ and $\{\psi_i\beta\}$, in terms of h_{ii}, coulomb (J_{ij}), and exchange (K_{ij}) integrals.

We begin with the one-electron contributions to the energy. Recall that an electron in spin orbital χ_i contributed the term $\langle i|h|i \rangle$ to the energy. If $\chi_i = \psi_i \alpha$, then $\langle i|h|i \rangle = \langle \psi_i \alpha | h | \psi_i \alpha \rangle = (\psi_i | h | \psi_i) = h_{ii}$. Similarly, if $\chi_i = \psi_i \beta$, then $\langle i|h|i \rangle = h_{ii}$, *Therefore, an electron (irrespective of its spin) in spatial orbital ψ_i contributes the term h_{ii} to the energy.*

Next we consider the two-electron contributions to the energy. Recall that each unique pair of electrons in spin orbitals χ_i and χ_j contributes the term $\langle ij||ij \rangle$ to the energy. A pair of electrons can have either parallel or opposite spins. If they have opposite spins, say $\chi_i = \psi_i \alpha$ and $\chi_j = \psi_j \beta$, then

$$\langle ij||ij \rangle = [\psi_i \psi_i | \bar{\psi}_j \bar{\psi}_j] - [\psi_i \bar{\psi}_j | \bar{\psi}_j \psi_i] = J_{ij} \qquad (2.188)$$

On the other hand, if they have parallel spins, say $\chi_i = \psi_i \beta$ and $\chi_j = \psi_j \beta$, then

$$\langle ij||ij \rangle = [\bar{\psi}_i \bar{\psi}_i | \bar{\psi}_j \bar{\psi}_j] - [\bar{\psi}_i \bar{\psi}_j | \bar{\psi}_j \bar{\psi}_i] = J_{ij} - K_{ij} \qquad (2.189)$$

Therefore, each unique pair of electrons (irrespective of their spin) in spatial orbitals ψ_i and ψ_j contributes the term J_{ij} to the energy, and each unique pair of electrons with parallel spins in spatial orbitals ψ_i and ψ_j contributes the term $-K_{ij}$ to the energy. The total energy of the determinant is the sum of all these contributions.

Thus we can think of the total energy of an N-electron system, which is described by a restricted determinant, as a sum of "one-electron-energies" (h_{ii} for an electron in spatial orbital ψ_i) plus all unique coulomb interaction energies (J_{ij} for a pair of electrons in spatial orbitals ψ_i and ψ_j) plus all unique exchange interaction energies between electrons with parallel spins ($-K_{ij}$ for a pair of electrons with parallel spin in spatial orbitals ψ_i and ψ_j). In using this language, it must be remembered that exchange interactions between electrons with parallel spin are not real physical interactions but a convenient way of representing the energy of a system described by a single determinant. The physical interaction between two electrons, as described by the coulomb repulsion term (r_{ij}^{-1}) in the Hamiltonian, does not depend on the spin of the electrons.

As an illustration of the above approach, consider the energy of the determinant

$$|\bar{\psi}_1 \psi_2 \bar{\psi}_2 \bar{\psi}_3 \rangle \equiv \begin{array}{c} \underline{\downarrow}3 \\ \underline{\uparrow\downarrow}2 \\ \underline{\downarrow}1 \end{array}$$

The one-electron contributions to the energy are h_{11}, $2h_{22}$, and h_{33}. The coulomb contributions are J_{22}, J_{13}, $2J_{12}$, and $2J_{23}$. The exchange contributions are $-K_{23}$, $-K_{12}$, and $-K_{13}$. Thus the total energy is $h_{11} + 2h_{22} + h_{33} + J_{22} + J_{13} + 2J_{12} + 2J_{23} - K_{23} - K_{12} - K_{13}$.

Exercise 2.23 Verify the energies of the following determinants by inspection.

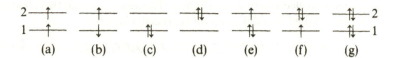

a. $h_{11} + h_{22} + J_{12} - K_{12}$.
b. $h_{11} + h_{22} + J_{12}$.
c. $2h_{11} + J_{11}$.
d. $2h_{22} + J_{22}$.
e. $2h_{11} + h_{22} + J_{11} + 2J_{12} - K_{12}$.
f. $2h_{22} + h_{11} + J_{22} + 2J_{12} - K_{12}$.
g. $2h_{11} + 2h_{22} + J_{11} + J_{22} + 4J_{12} - 2K_{12}$.

2.4 SECOND QUANTIZATION

The antisymmetry principle is an axiom of quantum mechanics quite apart from the Schrödinger equation. We have insured that this principle is satisfied by using Slater determinants and linear combinations of such determinants for wave functions. Can we satisfy the antisymmetry principle without using Slater determinants? Second quantization is a formalism in which the antisymmetry property of the wave function has been transferred onto the algebraic properties of certain operators. Second quantization introduces no new physics. It is just another, although very elegant, way of treating many-electron systems, which shifts much of the emphasis away from N-electron wave functions to the one- and two-electron integrals $\langle i|h|j \rangle$ and $\langle ij|kl \rangle$ that were discussed in the preceding section. The formalism of second quantization is widely used in the literature dealing with many-electron problems. We introduce it here not only as an interesting way of rederiving some of our previous results but also as a background for approaching such literature. Since we will not make general use of second quantization in the remaining chapters, this section can be considered optional.

2.4.1 Creation and Annihilation Operators and Their Anticommutation Relations

We shall gradually construct the formalism of second quantization by showing how the properties of determinants can be transferred onto the algebraic properties of operators. We begin by associating a *creation* operator $a_i^\dagger$ with each spin orbital χ_i. We define $a_i^\dagger$ by its action on an arbitrary Slater determinant $|\chi_k \cdots \chi_l\rangle$, as

$$a_i^\dagger |\chi_k \cdots \chi_l\rangle = |\chi_i \chi_k \cdots \chi_l\rangle \qquad (2.190)$$

Thus $a_i^\dagger$ creates an electron in spin orbital χ_i. The order in which two creation operators are applied to a determinant is crucial. Consider

$$a_i^\dagger a_j^\dagger |\chi_k \cdots \chi_l\rangle = a_i^\dagger |\chi_j \chi_k \cdots \chi_l\rangle = |\chi_i \chi_j \chi_k \cdots \chi_l\rangle \qquad (2.191)$$

On the other hand,

$$a_j^\dagger a_i^\dagger |\chi_k \cdots \chi_l\rangle = a_j^\dagger |\chi_i \chi_k \cdots \chi_l\rangle = |\chi_j \chi_i \chi_k \cdots \chi_l\rangle$$
$$= -|\chi_i \chi_j \chi_k \cdots \chi_l\rangle \qquad (2.192)$$

where we have used the antisymmetry property of Slater determinants (see Eq. (2.40)). Adding Eqs. (2.191) and (2.192), we have

$$(a_i^\dagger a_j^\dagger + a_j^\dagger a_i^\dagger)|\chi_k \cdots \chi_l\rangle = 0 \qquad (2.193)$$

Because $|\chi_k \cdots \chi_l\rangle$ is an arbitrary determinant, we have discovered the operator relation

$$a_i^\dagger a_j^\dagger + a_j^\dagger a_i^\dagger = 0 = \{a_i^\dagger, a_j^\dagger\} \qquad (2.194)$$

where we have used the notation for the *anticommutator* of two operators introduced in Eq. (1.19a). Since,

$$a_i^\dagger a_j^\dagger = -a_j^\dagger a_i^\dagger \qquad (2.195)$$

we can interchange the order of two creation operators provided we change the sign. If $i = j$, we have

$$a_i^\dagger a_i^\dagger = -a_i^\dagger a_i^\dagger = 0 \qquad (2.196)$$

which states that we cannot create two electrons in the same spin orbital χ_i (Pauli exclusion principle). Thus

$$a_1^\dagger a_1^\dagger |\chi_2 \chi_3\rangle = a_1^\dagger |\chi_1 \chi_2 \chi_3\rangle = |\chi_1 \chi_1 \chi_2 \chi_3\rangle = 0 \qquad (2.197)$$

In general,

$$a_i^\dagger |\chi_k \cdots \chi_l\rangle = 0 \qquad \text{if } i \in \{k, \ldots, l\} \qquad (2.198)$$

which states we cannot create an electron in spin orbital χ_i if there is one already there.

Exercise 2.24 Show, using the properties of determinants, that

$$(a_1^\dagger a_2^\dagger + a_2^\dagger a_1^\dagger)|K\rangle = 0$$

for every $|K\rangle$ in the set $\{|\chi_1 \chi_2\rangle, |\chi_1 \chi_3\rangle, |\chi_1 \chi_4\rangle, |\chi_2 \chi_3\rangle, |\chi_2 \chi_4\rangle, |\chi_3 \chi_4\rangle\}$.

We now introduce the *annihilation* operator a_i, which is the adjoint of the creation operator $a_i^\dagger$ (i.e., $(a_i^\dagger)^\dagger = a_i$). In analogy with Eq. (2.190), a_i is defined by

$$a_i|\chi_i\chi_k\cdots\chi_l\rangle = |\chi_k\cdots\chi_l\rangle \tag{2.199}$$

Thus a_i annihilates or destroys an electron in spin orbital χ_i. Note that the annihilation operator can only act on a determinant if the spin orbital, which will disappear, is immediately to the left. If a spin orbital is not in the proper position, it must be placed there by interchanging the columns of the determinant, e.g.,

$$a_i|\chi_k\chi_l\chi_i\rangle = -a_i|\chi_i\chi_l\chi_k\rangle = -|\chi_l\chi_k\rangle = |\chi_k\chi_l\rangle \tag{2.200}$$

Why is the annihilation operator defined as the adjoint of the creation operator? Consider the determinant

$$|K\rangle = |\chi_i\chi_j\rangle \tag{2.201}$$

Clearly,

$$|K\rangle = a_i^\dagger|\chi_j\rangle \tag{2.202}$$

The adjoint of this equation (see Eqs. (1.52) and (1.57)) is

$$\langle K| = \langle\chi_j|(a_i^\dagger)^\dagger = \langle\chi_j|a_i \tag{2.203}$$

Multiplying (2.203) on the right by $|K\rangle$, we have

$$\langle K|K\rangle = \langle\chi_j|a_i|K\rangle \tag{2.204}$$

Since $\langle K|K\rangle = 1 = \langle\chi_j|\chi_j\rangle$, our formalism is consistent when

$$a_i|K\rangle \equiv a_i|\chi_i\chi_j\rangle = |\chi_j\rangle \tag{2.205}$$

in agreement with the definition of (2.199) of the annihilation operator. From Eq. (2.203) we see that a_i acts like a creation operator if it operates on a determinant to the left. Similarly, $a_i^\dagger$ acts like an annihilation operator if it operates to the left. For example, the adjoint of Eq. (2.205) is

$$\langle K|a_i^\dagger = \langle\chi_j| \tag{2.206}$$

To obtain the anticommutation relation satisfied by annihilation operators we take the adjoint of (2.194). Since (c.f. Exercise 1.3)

$$(\mathscr{A}\mathscr{B})^\dagger = \mathscr{B}^\dagger\mathscr{A}^\dagger \tag{2.207}$$

we have

$$a_ja_i + a_ia_j = 0 = \{a_j, a_i\} \tag{2.208}$$

Since

$$a_ia_j = -a_ja_i \tag{2.209}$$

we can interchange the order of two annihilation operators provided we change the sign. If $i = j$, we have

$$a_ia_i = -a_ia_i = 0 \tag{2.210}$$

which states that we cannot destroy an electron twice. A consequence of this is that we cannot remove an electron from a spin orbital χ_i, if it is not already there,

$$a_i|\chi_k \cdots \chi_l\rangle = 0 \qquad \text{if } i \notin \{k, \ldots, l\} \tag{2.211}$$

It remains for us to discover how we can interchange creation and annihilation operators. Consider the operator $a_i a_i^\dagger + a_i^\dagger a_i$ acting on an arbitrary determinant, $|\chi_k \cdots \chi_l\rangle$. If spin orbital χ_i is *not* occupied in this determinant, we have

$$\begin{aligned}
(a_i a_i^\dagger + a_i^\dagger a_i)|\chi_k \cdots \chi_l\rangle &= a_i a_i^\dagger|\chi_k \cdots \chi_l\rangle \\
&= a_i|\chi_i\chi_k \cdots \chi_l\rangle \\
&= |\chi_k \cdots \chi_l\rangle
\end{aligned} \tag{2.212}$$

on the other hand, if χ_i is occupied, we have

$$\begin{aligned}
(a_i a_i^\dagger + a_i^\dagger a_i)|\chi_k \cdots \chi_i \cdots \chi_l\rangle &= a_i^\dagger a_i|\chi_k \cdots \chi_i \cdots \chi_l\rangle \\
&= -a_i^\dagger a_i|\chi_i \cdots \chi_k \cdots \chi_l\rangle \\
&= -a_i^\dagger|\cdots \chi_k \cdots \chi_l\rangle \\
&= -|\chi_i \cdots \chi_k \cdots \chi_l\rangle \\
&= |\chi_k \cdots \chi_i \cdots \chi_l\rangle
\end{aligned} \tag{2.213}$$

Since we recover the same determinant in both cases, we have discovered the operator relation

$$a_i a_i^\dagger + a_i^\dagger a_i = 1 = \{a_i, a_i^\dagger\} \tag{2.214}$$

Finally, consider $(a_i^\dagger a_i + a_i a_i^\dagger)|\chi_k \cdots \chi_l\rangle$ when $i \neq j$. This expression can be nonzero only if the spin orbital χ_i appears and the spin orbital χ_j does not appear in $|\chi_k \cdots \chi_l\rangle$. Otherwise, we obtain zero either because $a_j^\dagger$ tries to create an electron that is already there or a_i tries to destroy one that is not there. However, even when $i \in \{k, \ldots, l\}$ and $j \notin \{k, \ldots, l\}$ we obtain zero as a result of the antisymmetry property of determinants,

$$\begin{aligned}
(a_i a_j^\dagger + a_j^\dagger a_i)|\chi_k \cdots \chi_i \cdots \chi_l\rangle &= -(a_i a_j^\dagger + a_j^\dagger a_i)|\chi_i \cdots \chi_k \cdots \chi_l\rangle \\
&= -a_i|\chi_j\chi_i \cdots \chi_k \cdots \chi_l\rangle - a_j^\dagger|\cdots \chi_k \cdots \chi_l\rangle \\
&= a_i|\chi_i\chi_j \cdots \chi_k \cdots \chi_l\rangle - |\chi_j \cdots \chi_k \cdots \chi_l\rangle \\
&= |\chi_j \cdots \chi_k \cdots \chi_l\rangle - |\chi_j \cdots \chi_k \cdots \chi_l\rangle \\
&= 0
\end{aligned} \tag{2.215}$$

Thus we have

$$a_i a_j^\dagger + a_j^\dagger a_i = 0 = \{a_i, a_j^\dagger\} \qquad i \neq j \tag{2.216}$$

Combining this with (2.214), the anticommutation relation between a creation and an annihilation operator is

$$a_i a_j^\dagger + a_j^\dagger a_i = \delta_{ij} = \{a_i, a_j^\dagger\} \tag{2.217}$$

Thus we can interchange a creation and an annihilation operator, which refer to different spin orbitals, provided we change the sign, i.e.,

$$a_i a_j^\dagger = -a_j^\dagger a_i \qquad i \neq j \tag{2.218a}$$

However, if the operators refer to the same spin orbital, we have

$$a_i a_i^\dagger = 1 - a_i^\dagger a_i \tag{2.218b}$$

Exercise 2.25 Show, using the properties of determinants, that

$$(a_1 a_2^\dagger + a_2^\dagger a_1)|K\rangle = 0$$

$$(a_1 a_1^\dagger + a_1^\dagger a_1)|K\rangle = |K\rangle$$

for every $|K\rangle$ in the set $\{|\chi_1\chi_2\rangle, |\chi_1\chi_3\rangle, |\chi_1\chi_4\rangle, |\chi_2\chi_3\rangle, |\chi_2\chi_4\rangle, |\chi_3\chi_4\rangle\}$.

All the properties of Slater determinants are contained in the anti-commutation relations between two creation operators (Eq. (2.194)), between two annihilation operators (Eq. (2.208)), and between a creation and an annihilation operator (Eq. (2.217)). In order to define a Slater determinant in the formalism of second quantization, we introduce a vacuum state denoted by $|\ \rangle$. The vacuum state represents a state of the system that contains no electrons. It is normalized,

$$\langle\ |\ \rangle = 1 \tag{2.219}$$

and has the property that

$$a_i|\ \rangle = 0 = \langle\ |a_i^\dagger \tag{2.220}$$

that is, since the vacuum state contains no electrons, we cannot remove an electron from it. We can construct any state of the system by applying a succession of creation operators to the vacuum state. For example,

$$|\chi_i\rangle = a_i^\dagger|\ \rangle \tag{2.221}$$

or, in general,

$$a_i^\dagger a_k^\dagger \cdots a_l^\dagger|\ \rangle = |\chi_i\chi_k \cdots \chi_l\rangle \tag{2.222}$$

This relation is the second-quantized representation of a Slater determinant. Any result that can be obtained using the properties of determinants can also be proved using only the algebraic properties of creation and annihilation operators.

In Exercise 2.5 we evaluated the overlap between the two determinants

$$|K\rangle = |\chi_i\chi_j\rangle = a_i^\dagger a_j^\dagger|\ \rangle \tag{2.223}$$

$$|L\rangle = |\chi_k\chi_l\rangle = a_k^\dagger a_l^\dagger|\ \rangle \tag{2.224}$$

by expanding out the determinants, integrating over the space and spin coordinates of the two electrons, and using the orthonormality relation of spin orbitals. Here we evaluate the overlap by using the formalism of second quantization. Since the adjoint of Eq. (2.223) is

$$\langle K| = \langle \ |(a_i^\dagger a_j^\dagger)^\dagger = \langle \ |a_j a_i \qquad (2.225)$$

we have

$$\langle K|L\rangle = \langle \ |a_j a_i a_k^\dagger a_l^\dagger| \ \rangle \qquad (2.226)$$

The general strategy for evaluating such matrix elements is to move, using the anticommutation relations, the annihilation operators to the right until they operate directly on the vacuum state. We begin with a_i. Since

$$a_i a_k^\dagger = \delta_{ik} - a_k^\dagger a_i \qquad (2.227)$$

we have

$$\begin{aligned}
\langle K|L\rangle &= \langle \ |a_j(\delta_{ik} - a_k^\dagger a_i)a_l^\dagger| \ \rangle \\
&= \delta_{ik}\langle \ |a_j a_l^\dagger| \ \rangle - \langle \ |a_j a_k^\dagger a_i a_l^\dagger| \ \rangle \qquad (2.228)
\end{aligned}$$

To continue, we move a_j to the right in the first term and keep moving a_i to the right in the second,

$$\langle K|L\rangle = \delta_{ik}\delta_{jl}\langle \ | \ \rangle - \delta_{ik}\langle \ |a_l^\dagger a_j| \ \rangle - \delta_{il}\langle \ |a_j a_k^\dagger| \ \rangle + \langle \ |a_j a_k^\dagger a_l^\dagger a_i| \ \rangle \quad (2.229)$$

The second and last terms now have an annihilation operator acting on the vacuum and hence are zero. Finally, moving a_j to the right in the third term, we have

$$\begin{aligned}
\langle K|L\rangle &= \delta_{ik}\delta_{jl}\langle \ | \ \rangle - \delta_{il}\delta_{jk}\langle \ | \ \rangle + \delta_{il}\langle \ |a_k^\dagger a_j| \ \rangle \\
&= \delta_{ik}\delta_{jl} - \delta_{il}\delta_{jk} \qquad (2.230)
\end{aligned}$$

since the vacuum is normalized. This result is the same as found in Exercise 2.5.

Exercise 2.26 Show using second quantization that $\langle \chi_i|\chi_j\rangle = \delta_{ij}$

Exercise 2.27 Given a state

$$|K\rangle = |\chi_1\chi_2\cdots\chi_N\rangle = a_1^\dagger a_2^\dagger\cdots a_N^\dagger| \ \rangle$$

show that $\langle K|a_i^\dagger a_j|K\rangle = 1$ if $i = j$ and $i \in \{1, 2, \ldots, N\}$, but is zero otherwise.

Exercise 2.28 Let $|\Psi_0\rangle = |\chi_1\cdots\chi_a\chi_b\cdots\chi_N\rangle$ be the Hartree-Fock ground state wave function. Show that

a. $a_r|\Psi_0\rangle = 0 = \langle\Psi_0|a_r^\dagger$.
b. $a_a^\dagger|\Psi_0\rangle = 0 = \langle\Psi_0|a_a$.
c. $|\Psi_a^r\rangle = a_r^\dagger a_a|\Psi_0\rangle$.

d. $\langle \Psi_a^r | = \langle \Psi_0 | a_a^\dagger a_r$.

e. $| \Psi_{ab}^{rs} \rangle = a_s^\dagger a_b a_r^\dagger a_a | \Psi_0 \rangle = a_r^\dagger a_s^\dagger a_b a_a | \Psi_0 \rangle$.

f. $\langle \Psi_{ab}^{rs} | = \langle \Psi_0 | a_a^\dagger a_r a_b^\dagger a_s = \langle \Psi_0 | a_a^\dagger a_b^\dagger a_s a_r$.

2.4.2 Second-Quantized Operators and Their Matrix Elements

We have seen that we can represent determinants by using creation and annihilation operators, which obey a set of anticommutation relations and a vacuum state. Thus we have found a representation of a many-electron wave function that satisfies the requirement of the antisymmetry principle, but which can be manipulated without any knowledge of the properties of determinants. To be able to develop the entire theory of many-electron systems without using determinants, we must express the many-particle operators, $\mathcal{O}_1$ and $\mathcal{O}_2$, in terms of creation and annihilation operators. We can then evaluate matrix elements of these operators using only the algebraic properties of creation and annihilation operators. Clearly, the expression for an operator $\mathcal{O}$ in second quantization must be such that the value of the matrix element $\langle K | \mathcal{O} | L \rangle$ is the same irrespective of whether we obtained it using the properties of determinants or using the algebra of creation and annihilation operators. The appropriate expressions for $\mathcal{O}_1$ (our sum of one-electron operators) and $\mathcal{O}_2$ (the operator describing the total coulomb repulsion between electrons) in second quantization are

$$\mathcal{O}_1 = \sum_{ij} \langle i | h | j \rangle a_i^\dagger a_j \tag{2.231}$$

$$\mathcal{O}_2 = \frac{1}{2} \sum_{ijkl} \langle ij | kl \rangle a_i^\dagger a_j^\dagger a_l a_k \tag{2.232}$$

where the sums run over the set spin orbitals $\{\chi_i\}$. Note that the one- and two-electron integrals appear explicitly and that the form of these operators is independent of the number of electrons. One of the advantages of second quantization is that it treats systems with different numbers of particles on an equal footing. This is particularly convenient when one is dealing with infinite systems such as solids.

Exercise 2.29 Let $| \Psi_0 \rangle = | \chi_1 \chi_2 \rangle = a_1^\dagger a_2^\dagger | \; \rangle$ be the Hartree-Fock wave function for minimal basis H_2. Show using second quantization that

$$\langle \Psi_0 | \mathcal{O}_1 | \Psi_0 \rangle = \sum_{ij} \langle i | h | j \rangle \langle \; | a_2 a_1 a_i^\dagger a_j a_1^\dagger a_2^\dagger | \; \rangle$$

$$= \langle 1 | h | 1 \rangle + \langle 2 | h | 2 \rangle$$

As an illustration of the equivalence of second quantization with our previous development, based on Slater determinants, we calculate the energy of the Hartree-Fock ground state, $| \Psi_0 \rangle = | \chi_1 \cdots \chi_a \chi_b \cdots \chi_N \rangle$, using second quantization. For the sum of one-electron operators, we have

$$\langle\Psi_0|\mathcal{O}_1|\Psi_0\rangle = \sum_{ij} \langle i|h|j\rangle\langle\Psi_0|a_i^\dagger a_j|\Psi_0\rangle \tag{2.233}$$

Since both a_j and $a_i^\dagger$ are trying to destroy an electron (a_j to the right and $a_i^\dagger$ to the left), the indices i and j must belong to the set $\{a, b, \ldots\}$ and thus

$$\langle\Psi_0|\mathcal{O}_1|\Psi_0\rangle = \sum_{ab} \langle a|h|b\rangle\langle\Psi_0|a_a^\dagger a_b|\Psi_0\rangle \tag{2.234}$$

Using

$$a_a^\dagger a_b = \delta_{ab} - a_b a_a^\dagger$$

to move $a_a^\dagger$ to the right, we have

$$\langle\Psi_0|a_a^\dagger a_b|\Psi_0\rangle = \delta_{ab}\langle\Psi_0|\Psi_0\rangle - \langle\Psi_0|a_b a_a^\dagger|\Psi_0\rangle \tag{2.235}$$

The second term on the right is zero since $a_a^\dagger$ is trying to create an electron in χ_a, which is already occupied in $|\Psi_0\rangle$. Since $\langle\Psi_0|\Psi_0\rangle = 1$, we finally have

$$\langle\Psi_0|\mathcal{O}_1|\Psi_0\rangle = \sum_{ab} \langle a|h|b\rangle\delta_{ab} = \sum_a \langle a|h|a\rangle \tag{2.236}$$

in agreement with our previous result in Table 2.5.

For the sum of two-electron operators, we have

$$\langle\Psi_0|\mathcal{O}_2|\Psi_0\rangle = \frac{1}{2}\sum_{ijkl} \langle ij|kl\rangle\langle\Psi_0|a_i^\dagger a_j^\dagger a_l a_k|\Psi_0\rangle \tag{2.237}$$

By the same argument we used for one-electron operators, the indices i, j, k, l must belong to the set $\{a, b, \ldots\}$,

$$\langle\Psi_0|\mathcal{O}_2|\Psi_0\rangle = \frac{1}{2}\sum_{abcd} \langle ab|cd\rangle\langle\Psi_0|a_a^\dagger a_b^\dagger a_d a_c|\Psi_0\rangle \tag{2.238}$$

Our strategy, as before, is to move $a_a^\dagger$ and $a_b^\dagger$ to the right until they operate on $|\Psi_0\rangle$,

$$\begin{aligned}
\langle\Psi_0|a_a^\dagger a_b^\dagger a_d a_c|\Psi_0\rangle &= \delta_{bd}\langle\Psi_0|a_a^\dagger a_c|\Psi_0\rangle - \langle\Psi_0|a_a^\dagger a_d a_b^\dagger a_c|\Psi_0\rangle \\
&= \delta_{bd}\delta_{ac}\langle\Psi_0|\Psi_0\rangle - \delta_{bd}\langle\Psi_0|a_c a_a^\dagger|\Psi_0\rangle \\
&\quad - \delta_{bc}\langle\Psi_0|a_a^\dagger a_d|\Psi_0\rangle + \langle\Psi_0|a_a^\dagger a_d a_c a_b^\dagger|\Psi_0\rangle \\
&= \delta_{bd}\delta_{ac} - \delta_{bc}\delta_{ad}\langle\Psi_0|\Psi_0\rangle + \delta_{bc}\langle\Psi_0|a_d a_a^\dagger|\Psi_0\rangle \\
&= \delta_{bd}\delta_{ac} - \delta_{bc}\delta_{ad}
\end{aligned}$$

We thus get two terms; in the first term we set $c = a$ and $d = b$, and in the second term we set $c = b$ and $d = a$,

$$\langle\Psi_0|\mathcal{O}_2|\Psi_0\rangle = \frac{1}{2}\sum_{ab} \langle ab|ab\rangle - \langle ab|ba\rangle \tag{2.239}$$

in agreement with our previous result in Table 2.6.

Exercise 2.30 Show that

$$\langle \Psi_a^r | \mathcal{O}_1 | \Psi_0 \rangle = \sum_{ij} \langle i | h | j \rangle \langle \Psi_0 | a_a^\dagger a_r a_i^\dagger a_j | \Psi_0 \rangle$$

$$= \langle r | h | a \rangle$$

by moving $a_a^\dagger$ and a_r to the right.

Exercise 2.31 Show that

$$\langle \Psi_a^r | \mathcal{O}_2 | \Psi_0 \rangle = \sum_b^N \langle rb | | ab \rangle$$

Hint: first show that

$$\langle \Psi_0 | a_a^\dagger a_r a_i^\dagger a_j^\dagger a_l a_k | \Psi_0 \rangle = \delta_{rj} \delta_{al} \langle \Psi_0 | a_i^\dagger a_k | \Psi_0 \rangle - \delta_{rj} \delta_{ak} \langle \Psi_0 | a_i^\dagger a_l | \Psi_0 \rangle$$
$$+ \delta_{ri} \delta_{ak} \langle \Psi_0 | a_j^\dagger a_l | \Psi_0 \rangle - \delta_{ri} \delta_{al} \langle \Psi_0 | a_j^\dagger a_k | \Psi_0 \rangle$$

then refer to Exercise 2.27.

2.5 SPIN-ADAPTED CONFIGURATIONS

We have described the spin of a single electron by the two spin functions $\alpha(\omega) \equiv \alpha$ and $\beta(\omega) \equiv \beta$. In this section we will discuss spin in more detail and consider the spin states of many-electron systems. We will describe *restricted* Slater determinants that are formed from spin orbitals whose spatial parts are restricted to be the same for α and β spins (i.e., $\{\chi_i\} = \{\psi_i \alpha, \psi_i \beta\}$). Restricted determinants, except in special cases, are not eigenfunctions of the total electron spin operator. However, by taking appropriate linear combinations of such determinants we can form *spin-adapted configurations*, which are proper eigenfunctions. Finally, we will describe *unrestricted* determinants, which are formed from spin orbitals that have different spatial parts for different spins (i.e., $\{\chi_i\} = \{\psi_i^\alpha \alpha, \psi_i^\beta \beta\}$).

2.5.1 Spin Operators

The spin angular momentum of a particle is a vector operator $\vec{s}$,

$$\vec{s} = s_x \vec{i} + s_y \vec{j} + s_z \vec{k} \tag{2.240}$$

where $\vec{i}, \vec{j}$, and $\vec{k}$ are unit vectors along the x, y, and z directions. The squared magnitude of $\vec{s}$ is a scalar operator

$$s^2 = \vec{s} \cdot \vec{s} = s_x^2 + s_y^2 + s_z^2 \tag{2.241}$$

The components of the spin angular momentum satisfy the commutation relations

$$[s_x, s_y] = is_z, \qquad [s_y, s_z] = is_x, \qquad [s_z, s_x] = is_y \qquad (2.242)$$

The complete set of states describing the spin of a single particle can be taken to be the simultaneous eigenfunctions of s^2 and a single component of $\vec{s}$, usually chosen to be s_z,

$$s^2|s, m_s\rangle = s(s + 1)|s, m_s\rangle \qquad (2.243a)$$

$$s_z|s, m_s\rangle = m_s|s, m_s\rangle \qquad (2.243b)$$

where s is a quantum number describing the total spin and m_s is a quantum number describing the z component of the spin. The possible values of s are $0, \frac{1}{2}, 1, \frac{3}{2}, \ldots$ and m_s has $2s + 1$ possible values $-s, -s + 1, -s + 2, \ldots,$ $s - 1, s$. An electron is a particle with $s = \frac{1}{2}$ and $m_s = \pm\frac{1}{2}$. Thus the complete set of states describing the spin of the electron are

$$|\tfrac{1}{2}, \tfrac{1}{2}\rangle \equiv |\alpha\rangle \qquad (2.244a)$$

$$|\tfrac{1}{2}, -\tfrac{1}{2}\rangle \equiv |\beta\rangle \qquad (2.244b)$$

These spin states are eigenfunctions of s^2 and s_z,

$$s^2|\alpha\rangle = \tfrac{3}{4}|\alpha\rangle, \qquad s^2|\beta\rangle = \tfrac{3}{4}|\beta\rangle \qquad (2.245a)$$

$$s_z|\alpha\rangle = \tfrac{1}{2}|\alpha\rangle, \qquad s_z|\beta\rangle = -\tfrac{1}{2}|\beta\rangle \qquad (2.245b)$$

but are not eigenfunctions of s_x and s_y,

$$s_x|\alpha\rangle = \tfrac{1}{2}|\beta\rangle, \qquad s_x|\beta\rangle = \tfrac{1}{2}|\alpha\rangle \qquad (2.245c)$$

$$s_y|\alpha\rangle = \frac{i}{2}|\beta\rangle, \qquad s_y|\beta\rangle = -\frac{i}{2}|\alpha\rangle \qquad (2.245d)$$

Instead of using s_x and s_y, it is often more convenient to work with the "step-up" and "step-down" ladder operators, s_+ and s_-, defined as

$$s_+ = s_x + is_y \qquad (2.246a)$$

$$s_- = s_x - is_y \qquad (2.246b)$$

These operators increase or decrease the value of m_s by one,

$$s_+|\alpha\rangle = 0, \qquad s_+|\beta\rangle = |\alpha\rangle \qquad (2.247a)$$

$$s_-|\alpha\rangle = |\beta\rangle, \qquad s_-|\beta\rangle = 0 \qquad (2.247b)$$

Using the commutation relations (2.242), the expression (2.241) for s^2 can be rewritten as

$$s^2 = s_+s_- - s_z + s_z^2 \qquad (2.248a)$$

$$s^2 = s_-s_+ + s_z + s_z^2 \qquad (2.248b)$$

Exercise 2.32 a) Derive (2.247) from (2.245); b) Derive (2.248).

Exercise 2.33 Find the 2×2 matrix representations of s^2, s_z, s_+, and s_- in the basis $|\alpha\rangle$, $|\beta\rangle$. Verify the identities analogous to (2.248a,b) for these matrix representations.

Exercise 2.34 Using the commutation relations (2.242), show that $[s^2, s_z] = 0$.

In a many-electron system, the total spin angular momentum operator is simply the vector sum of the spin vectors of each of the electrons

$$\vec{\mathscr{S}} = \sum_{i=1}^{N} \vec{s}(i) \tag{2.249}$$

From this relation it is evident that the components of the total spin and the ladder operators are analogous sums of one-electron operators

$$\mathscr{S}_I = \sum_{i=1}^{N} s_I(i) \qquad I = x, y, z \tag{2.250a}$$

$$\mathscr{S}_\pm = \sum_{i=1}^{N} s_\pm(i) \tag{2.250b}$$

The total squared-magnitude of the spin,

$$\mathscr{S}^2 = \vec{\mathscr{S}} \cdot \vec{\mathscr{S}} = \sum_{i=1}^{N} \sum_{j=1}^{N} \vec{s}(i) \cdot \vec{s}(j)$$

$$= \mathscr{S}_+ \mathscr{S}_- - \mathscr{S}_z + \mathscr{S}_z^2$$

$$= \mathscr{S}_- \mathscr{S}_+ + \mathscr{S}_z + \mathscr{S}_z^2 \tag{2.251}$$

is the sum of one-electron operators (the diagonal terms $i = j$) plus the sum of two-electron operators (the cross-terms $i \neq j$).

In the usual nonrelativistic treatment, such as considered in this book, the Hamiltonian does not contain any spin coordinates and hence both $\mathscr{S}^2$ and $\mathscr{S}_z$ commute with the Hamiltonian,

$$[\mathscr{H}, \mathscr{S}^2] = 0 = [\mathscr{H}, \mathscr{S}_z] \tag{2.252}$$

Consequently, the exact eigenfunctions of the Hamiltonian are also eigenfunctions of the two spin operators,

$$\mathscr{S}^2 |\Phi\rangle = S(S+1)|\Phi\rangle \tag{2.253a}$$

$$\mathscr{S}_z |\Phi\rangle = M_S|\Phi\rangle \tag{2.253b}$$

where S and M_S are the spin quantum numbers describing the total spin and its z component of an N-electron state $|\Phi\rangle$. States with $S = 0, \frac{1}{2}, 1, \frac{3}{2}, \ldots$

have multiplicity $(2S + 1) = 1, 2, 3, 4, \ldots$ and are called singlets, doublets, triplets, quartets, etc. Approximate solutions of the Schrödinger equation are not necessarily pure spin states. However, it is often convenient to constrain approximate wave functions to be pure singlets, doublets, triplets, etc. Any single determinant is an eigenfunction of $\mathscr{S}_z$ (see Exercise 2.37). In particular

$$\mathscr{S}_z |\chi_i \chi_j \cdots \chi_k\rangle = \tfrac{1}{2}(N^\alpha - N^\beta)|\chi_i \chi_j \cdots \chi_k\rangle = M_S |\chi_i \chi_j \cdots \chi_k\rangle \quad (2.254)$$

where N^α is the number of spin orbitals with α spin and N^β is the number of spin orbitals with β spin. However, single determinants are not necessarily eigenfunctions of $\mathscr{S}^2$. As we will discuss in the next subsection, by combining a small number of single determinants it is possible to form spin-adapted configurations that are correct eigenfunctions of $\mathscr{S}^2$.

Exercise 2.35 Consider an operator $\mathscr{A}$ that commutes with the Hamiltonian. Suppose $|\Phi\rangle$ is an eigenfunction of $\mathscr{H}$ with eigenvalue E. Show that $\mathscr{A}|\Phi\rangle$ is also an eigenfunction of $\mathscr{H}$ with eigenvalue E. Thus if $|\Phi\rangle$ is (energetically) nondegenerate, then $\mathscr{A}|\Phi\rangle$ is at most a constant multiple of $|\Phi\rangle$ (i.e., $\mathscr{A}|\Phi\rangle = a|\Phi\rangle$) and hence $|\Phi\rangle$ is an eigenfunction of $\mathscr{A}$. In case of degeneracies, we can always construct appropriate linear combinations of the degenerate eigenfunctions of $\mathscr{H}$ that are also eigenfunctions of $\mathscr{A}$.

Exercise 2.36 Given two nondegenerate eigenfunctions of a hermitian operator $\mathscr{A}$ that commutes with $\mathscr{H}$, i.e., $\mathscr{A}|\Psi_1\rangle = a_1|\Psi_1\rangle$, $\mathscr{A}|\Psi_2\rangle = a_2|\Psi_2\rangle$, $a_1 \neq a_2$, show that $\langle\Psi_1|\mathscr{H}|\Psi_2\rangle = 0$. Thus the matrix element of the Hamiltonian between, say, singlet and triplet spin-adapted configurations is zero.

Exercise 2.37 Prove Eq. (2.254). *Hint*: Use expansion (2.115) for a Slater determinant and note that $\mathscr{S}_z$, since it is invariant to any permutation of the electron labels, commutes with $\mathscr{P}_n$.

2.5.2 Restricted Determinants and Spin-Adapted Configurations

As we have seen in Subsection 2.2.1, given a set of K orthonormal spatial orbitals $\{\psi_i | i = 1, 2, \ldots, K\}$ we can form a set of $2K$ spin orbitals $\{\chi_i | i = 1, 2, \ldots, 2K\}$ by multiplying each spatial orbital by either the α or β spin function

$$\chi_{2i-1}(\mathbf{x}) = \psi_i(\mathbf{r})\alpha(\omega)$$
$$\chi_{2i}(\mathbf{x}) = \psi_i(\mathbf{r})\beta(\omega) \qquad i = 1, 2, \ldots, K \qquad (2.255)$$

Such spin orbitals are called *restricted* spin orbitals, and determinants formed from them are restricted determinants. In such a determinant a given spatial

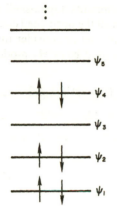

$|{}^1\Psi\rangle \;=\; |\,\psi_1\,\bar{\psi}_1\,\psi_2\,\bar{\psi}_2\,\psi_4\,\bar{\psi}_4\,\rangle$ **Figure 2.11** A singlet closed-shell determinant.

orbital ψ_i can be occupied either by a single electron (spin up or down) or by two electrons (one with spin up and the other with spin down). It is convenient to classify the types of restricted determinants according to the number of spatial orbitals that are singly occupied. A determinant in which each spatial orbital is doubly occupied is called a *closed-shell* determinant (see Fig. 2.11). An *open shell* refers to a spatial orbital that contains a single electron. One refers to determinants by the number of open shells they contain.

All the electron spins are paired in a closed-shell determinant, and it is not surprising that a closed-shell determinant is a pure singlet. That is, it is an eigenfunction of $\mathscr{S}^2$ with eigenvalue zero,

$$\mathscr{S}^2|\psi_i\bar{\psi}_i\psi_j\bar{\psi}_j\cdots\rangle = 0(0+1)|\psi_i\bar{\psi}_i\psi_j\bar{\psi}_j\cdots\rangle = 0 \qquad (2.256)$$

as shown in Exercise 2.38. The simplest example of a closed-shell determinant is the Hartree-Fock ground state wave function of minimal basis H_2,

$$|\Psi_0\rangle = |\psi_1\bar{\psi}_1\rangle = [\psi_1(1)\psi_1(2)]2^{-1/2}(\alpha(1)\beta(2) - \beta(1)\alpha(2)) \qquad (2.257)$$

where we have expanded out the determinant. The spin part of this wave function is just the singlet spin function of a two-electron system. The doubly excited state $|\Psi_{1\bar{1}}^{2\bar{2}}\rangle = |2\bar{2}\rangle$ is, of course, also a singlet.

Exercise 2.38 Prove Eq. (2.256). *Hints:* 1) $\mathscr{S}^2 = \mathscr{S}_-\mathscr{S}_+ + \mathscr{S}_z + \mathscr{S}_z^2$, 2) as a result of Eq. (2.254) it is sufficient to show $\mathscr{S}_+|\psi_i\bar{\psi}_i\cdots\rangle = 0$, 3) use expansion (2.115) for the determinant, and note the $\mathscr{S}_+$ commutes with the permutation operator, 4) $s_+\psi\alpha = 0$, 5) finally, $s_+\psi\beta = \psi\alpha$, but the determinant vanishes because it has two indentical columns.

We now consider open-shell restricted determinants. Open-shell determinants are *not* eigenfunctions of $\mathscr{S}^2$, except when all the open-shell electrons have parallel spin, as in Fig. 2.12. As an illustration, let us consider the four singly excited determinants that arise in the minimal basis H_2 model (see Eq. (2.76)). The open-shell determinants

$$|\Psi_1^2\rangle = |\bar{2}\,\bar{1}\rangle = -2^{-1/2}[\psi_1(1)\psi_2(2) - \psi_2(1)\psi_1(2)]\beta(1)\beta(2) \quad (2.258a)$$

$$|\Psi_{\bar{1}}^2\rangle = |1\,2\rangle = 2^{-1/2}[\psi_1(1)\psi_2(2) - \psi_2(1)\psi_1(2)]\alpha(1)\alpha(2) \quad (2.258b)$$

are eigenfunctions of $\mathscr{S}^2$ with eigenvalue $1(1+1) = 2$ and thus are both triplets. On the other hand, the determinants

$$|\Psi_1^2\rangle = |2\,\bar{1}\rangle \quad (2.259a)$$

$$|\Psi_{\bar{1}}^2\rangle = |1\,\bar{2}\rangle \quad (2.259b)$$

are not pure spin states. However, by taking appropriate linear combinations of these determinants we can form spin-adapted configurations, which are eigenfunctions of $\mathscr{S}^2$. In particular, the singlet spin-adapted configuration is

$$
\begin{aligned}
|^1\Psi_1^2\rangle &= 2^{-1/2}(|\Psi_{\bar{1}}^2\rangle + |\Psi_1^2\rangle) \\
&= 2^{-1/2}(|1\,\bar{2}\rangle + |2\,\bar{1}\rangle) \\
&= 2^{-1/2}[\psi_1(1)\psi_2(2) + \psi_1(2)\psi_2(1)]2^{-1/2}(\alpha(1)\beta(2) - \beta(1)\alpha(2)) \quad (2.260)
\end{aligned}
$$

and the triplet spin-adapted configuration is

$$
\begin{aligned}
|^3\Psi_1^2\rangle &= 2^{-1/2}(|\Psi_{\bar{1}}^2\rangle - |\Psi_1^2\rangle) \\
&= 2^{-1/2}(|1\,\bar{2}\rangle - |2\,\bar{1}\rangle) \\
&= 2^{-1/2}[\psi_1(1)\psi_2(2) - \psi_1(2)\psi_2(1)]2^{-1/2}(\alpha(1)\beta(2) + \beta(1)\alpha(2)) \quad (2.261)
\end{aligned}
$$

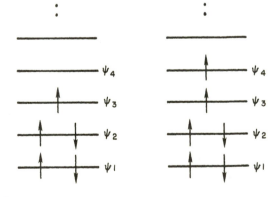

$$|^2\Psi\rangle = |\psi_1\,\bar{\psi}_1\,\psi_2\,\bar{\psi}_2\,\psi_3\rangle \qquad |^3\Psi\rangle = |\psi_1\,\bar{\psi}_1\,\psi_2\,\bar{\psi}_2\,\psi_3\,\psi_4\rangle$$

Figure 2.12 Doublet and triplet restricted single determinants.

As expected, the spin part of $|^1\Psi_1^2\rangle$ is identical to the spin part of the closed-shell wave function (2.257) since both wave functions are singlets.

Exercise 2.39 Using $\mathscr{S}^2 = \mathscr{S}_-\mathscr{S}_+ + \mathscr{S}_z + \mathscr{S}_z^2$, show that $|^1\Psi_1^2\rangle$ is a singlet while $|^3\Psi_1^2\rangle$, $|\Psi_1^2\rangle$ and $|\Psi_{\bar{1}}^2\rangle$ are triplets.

Exercise 2.40 Show that

$$\langle ^1\Psi_1^2|\mathscr{H}|^1\Psi_1^2\rangle = h_{11} + h_{22} + J_{12} + K_{12}$$

$$\langle ^3\Psi_1^2|\mathscr{H}|^3\Psi_1^2\rangle = h_{11} + h_{22} + J_{12} - K_{12}$$

Note that the energy of the triplet is lower than the energy of the singlet. Why is this to be expected from the space parts of the two wave functions?

Let us generalize the above results for minimal basis H_2. In Chapters 4 and 5 we use singlet spin-adapted configurations that arise as a result of single and double excitations from a closed-shell Hartree-Fock ground state,

$$|\Psi_0\rangle = |1\bar{1}\cdots a\bar{a}\, b\bar{b}\cdots\rangle \tag{2.262}$$

The procedure for finding the appropriate linear combinations of singly and doubly excited determinants to form spin-adapted configurations is beyond the scope of this book; we shall merely quote the results. A variety of methods are available for constructing spin eigenfunctions. An authoritative and clear description of many of these methods has been given by Paunz.[2]

The singlet spin-adapted configuration corresponding to the single excitation in which an electron has been promoted from spatial orbital ψ_a to spatial orbital ψ_r is

$$|^1\Psi_a^r\rangle = 2^{-1/2}(|\Psi_a^{\bar{r}}\rangle + |\Psi_{\bar{a}}^r\rangle) \tag{2.263}$$

Note that if $a = 1$ and $r = 2$, this expression reduces to the minimal basis result (2.260).

For double excitations, a number of different types of singlet spin-adapted configurations can occur. They are presented in Table 2.7. The spin-adapted configuration corresponding to the situation that both electrons come from the same spatial orbital and go into the same spatial orbital is $|^1\Psi_{aa}^{rr}\rangle$. This is the generalization of the doubly excited state ($|\Psi_{1\bar{1}}^{2\bar{2}}\rangle = |2\bar{2}\rangle$) of minimal basis H_2. If two electrons come from the same spatial orbital but go to different spatial orbitals, the appropriate spin-adapted configuration is $|^1\Psi_{aa}^{rs}\rangle$. If two electrons come from the different spatial orbitals but go to the same spatial orbital, the appropriate spin-adapted configuration is $|^1\Psi_{ab}^{rr}\rangle$. Finally, for the situation that both electrons come from different spatial orbitals and go to different spatial orbitals there are two linearly independent spin-adapted configurations, $|^A\Psi_{ab}^{rs}\rangle$ and $|^B\Psi_{ab}^{rs}\rangle$.

Table 2.7 Doubly-excited singlet spin-adapted configurations

$$|^1\Psi_{aa}^{rr}\rangle = |\Psi_{a\bar{a}}^{r\bar{r}}\rangle$$

$$|^1\Psi_{aa}^{rs}\rangle = 2^{-1/2}(|\Psi_{a\bar{a}}^{r\bar{s}}\rangle + |\Psi_{a\bar{a}}^{s\bar{r}}\rangle)$$

$$|^1\Psi_{ab}^{rr}\rangle = 2^{-1/2}(|\Psi_{\bar{a}b}^{\bar{r}r}\rangle + |\Psi_{a\bar{b}}^{r\bar{r}}\rangle)$$

$$|^A\Psi_{ab}^{rs}\rangle = (12)^{-1/2}(2|\Psi_{ab}^{rs}\rangle + 2|\Psi_{\bar{a}\bar{b}}^{\bar{r}\bar{s}}\rangle - |\Psi_{\bar{a}b}^{\bar{s}r}\rangle + |\Psi_{\bar{a}b}^{\bar{r}s}\rangle + |\Psi_{a\bar{b}}^{r\bar{s}}\rangle - |\Psi_{a\bar{b}}^{s\bar{r}}\rangle)$$

$$|^B\Psi_{ab}^{rs}\rangle = \tfrac{1}{2}(|\Psi_{\bar{a}b}^{\bar{s}r}\rangle + |\Psi_{\bar{a}b}^{\bar{r}s}\rangle + |\Psi_{a\bar{b}}^{r\bar{s}}\rangle + |\Psi_{a\bar{b}}^{s\bar{r}}\rangle)$$

2.5.3 Unrestricted Determinants

With restricted spin orbitals and restricted determinants, the spatial orbitals are constrained to be identical for α and β spins. For example, the restricted Hartree-Fock (RHF) ground state of the Li atom is

$$|^2\Psi_{\text{RHF}}\rangle = |\psi_{1s}\bar{\psi}_{1s}\psi_{2s}\rangle \qquad (2.264)$$

as shown in Fig. 2.13. The spatial description of the $1s\alpha$ electron is forced to be identical to that of the $1s\beta$ electron. This is a real constraint since the $1s\alpha$ electron has an exchange interaction with the $2s\alpha$ electron, whereas the $1s\beta$ electron does not. The $2s\alpha$ electron spin "polarizes" the $1s$ shell. The $1s\alpha$ and $1s\beta$ electrons will experience different effective potentials and would "prefer" not to be described by the same spatial function. Intuitively, we expect that if this constraint is relaxed by using different orbitals for different spins,

$$|\Psi_{\text{UHF}}\rangle = |\psi_{1s}^{\alpha}\bar{\psi}_{1s}^{\beta}\psi_{2s}^{\alpha}\rangle \qquad (2.265)$$

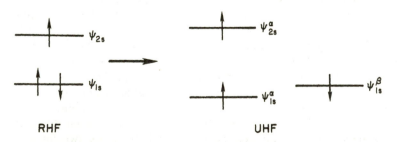

Figure 2.13 Relaxation of a restricted single determinant to an unrestricted single determinant for the Li atom.

we will obtain a lower energy. This is indeed the case. The wave function (2.265) is an example of an *unrestricted* determinant. It is, in fact, the unrestricted Hartree-Fock (UHF) ground state wave function of the Li atom.

Unrestricted determinants are formed from unrestricted spin orbitals. Unrestricted spin orbitals have different spatial orbitals for different spins. Given a set of K orthonormal spatial orbitals $\{\psi_i^\alpha\}$,

$$\langle \psi_i^\alpha | \psi_j^\alpha \rangle = \delta_{ij} \tag{2.266}$$

and a different set of K orthonormal spatial orbitals $\{\psi_i^\beta\}$

$$\langle \psi_i^\beta | \psi_j^\beta \rangle = \delta_{ij} \tag{2.267}$$

such that the two sets are not orthogonal,

$$\langle \psi_i^\alpha | \psi_j^\beta \rangle = S_{ij}^{\alpha\beta} \tag{2.268}$$

where $\mathbf{S}^{\alpha\beta}$ is an overlap matrix, we can form $2K$ *unrestricted* spin orbitals as

$$\chi_{2i-1}(\mathbf{x}) = \psi_i^\alpha(\mathbf{r})\alpha(\omega)$$
$$\chi_{2i}(\mathbf{x}) = \psi_i^\beta(\mathbf{r})\beta(\omega) \qquad i = 1, 2, \ldots, K \tag{2.269}$$

As shown in Exercise 2.1, the $2K$ unrestricted spin orbitals form an orthonormal set, in spite of the fact that the α and β spatial orbitals are not orthogonal.

Unrestricted determinants are not eigenfunctions of $\mathscr{S}^2$. Moreover, they cannot be spin-adapted by combining a small number of unrestricted determinants as is the case for restricted determinants. Thus the UHF ground state (2.265) of the Li atom is not a pure doublet as is the RHF ground state (2.264). Nevertheless, unrestricted wave functions are commonly used as a first approximation to doublet and triplet states.

Figure 2.14 shows a representation of an unrestricted wave function, which is approximately a singlet. Note that $N^\alpha = N^\beta$. The α and β orbitals

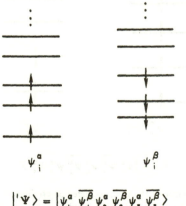

$$|'\Psi\rangle = |\psi_1^\alpha \,\overline{\psi_1^\beta}\, \psi_2^\alpha \,\overline{\psi_2^\beta}\, \psi_3^\alpha \,\overline{\psi_3^\beta}\rangle$$

Figure 2.14 An unrestricted determinant that is *approximately* a singlet.

are drawn as nondegenerate for emphasis. Unrestricted singlets frequently collapse to the corresponding restricted singlets, i.e., to a closed-shell state. In our minimal basis H_2 problem, for example, the closed-shell ground state is $|\psi_1\bar{\psi}_1\rangle$ and at normal bond lengths the energy is raised rather than lowered by using different spatial orbitals for the two electrons. However, when the bond length is very large, one electron is effectively around one hydrogen atom, and the other electron, around the other hydrogen atom, should have a very different spatial description. Thus at large bond lengths, the energy is lowered by using an unrestricted rather than a restricted description, as we shall see in the next chapter.

If $N_\alpha = N_\beta + 1$, then an unrestricted determinant is approximately a doublet (see Fig. 2.15). An unrestricted doublet is often used as the first description of free radicals with one unpaired electron such as CH_3. An approximate triplet determinant has two more α electrons than β electrons as shown in Fig. 2.16.

If $|1\rangle$, $|2\rangle$, $|3\rangle$, etc. are exact singlet, doublet, triplet states etc., then the unrestricted states in Figs. 2.14, 2.15, and 2.16, can be expanded as

$$|^1\Psi\rangle = c_1^1|1\rangle + c_3^1|3\rangle + c_5^1|5\rangle + \cdots \tag{2.270a}$$

$$|^2\Psi\rangle = c_2^2|2\rangle + c_4^2|4\rangle + c_6^2|6\rangle + \cdots \tag{2.270b}$$

$$|^3\Psi\rangle = c_3^3|3\rangle + c_5^3|5\rangle + c_7^3|7\rangle + \cdots \tag{2.270c}$$

Thus an unrestricted wave function is contaminated by higher, not lower, multiplicity components. If the leading term in the above expansion is dominant, then one can describe, to a good approximation, unrestricted

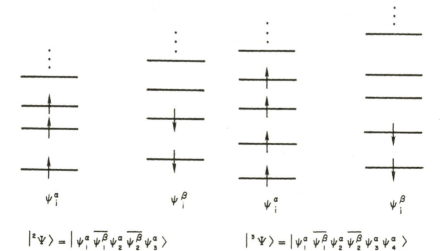

$$|^2\Psi\rangle = |\psi_1^\alpha \overline{\psi_1^\beta} \psi_2^\alpha \overline{\psi_2^\beta} \psi_3^\alpha\rangle \qquad\qquad |^3\Psi\rangle = |\psi_1^\alpha \overline{\psi_1^\beta} \psi_2^\alpha \overline{\psi_2^\beta} \psi_3^\alpha \psi_4^\alpha\rangle$$

Figure 2.15 An unrestricted determinant that is *approximately* a doublet.

Figure 2.16 An unrestricted determinant that is *approximately* a triplet.

determinants as doublet, triplet, etc. The expectation value of $\mathscr{S}^2$ for an unrestricted determinant is always too large because the contaminants have larger values of S. In particular, it can be shown that

$$\langle \mathscr{S}^2 \rangle_{\text{UHF}} = \langle \mathscr{S}^2 \rangle_{\text{Exact}} + N^\beta - \sum_i^N \sum_j^N |S_{ij}^{\alpha\beta}|^2 \qquad (2.271)$$

where we have assumed, as always, that $N^\alpha \geq N^\beta$ and where

$$\langle \mathscr{S}^2 \rangle_{\text{Exact}} = \left(\frac{N^\alpha - N^\beta}{2} \right)\left(\frac{N^\alpha - N^\beta}{2} + 1 \right) \qquad (2.272)$$

In spite of spin contamination, an unrestricted determinant is often used as a first approximation to the wave function for doublets and triplets because unrestricted wave functions have lower energies than the corresponding restricted wave functions.

Exercise 2.41 Consider the determinant $|K\rangle = |\psi_1^\alpha \bar{\psi}_1^\beta\rangle$ formed from *nonorthogonal* spatial orbitals, $\langle \psi_1^\alpha | \psi_1^\beta \rangle = S_{11}^{\alpha\beta}$. a. Show that $|K\rangle$ is an eigenfunction of $\mathscr{S}^2$ only if $\psi_1^\alpha = \psi_1^\beta$. b. Show that $\langle K | \mathscr{S}^2 | K \rangle = 1 - |S_{11}^{\alpha\beta}|^2$ in agreement with Eq. (2.271).

NOTES

1. B. T. Sutcliffe, Fundamentals of computational quantum chemistry, in *Computational Techniques in Quantum Chemistry*, G. H. F. Diercksen, B. T. Sutcliffe, and A. Veillard (Eds.), Reidel, Boston, 1975, p. 1.
2. R. Paunz, *Spin Eigenfunctions*, Plenum, New York, 1979.

FURTHER READING

Avery, J., *Creation and Annihilation Operators*, McGraw-Hill, New York, 1976. Chapter 2 formulates a number of quantum mechanical approximations, including the Hartree-Fock approximation, in the language of second quantization.

Flurry, R. L. Jr., *Symmetry Groups*, Prentice-Hall, Englewood Cliffs, New Jersey, 1980. An excellent introduction to chemical applications of group theory. Our text assumes familiarity with only the most elementary group theoretical ideas and notations.

Mattuck, R. D., 2nd ed., *A Guide to Feynman Diagrams in the Many-Body Problem*, McGraw-Hill, New York, 1976. Chapter 7 introduces second quantization using a more sophisticated point of view than adopted here.

McWeeny, R. and Sutcliffe, B. T., *Methods of Molecular Quantum Mechanics*, 2nd ed., Academic Press, New York, 1976. Discusses the rules for evaluating matrix elements between Slater determinants formed from *nonorthogonal* spin orbitals. This book also contains a concise introduction to methods for obtaining spin eigenfunctions.

Slater, J. C., *Quantum Theory of Matter*, 2nd ed., McGraw-Hill, New York, 1968. Chapter 11 discusses determinantal wave functions and derives expressions for matrix elements between determinants in a somewhat different way than we have done.

THREE

THE HARTREE-FOCK APPROXIMATION

The Hartree-Fock approximation, which is equivalent to the molecular orbital approximation, is central to chemistry. The simple picture, that chemists carry around in their heads, of electrons occupying orbitals is in reality an approximation, sometimes a very good one but, nevertheless, an approximation—the Hartree-Fock approximation. In this chapter we describe, in detail, Hartree-Fock theory and the principles of *ab initio* Hartree-Fock calculations. The length of this chapter testifies to the important role Hartree-Fock theory plays in quantum chemistry. The Hartree-Fock approximation is important not only for its own sake but as a starting point for more accurate approximations, which include the effects of electron correlation. A few of the computational methods of quantum chemistry bypass the Hartree-Fock approximation, but most do not, and all the methods described in the subsequent chapters of this book use the Hartree-Fock approximation as a starting point. Chapters 1 and 2 introduced the basic concepts and mathematical tools important for an indepth understanding of the structure of many-electron theory. We are now in a position to tackle and understand the formalism and computational procedures associated with the Hartree-Fock approximation, at other than a superficial level.

In addition to the basic formalisms of Hartree-Fock theory, this chapter includes a number of *ab initio* calculations. These calculations are not included as a review of available computational results, but as a means of illustrating fundamental ideas. The importance of these calculations to an understanding of the formalisms of this and later chapters cannot be overemphasized. To illustrate the Hartree-Fock approximation, we have performed calculations of each of the quantities discussed in the text (total energies, ionization potentials, equilibrium geometries, dipole moments, etc.), using a standard hierarchy of basis sets (STO-3G, 4-31G, 6-31G* and 6-31G**) and a standard collection of molecules (H_2, CO, N_2, CH_4, NH_3, H_2O, and FH). We thus illustrate the formalism and the accuracy or in-accuracy of the Hartree-Fock approximation with these calculations. In later chapters, we use the same molecules and the same basis sets to illustrate the formalisms of those chapters. In this way we can compare calculations from chapter to chapter to see how a perturbation calculation improves the Hartree-Fock dipole moment of CO, to see how a Green's function calculation improves the Hartree-Fock ionization potentials of N_2, etc. The calculations are intimately related to our discussion of the formal methods of quantum chemistry.

In addition to these larger calculations, we have used two smaller *ab initio* models to illustrate theory. The minimal basis model of H_2, which we introduced in the previous chapter, is applied ubiquitously throughout the book. To specifically illustrate the machinery of Hartree-Fock self-consistent-field (SCF) calculations, we use the minimal basis model of HeH^+. This model is perhaps our most important means of describing the SCF procedure. Appendix A contains a derivation of formulas for all the integrals required in this HeH^+ calculation, and Appendix B contains a short FORTRAN program for performing *ab initio* Hartree-Fock calculations on any two-electron diatomic molecule using the STO-3G basis set. Included is the detailed output for the calculation on HeH^+. The program is written so that it can be easily understood by anyone who has followed the text and has a minimal knowledge of FORTRAN. While it is simple, this program contains the essential ideas (but not the details) of large *ab initio* packages such as Gaussian 80.[1] Appendices A and B and Subsection 3.5.3 are intended to make explicit the basic manipulations of the SCF procedure and take some of the mystery out of such calculations.

In Section 3.1 we present the Hartree-Fock eigenvalue equations and define and discuss associated quantities such as the coulomb, exchange, and Fock operators. The results of this section are presented without derivation as summary of the main equations of Hartree-Fock theory.

Section 3.2 constitutes a derivation of the results of the previous section. The order of presentation of these two sections is such that the derivations of Section 3.2 can be skipped if necessary. For a fuller appreciation of Hartree-Fock theory, however, it is recommended that the derivations be followed. We first present the elements of functional variation and then use this technique to minimize the energy of a single Slater determinant. A unitary transformation of the spin orbitals then leads to the canonical Hartree-Fock equations.

Section 3.3 continues with formal aspects of Hartree-Fock theory. We derive and discuss two important theorems associated with the Hartree-Fock equations: Koopmans' theorem and Brillouin's theorem. The first theorem constitutes an interpretation of the Hartree-Fock orbital energies as ionization potentials and electron affinities. The second theorem states that the matrix element between a Hartree-Fock single determinant and determinants which differ by a single excitation is zero. This theorem is important in multideterminantal theories. Finally, in preparation for perturbation theory (Chapter 6), we define a Hartree-Fock Hamiltonian such that determinants formed from the Hartree-Fock spin orbitals become exact eigenfunctions of this Hamiltonian.

Section 3.4 is the most important section of this chapter. Here we derive the Roothaan equations, which allow one to calculate Hartree-Fock solutions for the ground state of closed-shell molecules. To solve the Hartree-Fock equations, it is necessary to give explicit form to the spin orbitals. This book uses two sets of spin orbitals. The restricted closed-shell set of spin orbitals leads to restricted closed-shell wave functions via the Roothaan equations. An unrestricted open-shell set of spin orbitals leads to unrestricted open-shell wave functions via the Pople-Nesbet equations as discussed in a later section. In this section, the general spin orbital formulation of the Hartree-Fock equations is first reduced to a spatially restricted closed-shell formulation, by replacing the general spin orbitals by a set of restricted closed-shell spin orbitals. A basis set is then introduced; this converts the spatial integro-differential closed-shell Hartree-Fock equations to a set of algebraic equations, the Roothaan equations. The rest of the section then constitutes a detailed discussion of the Roothaan equations, their method of solution (the self-consistent-field (SCF) procedure), and the interpretation of the resulting wave functions.

Section 3.5 contains a detailed illustration of the closed-shell *ab initio* SCF procedure using two simple systems: the minimal basis set descriptions of the homonuclear (H_2) and heteronuclear (HeH^+) two-electron molecules. We first describe the STO-3G minimal basis set used in calculations on these two molecules. We then describe the application of closed-shell Hartree-Fock theory to H_2. This is a very simple model system, which allows one to examine the results of calculations in explicit analytical form. Finally, we apply the Roothaan SCF procedure to HeH^+. Unlike H_2, the final SCF wave function for minimal basis HeH^+ is not symmetry determined and the HeH^+ example provides the simplest possible illustration of the iterative SCF procedure. The description of the *ab initio* HeH^+ calculation given in the text is based on a simple FORTRAN program and the output of a HeH^+ calculation found in Appendix B. By following the details of this simple but, nevertheless, real calculation, the formalism of closed-shell *ab initio* SCF calculations is made concrete.

Section 3.6 describes general aspects of the polyatomic basis sets used in many current calculations. The choice of a basis set for quantum chemical calculations is mainly an art rather than a science, but the principal unifying concepts involved in the choice of a basis set are described. In addition, the basis sets of Pople and co-workers, used in the calculations of this book, are explicitly defined.

In Section 3.7 we perform a number of *ab initio* calculations to illustrate the application and the results of the closed-shell SCF procedure. Our principal aim is to give the reader a feeling for a few of the problems to which *ab initio* SCF calculations can be applied, and the accuracy that can be expected of such Hartree-Fock calculations. To systematize these applications, we apply a standard hierarchy of basis sets to each problem.

In the final Section 3.8, we leave the restricted closed-shell formalism and derive and illustrate unrestricted open-shell calculations. We do not discuss restricted open-shell calculations. By procedures that are strictly analogous to those used in deriving the Roothaan equations of Section 3.4, we derive the corresponding unrestricted open-shell equations of Pople and Nesbet. To illustrate the formalism and the results of unrestricted calculations, we apply our standard basis sets to a description of the electronic structure and ESR spectra of the methyl radical, the ionization potential of N_2, and the orbital structure of the triplet ground state of O_2. Finally, we describe in some detail the application of unrestricted wave functions to the improper behavior of restricted closed-shell wave functions upon dissociation. We again use our minimal basis H_2 model to make the discussion concrete.

3.1 THE HARTREE-FOCK EQUATIONS

In this section we summarize the main results obtained in a derivation of the Hartree-Fock equations. We do this so that the somewhat involved details of the derivation given in the next section (Section 3.2) can be skipped, if desired.

For our purposes, we can equate Hartree-Fock theory to single determinant theory,[2] and we are thus interested in finding a set of spin orbitals $\{\chi_a\}$ such that the single determinant formed from these spin orbitals

$$|\Psi_0\rangle = |\chi_1\chi_2\cdots\chi_a\chi_b\cdots\chi_N\rangle \qquad (3.1)$$

is the best possible approximation to the ground state of the N-electron system described by an electronic Hamiltonian $\mathscr{H}$. According to the variational principle, the "best" spin orbitals are those which minimize the electronic energy

$$E_0 = \langle\Psi_0|\mathscr{H}|\Psi_0\rangle = \sum_a \langle a|h|a\rangle + \frac{1}{2}\sum_{ab} \langle ab||ab\rangle$$

$$= \sum_a \langle a|h|a\rangle + \frac{1}{2}\sum_{ab} [aa|bb] - [ab|ba] \qquad (3.2)$$

By a procedure outlined in Section 3.2, we can systematically vary the spin orbitals $\{\chi_a\}$, constraining them only to the extend that they remain orthonormal,

$$\langle \chi_a | \chi_b \rangle = \delta_{ab} \tag{3.3}$$

until the energy E_0 is a minimum. In doing so (in a formal way) one obtains an equation that defines the best spin orbitals, the ones that minimize E_0. This equation for the best (Hartree-Fock) spin orbitals is the Hartree-Fock integro-differential equation

$$h(1)\chi_a(1) + \sum_{b \neq a} \left[\int d\mathbf{x}_2 \, |\chi_b(2)|^2 r_{12}^{-1} \right] \chi_a(1) - \sum_{b \neq a} \left[\int d\mathbf{x}_2 \, \chi_b^*(2)\chi_a(2) r_{12}^{-1} \right] \chi_b(1)$$

$$= \varepsilon_a \chi_a(1) \tag{3.4}$$

where

$$h(1) = -\frac{1}{2}\nabla_1^2 - \sum_A \frac{Z_A}{r_{1A}} \tag{3.5}$$

is the kinetic energy and potential energy for attraction to the nuclei, of a single electron chosen to be electron-one. The orbital energy of the spin orbital χ_a is ε_a.

3.1.1 The Coulomb and Exchange Operators

The two terms in Eq. (3.4) involving sums over b are those that in single determinant Hartree-Fock theory represent electron-electron interactions. Without these terms,

$$h(1)\chi_a(1) = \varepsilon_a \chi_a(1) \tag{3.6}$$

would simply be a one-electron Schrödinger equation for the spin orbital states of a single electron in the field of the nuclei. The first of the two-electron terms is the *coulomb* term, which is also present in Hartree theory—a theory which uses a Hartree product wave function rather than an antisymmetrized Hartree product (Slater determinant) wave function. The second two-electron term is the *exchange* term, which arises because of the antisymmetric nature of the determinantal wave function.

The coulomb term has a simple interpretation. In an exact theory, the coulomb interaction is represented by the two-electron operator r_{ij}^{-1}. In the Hartree or Hartree-Fock approximation, as Eq. (3.4) shows, electron-one in χ_a experiences a one-electron coulomb potential

$$v_a^{coul}(1) = \sum_{b \neq a} \int d\mathbf{x}_2 |\chi_b(2)|^2 r_{12}^{-1} \tag{3.7}$$

Let us consider this potential. Suppose electron 2 occupies χ_b. The *two-electron potential* r_{12}^{-1} felt by electron 1 and associated with the instantaneous position of electron 2 is thus replaced by a *one-electron potential*, obtained by averaging the interaction r_{12}^{-1} of electron 1 and electron 2, over all space and spin coordinates $\mathbf{x}_2$ of electron 2, weighted by the probability $d\mathbf{x}_2 |\chi_b(2)|^2$ that electron 2 occupies the volume element $d\mathbf{x}_2$ at $\mathbf{x}_2$. By summing over all $b \neq a$, one obtains the total averaged potential acting on the electron in χ_a, arising from the $N - 1$ electrons in the other spin orbitals. Associated with this interpretation it is convenient to define a *coulomb operator*

$$\mathscr{J}_b(1) = \int d\mathbf{x}_2 \, |\chi_b(2)|^2 r_{12}^{-1} \tag{3.8}$$

which represents the average local potential at $\mathbf{x}_1$ arising from an electron in χ_b.

The exchange term in (3.4), arising from the antisymmetric nature of the single determinant, has a somewhat strange form and does not have a simple classical interpretation like the coulomb term. We can, however, write the Hartree-Fock equation (3.4) as an eigenvalue equation

$$\left[h(1) + \sum_{b \neq a} \mathscr{J}_b(1) - \sum_{b \neq a} \mathscr{K}_b(1) \right] \chi_a(1) = \varepsilon_a \chi_a(1) \tag{3.9}$$

provided we introduce an *exchange operator* $\mathscr{K}_b(1)$, defined by its effect when operating on a spin orbital $\chi_a(1)$,

$$\mathscr{K}_b(1)\chi_a(1) = \left[\int d\mathbf{x}_2 \, \chi_b^*(2) r_{12}^{-1} \chi_a(2) \right] \chi_b(1) \tag{3.10}$$

This is to be compared with the previous result (3.8) for the coulomb operator,

$$\mathscr{J}_b(1)\chi_a(1) = \left[\int d\mathbf{x}_2 \, \chi_b^*(2) r_{12}^{-1} \chi_b(2) \right] \chi_a(1) \tag{3.11}$$

Operating with $\mathscr{K}_b(1)$ on $\chi_a(1)$ involves an "exchange" of electron 1 and electron 2 to the right of r_{12}^{-1} in (3.10), relative to (3.11). Unlike the *local* coulomb operator, the exchange operator is said to be a *nonlocal* operator, since there does not exist a simple potential $\mathscr{K}_b(\mathbf{x}_1)$ uniquely defined at a local point in space $\mathbf{x}_1$. The result of operating with $\mathscr{K}_b(\mathbf{x}_1)$ on $\chi_a(\mathbf{x}_1)$ depends on the value of χ_a throughout all space, not just at $\mathbf{x}_1$, as is evident from (3.10). One could not, for example, draw contour plots of the exchange potential as one can for the coulomb potential. For an electron in χ_a the expectation values of the coulomb and exchange potentials $\mathscr{J}_b$ and $\mathscr{K}_b$ are just the coulomb and exchange integrals described in the last chapter, i.e.,

$$\langle \chi_a(1) | \mathscr{J}_b(1) | \chi_a(1) \rangle = \int d\mathbf{x}_1 \, d\mathbf{x}_2 \, \chi_a^*(1) \chi_a(1) r_{12}^{-1} \chi_b^*(2) \chi_b(2) = [aa|bb] \tag{3.12}$$

$$\langle \chi_a(1) | \mathscr{K}_b(1) | \chi_a(1) \rangle = \int d\mathbf{x}_1 \, d\mathbf{x}_2 \, \chi_a^*(1) \chi_b(1) r_{12}^{-1} \chi_b^*(2) \chi_a(2) = [ab|ba] \tag{3.13}$$

3.1.2 The Fock Operator

The Hartree-Fock equation, as we have written it up to this point, is

$$\left[h(1) + \sum_{b \neq a} \mathcal{J}_b(1) - \sum_{b \neq a} \mathcal{K}_b(1) \right] \chi_a(1) = \varepsilon_a \chi_a(1) \tag{3.14}$$

This is of the eigenvalue form. However, the operator in square brackets appears to be different for every spin orbital χ_a on which it operates (because of the restricted summation over $b \neq a$). Inspecting Eqs. (3.10) and (3.11), it is obvious, however, that

$$[\mathcal{J}_a(1) - \mathcal{K}_a(1)] \chi_a(1) = 0 \tag{3.15}$$

It is thus possible to add this term to (3.14), eliminate the restriction on the summation, and define a *Fock operator* f by

$$f(1) = h(1) + \sum_b \mathcal{J}_b(1) - \mathcal{K}_b(1) \tag{3.16}$$

so that the Hartree-Fock equations become

$$f | \chi_a \rangle = \varepsilon_a | \chi_a \rangle \tag{3.17}$$

This is the usual form of the Hartree-Fock equations. The Fock operator $f(1)$ is the sum of a *core-Hamiltonian operator* $h(1)$ and an effective one-electron potential operator called the *Hartree-Fock potential* $v^{HF}(1)$,

$$v^{HF}(1) = \sum_b \mathcal{J}_b(1) - \mathcal{K}_b(1) \tag{3.18}$$

That is,

$$f(1) = h(1) + v^{HF}(1) \tag{3.19}$$

Sometimes it is convenient to write the exchange potential in terms of an operator $\mathcal{P}_{12}$, which, operating to the right, interchanges electron 1 and electron 2. Thus

$$\mathcal{K}_b(1) \chi_a(1) = \left[\int d\mathbf{x}_2 \, \chi_b^*(2) r_{12}^{-1} \chi_a(2) \right] \chi_b(1)$$

$$= \left[\int d\mathbf{x}_2 \, \chi_b^*(2) r_{12}^{-1} \mathcal{P}_{12} \chi_b(2) \right] \chi_a(1) \tag{3.20}$$

The Fock operator is thus written, using $\mathcal{P}_{12}$, as

$$f(1) = h(1) + v^{HF}(1)$$

$$= h(1) + \sum_b \int d\mathbf{x}_2 \, \chi_b^*(2) r_{12}^{-1} (1 - \mathcal{P}_{12}) \chi_b(2) \tag{3.21}$$

The Hartree-Fock equation

$$f | \chi_a \rangle = \varepsilon_a | \chi_a \rangle \tag{3.22}$$

is an eigenvalue equation with the spin orbitals as eigenfunctions and the energy of the spin orbitals as eigenvalues. The exact solutions to this integro-differential equation correspond to the "exact" Hartree-Fock spin orbitals. In practice it is only possible to solve this equation exactly (i.e., as an integro-differential equation) for atoms. One normally, instead, introduces a set of basis functions for expansion of the spin orbitals and solves a set of matrix equations, as will be described subsequently. Only as the basis set approaches completeness, i.e., as one approaches the Hartree-Fock limit, will the spin orbitals that one obtains approach the exact Hartree-Fock spin orbitals.

While (3.22) is written as a linear eigenvalue equation, it might best be described as a pseudo-eigenvalue equation since the Fock operator has a functional dependence, through the coulomb and exchange operators, on the solutions $\{\chi_a\}$ of the pseudo-eigenvalue equation. Thus the Hartree-Fock equations are really nonlinear equations and will need to be solved by iterative procedures.

Exercise 3.1 Show that the general matrix element of the Fock operator has the form

$$\langle \chi_i | f | \chi_j \rangle = \langle i|h|j \rangle + \sum_b \left[ij|bb \right] - \left[ib|bj \right] = \langle i|h|j \rangle + \sum_b \langle ib||jb \rangle \quad (3.23)$$

3.2 DERIVATION OF THE HARTREE-FOCK EQUATIONS

In this section we derive the Hartree-Fock equations in their general spin orbital form, i.e., we obtain the eigenvalue equation (3.17) by minimizing the energy expression for a single Slater determinant. The derivation makes no assumptions about the spin orbitals. Later, we will specialize to restricted and unrestricted spin orbitals and introduce a basis set, in order to generate algebraic equations (matrix equations) that can be conveniently solved on a computer. In the meantime, we are concerned only with the derivation of the general integro-differential equations (the Hartree-Fock eigenvalue equations), the nature of these equations, and the nature of their formal solution. To derive the equations we will use the general and useful technique of functional variation.

3.2.1 Functional Variation

Given any trial function $\tilde{\Phi}$, the expectation value $E[\tilde{\Phi}]$ of the Hamiltonian operator $\mathscr{H}$ is a number given by

$$E[\tilde{\Phi}] = \langle \tilde{\Phi} | \mathscr{H} | \tilde{\Phi} \rangle \quad (3.24)$$

We say that $E[\tilde{\Phi}]$ is a functional of $\tilde{\Phi}$ since its value depends on the form of a function, i.e., the function $\tilde{\Phi}$, rather than any single independent variable. Suppose we vary $\tilde{\Phi}$ by an arbitrarily small amount, by changing the parameters upon which $\tilde{\Phi}$ depends, for example. That is,

$$\tilde{\Phi} \to \tilde{\Phi} + \delta\tilde{\Phi} \tag{3.25}$$

The energy then becomes

$$
\begin{aligned}
E[\tilde{\Phi} + \delta\tilde{\Phi}] &= \langle \tilde{\Phi} + \delta\tilde{\Phi} | \mathscr{H} | \tilde{\Phi} + \delta\tilde{\Phi} \rangle \\
&= E[\tilde{\Phi}] + \{\langle \delta\tilde{\Phi} | \mathscr{H} | \tilde{\Phi} \rangle + \langle \tilde{\Phi} | \mathscr{H} | \delta\tilde{\Phi} \rangle\} + \cdots \\
&= E[\tilde{\Phi}] + \delta E + \cdots
\end{aligned} \tag{3.26}
$$

where δE, which is called the first variation in E, includes all terms that are linear, i.e., first-order, in the variation $\delta\tilde{\Phi}$. Notice that we can treat "δ" just like a differential operator, i.e., $\delta\langle \tilde{\Phi} | \mathscr{H} | \tilde{\Phi} \rangle = \langle \delta\tilde{\Phi} | \mathscr{H} | \tilde{\Phi} \rangle + \langle \tilde{\Phi} | \mathscr{H} | \delta\tilde{\Phi} \rangle$. In the variation method, we are looking for that $\tilde{\Phi}$ for which $E[\tilde{\Phi}]$ is a minimum. In other words, we wish to find that $\tilde{\Phi}$ for which the first variation in $E[\tilde{\Phi}]$ is zero, i.e.,

$$\delta E = 0 \tag{3.27}$$

This condition only ensures that E is *stationary* with respect to any variation in $\tilde{\Phi}$. Normally, however, the stationary point will also be a minimum.

We will illustrate the variational technique by rederiving the matrix eigenvalue equation of the linear variational problem given in Subsection 1.3.2. Given a linear variational trial wave function,

$$|\tilde{\Phi}\rangle = \sum_{i=1}^{N} c_i |\Psi_i\rangle \tag{3.28}$$

we want to minimize the energy

$$E = \langle \tilde{\Phi} | \mathscr{H} | \tilde{\Phi} \rangle = \sum_{ij} c_i^* c_j \langle \Psi_i | \mathscr{H} | \Psi_j \rangle \tag{3.29}$$

subject to the constraint that the trial wave function remains normalized, i.e.,

$$\langle \tilde{\Phi} | \tilde{\Phi} \rangle - 1 = \sum_{ij} c_i^* c_j \langle \Psi_i | \Psi_j \rangle - 1 = 0 \tag{3.30}$$

Using Lagrange's method of undetermined multipliers described in Chapter 1, we therefore minimize, with respect to the coefficients c_i, the following functional

$$
\begin{aligned}
\mathscr{L} &= \langle \tilde{\Phi} | \mathscr{H} | \tilde{\Phi} \rangle - E(\langle \tilde{\Phi} | \tilde{\Phi} \rangle - 1) \\
&= \sum_{ij} c_i^* c_j \langle \Psi_i | \mathscr{H} | \Psi_j \rangle - E\left(\sum_{ij} c_i^* c_j \langle \Psi_i | \Psi_j \rangle - 1 \right)
\end{aligned} \tag{3.31}
$$

where E is the Lagrange multiplier. Therefore, we set the first variation in $\mathscr{L}$ equal to zero.

$$\delta\mathscr{L} = \sum_{ij} \delta c_i^* c_j \langle \Psi_i | \mathscr{H} | \Psi_j \rangle - E \sum_{ij} \delta c_i^* c_j \langle \Psi_i | \Psi_j \rangle$$
$$+ \sum_{ij} c_i^* \delta c_j \langle \Psi_i | \mathscr{H} | \Psi_j \rangle - E \sum_{ij} c_i^* \delta c_j \langle \Psi_i | \Psi_j \rangle = 0 \qquad (3.32)$$

Since E is real ($\mathscr{L}$ is real), after collecting terms and interchanging indices, we get

$$\sum_i \delta c_i^* \left[\sum_j H_{ij} c_j - E S_{ij} c_j \right] + \text{complex conjugate} = 0 \qquad (3.33)$$

where $H_{ij} = \langle \Psi_i | \mathscr{H} | \Psi_j \rangle$. The linear expansion functions $|\Psi_i\rangle$ are not assumed to be orthonormal, but are assumed to overlap according to

$$\langle \Psi_i | \Psi_j \rangle = S_{ij} \qquad (3.34)$$

Since δc_i^* is arbitrary (c_i^* and c_i are both independent variables), the quantity in square brackets in (3.33) must be zero, or

$$\sum_j H_{ij} c_j = E \sum_j S_{ij} c_j$$

$$\mathbf{Hc} = E\mathbf{Sc} \qquad (3.35)$$

Essentially the same result (with $\mathbf{S} = \mathbf{1}$ and real coefficients) was previously obtained in Subsection 1.3.2. The functional variation technique thus leads to the same result as is obtained by differentiating with respect to the coefficients. Functional variation is a more general technique, however, and we now proceed to derive the Hartree-Fock equations using it.

3.2.2 Minimization of the Energy of a Single Determinant

Given the single determinant $|\Psi_0\rangle = |\chi_1 \chi_2 \cdots \chi_a \chi_b \cdots \chi_N\rangle$, the energy $E_0 = \langle \Psi_0 | \mathscr{H} | \Psi_0 \rangle$ is a functional of the spin orbitals $\{\chi_a\}$. To derive the Hartree-Fock equations we need to minimize $E_0[\{\chi_a\}]$ with respect to the spin orbitals, subject to the constraint that the spin orbitals remain orthonormal,

$$\int d\mathbf{x}_1 \, \chi_a^*(1) \chi_b(1) = [a|b] = \delta_{ab} \qquad (3.36)$$

That is, the constraints are of the form

$$[a|b] - \delta_{ab} = 0 \qquad (3.37)$$

We therefore consider the functional $\mathscr{L}[\{\chi_a\}]$ of the spin orbitals

$$\mathscr{L}[\{\chi_a\}] = E_0[\{\chi_a\}] - \sum_{a=1}^{N} \sum_{b=1}^{N} \varepsilon_{ba}([a|b] - \delta_{ab}) \qquad (3.38)$$

where E_0 is the expectation value of the single determinant $|\Psi_0\rangle$,

$$E_0[\{\chi_a\}] = \sum_{a=1}^{N} [a|h|a] + \frac{1}{2} \sum_{a=1}^{N} \sum_{b=1}^{N} [aa|bb] - [ab|ba] \qquad (3.39)$$

and the ε_{ba} constitute a set of Lagrange multipliers. Because $\mathscr{L}$ is real and $[a|b] = [b|a]^*$, the Lagrange multipliers must be elements of a Hermitian matrix

$$\varepsilon_{ba} = \varepsilon_{ab}^* \qquad (3.40)$$

Exercise 3.2 Prove Eq. (3.40).

Minimization of E_0, subject to the constraints, is thus obtained by minimizing $\mathscr{L}$. We therefore vary the spin orbitals an arbitrary infinitesimal amount, i.e.,

$$\chi_a \rightarrow \chi_a + \delta\chi_a \qquad (3.41)$$

and set the first variation in $\mathscr{L}$ equal to zero,

$$\delta\mathscr{L} = \delta E_0 - \sum_{a=1}^{N} \sum_{b=1}^{N} \varepsilon_{ba}\delta[a|b] = 0 \qquad (3.42)$$

This follows directly from Eq. (3.38) since the variation in a constant (δ_{ab}) is zero. Now

$$\delta[a|b] = [\delta\chi_a|\chi_b] + [\chi_a|\delta\chi_b] \qquad (3.43)$$

and

$$\delta E_0 = \sum_{a=1}^{N} [\delta\chi_a|h|\chi_a] + [\chi_a|h|\delta\chi_a]$$

$$+ \frac{1}{2} \sum_{a=1}^{N} \sum_{b=1}^{N} [\delta\chi_a\chi_a|\chi_b\chi_b] + [\chi_a\delta\chi_a|\chi_b\chi_b] + [\chi_a\chi_a|\delta\chi_b\chi_b] + [\chi_a\chi_a|\chi_b\delta\chi_b]$$

$$- \frac{1}{2} \sum_{a=1}^{N} \sum_{b=1}^{N} [\delta\chi_a\chi_b|\chi_b\chi_a] + [\chi_a\delta\chi_b|\chi_b\chi_a] + [\chi_a\chi_b|\delta\chi_b\chi_a] + [\chi_a\chi_b|\chi_b\delta\chi_a]$$

$$(3.44)$$

Exercise 3.3 Manipulate Eq. (3.44) to show that

$$\delta E_0 = \sum_{a=1}^{N} [\delta\chi_a|h|\chi_a] + \sum_{a=1}^{N} \sum_{b=1}^{N} [\delta\chi_a\chi_a|\chi_b\chi_b] - [\delta\chi_a\chi_b|\chi_b\chi_a]$$

+ complex conjugate

Also

$$\sum_{ab} \varepsilon_{ba}([\delta\chi_a|\chi_b] + [\chi_a|\delta\chi_b]) = \sum_{ab} \varepsilon_{ba}[\delta\chi_a|\chi_b] + \sum_{ab} \varepsilon_{ab}[\chi_b|\delta\chi_a]$$

$$= \sum_{ab} \varepsilon_{ba}[\delta\chi_a|\chi_b] + \sum_{ab} \varepsilon_{ba}^*[\delta\chi_a|\chi_b]^*$$

$$= \sum_{ab} \varepsilon_{ba}[\delta\chi_a|\chi_b] + \text{complex conjugate} \quad (3.45)$$

As a result of the above exercise and Eq. (3.45), the first variation in $\mathscr{L}$ of Eq. (3.42) becomes

$$\delta\mathscr{L} = \sum_{a=1}^{N} [\delta\chi_a|h|\chi_a] + \sum_{a=1}^{N} \sum_{b=1}^{N} [\delta\chi_a\chi_a|\chi_b\chi_b] - [\delta\chi_a\chi_b|\chi_b\chi_a]$$

$$- \sum_{a=1}^{N} \sum_{b=1}^{N} \varepsilon_{ba}[\delta\chi_a|\chi_b] + \text{complex conjugate}$$

$$= 0 \quad (3.46)$$

We can use definitions (3.10) and (3.11) for the coulomb and exchange operators to write this result in the form

$$\delta\mathscr{L} = \sum_{a=1}^{N} \int d\mathbf{x}_1 \, \delta\chi_a^*(1)\left[h(1)\chi_a(1) + \sum_{b=1}^{N} (\mathscr{J}_b(1) - \mathscr{K}_b(1))\chi_a(1) - \sum_{b=1}^{N} \varepsilon_{ba}\chi_b(1)\right]$$

$$+ \text{complex conjugate} = 0 \quad (3.47)$$

Since $\delta\chi_a^*(1)$ is arbitrary, it must be that the quantity in square brackets is zero for all a. Therefore,

$$\left[h(1) + \sum_{b=1}^{N} \mathscr{J}_b(1) - \mathscr{K}_b(1)\right]\chi_a(1) = \sum_{b=1}^{N} \varepsilon_{ba}\chi_b(1) \qquad a = 1, 2, \ldots, N \quad (3.48)$$

The quantity in square brackets above is just our definition of the Fock operator $f(1)$; therefore, the equation for the spin orbitals takes the form

$$f|\chi_a\rangle = \sum_{b=1}^{N} \varepsilon_{ba}|\chi_b\rangle \quad (3.49)$$

This result is perhaps surprising at first glance since it is not in the canonical (standard) eigenvalue form of Eq. (3.17). The reason is that any single determinant wave function $|\Psi_0\rangle$ formed from a set of spin orbitals $\{\chi_a\}$ retains a certain degree of flexibility in the spin orbitals; the spin orbitals can be mixed among themselves without changing the expectation value $E_0 = \langle\Psi_0|\mathscr{H}|\Psi_0\rangle$. Before obtaining the canonical form of the Hartree-Fock equations, we need to consider unitary transformations of the spin orbitals among themselves.

3.2.3 The Canonical Hartree-Fock Equations

Let us consider a new set of spin orbitals $\{\chi'_a\}$ that are obtained from an old set $\{\chi_a\}$ (those of Eq. (3.49)) by a unitary transformation,

$$\chi'_a = \sum_b \chi_b U_{ba} \tag{3.50}$$

A unitary transformation, which satisfies the relation

$$\mathbf{U}^\dagger = \mathbf{U}^{-1} \tag{3.51}$$

is one which preserves the orthonormality property. That is, if we start with a set $\{\chi_a\}$ of orthonormal spin orbitals, the new set $\{\chi'_a\}$ will also be orthonormal. Let us define a square matrix $\mathbf{A}$

$$\mathbf{A} = \begin{pmatrix} \chi_1(1) & \chi_2(1) & \cdots & \chi_a(1) & \cdots & \chi_N(1) \\ \chi_1(2) & \chi_2(2) & \cdots & \chi_a(2) & \cdots & \chi_N(2) \\ \vdots & \vdots & & \vdots & & \vdots \\ \chi_1(N) & \chi_2(N) & \cdots & \chi_a(N) & \cdots & \chi_N(N) \end{pmatrix} \tag{3.52}$$

such that the wave function $|\Psi_0\rangle$ is just the normalized determinant of this matrix

$$|\Psi_0\rangle = (N!)^{-1/2} \det(\mathbf{A}) \tag{3.53}$$

Using definition (3.50) for the transformed orbitals and the rules for ordinary multiplication, it becomes clear that the matrix $\mathbf{A}'$ which corresponds to $\mathbf{A}$ but contains the transformed spin orbitals is

$$\mathbf{A}' = \mathbf{AU} = \begin{pmatrix} \chi_1(1) & \chi_2(1) & \cdots & \chi_N(1) \\ \chi_1(2) & \chi_2(2) & \cdots & \chi_N(2) \\ \vdots & \vdots & & \vdots \\ \chi_1(N) & \chi_2(N) & \cdots & \chi_N(N) \end{pmatrix} \begin{pmatrix} U_{11} & U_{12} & \cdots & U_{1N} \\ U_{21} & U_{22} & \cdots & U_{2N} \\ \vdots & \vdots & & \vdots \\ U_{N1} & U_{N2} & \cdots & U_{NN} \end{pmatrix}$$

$$= \begin{pmatrix} \chi'_1(1) & \chi'_2(1) & \cdots & \chi'_N(1) \\ \chi'_1(2) & \chi'_2(2) & \cdots & \chi'_N(2) \\ \vdots & \vdots & & \vdots \\ \chi'_1(N) & \chi'_2(N) & \cdots & \chi'_N(N) \end{pmatrix} \tag{3.54}$$

Therefore, since

$$\det(\mathbf{AB}) = \det(\mathbf{A})\det(\mathbf{B}) \tag{3.55}$$

the determinant of transformed spin orbitals is related to the determinant of the original spin orbitals by

$$\det(\mathbf{A}') = \det(\mathbf{U})\det(\mathbf{A}) \tag{3.56}$$

or, equivalently

$$|\Psi'_0\rangle = \det(\mathbf{U})|\Psi_0\rangle \tag{3.57}$$

Now, since

$$U^\dagger U = 1 \qquad (3.58)$$

we have

$$\det(U^\dagger U) = \det(U^\dagger)\det(U) = (\det(U))^* \det(U) = |\det(U)|^2 = \det(1) = 1 \qquad (3.59)$$

Therefore,

$$\det(U) = e^{i\phi} \qquad (3.60)$$

and the transformed single determinant $|\Psi_0'\rangle$ of Eq. (3.57) can at most differ from the original determinant $|\Psi_0\rangle$ by a phase factor. If U is a real matrix then this phase factor is just ± 1. Because any observable property depends on $|\Psi|^2$, for all intents and purposes, the original wave function in terms of the spin orbitals $\{\chi_a\}$ and the transformed wave function in terms of the spin orbitals $\{\chi_a'\}$ are identical. For a single determinant wave function, any expectation value is therefore invariant to an arbitrary unitary transformation of the spin orbitals. Thus the spin orbitals that make the total energy stationary are not unique, and no particular physical significance can be given to a particular set of spin orbitals. Localized spin orbitals, for example, are not more "physical" than delocalized spin orbitals.

We can use the invariance of a single determinant to a unitary transformation of the spin orbitals to simplify Eq. (3.49) and put it in the form of an eigenvalue equation for a particular set of spin orbitals. First, however, we need to determine the effect of the above unitary transformation on the Fock operator f and the Lagrange multipliers ε_{ab}. The only parts of the Fock operator that depend on the spin orbitals are the coulomb and exchange terms. The transformed sum of the coulomb operators is

$$\sum_a \mathscr{J}_a'(1) = \sum_a \int d\mathbf{x}_2\, \chi_a'^*(2) r_{12}^{-1} \chi_a'(2)$$

$$= \sum_{bc} \left[\sum_a U_{ba}^* U_{ca}\right] \int d\mathbf{x}_2\, \chi_b^*(2) r_{12}^{-1} \chi_c(2) \qquad (3.61)$$

But

$$\sum_a U_{ba}^* U_{ca} = (UU^\dagger)_{cb} = \delta_{cb} \qquad (3.62)$$

so that

$$\sum_a \mathscr{J}_a'(1) = \sum_b \int d\mathbf{x}_2\, \chi_b^*(2) r_{12}^{-1} \chi_b(2) = \sum_b \mathscr{J}_b(1) \qquad (3.63)$$

Thus the sum of coulomb operators is invariant to a unitary transformation of the spin orbitals. In an identical manner it is easy to show that the sum of exchange operators, and hence the Fock operator itself, is invariant to an arbitrary unitary transformation of the spin orbitals, i.e.,

$$f'(1) = f(1) \tag{3.64}$$

We now need to determine the effect of the unitary transformation on the Lagrange multipliers ε_{ba}. Multiplying Equation (3.49) by $\langle \chi_c |$ shows that the Lagrange multipliers are matrix elements of the Fock operator

$$\langle \chi_c | f | \chi_a \rangle = \sum_{b=1}^{N} \varepsilon_{ba} \langle \chi_c | \chi_b \rangle = \varepsilon_{ca} \tag{3.65}$$

Therefore,

$$
\begin{aligned}
\varepsilon'_{ab} &= \int d\mathbf{x}_1 \, \chi_a'^*(1) f(1) \chi_b'(1) \\
&= \sum_{cd} U_{ca}^* U_{db} \int d\mathbf{x}_1 \chi_c^*(1) f(1) \chi_d(1) \\
&= \sum_{cd} U_{ca}^* \varepsilon_{cd} U_{db} \tag{3.66}
\end{aligned}
$$

or in matrix form

$$\boldsymbol{\varepsilon}' = \mathbf{U}^\dagger \boldsymbol{\varepsilon} \mathbf{U} \tag{3.67}$$

From (3.40), $\boldsymbol{\varepsilon}$ is a Hermitian matrix. It is always possible, therefore, to find a unitary matrix $\mathbf{U}$ such that the transformation (3.67) diagonalizes $\boldsymbol{\varepsilon}$. We are not concerned with how to obtain such a matrix, only that such a matrix exists and is unique. There must exist, then, a set of spin orbitals $\{\chi_a'\}$ for which the matrix of Lagrange multipliers is diagonal.

$$f | \chi_a' \rangle = \varepsilon_a' | \chi_a' \rangle \tag{3.68}$$

The unique set of spin orbitals $\{\chi_a'\}$ obtained from a solution of this eigenvalue equation is called the set of *canonical spin orbitals*. We henceforth drop the primes and write the Hartree-Fock equations as

$$f | \chi_a \rangle = \varepsilon_a | \chi_a \rangle \tag{3.69}$$

The canonical spin orbitals, which are a solution to this equation, will generally be delocalized and form a basis for an irreducible representation of the point group of the molecule, i.e., they will have certain symmetry properties characteristic of the symmetry of the molecule or, equivalently, of the Fock operator. Once the canonical spin orbitals have been obtained it would be possible to obtain an infinite number of equivalent sets by a unitary transformation of the canonical set. In particular, there are various criteria (see Further Reading) for choosing a unitary transformation so that the transformed set of spin orbitals is in some sense localized, more in line with our intuitive feeling for chemical bonds.

3.3 INTERPRETATION OF SOLUTIONS TO THE HARTREE-FOCK EQUATIONS

In order to solve the Hartree-Fock equations it is necessary to introduce a basis set and solve a set of matrix equations. Before doing so, however, there are certain aspects of the eigenvalue equation and its solutions that are independent of any basis, and it is appropriate to discuss them at this point.

3.3.1 Orbital Energies and Koopmans' Theorem

For an N-electron system, minimization of the energy of the determinant $|\Psi_0\rangle = |\chi_1\chi_2 \cdots \chi_a\chi_b \cdots \chi_N\rangle$ leads to an eigenvalue equation $f|\chi_a\rangle = \varepsilon_a|\chi_a\rangle$ for the N occupied spin orbitals $\{\chi_a\}$. The Fock operator has a functional dependence on these occupied spin orbitals, but once the occupied spin orbitals are known the Fock operator becomes a well-defined Hermitian operator, which will have an infinite number of eigenfunctions, i.e.,

$$f|\chi_j\rangle = \varepsilon_j|\chi_j\rangle \qquad j = 1, 2, \ldots, \infty \tag{3.70}$$

Exercise 3.4 Use the result of Exercise 3.1 to show that the Fock operator is a Hermitian operator, by showing that $f_{ij} = \langle\chi_i|f|\chi_j\rangle$ is an element of a Hermitian matrix.

Each of the solutions $|\chi_j\rangle$ of (3.70) has a spin orbital energy ε_j. The N spin orbitals with the lowest orbital energies are just the spin orbitals occupied in $|\Psi_0\rangle$ for which we use the indices a, b, ... The remaining infinite number of spin orbitals with higher energies are the *virtual* or unoccupied spin orbitals, which we label with the indices r, s, ... Our main interest here is to obtain expressions for the orbital energies ε_a and ε_r and to investigate what physical significance we can attach to these orbital energies.

Multiplying (3.70) by $\langle\chi_i|$, shows that the matrix representation of the Fock operator in the basis of spin orbital eigenfunctions is diagonal with diagonal elements equal to the orbital energies.

$$\langle\chi_i|f|\chi_j\rangle = \varepsilon_j\langle\chi_i|\chi_j\rangle = \varepsilon_j\delta_{ij} \tag{3.71}$$

Using expression (3.16) for the Fock operator, the orbital energies can be expressed as

$$\begin{aligned}
\varepsilon_i &= \langle\chi_i|f|\chi_i\rangle = \langle\chi_i|h + \sum_b (\mathscr{J}_b - \mathscr{K}_b)|\chi_i\rangle \\
&= \langle\chi_i|h|\chi_i\rangle + \sum_b \langle\chi_i|\mathscr{J}_b|\chi_i\rangle - \langle\chi_i|\mathscr{K}_b|\chi_i\rangle \\
&= \langle i|h|i\rangle + \sum_b \langle ib|ib\rangle - \langle ib|bi\rangle \\
&= \langle i|h|i\rangle + \sum_b \langle ib||ib\rangle
\end{aligned} \tag{3.72}$$

where, from definitions (3.10) and (3.11) of the exchange and coulomb operators, we have used

$$\langle \chi_i | \mathcal{J}_k | \chi_j \rangle = \langle ik | jk \rangle = [ij | kk] \tag{3.73}$$

$$\langle \chi_i | \mathcal{K}_k | \chi_j \rangle = \langle ik | kj \rangle = [ik | kj] \tag{3.74}$$

In particular then

$$\varepsilon_a = \langle a | h | a \rangle + \sum_{b=1}^{N} \langle ab | | ab \rangle \tag{3.75}$$

$$\varepsilon_r = \langle r | h | r \rangle + \sum_{b=1}^{N} \langle rb | | rb \rangle \tag{3.76}$$

Now, since,

$$\langle aa | | aa \rangle = 0 \tag{3.77}$$

we can rewrite these results as

$$\varepsilon_a = \langle a | h | a \rangle + \sum_{b \neq a} \langle ab | ab \rangle - \langle ab | ba \rangle \tag{3.78}$$

$$\varepsilon_r = \langle r | h | r \rangle + \sum_{b} \langle rb | rb \rangle - \langle rb | br \rangle \tag{3.79}$$

Let us examine these last two expressions. The orbital energy ε_a represents the energy of an electron in the spin orbital $|\chi_a\rangle$. From (3.78) this energy is the kinetic energy and attraction to the nuclei ($\langle a|h|a\rangle$) plus a coulomb ($\langle ab|ab\rangle$) and exchange ($-\langle ab|ba\rangle$) interaction with each of the remaining $N-1$ electrons in the $N-1$ spin orbitals $|\chi_b\rangle$, where $b \neq a$. As we have seen before, the integral $\langle ab|ba\rangle$ is nonzero only if the spins of the electrons in $|\chi_a\rangle$ and $|\chi_b\rangle$ are parallel. In the general spin orbital formulation given here, we have not specified the spins of the electrons, so the general term $\langle ab|ba\rangle$ remains for all electron-electron interactions, even though some of these integrals will be zero.

The result for ε_a is as might be expected, but the formula (3.79) for the virtual spin orbital energy ε_r has a different character. It includes the kinetic energy and nuclear attraction of an electron in $|\chi_r\rangle$, i.e., $\langle r|h|r\rangle$, as expected, but includes coulomb ($\langle rb|rb\rangle$) and exchange ($-\langle rb|br\rangle$) interactions *with all N electrons* of the Hartree-Fock ground state $|\Psi_0\rangle$, i.e., interactions with all N electrons in the spin orbitals $\{\chi_b | b = 1, 2, \ldots, N\}$. It is as if an electron had been added to $|\Psi_0\rangle$ to produce an $(N+1)$-electron state and ε_r represented the energy of this extra electron. This is exactly the case. We will return to this point when we describe Koopmans' theorem. First, we want to relate the occupied orbital energies ε_a to the total energy E_0.

If we simply add up the orbital energies ε_a of Eq. (3.75) for each of the N electrons in the ground state $|\Psi_0\rangle$, we get

$$\sum_a^N \varepsilon_a = \sum_a^N \langle a|h|a\rangle + \sum_a^N \sum_b^N \langle ab||ab\rangle \tag{3.80}$$

The correct expectation value $E_0 = \langle \Psi_0|\mathscr{H}|\Psi_0\rangle$ for this state, from Eq. (2.112), for example, is

$$E_0 = \sum_a^N \langle a|h|a\rangle + \frac{1}{2}\sum_a^N \sum_b^N \langle ab||ab\rangle \tag{3.81}$$

It is thus apparent that

$$E_0 \neq \sum_a^N \varepsilon_a \tag{3.82}$$

and the total energy of the state $|\Psi_0\rangle$ is not just the sum of the orbital energies. The reason is as follows. The energy ε_a includes coulomb and exchange interactions between an electron in χ_a and electrons in all other occupied spin orbitals (in particular, χ_b). But ε_b includes coulomb and exchange interactions between an electron in χ_b and electrons in all other occupied spin orbitals (in particular, χ_a). Thus when we add ε_a and ε_b we include the electron-electron interactions between an electron in χ_a and one in χ_b, twice. The sum of orbital energies counts the electron-electron interactions twice. This is the reason for the factor $\frac{1}{2}$ in the correct expression (3.81) for the total energy E_0 relative to the sum of orbital energies (3.80).

If the total energy is not the sum of orbital energies what physical significance can we attach to orbital energies? The answer is provided by investigating the process of adding or subtracting an electron to the N-electron state $|\Psi_0\rangle = |^N\Psi_0\rangle = |\chi_1\chi_2 \cdots \chi_c \cdots \chi_N\rangle$. Suppose we consider removing an electron from the spin orbital χ_c to produce the $(N-1)$-electron single determinant state $|^{N-1}\Psi_c\rangle = |\chi_1\chi_2 \cdots \chi_{c-1}\chi_{c+1} \cdots \chi_N\rangle$, where the remaining $N-1$ spin orbitals in $|^{N-1}\Psi_c\rangle$ are identical to those in $|^N\Psi_0\rangle$. In second quantization, this would be accomplished by annihilating an electron in χ_c, so that to within a sign,

$$|^{N-1}\Psi_c\rangle = a_c|^N\Psi_0\rangle \tag{3.83}$$

The ionization potential of $|^N\Psi_0\rangle$ for this process is

$$\text{IP} = {}^{N-1}E_c - {}^NE_0 \tag{3.84}$$

where ${}^{N-1}E_c$ and ${}^NE_0$ are the expectation values of the energy of the two relevant single determinants.

$$^NE_0 = \langle ^N\Psi_0|\mathscr{H}|^N\Psi_0\rangle \tag{3.85}$$

$$^{N-1}E_c = \langle ^{N-1}\Psi_c|\mathscr{H}|^{N-1}\Psi_c\rangle \tag{3.86}$$

Depending from which spin orbital χ_c we remove an electron, the state $|^{N-1}\Psi_c\rangle$ may or may not represent the ground state of the ionized species. Since $|^{N-1}\Psi_c\rangle$ is a different state from $|^N\Psi_0\rangle$, one could not in general expect its optimum spin orbitals to be identical with those of $|^N\Psi_0\rangle$. With our assumption that the spin orbitals are identical, however, we can calculate the energy difference between the two states. From the rules of the last chapter, the energy of a single determinant is

$$E = \sum_i^{occ} \langle i|h|i\rangle + \frac{1}{2} \sum_i^{occ} \sum_j^{occ} \langle ij||ij\rangle \tag{3.87}$$

where the sums go over all spin orbitals occupied in the determinant. Thus

$$^N E_0 = \sum_a \langle a|h|a\rangle + \frac{1}{2} \sum_a \sum_b \langle ab||ab\rangle \tag{3.88}$$

where the indices $a, b, \ldots$ refer to the spin orbitals occupied in $|^N\Psi_0\rangle$. With this convention, we have

$$^{N-1} E_c = \sum_{a \neq c} \langle a|h|a\rangle + \frac{1}{2} \sum_{a \neq c} \sum_{b \neq c} \langle ab||ab\rangle \tag{3.89}$$

The ionization potential is the difference between these two results

$$\begin{aligned}
\text{IP} &= {}^{N-1}E_c - {}^N E_0 \\
&= -\langle c|h|c\rangle - \frac{1}{2} \sum_{a[b\equiv c]} \langle ab||ab\rangle - \frac{1}{2} \sum_{b[a\equiv c]} \langle ab||ab\rangle \\
&= -\langle c|h|c\rangle - \frac{1}{2} \sum_a \langle ac||ac\rangle - \frac{1}{2} \sum_b \langle cb||cb\rangle \\
&= -\langle c|h|c\rangle - \sum_b \langle cb||cb\rangle
\end{aligned} \tag{3.90}$$

Comparing this with the definition (3.75) of an occupied spin orbital energy, we see that the ionization potential for removing an electron from χ_c is just the negative of the orbital energy ε_c

$$\text{IP} = {}^{N-1}E_c - {}^N E_0 = -\varepsilon_c \tag{3.91}$$

Thus occupied spin orbital energies in the single determinant approximation represent the energy (with opposite sign) required to remove an electron from that spin orbital. Orbital energies ε_a are generally negative and ionization potentials are positive.

Exercise 3.5 Show that the energy required to remove an electron from χ_c and one from χ_d to produce the $(N-2)$-electron single determinant $|^{N-2}\Psi_{cd}\rangle$ is $-\varepsilon_c - \varepsilon_d + \langle cd|cd\rangle - \langle cd|dc\rangle$.

Now let us consider the process of adding an electron to one of the virtual spin orbitals χ_r to produce the $(N+1)$-electron single determinant $|^{N+1}\Psi^r\rangle = |\chi_r\chi_1\chi_2\cdots\chi_N\rangle$, where again the remaining spin orbitals are identical to those in $|^N\Psi_0\rangle$. In second quantization, this would be accomplished by creating an electron in χ_r

$$|^{N+1}\Psi^r\rangle = a_r^\dagger|^N\Psi_0\rangle \tag{3.92}$$

The electron affinity of $|^N\Psi_0\rangle$ for this process is

$$\text{EA} = {}^NE_0 - {}^{N+1}E^r \tag{3.93}$$

where ${}^{N+1}E^r$ is the energy of the single determinant $|^{N+1}\Psi^r\rangle$,

$${}^{N+1}E^r = \langle {}^{N+1}\Psi^r|\mathcal{H}|{}^{N+1}\Psi^r\rangle \tag{3.94}$$

As in the ionization process, the optimum spin orbitals of the $(N+1)$-electron single determinant will not, in general, be identical with those of $|^N\Psi_0\rangle$. However, with the assumption that they are identical, the electron affinity is readily calculated.

Exercise 3.6 Use Eq. (3.87) to obtain an expression for ${}^{N+1}E^r$ and then subtract it from ${}^NE_0$ (Eq. (3.88)) to show that

$${}^NE_0 - {}^{N+1}E^r = -\langle r|h|r\rangle - \sum_b \langle rb||rb\rangle$$

With the result of the above exercise and Eq. (3.76), we see that the electron affinity for adding an electron to the virtual spin orbital χ_r is just the negative of the orbital energy of that virtual spin orbital, i.e.,

$$\text{EA} = {}^NE_0 - {}^{N+1}E^r = -\varepsilon_r \tag{3.95}$$

This result is consistent with our previous observation that ε_r included interactions with all N other electrons of the ground state $|^N\Psi_0\rangle$ and thus describes an $(N+1)$th electron. If ε_r is negative (i.e., if $|^{N+1}\Psi^r\rangle$ is more stable than $|^N\Psi_0\rangle$), the electron affinity is positive.

The above results were first obtained by Koopmans. We are now in a position to state Koopmans' theorem.

Koopmans' Theorem Given an N-electron Hartree-Fock single determinant $|^N\Psi_0\rangle$ with occupied and virtual spin orbital energies ε_a and ε_r, then the ionization potential to produce an $(N-1)$-electron single determinant $|^{N-1}\Psi_a\rangle$ with identical spin orbitals, obtained by removing an electron from spin orbital χ_a, and the electron affinity to produce an $(N+1)$-electron single determinant $|^{N+1}\Psi^r\rangle$ with identical spin orbitals, obtained by adding an electron to spin orbital χ_r, are just $-\varepsilon_a$ and $-\varepsilon_r$, respectively.

Koopmans' theorem thus gives us a way of calculating approximate ionization potentials and electron affinities. This "frozen orbital" approximation assumes that the spin orbitals in the $(N \pm 1)$-electron states, i.e., the positive and negative ions if $|{}^N\Psi_0\rangle$ is a neutral species, are identical with those of the N-electron state. This approximation neglects relaxation of the spin orbitals in the $(N \pm 1)$-electron states, i.e., the spin orbitals of $|{}^N\Psi_0\rangle$ are not the optimum spin orbitals for $|{}^{N-1}\Psi_a\rangle$ or $|{}^{N+1}\Psi^r\rangle$. Optimizing the spin orbitals of the $(N \pm 1)$-electron single determinants by performing a separate Hartree-Fock calculation on these states would lower the energies ${}^{N-1}E_a$ and ${}^{N+1}E^r$ and thus the neglect of relaxation in Koopmans' theorem calculations tends to produce too positive an ionization potential and too negative an electron affinity. In addition, of course, the approximation of a single determinant wave function leads to errors, and the *correlation effects*, which one obtains in going beyond the Hartree-Fock approximation, will produce further corrections to Koopmans' theorem results. In particular, correlation energies are largest for the system with the highest number of electrons. Therefore, correlation effects tend to cancel the relaxation error for ionization potentials, but add to the relaxation error for electron affinities. In general, Koopmans' ionization potentials are reasonable first approximations to experimental ionization potentials and we shall be discussing a number of such calculations later in this chapter. Koopmans' electron affinities are unfortunately often bad. Many neutral molecules will add an electron to form a stable negative ion. Hartree-Fock calculations on neutral molecules, however, almost always give positive orbital energies for all the virtual orbitals. Electron affinities are considerably more difficult to calculate than ionization potentials and we will not be concerned, to any extent, with electron affinities in this book.

3.3.2 Brillouin's Theorem

The Hartree-Fock equation (3.70) produces a set $\{\chi_i\}$ of spin orbitals. The single determinant $|\Psi_0\rangle$, formed from the N spin orbitals $\{\chi_a\}$ with the lowest orbital energies, is the Hartree-Fock approximation to the ground state. As discussed in the last chapter, there are many other determinants that can be formed from the set $\{\chi_i\}$. Having derived the form of the Fock operator, we are now in a position to prove a theorem about a subset of these determinants. This subset is the set of singly excited determinants $|\Psi_a^r\rangle$ obtained from $|\Psi_0\rangle$ by a single replacement of χ_a with χ_r (Fig. 2.7). In a multideterminantal representation of the exact ground state $|\Phi_0\rangle$, it is these determinants which we might expect, *a priori*, to give the leading correction to the Hartree-Fock ground state $|\Psi_0\rangle$,

$$|\Phi_0\rangle = c_0|\Psi_0\rangle + \sum_{ra} c_a^r|\Psi_a^r\rangle + \cdots \tag{3.96}$$

If we consider only the singly excited determinants as corrections, then the coefficients c_a^r are determined from the variational principle by diagonalizing the Hamiltonian matrix in the basis of the states $\{\Psi_0, \{\Psi_a^r\}\}$. Consider for a moment the matrix eigenvalue problem involving one singly excited state

$$\begin{pmatrix} \langle \Psi_0 | \mathscr{H} | \Psi_0 \rangle & \langle \Psi_0 | \mathscr{H} | \Psi_a^r \rangle \\ \langle \Psi_a^r | \mathscr{H} | \Psi_0 \rangle & \langle \Psi_a^r | \mathscr{H} | \Psi_a^r \rangle \end{pmatrix} \begin{pmatrix} c_0 \\ c_a^r \end{pmatrix} = \mathscr{E}_0 \begin{pmatrix} c_0 \\ c_a^r \end{pmatrix} \tag{3.97}$$

The mixing of the two states depends on the off-diagonal element $\langle \Psi_0 | \mathscr{H} | \Psi_a^r \rangle$. This matrix element is obtained by using the rules for evaluating matrix elements between determinants, and the result can be read directly from Tables 2.5 and 2.6.

$$\langle \Psi_0 | \mathscr{H} | \Psi_a^r \rangle = \langle a | h | r \rangle + \sum_b \langle ab | | rb \rangle \tag{3.98}$$

The right-hand side of this equation can be simplified; as Exercise 3.1 shows, matrix elements of the Fock operator are given by

$$\langle \chi_i | f | \chi_j \rangle = \langle i | h | j \rangle + \sum_b \langle ib | | jb \rangle \tag{3.99}$$

Therefore,

$$\langle \Psi_0 | \mathscr{H} | \Psi_a^r \rangle = \langle \chi_a | f | \chi_r \rangle \tag{3.100}$$

The matrix element that mixes singly excited determinants with $|\Psi_0\rangle$ is thus equal to an off-diagonal element of the Fock matrix. Now, by definition, solving the Hartree-Fock eigenvalue problem requires the off-diagonal elements to satisfy $\langle \chi_i | f | \chi_j \rangle = 0$, $(i \neq j)$. One can then say that solving the Hartree-Fock eigenvalue equation is equivalent to ensuring that $|\Psi_0\rangle$ will not mix with any singly excited determinants. The lowest solution to (3.97) is thus

$$\begin{pmatrix} E_0 & 0 \\ 0 & \langle \Psi_a^r | \mathscr{H} | \Psi_a^r \rangle \end{pmatrix} \begin{pmatrix} 1 \\ 0 \end{pmatrix} = E_0 \begin{pmatrix} 1 \\ 0 \end{pmatrix} \tag{3.101}$$

The Hartree-Fock ground state is in this sense "stable" since it cannot be improved by mixing it with singly excited determinants. One then expects doubly excited determinants $|\Psi_{ab}^{rs}\rangle$ to provide the leading and most important corrections to $|\Psi_0\rangle$. This does not mean that there are no singly excited determinants $|\Psi_a^r\rangle$ in an exact ground state $|\Phi_0\rangle$. They can mix indirectly with $|\Psi_0\rangle$ through the doubly excited determinants by way of the matrix elements $\langle \Psi_a^r | \mathscr{H} | \Psi_{ab}^{rs} \rangle$ and $\langle \Psi_{ab}^{rs} | \mathscr{H} | \Psi_0 \rangle$. The important result we have just derived is termed Brillouin's theorem.

Brillouin's Theorem Singly excited determinants $|\Psi_a^r\rangle$ will not interact directly with a reference Hartree-Fock determinant $|\Psi_0\rangle$, i.e., $\langle \Psi_0 | \mathscr{H} | \Psi_a^r \rangle = 0$.

We will have the opportunity to use this theorem many times in later chapters.

3.3.3 The Hartree-Fock Hamiltonian

Until now the Hartree-Fock approximation has been viewed as an approximation in which the Hamiltonian is exact but the wave function is approximated as a single Slater determinant. For later use in the perturbation theory of Chapter 6, we now preview a different but equivalent view of Hartree-Fock theory that focuses on the Hamiltonian.

We have not solved the exact electronic Schrödinger equation

$$\mathscr{H}|\Phi_0\rangle = \mathscr{E}_0|\Phi_0\rangle \tag{3.102}$$

but rather we have used the variational principle to find an approximation $|\Psi_0\rangle$ to $|\Phi_0\rangle$. We now ask the question, "Is there some approximate N-electron Hamiltonian and eigenvalue equation that we have solved exactly, i.e., is there an approximate Hamiltonian for which $|\Psi_0\rangle$ is an exact eigenfunction?" The answer is "Yes." The *Hartree-Fock Hamiltonian* is

$$\mathscr{H}_0 = \sum_{i=1}^{N} f(i) \tag{3.103}$$

where $f(i)$ is a Fock operator for the ith electron.

Exercise 3.7 Use definition (2.115) of a Slater determinant and the fact that $\mathscr{H}_0$ commutes with any operator that permutes the electron labels, to show that $|\Psi_0\rangle$ is an eigenfunction of $\mathscr{H}_0$ with eigenvalue $\sum_a \varepsilon_a$. Why does $\mathscr{H}_0$ commute with the permutation operator?

As the above exercise shows, $|\Psi_0\rangle$ is an eigenfunction of a Hartree-Fock Hamiltonian with an eigenvalue that is not the Hartree-Fock energy E_0, but the sum of orbital energies $\sum_a \varepsilon_a$. We can in fact show that any single determinant formed from the set $\{\chi_i\}$ of eigenfunctions of the Fock operator, f, is an eigenfunction of $\mathscr{H}_0$ with eigenvalue equal to the sum of the orbital energies of the spin orbitals included in the determinant. In the context of perturbation theory, which is extensively discussed in Chapter 6, we have obtained a complete set of eigenfunctions to an unperturbed Hamiltonian $\mathscr{H}_0$, which can form the basis for a perturbation expansion of the exact energy,

$$\mathscr{E}_0 = E_0^{(0)} + E_0^{(1)} + E_0^{(2)} + \cdots \tag{3.104}$$

The unperturbed zeroth-order energy is just

$$E_0^{(0)} = \sum_a \varepsilon_a \tag{3.105}$$

where

$$\mathscr{H}_0|\Psi_0\rangle = E_0^{(0)}|\Psi_0\rangle \tag{3.106}$$

If

$$\mathcal{H} = \mathcal{H}_0 + \mathcal{V} \tag{3.107}$$

then the perturbation $\mathcal{V}$ is

$$
\begin{aligned}
\mathcal{V} &= \mathcal{H} - \mathcal{H}_0 \\
&= \sum_{i=1}^{N} h(i) + \sum_{i=1}^{N} \sum_{j>i}^{N} r_{ij}^{-1} - \sum_{i=1}^{N} f(i) \\
&= \sum_{i=1}^{N} \sum_{j>i}^{N} r_{ij}^{-1} - \sum_{i=1}^{N} v^{HF}(i)
\end{aligned}
\tag{3.108}
$$

or just the difference between the exact electron-electron interaction and the sum of the Hartree-Fock coulomb and exchange potentials. We can now evaluate the Hartree-Fock energy as

$$
\begin{aligned}
E_0 &= \langle \Psi_0 | \mathcal{H} | \Psi_0 \rangle = \langle \Psi_0 | \mathcal{H}_0 | \Psi_0 \rangle + \langle \Psi_0 | \mathcal{V} | \Psi_0 \rangle \\
&= \sum_a \varepsilon_a + \langle \Psi_0 | \mathcal{V} | \Psi_0 \rangle = E_0^{(0)} + E_0^{(1)}
\end{aligned}
\tag{3.109}
$$

where $\langle \Psi_0 | \mathcal{V} | \Psi_0 \rangle$ has been defined as the first-order energy in the expansion (3.104) for the exact energy. In Chapter 6 we will primarily be concerned with finding the second-order energy $E_0^{(2)}$ and other higher-order energies.

Exercise 3.8 Use expression (3.108) for $\mathcal{V}$, expression (3.18) for the Hartree-Fock potential $v^{HF}(i)$, and the rules for evaluating matrix elements to explicitly show that $\langle \Psi_0 | \mathcal{V} | \Psi_0 \rangle = -\frac{1}{2} \sum_a \sum_b \langle ab || ab \rangle$ and hence that $E_0^{(1)}$ cancels the double counting of electron-electron repulsions in $E_0^{(0)} = \sum_a \varepsilon_a$ to give the correct Hartree-Fock energy E_0.

3.4 RESTRICTED CLOSED-SHELL HARTREE-FOCK: THE ROOTHAAN EQUATIONS

So far in this chapter we have discussed the Hartree-Fock equations from a formal point of view in terms of a general set of spin orbitals $\{\chi_i\}$. We are now in a position to consider the actual calculation of Hartree-Fock wave functions, and we must be more specific about the form of the spin orbitals. In the last chapter we briefly discussed two types of spin orbitals: restricted spin orbitals, which are constrained to have the same spatial function for α (spin up) and β (spin down) spin functions; and unrestricted spin orbitals, which have different spatial functions for α and β spins. Later in this chapter we will discuss the unrestricted Hartree-Fock formalism and unrestricted calculations. In this section we are concerned with procedures for calculating restricted Hartree-Fock wave functions and, specifically, we consider here only closed-shell calculations. Our molecular states are thus allowed to have only an even number N of electrons, with all electrons paired such that $n = N/2$ spatial orbitals are doubly occupied. In essence this restricts our discussion to closed-shell ground states.[3] For describing open-shell ground states we will use the unrestricted formalism of the last section of this chapter. To describe open-shell excited states we also use unrestricted Hartree-Fock theory. The restricted open-shell formalism is somewhat more involved than our restricted closed-shell or unrestricted open-shell formalism, and we do not describe restricted open-shell Hartree-Fock calculations in this book. An excellent introduction to such calculations is contained in the book by Hurley, suggested for further reading at the end of this chapter.

3.4.1 Closed-Shell Hartree-Fock: Restricted Spin Orbitals

A restricted set of spin orbitals has the form

$$\chi_i(\mathbf{x}) = \begin{cases} \psi_j(\mathbf{r})\alpha(\omega) \\ \psi_j(\mathbf{r})\beta(\omega) \end{cases} \tag{3.110}$$

and the closed-shell restricted ground state is

$$|\Psi_0\rangle = |\chi_1\chi_2\cdots\chi_{N-1}\chi_N\rangle = |\psi_1\bar{\psi}_1\cdots\psi_a\bar{\psi}_a\cdots\psi_{N/2}\bar{\psi}_{N/2}\rangle \tag{3.111}$$

We now want to convert the general spin orbital Hartree-Fock equation $f(1)\chi_i(1) = \varepsilon_i\chi_i(1)$ to a spatial eigenvalue equation where each of the occupied spatial molecular orbitals $\{\psi_a | a = 1, 2, \ldots, N/2\}$ is doubly occupied. The procedure for converting from spin orbitals to spatial orbitals was described in Subsection 2.3.5; we must integrate out the spin functions. Let us first apply this technique to the Hartree-Fock equation

$$f(\mathbf{x}_1)\chi_i(\mathbf{x}_1) = \varepsilon_i\chi_i(\mathbf{x}_1) \tag{3.112}$$

The spin orbital $\chi_i(\mathbf{x}_1)$ will have either the α or β spin function. Let us assume α; identical results will be obtained by assuming β,

$$f(\mathbf{x}_1)\psi_j(\mathbf{r}_1)\alpha(\omega_1) = \varepsilon_j\psi_j(\mathbf{r}_1)\alpha(\omega_1) \tag{3.113}$$

where ε_j, the energy of the spatial orbital ψ_j is identical with ε_i, the energy of the spin orbital χ_i. Multiplying on the left by $\alpha^*(\omega_1)$ and integrating over spin gives

$$\left[\int d\omega_1 \, \alpha^*(\omega_1)f(\mathbf{x}_1)\alpha(\omega_1)\right]\psi_j(\mathbf{r}_1) = \varepsilon_j\psi_j(\mathbf{r}_1) \tag{3.114}$$

To proceed we must evaluate the left-hand side of (3.114). Let us write the spin orbital Fock operator as

$$f(\mathbf{x}_1) = h(\mathbf{r}_1) + \sum_c^N \int d\mathbf{x}_2 \, \chi_c^*(\mathbf{x}_2)r_{12}^{-1}(1 - \mathscr{P}_{12})\chi_c(\mathbf{x}_2) \tag{3.115}$$

so that (3.114) becomes

$$\left[\int d\omega_1 \, \alpha^*(\omega_1)f(\mathbf{x}_1)\alpha(\omega_1)\right]\psi_j(\mathbf{r}_1) = \left[\int d\omega_1 \, \alpha^*(\omega_1)h(\mathbf{r}_1)\alpha(\omega_1)\right]\psi_j(\mathbf{r}_1)$$

$$+ \left[\sum_c \int d\omega_1 \, d\mathbf{x}_2 \, \alpha^*(\omega_1)\chi_c^*(\mathbf{x}_2)r_{12}^{-1}(1-\mathscr{P}_{12})\chi_c(\mathbf{x}_2)\alpha(\omega_1)\right]\psi_j(\mathbf{r}_1)$$

$$= \varepsilon_j \psi_j(\mathbf{r}_1) \tag{3.116}$$

If we let $f(\mathbf{r}_1)$ be the closed-shell Fock operator

$$f(\mathbf{r}_1) = \int d\omega_1 \, \alpha^*(\omega_1)f(\mathbf{x}_1)\alpha(\omega_1) \tag{3.117}$$

then

$$f(\mathbf{r}_1)\psi_j(\mathbf{r}_1) = h(\mathbf{r}_1)\psi_j(\mathbf{r}_1) + \sum_c \int d\omega_1 \, d\mathbf{x}_2 \, \alpha^*(\omega_1)\chi_c^*(\mathbf{x}_2)r_{12}^{-1}\chi_c(\mathbf{x}_2)\alpha(\omega_1)\psi_j(\mathbf{r}_1)$$

$$- \sum_c \int d\omega_1 \, d\mathbf{x}_2 \, \alpha^*(\omega_1)\chi_c^*(\mathbf{x}_2)r_{12}^{-1}\chi_c(\mathbf{x}_1)\alpha(\omega_2)\psi_j(\mathbf{r}_2)$$

$$= \varepsilon_j \psi_j(\mathbf{r}_1) \tag{3.118}$$

where we have performed the integration over $d\omega_1$ in the expression involving $h(\mathbf{r}_1)$ and used $\mathscr{P}_{12}$ to generate the explicit exchange term. Now, if we have a closed-shell, the sum over occupied spin orbitals includes an equal sum over those with the α spin function and those with the β spin function

$$\sum_c^N \rightarrow \sum_c^{N/2} + \sum_{\bar{c}}^{N/2} \tag{3.119}$$

and therefore

$$f(\mathbf{r}_1)\psi_j(\mathbf{r}_1) = h(\mathbf{r}_1)\psi_j(\mathbf{r}_1)$$

$$+ \sum_c^{N/2} \int d\omega_1 \, d\omega_2 \, d\mathbf{r}_2 \, \alpha^*(\omega_1)\psi_c^*(\mathbf{r}_2)\alpha^*(\omega_2)r_{12}^{-1}\psi_c(\mathbf{r}_2)\alpha(\omega_2)\alpha(\omega_1)\psi_j(\mathbf{r}_1)$$

$$+ \sum_c^{N/2} \int d\omega_1 \, d\omega_2 \, d\mathbf{r}_2 \, \alpha^*(\omega_1)\psi_c^*(\mathbf{r}_2)\beta^*(\omega_2)r_{12}^{-1}\psi_c(\mathbf{r}_2)\beta(\omega_2)\alpha(\omega_1)\psi_j(\mathbf{r}_1)$$

$$- \sum_c^{N/2} \int d\omega_1 \, d\omega_2 \, d\mathbf{r}_2 \, \alpha^*(\omega_1)\psi_c^*(\mathbf{r}_2)\alpha^*(\omega_2)r_{12}^{-1}\psi_c(\mathbf{r}_1)\alpha(\omega_1)\alpha(\omega_2)\psi_j(\mathbf{r}_2)$$

$$- \sum_c^{N/2} \int d\omega_1 \, d\omega_2 \, d\mathbf{r}_2 \, \alpha^*(\omega_1)\psi_c^*(\mathbf{r}_2)\beta^*(\omega_2)r_{12}^{-1}\psi_c(\mathbf{r}_1)\beta(\omega_1)\alpha(\omega_2)\psi_j(\mathbf{r}_2)$$

$$= \varepsilon_j \psi_j(\mathbf{r}_1) \tag{3.120}$$

We can now perform the integrations over $d\omega_1$ and $d\omega_2$. The last term of (3.120) disappears because of spin orthogonality. This reflects the fact that there is only an exchange interaction between electrons of parallel spin. The two coulomb terms are equal and thus one obtains

$$f(\mathbf{r}_1)\psi_j(\mathbf{r}_1) = h(\mathbf{r}_1)\psi_j(\mathbf{r}_1) + \left[2 \sum_c^{N/2} \int d\mathbf{r}_2 \, \psi_c^*(\mathbf{r}_2) r_{12}^{-1} \psi_c(\mathbf{r}_2) \right] \psi_j(\mathbf{r}_1)$$

$$- \left[\sum_c^{N/2} \int d\mathbf{r}_2 \, \psi_c^*(\mathbf{r}_2) r_{12}^{-1} \psi_j(\mathbf{r}_2) \right] \psi_c(\mathbf{r}_1)$$

$$= \varepsilon_j \psi_j(\mathbf{r}_1) \tag{3.121}$$

The closed-shell Fock operator thus has the form,

$$f(\mathbf{r}_1) = h(\mathbf{r}_1) + \sum_a^{N/2} \int d\mathbf{r}_2 \, \psi_a^*(\mathbf{r}_2)(2 - \mathscr{P}_{12}) r_{12}^{-1} \psi_a(\mathbf{r}_2) \tag{3.122}$$

or, equivalently,

$$f(1) = h(1) + \sum_a^{N/2} 2J_a(1) - K_a(1) \tag{3.123}$$

where the closed-shell coulomb and exchange operators are defined by

$$J_a(1) = \int d\mathbf{r}_2 \, \psi_a^*(2) r_{12}^{-1} \psi_a(2) \tag{3.124}$$

$$K_a(1)\psi_i(1) = \left[\int d\mathbf{r}_2 \, \psi_a^*(2) r_{12}^{-1} \psi_i(2) \right] \psi_a(1) \tag{3.125}$$

These equations are quite analogous to those for spin orbitals, except for the factor of 2 occurring with the coulomb operator. The sum in (3.122) is, of course, over the $N/2$ occupied orbitals $\{\psi_a\}$. The closed-shell spatial Hartree-Fock equation is just

$$f(1)\psi_j(1) = \varepsilon_j \psi_j(1) \tag{3.126}$$

The closed-shell Hartree-Fock energy was derived in Subsection 2.3.5 as an example of the transition from spin orbitals to spatial orbitals. For the closed-shell determinant, $|\Psi_0\rangle = |\psi_1 \bar{\psi}_1 \cdots \psi_a \bar{\psi}_a \cdots \psi_{N/2} \bar{\psi}_{N/2}\rangle$, it is

$$E_0 = \langle \Psi_0 | \mathscr{H} | \Psi_0 \rangle = 2 \sum_a (a|h|a) + \sum_a \sum_b 2(aa|bb) - (ab|ba)$$

$$= 2 \sum_a h_{aa} + \sum_a \sum_b 2J_{ab} - K_{ab} \tag{3.127}$$

It remains to convert the expression for orbital energies in Eq. (3.72) to the closed-shell spatial orbital form.

Exercise 3.9 Convert the spin orbital expression for orbital energies

$$\varepsilon_i = \langle \chi_i | h | \chi_i \rangle + \sum_b^N \langle \chi_i \chi_b | | \chi_i \chi_b \rangle$$

to the closed-shell expression

$$\varepsilon_i = (\psi_i|h|\psi_i) + \sum_b^{N/2} 2(ii|bb) - (ib|bi) = h_{ii} + \sum_b^{N/2} 2J_{ib} - K_{ib} \quad (3.128)$$

With the results of this last exercise we now have closed-shell expressions for most quantities of interest. Let us examine these, for a moment, in the context of our minimal basis H_2 model.

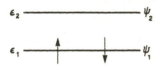

We can evaluate the total energy by inspection. Each of the two electrons has kinetic energy plus attraction to the nuclei of $h_{11} = (\psi_1|h|\psi_1)$. In addition there is the coulomb repulsion between the two electrons $J_{11} = (\psi_1\psi_1|\psi_1\psi_1)$. There are no exchange interactions since the two electrons have antiparallel spins. The Hartree-Fock energy is thus

$$E_0 = 2h_{11} + J_{11} \quad (3.129)$$

This is an agreement with (3.127) since $J_{ii} = K_{ii}$.

One can evaluate the orbital energies similarly.

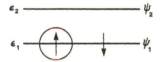

To evaluate ε_1 we need only add up the interactions of the circled electron. It has kinetic energy and nuclear attraction h_{11} and a coulomb interaction J_{11} and, therefore,

$$\varepsilon_1 = h_{11} + J_{11} \quad (3.130)$$

We could do the same for any occupied orbital energy. For virtual orbitals, as we have seen before, the orbital energy corresponds to the interactions of an extra $(N + 1)$th electron, in agreement with Koopmans' theorem. For the minimal basis model, we must keep the two electrons of $|\Psi_0\rangle$ and evaluate the interactions of the extra electron in the virtual orbital ψ_2, as shown below.

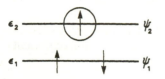

The circled electron has kinetic energy and nuclear attraction h_{22}. It has two coulomb interactions J_{12}, with each of the other two electrons, and an exchange interaction $-K_{12}$, with the electron of parallel spin. Thus

$$\varepsilon_2 = h_{22} + 2J_{12} - K_{12} \tag{3.131}$$

Both of the results are in agreement with the general expression for closed-shell orbital energies obtained in Exercise 3.9.

3.4.2 Introduction of a Basis: The Roothaan Equations

Now that we have eliminated spin, the calculation of molecular orbitals becomes equivalent to the problem of solving the spatial integro-differential equation

$$f(\mathbf{r}_1)\psi_i(\mathbf{r}_1) = \varepsilon_i \psi_i(\mathbf{r}_1) \tag{3.132}$$

One might attempt to solve this equation numerically; numerical solutions are common in atomic calculations. No practical procedures are presently available, however, for obtaining numerical solutions for molecules. The contribution of Roothaan[4] was to show how, by introducing a set of known spatial basis functions, the differential equation could be converted to a set of algebraic equations and solved by standard matrix techniques.

We, therefore, introduce a set of K known basis functions $\{\phi_\mu(\mathbf{r}) | \mu = 1, 2, \ldots, K\}$ and expand the unknown molecular orbitals in the linear expansion

$$\psi_i = \sum_{\mu=1}^{K} C_{\mu i} \phi_\mu \qquad i = 1, 2, \ldots, K \tag{3.133}$$

If the set $\{\phi_\mu\}$ was complete, this would be an exact expansion, and any complete set $\{\phi_\mu\}$ could be used. Unfortunately, one is always restricted, for practical computational reasons, to a finite set of K basis functions. As such, it is important to choose a basis that will provide, as far as is possible, a reasonably accurate expansion for the exact molecular orbitals $\{\psi_i\}$, particularly, for those molecular orbitals $\{\psi_a\}$ which are occupied in $|\Psi_0\rangle$ and determine the ground state energy E_0. A later section of this chapter discusses the questions involved in the choice of a basis set and describes some of the art of choosing a basis set. For our purposes here, we need only assume that $\{\phi_\mu\}$ is a set of known functions. As the basis set becomes more and more complete, the expansion (3.133) leads to more and more accurate representations of the "exact" molecular orbitals, i.e., the molecular orbitals converge to those of Eq. (3.132), the true eigenfunctions of the Fock operator. For any finite basis set we will obtain molecular orbitals from the truncated expansion (3.133), which are exact only in the space spanned by the basis functions $\{\phi_\mu\}$.

From (3.133), the problem of calculating the Hartree-Fock molecular orbitals reduces to the problem of calculating the set of expansion coefficients

$C_{\mu i}$. We can obtain a matrix equation for the $C_{\mu i}$ by substituting the linear expansion (3.133) into the Hartree-Fock equation (3.132). Using the index v, gives

$$f(1) \sum_v C_{vi}\phi_v(1) = \varepsilon_i \sum_v C_{vi}\phi_v(1) \qquad (3.134)$$

By multiplying by $\phi_\mu^*(1)$ on the left and integrating, we turn the integro-differential equation into a matrix equation,

$$\sum_v C_{vi} \int d\mathbf{r}_1 \, \phi_\mu^*(1)f(1)\phi_v(1) = \varepsilon_i \sum_v C_{vi} \int d\mathbf{r}_1 \, \phi_\mu^*(1)\phi_v(1) \qquad (3.135)$$

We now define two matrices.
1. The *overlap matrix* **S** has elements

$$S_{\mu v} = \int d\mathbf{r}_1 \, \phi_\mu^*(1)\phi_v(1) \qquad (3.136)$$

and is a $K \times K$ Hermitian (although usually real and symmetric) matrix. The basis functions $\{\phi_\mu\}$, although assumed to be normalized and linearly independent, are not in general orthogonal to each other and, hence, overlap with a magnitude $0 \le |S_{\mu v}| \le 1$, i.e., the diagonal elements of **S** are unity and the off-diagonal elements are numbers less than one in magnitude. The sign of the off-diagonal elements depends on the relative sign of the two basis functions, and their relative orientation and separation in space. If two off-diagonal elements approach unity (in magnitude) i.e., approach complete overlap, then the two basis functions approach linear dependence. Because the overlap matrix is Hermitian, it can be diagonalized by a unitary matrix, as we will have occasion to do, later. The eigenvalues of the overlap matrix can be shown to be necessarily positive numbers and, hence, the overlap matrix is said to be a positive-definite matrix. Linear dependence in the basis set is associated with eigenvalues of the overlap matrix approaching zero. The overlap matrix is sometimes called the metric matrix.
2. The *Fock matrix* **F** has elements

$$F_{\mu v} = \int d\mathbf{r}_1 \, \phi_\mu^*(1)f(1)\phi_v(1) \qquad (3.137)$$

and is also a $K \times K$ Hermitian (although usually real and symmetric) matrix. The Fock operator $f(1)$ is a one-electron operator, and any set of one-electron functions defines a matrix representation of this operator. We have previously discussed matrix elements of the Fock operator with spin orbitals. The Fock matrix **F** is the matrix representation of the Fock operator with the set of basis functions $\{\phi_\mu\}$.

With these definitions of **F** and **S** we can now write the integrated Hartree-Fock equation (3.135) as

$$\sum_v F_{\mu v}C_{vi} = \varepsilon_i \sum_v S_{\mu v}C_{vi} \qquad i = 1, 2, \ldots, K \qquad (3.138)$$

These are the *Roothaan equations*, which can be written more compactly as the single matrix equation

$$\mathbf{FC} = \mathbf{SC}\boldsymbol{\varepsilon} \tag{3.139}$$

where $\mathbf{C}$ is a $K \times K$ square matrix of the expansion coefficients $C_{\mu i}$

$$\mathbf{C} = \begin{pmatrix} C_{11} & C_{12} & \cdots & C_{1K} \\ C_{21} & C_{22} & \cdots & C_{2K} \\ \vdots & \vdots & & \vdots \\ C_{K1} & C_{K2} & \cdots & C_{KK} \end{pmatrix} \tag{3.140}$$

and $\boldsymbol{\varepsilon}$ is a diagonal matrix of the orbital energies ε_i,

$$\boldsymbol{\varepsilon} = \begin{pmatrix} \varepsilon_1 & & & \\ & \varepsilon_2 & & \mathbf{0} \\ & & \ddots & \\ \mathbf{0} & & & \varepsilon_K \end{pmatrix} \tag{3.141}$$

Note that from (3.133) and (3.140) it is the columns of $\mathbf{C}$ which describe the molecular orbitals, i.e., the coefficients describing ψ_1 are in the first column of $\mathbf{C}$, those describing ψ_2 are in the second column of $\mathbf{C}$, etc.

Exercise 3.10 Show that $\mathbf{C}^\dagger \mathbf{SC} = \mathbf{1}$. *Hint*: Use the fact that the molecular orbitals $\{\psi_i\}$ are orthonormal.

At this point the problem of determining the Hartree-Fock molecular orbitals $\{\psi_i\}$ and orbital energies ε_i involves solving the matrix equation $\mathbf{FC} = \mathbf{SC}\boldsymbol{\varepsilon}$. To proceed, however, we need an explicit expression for the Fock matrix. It is first of all necessary, however, to introduce the concept of a density matrix.

3.4.3 The Charge Density

If we have an electron described by the spatial wave function $\psi_a(\mathbf{r})$, then the probability of finding that electron in a volume element $d\mathbf{r}$ at a point $\mathbf{r}$ is $|\psi_a(\mathbf{r})|^2 \, d\mathbf{r}$. The probability distribution function (charge density) is $|\psi_a(\mathbf{r})|^2$. If we have a closed-shell molecule described by a single determinant wave function with each occupied molecular orbital ψ_a containing two electrons, then the total charge density is just

$$\rho(\mathbf{r}) = 2 \sum_a^{N/2} |\psi_a(\mathbf{r})|^2 \tag{3.142}$$

such that $\rho(\mathbf{r}) \, d\mathbf{r}$ is the probability of finding an electron (any electron) in $d\mathbf{r}$ at $\mathbf{r}$. The integral of this charge density is just the total number of electrons,

$$\int d\mathbf{r}\, \rho(\mathbf{r}) = 2 \sum_a^{N/2} \int d\mathbf{r}\, |\psi_a(\mathbf{r})|^2 = 2 \sum_a^{N/2} 1 = N \tag{3.143}$$

For a single determinant, these equations show that the total charge density is just a sum of charge densities for each of the electrons.

Exercise 3.11 Use the density operator $\hat{\rho}(\mathbf{r}) = \sum_{i=1}^{N} \delta(\mathbf{r}_i - \mathbf{r})$, the rules for evaluating matrix elements in Chapter 2, and the rules for converting from spin orbitals to spatial orbitals, to derive (3.142) from $\rho(\mathbf{r}) = \langle \Psi_0 | \hat{\rho}(\mathbf{r}) | \Psi_0 \rangle$.

Let us now insert the molecular orbital expansion (3.133) into the expression (3.142) for the charge density,

$$
\begin{aligned}
\rho(\mathbf{r}) &= 2 \sum_{a}^{N/2} \psi_a^*(\mathbf{r}) \psi_a(\mathbf{r}) \\
&= 2 \sum_{a}^{N/2} \sum_{\nu} C_{\nu a}^* \phi_\nu^*(\mathbf{r}) \sum_{\mu} C_{\mu a} \phi_\mu(\mathbf{r}) \\
&= \sum_{\mu\nu} \left[2 \sum_{a}^{N/2} C_{\mu a} C_{\nu a}^* \right] \phi_\mu(\mathbf{r}) \phi_\nu^*(\mathbf{r}) \\
&= \sum_{\mu\nu} P_{\mu\nu} \phi_\mu(\mathbf{r}) \phi_\nu^*(\mathbf{r})
\end{aligned}
\tag{3.144}
$$

where we have defined a *density matrix* or, as it is sometimes called, a *charge-density bond-order matrix*

$$
P_{\mu\nu} = 2 \sum_{a}^{N/2} C_{\mu a} C_{\nu a}^*
\tag{3.145}
$$

From (3.144), given a set of known basis functions $\{\phi_\mu\}$, the matrix **P** specifies completely the charge density $\rho(\mathbf{r})$. It is directly related to the expansion coefficients **C** by (3.145), and we can characterize the results of closed-shell Hartree-Fock calculations either by the $C_{\mu i}$ or by the $P_{\mu\nu}$.

Exercise 3.12 A matrix **A** is said to be idempotent if $\mathbf{A}^2 = \mathbf{A}$. Use the result of Exercise 3.10 to show that $\mathbf{PSP} = 2\mathbf{P}$, i.e., show that $\frac{1}{2}\mathbf{P}$ would be idempotent in an orthonormal basis.

Exercise 3.13 Use the expression (3.122) for the closed-shell Fock operator to show that

$$
\begin{aligned}
f(\mathbf{r}_1) &= h(\mathbf{r}_1) + v^{\text{HF}}(\mathbf{r}_1) \\
&= h(\mathbf{r}_1) + \frac{1}{2} \sum_{\lambda\sigma} P_{\lambda\sigma} \left[\int d\mathbf{r}_2 \, \phi_\sigma^*(\mathbf{r}_2)(2 - \mathscr{P}_{12}) r_{12}^{-1} \phi_\lambda(\mathbf{r}_2) \right]
\end{aligned}
\tag{3.146}
$$

The result of the above exercise expresses the Fock operator in terms of the density matrix. We can use this expression to indicate in an intuitive way how the Hartree-Fock procedure operates. We first guess a density matrix

P, i.e., we guess the charge density $\rho(\mathbf{r})$ describing the positions of the electrons. Later we will say something about how to obtain such a guess. We then use this charge density to calculate an effective one-electron potential $v^{HF}(\mathbf{r}_1)$ for the electrons according to (3.146). We thus have an effective one-electron Hamiltonian (the Fock operator), and we can solve a one-electron Schrödinger-like equation to determine the states $\{\psi_i\}$ of an electron in the effective potential. The new one-electron states (molecular orbitals ψ_i) can then be used to obtain a better approximation to the density, using (3.142), for example. With this new charge density we can calculate a new Hartree-Fock potential and repeat the procedure until the Hartree-Fock potential (and, consequently, an effective electrostatic field) no longer changes, i.e., until the field which produced a particular charge density (by solving a one-electron Schrödinger-like equation, the Hartree-Fock eigenvalue equation) is consistent (identical) with the field which would be calculated from that charge density (using (3.146)). This is why the Hartree-Fock equations are commonly called the *self-consistent-field* (SCF) equations. This is a way of viewing the physics involved in solving the Roothaan equations. To return to the actual algebraic procedure, we need an explicit expression for the Fock matrix **F**.

3.4.4 Expression for the Fock Matrix

The Fock matrix **F** is the matrix representation of the Fock operator

$$f(1) = h(1) + \sum_{a}^{N/2} 2J_a(1) - K_a(1) \tag{3.147}$$

in the basis $\{\phi_\mu\}$, i.e.,

$$
\begin{aligned}
F_{\mu\nu} &= \int d\mathbf{r}_1 \; \phi_\mu^*(1) f(1) \phi_\nu(1) \\
&= \int d\mathbf{r}_1 \; \phi_\mu^*(1) h(1) \phi_\nu(1) + \sum_a^{N/2} \int d\mathbf{r}_1 \; \phi_\mu^*(1)[2J_a(1) - K_a(1)]\phi_\nu(1) \\
&= H_{\mu\nu}^{\text{core}} + \sum_a^{N/2} 2(\mu\nu|aa) - (\mu a|a\nu) \tag{3.148}
\end{aligned}
$$

where we have defined a *core-Hamiltonian* matrix

$$H_{\mu\nu}^{\text{core}} = \int d\mathbf{r}_1 \; \phi_\mu^*(1) h(1) \phi_\nu(1) \tag{3.149}$$

The elements of the core-Hamiltonian matrix are integrals involving the one-electron operator $h(1)$, describing the kinetic energy and nuclear attraction of an electron, i.e.,

$$h(1) = -\frac{1}{2}\nabla_1^2 - \sum_A \frac{Z_A}{|\mathbf{r}_1 - \mathbf{R}_A|} \tag{3.150}$$

Calculating the elements of the core-Hamiltonian matrix thus involves the kinetic energy integrals

$$T_{\mu\nu} = \int d\mathbf{r}_1 \, \phi_\mu^*(1)[-\tfrac{1}{2}\nabla_1^2]\phi_\nu(1) \tag{3.151}$$

and the nuclear attraction integrals

$$V_{\mu\nu}^{\mathrm{nucl}} = \int d\mathbf{r}_1 \, \phi_\mu^*(1)\left[-\sum_A \frac{Z_A}{|\mathbf{r}_1 - \mathbf{R}_A|}\right]\phi_\nu(1) \tag{3.152}$$

where

$$H_{\mu\nu}^{\mathrm{core}} = T_{\mu\nu} + V_{\mu\nu}^{\mathrm{nucl}} \tag{3.153}$$

Given a particular basis set $\{\phi_\mu\}$, the integrals of $\mathbf{T}$ and $\mathbf{V}^{\mathrm{nucl}}$ need to be evaluated and the core-Hamiltonian matrix formed. The core-Hamiltonian matrix, unlike the full Fock matrix, needs only to be evaluated once as it remains constant during the iterative calculation. The calculation of kinetic energy and nuclear attraction integrals is described in Appendix A.

To return to expression (3.148) for the Fock matrix, we now insert the linear expansion for the molecular orbitals (3.133) into the two-electron terms to get

$$
\begin{aligned}
F_{\mu\nu} &= H_{\mu\nu}^{\mathrm{core}} + \sum_a^{N/2} \sum_{\lambda\sigma} C_{\lambda a} C_{\sigma a}^* [2(\mu\nu|\sigma\lambda) - (\mu\lambda|\sigma\nu)] \\
&= H_{\mu\nu}^{\mathrm{core}} + \sum_{\lambda\sigma} P_{\lambda\sigma}[(\mu\nu|\sigma\lambda) - \tfrac{1}{2}(\mu\lambda|\sigma\nu)] \\
&= H_{\mu\nu}^{\mathrm{core}} + G_{\mu\nu}
\end{aligned}
\tag{3.154}
$$

where $G_{\mu\nu}$ is the two-electron part of the Fock matrix. This is our final expression for the Fock matrix. It contains a one-electron part $\mathbf{H}^{\mathrm{core}}$ which is fixed, given the basis set, and a two-electron part $\mathbf{G}$ which depends on the density matrix $\mathbf{P}$ and a set of two-electron integrals

$$(\mu\nu|\lambda\sigma) = \int d\mathbf{r}_1 \, d\mathbf{r}_2 \, \phi_\mu^*(1)\phi_\nu(1)r_{12}^{-1}\phi_\lambda^*(2)\phi_\sigma(2) \tag{3.155}$$

Because of their large number, the evaluation and manipulation of these two-electron integrals is the major difficulty in a Hartree-Fock calculation.

Exercise 3.14 Assume that the basis functions are real and use the symmetry of the two-electron integrals $[(\mu\nu|\lambda\sigma) = (\nu\mu|\lambda\sigma) = (\lambda\sigma|\mu\nu)$, etc.] to show that for a basis set of size $K = 100$ there are $12{,}753{,}775 = O(K^4/8)$ unique two-electron integrals.

Because the Fock matrix depends on the density matrix,

$$\mathbf{F} = \mathbf{F}(\mathbf{P}) \tag{3.156}$$

or, equivalently, on the expansion coefficients,

$$\mathbf{F} = \mathbf{F(C)} \tag{3.157}$$

the Roothaan equations are nonlinear, i.e.,

$$\mathbf{F(C)C} = \mathbf{SC\varepsilon} \tag{3.158}$$

and they will need to be solved in an iterative fashion. Before considering how such iterations should proceed, we need to discuss the solution of the matrix equation

$$\mathbf{FC} = \mathbf{SC\varepsilon} \tag{3.159}$$

at each step in the iteration. If $\mathbf{S}$ were the unit matrix (i.e., if we had an orthonormal basis set), then we would have

$$\mathbf{FC} = \mathbf{C\varepsilon} \tag{3.160}$$

and Roothaan's equations would just have the form of the usual matrix eigenvalue problem, and we could find the eigenvectors $\mathbf{C}$ and eigenvalues ε by diagonalizing $\mathbf{F}$. Because of the nonorthogonal basis, we need to reformulate the eigenvalue problem $\mathbf{FC} = \mathbf{SC\varepsilon}$.

3.4.5 Orthogonalization of the Basis

The basis sets that are used in molecular calculations are not orthonormal sets. The basis functions are normalized, but they are not orthogonal to each other. This gives rise to the overlap matrix in Roothaan's equations. In order to put Roothaan's equations into the form of the usual matrix eigenvalue problem, we need to consider procedures for orthogonalizing the basis functions.

If we have a set of functions $\{\phi_\mu\}$ that are not orthogonal, i.e.,

$$\int d\mathbf{r}\, \phi_\mu^*(\mathbf{r})\phi_\nu(\mathbf{r}) = S_{\mu\nu} \tag{3.161}$$

then it will always be possible to find a transformation matrix $\mathbf{X}$ (not unitary) such that a transformed set of functions $\{\phi_\mu'\}$ given by

$$\phi_\mu' = \sum_\nu X_{\nu\mu}\phi_\nu \qquad \mu = 1, 2, \ldots, K \tag{3.162}$$

do form an orthonormal set, i.e.,

$$\int d\mathbf{r}\, \phi_\mu'^*(\mathbf{r})\phi_\nu'(\mathbf{r}) = \delta_{\mu\nu} \tag{3.163}$$

To derive the properties of $\mathbf{X}$, we substitute the transformation (3.162) into (3.163) to get

$$\int d\mathbf{r}\ \phi_\mu'^*(\mathbf{r})\phi_\nu'(\mathbf{r}) = \int d\mathbf{r}\left[\sum_\lambda X_{\lambda\mu}^*\phi_\lambda^*(\mathbf{r})\right]\left[\sum_\sigma X_{\sigma\nu}\phi_\sigma(\mathbf{r})\right]$$

$$= \sum_\lambda\sum_\sigma X_{\lambda\mu}^*\int d\mathbf{r}\ \phi_\lambda^*(\mathbf{r})\phi_\sigma(\mathbf{r})X_{\sigma\nu}$$

$$= \sum_\lambda\sum_\sigma X_{\lambda\mu}^* S_{\lambda\sigma} X_{\sigma\nu} = \delta_{\mu\nu} \tag{3.164}$$

This last equation can be written as the matrix equation

$$\mathbf{X}^\dagger\mathbf{S}\mathbf{X} = 1 \tag{3.165}$$

and defines the relation that the matrix $\mathbf{X}$ must satisfy if the transformed orbitals are to form an orthonormal set. As we shall see later, $\mathbf{X}$ must also be nonsingular, i.e., it must possess an inverse $\mathbf{X}^{-1}$. We now proceed to show how to obtain two different transformation matrices $\mathbf{X}$. Since $\mathbf{S}$ is Hermitian it can be diagonalized by a unitary matrix $\mathbf{U}$,

$$\mathbf{U}^\dagger\mathbf{S}\mathbf{U} = \mathbf{s} \tag{3.166}$$

where $\mathbf{s}$ is a diagonal matrix of the eigenvalues of $\mathbf{S}$.

Exercise 3.15 Use the definition of $S_{\mu\nu} = \int d\mathbf{r}\ \phi_\mu^*\phi_\nu$ to show that the eigenvalues of $\mathbf{S}$ are all positive. *Hint*: consider $\sum_\nu S_{\mu\nu}c_\nu^i = s_i c_\mu^i$, multiply by c_μ^{i*} and sum, where $\mathbf{c}^i$ is the ith column of $\mathbf{U}$.

There are two ways of orthogonalizing the basis set $\{\phi_\mu\}$ in common use. The first procedure, called *symmetric orthogonalization*, uses the inverse square root of $\mathbf{S}$ for $\mathbf{X}$

$$\mathbf{X} \equiv \mathbf{S}^{-1/2} = \mathbf{U}\mathbf{s}^{-1/2}\mathbf{U}^\dagger \tag{3.167}$$

If you will recall from the discussion of functions of a matrix in Chapter 1, we can form $\mathbf{S}^{-1/2}$ by diagonalizing $\mathbf{S}$ to form $\mathbf{s}$, then taking the inverse square root of each of the eigenvalues to form the diagonal matrix $\mathbf{s}^{-1/2}$ and then "undiagonalizing" by the transformation in (3.167). If $\mathbf{S}$ is Hermitian then $\mathbf{S}^{-1/2}$ is also Hermitian. Substituting (3.167) into (3.165),

$$\mathbf{S}^{-1/2}\mathbf{S}\mathbf{S}^{-1/2} = \mathbf{S}^{-1/2}\mathbf{S}^{1/2} = \mathbf{S}^0 = 1 \tag{3.168}$$

shows that $\mathbf{X} = \mathbf{S}^{-1/2}$ is indeed an orthogonalizing transformation matrix. Since the eigenvalues of $\mathbf{S}$ are all positive (Exercise 3.15), there is no difficulty in (3.167) of taking square roots. However, if there is linear dependence or near linear dependence in the basis set, then some of the eigenvalues will approach zero and (3.167) will involve dividing by quantities that are nearly zero. Thus symmetric orthogonalization will lead to problems in numerical precision for basis sets with near linear dependence.

A second way of obtaining an orthonormal set of basis functions is called *canonical orthogonalization*. It uses the transformation matrix

$$\mathbf{X} = \mathbf{U}\mathbf{s}^{-1/2} \tag{3.169}$$

that is, the columns of the unitary matrix $\mathbf{U}$ are divided by the square root of the corresponding eigenvalue

$$X_{ij} = U_{ij}/s_j^{1/2} \tag{3.170}$$

Substituting this definition of $\mathbf{X}$ into (3.165) gives

$$\mathbf{X}^\dagger\mathbf{S}\mathbf{X} = (\mathbf{U}\mathbf{s}^{-1/2})^\dagger\mathbf{S}\mathbf{U}\mathbf{s}^{-1/2} = \mathbf{s}^{-1/2}\mathbf{U}^\dagger\mathbf{S}\mathbf{U}\mathbf{s}^{-1/2} = \mathbf{s}^{-1/2}\mathbf{s}\mathbf{s}^{-1/2} = \mathbf{1} \tag{3.171}$$

showing that $\mathbf{X} = \mathbf{U}\mathbf{s}^{-1/2}$ is also an orthogonalizing transformation matrix. It appears, from (3.170), that this orthogonalization procedure will also entail difficulties if there is linear dependence in the basis set, i.e., if any of the eigenvalues s_i approach zero. We can circumvent this problem with canonical orthogonalization, however. In the matrix eigenvalue problem (3.166), we can order the eigenvalues in any way in the diagonal matrix $\mathbf{s}$, provided we order the columns of $\mathbf{U}$ in the same way. Suppose we order the positive eigenvalues s_i in the order $s_1 > s_2 > s_3 > \cdots$. Upon inspection we may decide that the last m of these are too small and will give numerical problems. We can then use as a transformation matrix, the truncated matrix $\tilde{\mathbf{X}}$,

$$\tilde{\mathbf{X}} = \begin{pmatrix} U_{1,1}/s_1^{1/2} & U_{1,2}/s_2^{1/2} & \cdots & U_{1,K-m}/s_{K-m}^{1/2} \\ U_{2,1}/s_1^{1/2} & U_{2,2}/s_2^{1/2} & \cdots & U_{2,K-m}/s_{K-m}^{1/2} \\ \vdots & \vdots & & \vdots \\ U_{K,1}/s_1^{1/2} & U_{K,2}/s_2^{1/2} & \cdots & U_{K,K-m}/s_{K-m}^{1/2} \end{pmatrix} \tag{3.172}$$

where we have eliminated the last m columns of $\mathbf{X}$ to give the $K \times (K - m)$ matrix $\tilde{\mathbf{X}}$. With this truncated transformation matrix, we get only $K - m$ transformed orthonormal basis functions

$$\phi'_\mu = \sum_{v=1}^{K} \tilde{X}_{v\mu}\phi_v, \qquad \mu = 1, 2, \ldots, K - m \tag{3.173}$$

These would span exactly the same region of space as the original set, provided the eliminated eigenvalues were exactly zero. In practice, one often finds linear dependence problems with eigenvalues in the region $s_i \le 10^{-4}$ (depending, of course, on the machine precision of the calculation). In eliminating the columns with these eigenvalues one is "throwing away" part of the basis set, but only a very small part.

One way of dealing with the problem of a nonorthogonal basis set would thus be to orthogonalize the functions $\{\phi_\mu\}$ to obtain the transformed basis functions $\{\phi'_\mu\}$ and work with these orthonormal functions throughout. This would eliminate the overlap matrix **S** from Roothaan's equations, which could then be solved just by diagonalizing the Fock matrix. This would mean, however, that we would have to calculate all our two-electron integrals using the new orbitals or else transform all the old integrals $(\mu\nu|\lambda\sigma)$ to set $(\mu'\nu'|\lambda'\sigma')$. In practice this is very time consuming, and we can solve the same problem in a more efficient way. Consider a new coefficient matrix **C'** related to the old coefficient matrix **C** by

$$\mathbf{C'} = \mathbf{X}^{-1}\mathbf{C} \qquad \mathbf{C} = \mathbf{X}\mathbf{C'} \tag{3.174}$$

where we have assumed that **X** possesses an inverse. This will be the case if we have eliminated linear dependencies. Substituting $\mathbf{C} = \mathbf{X}\mathbf{C'}$ into the Roothaan equations gives

$$\mathbf{FXC'} = \mathbf{SXC'\varepsilon} \tag{3.175}$$

Multiplying on the left by $\mathbf{X}^\dagger$ gives

$$(\mathbf{X}^\dagger\mathbf{FX})\mathbf{C'} = (\mathbf{X}^\dagger\mathbf{SX})\mathbf{C'\varepsilon} \tag{3.176}$$

If we define a new matrix **F'** by

$$\mathbf{F'} = \mathbf{X}^\dagger\mathbf{FX} \tag{3.177}$$

and use (3.165), then

$$\mathbf{F'C'} = \mathbf{C'\varepsilon} \tag{3.178}$$

These are the transformed Roothaan equations, which can be solved for **C'** by diagonalizing **F'**. Given **C'**, then **C** can be obtained from (3.174). Therefore, given **F**, we can use (3.177), (3.178), and (3.174) to solve the Roothaan equations $\mathbf{FC} = \mathbf{SC\varepsilon}$ for **C** and ε. The intermediate primed matrices are just the Fock matrix and expansion coefficients in the orthogonalized basis, i.e.,

$$\psi_i = \sum_{\mu=1}^{K} C'_{\mu i}\phi'_\mu \qquad i = 1, 2, \ldots, K \tag{3.179}$$

$$F'_{\mu\nu} = \int d\mathbf{r}_1\, \phi'^*_\mu(1)f(1)\phi'_\nu(1) \tag{3.180}$$

Exercise 3.16 Use (3.179), (3.180), and (3.162) to derive (3.174) and (3.177).

3.4.6 The SCF Procedure

With the background of the previous sections we are now in a position to describe the actual computational procedure for obtaining restricted closed-shell Hartree-Fock wave functions for molecules, i.e., wave functions $|\Psi_0\rangle$. Some authors restrict the term Hartree-Fock solution to one that is at the

Hartree-Fock limit, where the basis set is essentially complete, and use the term self-consistent-field (SCF) solution for one obtained with a finite, possibly small, basis set. We use the terms Hartree-Fock and SCF interchangeably, however, and specifically refer to the Hartree-Fock limit when necessary. The SCF procedure is as follows:

1. Specify a molecule (a set of nuclear coordinates $\{\mathbf{R}_A\}$, atomic numbers $\{Z_A\}$, and number of electrons N) and a basis set $\{\phi_\mu\}$.
2. Calculate all required molecular integrals, $S_{\mu v}$, $H_{\mu v}^{core}$, and $(\mu v | \lambda \sigma)$.
3. Diagonalize the overlap matrix $\mathbf{S}$ and obtain a transformation matrix $\mathbf{X}$ from either (3.167) or (3.169).
4. Obtain a guess at the density matrix $\mathbf{P}$.
5. Calculate the matrix $\mathbf{G}$ of equation (3.154) from the density matrix $\mathbf{P}$ and the two-electron integrals $(\mu v | \lambda \sigma)$.
6. Add $\mathbf{G}$ to the core-Hamiltonian to obtain the Fock matrix $\mathbf{F} = \mathbf{H}^{core} + \mathbf{G}$.
7. Calculate the transformed Fock matrix $\mathbf{F}' = \mathbf{X}^\dagger \mathbf{F} \mathbf{X}$.
8. Diagonalize $\mathbf{F}'$ to obtain $\mathbf{C}'$ and $\boldsymbol{\varepsilon}$.
9. Calculate $\mathbf{C} = \mathbf{X}\mathbf{C}'$.
10. Form a new density matrix $\mathbf{P}$ from $\mathbf{C}$ using Eq. (3.145).
11. Determine whether the procedure has converged, i.e., determine whether the new density matrix of step (10) is the same as the previous density matrix within a specified criterion. If the procedure has not converged, return to step (5) with the new density matrix.
12. If the procedure has converged, then use the resultant solution, represented by $\mathbf{C}, \mathbf{P}, \mathbf{F}$, etc., to calculate expectation values and other quantities of interest.

We will describe the calculation of expectation values like the energy, dipole moment, etc., and other quantities of interest like population analyses shortly (Subsection 3.4.7) but let us first consider some of the practical questions involved in each of the twelve steps.

Within the Born-Oppenheimer approximation, what we have done in the above procedure is to determine an electronic wave function $|\Psi_0\rangle$ (and hence an electronic energy E_0) for a collection of N electrons in the field of a set of M point charges (the M nuclei with charges Z_A). By adding the classical nuclear-nuclear repulsion to the electronic energy we will have a total energy as a function of a set of nuclear coordinates $\{\mathbf{R}_A\}$. By repeating the calculation for different nuclear coordinates we can then explore the potential energy surface for nuclear motion. A common calculation is to find the set $\{\mathbf{R}_A\}$ which minimize this total energy; this is a calculation of the equilibrium geometry of a molecule. The procedure is valid for any collection of point charges. In particular, "supermolecule" calculations, which use a set of nuclear charges representative of more than one molecule, are common for exploring, for example, intermolecular forces.

Having chosen a set of nuclear coordinates, the calculation of a restricted closed-shell single determinant wave function is then completely specified by the set of basis functions $\{\phi_\mu\}$. As such, this is an example of an *ab initio* calculation which makes no approximation to the integrals or the electronic Hamiltonian, but is completely specified by the choice of a basis set and the coordinates of the nuclei. The choice of a basis set is more of an art than a science. One is obviously limited by computer facilities, budget, etc. to a rather small, finite set of functions. One must, therefore, be rather judicious in the choice of a basis set. Only Slater- and Gaussian-type functions are currently in common use. If one uses a very small set of functions per atom, then Slater-type functions give definitely superior energies. As the number of functions per atom increases, the clear cut superiority of Slater-type functions is somewhat diminished. As well as the ability of the basis set to span the function space, however, one has to consider, for practical reasons, the time required to evaluate molecular integrals. Most polyatomic calculations now use Gaussian orbitals because of the speed with which integrals can be evaluated, and in this book we will emphasize Gaussian basis functions. Basis functions are discussed in Section 3.6 of this chapter, and the $1s$ STO-3G basis is discussed in Subsection 3.5.1, prior to its use in model calculations on H_2 and HeH^+.

Having defined a basis set, one then needs to calculate and store a number of different types of integrals. Appendix A describes molecular integral evaluation using Gaussian basis functions; we will only mention a few pertinent points here. The overlap integrals and the one-electron integrals that are needed for the core-Hamiltonian and one-electron expectation values, described later, are relatively trivial compared to the two-electron repulsion integrals, primarily because of the much smaller number of one-electron integrals. The major difficulty of a large calculation is the evaluation and handling of large numbers of two-electron integrals. If there are K basis functions, then there will be of the order of $K^4/8$ (see Exercise 3.14) unique two-electron integrals. These can quickly run into the millions even for small basis sets on moderately sized molecules. The problem is not quite this bad since many integrals will be effectively zero for large molecules, as the distance between basis functions becomes large. A number of integrals may also be zero because of molecular symmetry. There will almost always be, however, too many two-electrons integrals to store them all in main computer memory. One common procedure is to store all nonzero integrals in random order on an external magnetic disk or tape, associating with each integral a label identifying the indices μ, ν, λ, and σ.

In the last subsection we described two ways of orthogonalizing the basis set or deriving a transformation $\mathbf{X}$, which enables one to solve Roothaan's equations by a diagonalization. The use of $\mathbf{X} = \mathbf{S}^{-1/2}$ is conceptually simple, and only in unusual situations, where linear dependence in the basis set is a problem, does one need to use canonical orthogonalization. With canonical orthogonalization, columns of $\mathbf{X} = \mathbf{U}\mathbf{s}^{-1/2}$ can just be dropped to give a rectangular matrix. If m columns of $\mathbf{X}$ are deleted, one will effectively be using a basis set of size $K - m$ and one will obtain $K - m$ molecular orbitals ψ_i, i.e., $\mathbf{F}'$ will be a $(K - m) \times (K - m)$ matrix and $\mathbf{C}'$ will be a $K \times (K - m)$ matrix, with columns describing the $K - m$ molecular orbitals in terms of the original K basis functions.

The simplest possible guess at the density matrix $\mathbf{P}$ is to use a null (zero) matrix. This is equivalent to approximating $\mathbf{F}$ as $\mathbf{H}^{core}$ and neglecting all electron-electron interactions in the first iteration. This is a very convenient way of starting the iteration procedure. It corresponds to approximating the converged molecular orbitals by those describing a single electron in the field of the nuclear point charges. The molecular orbitals for the N-electron molecule may be quite different from those for the corresponding one-electron molecule, however, and the SCF procedure will often not converge with the core-Hamiltonian as an initial guess to the Fock matrix. A semi-empirical extended Hückel type calculation, with an "effective" $\mathbf{F}$, is often used for an initial guess at the wave function and, commonly, provides a better guess than just using the core-Hamiltonian. There are clearly many ways one could generate an initial guess.

The major time-consuming part of the actual iteration procedure is the assembling of the two-electron integrals and density matrix into the matrix $\mathbf{G}$ in step (5). If the integrals are stored in random order with associated labels on an external device, then as they are read into main memory one uses the label, i.e., the indices μ, ν, λ, and σ identifying the integral, to determine which elements of the density matrix to multiply by and to which elements of the $\mathbf{G}$ matrix the products must be added, according to the expression for the $\mathbf{G}$ matrix given by Eq. (3.154).

In most calculations, the matrix operations in steps (6) to (10) are not time consuming relative to the formation of the $\mathbf{G}$ matrix, provided one uses an efficient diagonalization procedure.

Because the Roothan equations are nonlinear equations, the simple iteration procedure we have outlined here will not always converge. It may oscillate or diverge, possibly because of a poor initial guess. If it oscillates, averaging successive density matrices may help. If it does converge, it may do so only slowly. With two or more successive density matrices, various extrapolation procedures can be devised. Convergence problems are not unusual, but also are not a major problem for many calculations. The iterative procedure we have described is perhaps the simplest procedure one might try, but it also is somewhat naive. A number of other techniques have been suggested for ensuring or accelerating convergence to the SCF solution.

One, of course, requires a criterion for establishing convergence, and it is not uncommon simply to observe the total electronic energy of each iteration and require that two successive values differ by no more than a small quantity δ. A value of $\delta = 10^{-6}$ Hartrees is adequate for most purposes. We will show shortly that the energy of each iteration can be calculated without due expense. Alternatively, one might require convergence for elements of the density matrix, by requiring the standard deviation of successive density matrix elements, i.e., the quantity

$$\left[K^{-2} \sum_{\mu} \sum_{\nu} \left[P_{\mu\nu}^{(i)} - P_{\mu\nu}^{(i-1)} \right]^2 \right]^{1/2}$$

to be less than δ. A value of $\delta = 10^{-4}$ for the error in the density matrix will usually give an error in the energy of less than 10^{-6} Hartrees.

We have only been able to touch on a few aspects of the SCF procedure. Research efforts by many groups and large numbers of man-years of programming have gone into the large computer programs, which are currently available, for performing *ab initio* SCF calculations.

3.4.7 Expectation Values and Population Analysis

Once we have a converged value for the density matrix, Fock matrix, etc., there are a number of ways we might use our wave function $|\Psi_0\rangle$ or analyze the results of our calculation. Only some of the more common quantities will be discussed.

The eigenvalues of $\mathbf{F}'$ are the orbital energies ε_i. As we discussed when describing Koopmans' theorem, the occupied orbital energies ε_a constitute a prediction of ionization potentials and the virtual orbital energies ε_r constitute a prediction of electron affinities. The values of $-\varepsilon_a$ are commonly a reasonable approximation to the observed ionization potentials, but $-\varepsilon_r$ is usually of little use, even for a qualitative understanding of electron affinities.

The total electronic energy is the expectation value $E_0 = \langle \Psi_0 | \mathcal{H} | \Psi_0 \rangle$ and, as we have seen a number of times now, it is given by

$$E_0 = 2 \sum_a^{N/2} h_{aa} + \sum_a^{N/2} \sum_b^{N/2} 2J_{ab} - K_{ab} \tag{3.181}$$

With definition (3.147) of the Fock operator, we have

$$\varepsilon_a = f_{aa} = h_{aa} + \sum_b^{N/2} 2J_{ab} - K_{ab} \tag{3.182}$$

and, therefore, we can write the energy as

$$E_0 = \sum_a^{N/2} (h_{aa} + f_{aa}) = \sum_a^{N/2} (h_{aa} + \varepsilon_a) \tag{3.183}$$

This is a convenient result; if we substitute the basis function expansion (3.133) for the molecular orbitals into this expression, we obtain a formula for the energy, which is readily evaluated from quantities available at any stage of the SCF iteration procedure, i.e.,

$$E_0 = \frac{1}{2} \sum_\mu \sum_\nu P_{\nu\mu}(H_{\mu\nu}^{\text{core}} + F_{\mu\nu}) \tag{3.184}$$

Exercise 3.17 Derive Equation (3.184) from (3.183)

If E_0 is calculated from (3.184) using the same matrix **P** as was used to form **F**, then E_0 will be an upper bound to the true energy at any stage of the iteration and will usually converge monotonically from above to the converged result. If one adds the nuclear-nuclear repulsion to the electronic energy E_0 one obtains the total energy E_{tot}

$$E_{\text{tot}} = E_0 + \sum_A \sum_{B>A} \frac{Z_A Z_B}{R_{AB}} \tag{3.185}$$

This is commonly the quantity of most interest, particularly in structure determinations, because the predicted equilibrium geometry of a molecule occurs when E_{tot} is a minimum.

Most of the properties of molecules that one might evaluate from a molecular wave function, such as the dipole moment, quadrupole moment, field gradient at a nucleus, diamagnetic susceptibility, etc., are described by sums of one-electron operators of the general form

$$\mathcal{O}_1 = \sum_{i=1}^N h(i) \tag{3.186}$$

where $h(i)$ is not necessarily the core-Hamiltonian here, but any operator depending only on the coordinates of a single electron. From the rules for matrix elements, expectation values for such operators will always have the form

$$\langle \mathcal{O}_1 \rangle = \langle \Psi_0 | \mathcal{O}_1 | \Psi_0 \rangle = \sum_a^{N/2} (\psi_a | h | \psi_a) = \sum_{\mu\nu} P_{\mu\nu}(\nu | h | \mu) \tag{3.187}$$

so that, in addition to the density matrix, we need only evaluate the set of one-electron integrals $(\mu | h | \nu)$ to calculate one-electron expectation values. We will use the dipole moment to illustrate such a calculation.

The classical definition of the dipole moment of a collection of charges q_i with position vectors $\mathbf{r}_i$ is

$$\vec{\mu} = \sum_i q_i \mathbf{r}_i \tag{3.188}$$

The corresponding definition for a quantum mechanical calculation on a molecule is

$$\vec{\mu} = \left\langle \Psi_0 \left| - \sum_{i=1}^N \mathbf{r}_i \right| \Psi_0 \right\rangle + \sum_A Z_A \mathbf{R}_A \tag{3.189}$$

where the first term is the contribution (quantum mechanical) of the electrons, of charge -1, and the second term is the contribution (classical) of the nuclei, of charge Z_A, to the dipole moment. The electronic dipole operator is $-\sum_{i=1}^{N} \mathbf{r}_i$, a sum of one-electron operators. Therefore, using (3.187), we have

$$\vec{\mu} = -\sum_{\mu}\sum_{\nu} P_{\mu\nu}(\nu|\mathbf{r}|\mu) + \sum_{A} Z_A \mathbf{R}_A \qquad (3.190)$$

this is a vector equation with components (for example the x component) given by

$$\mu_x = -\sum_{\mu}\sum_{\nu} P_{\mu\nu}(\nu|x|\mu) + \sum_{A} Z_A X_A \qquad (3.191)$$

and to calculate the dipole moment we need in addition to **P** only the dipole integrals

$$(\nu|x|\mu) = \int d\mathbf{r}_1 \; \phi_\nu^*(\mathbf{r}_1) x_1 \phi_\mu(\mathbf{r}_1) \qquad (3.192)$$

with corresponding values for the y and z components.

The charge density

$$\rho(\mathbf{r}) = \sum_{\mu}\sum_{\nu} P_{\mu\nu}\phi_\mu(\mathbf{r})\phi_\nu^*(\mathbf{r}) \qquad (3.193)$$

representing the probability of finding an electron in various regions of space, is commonly pictured by contour maps for various planes drawn through the molecule. There is no unique definition of the number of electrons to be associated with a given atom or nucleus in a molecule, but it is still sometimes useful to perform such population analyses. Since

$$N = 2\sum_{a}^{N/2} \int d\mathbf{r}|\psi_a(\mathbf{r})|^2 \qquad (3.194)$$

divides the total number of electrons into two electrons per molecular orbital, by substituting the basis expansion of ψ_a into (3.194), we have

$$N = \sum_{\mu}\sum_{\nu} P_{\mu\nu}S_{\nu\mu} = \sum_{\mu} (\mathbf{PS})_{\mu\mu} = \text{tr } \mathbf{PS} \qquad (3.195)$$

and it is possible to interpret $(\mathbf{PS})_{\mu\mu}$ as the number of electrons to be associated with ϕ_μ. This is called a *Mulliken population analysis*. Assuming the basis functions are centered on atomic nuclei, the corresponding number of electrons to be associated with a given atom in a molecule are obtained by summing over all basis functions centered on that atom. The net charge associated with an atom is then given by

$$q_A = Z_A - \sum_{\mu \in A} (\mathbf{PS})_{\mu\mu} \qquad (3.196)$$

where Z_A is the charge of atomic nucleus A; the index of summation indicates that we only sum over the basis functions centered on A.

The definition (3.195) is by no means unique. Since tr $\mathbf{AB}$ = tr $\mathbf{BA}$,

$$N = \sum_\mu (\mathbf{S}^\alpha \mathbf{P} \mathbf{S}^{1-\alpha})_{\mu\mu} \tag{3.197}$$

for any α. With $\alpha = 1/2$, we have

$$N = \sum_\mu (\mathbf{S}^{1/2} \mathbf{P} \mathbf{S}^{1/2})_{\mu\mu} = \sum_\mu \mathbf{P}'_{\mu\mu} \tag{3.198}$$

where we can show that $\mathbf{P}'$ is the density matrix in terms of a symmetrically orthogonalized basis set,

$$\rho(\mathbf{r}) = \sum_\mu \sum_\nu P'_{\mu\nu} \phi'_\mu(\mathbf{r}) \phi'^*_\nu(\mathbf{r}) \tag{3.199}$$

$$\phi'_\mu(\mathbf{r}) = \sum_\nu (\mathbf{S}^{-1/2})_{\nu\mu} \phi_\nu(\mathbf{r}) \tag{3.200}$$

The diagonal elements of $\mathbf{P}'$ are commonly used for a *Löwdin population analysis*

$$q_A = Z_A - \sum_{\mu \in A} (\mathbf{S}^{1/2} \mathbf{P} \mathbf{S}^{1/2})_{\mu\mu} \tag{3.201}$$

Exercise 3.18 Derive the right-hand side of Eq. (3.198), i.e., show that $\alpha = 1/2$ is equivalent to a population analysis based on the diagonal elements of $\mathbf{P}'$.

None of these population analysis schemes is unique, but they are often useful when comparing different molecules using the same type of basis set for each molecule. The basis sets must be "balanced" as can be illustrated by a simple example. It is possible to have a complete set of basis functions by placing them all at one center, on oxygen, for example, in an H_2O calculation. A population analysis would then suggest that the H atoms in water have a charge of $+1$ and that all the electrons reside on the oxygen. This example makes it obvious that care must be used in assigning physical significance to any population analysis.

3.5 MODEL CALCULATIONS ON H_2 AND HeH^+

We have discussed and will subsequently discuss a number of formal mathematical procedures associated with solutions to the many-electron problem. The ideas and concepts that we are presenting may appear to be rather formidable and difficult to the uninitiated. With only a formal presentation, it is unlikely that this situation would be radically altered. We want, for the reader's benefit, to avoid the burden of endless formalism, without hint of application. We particularly feel that what appears at the outset as a somewhat obscure formal theory is usually made clear by application to a simple, but nevertheless realistic, model system. In this section we apply the closed-shell Hartree-Fock procedure to the model systems H_2 and HeH^+.

The two-electron molecules H_2 and HeH^+ are prototypes for homonuclear and heteronuclear diatomic molecules. We will consider both molecules in the approximation of a *minimal basis set*, i.e., a basis set $\{\phi_\mu\}$ consisting of only two functions, one on each nucleus. The limitation of these models is only in the basis set (and the usual assumption of a nonrelativistic, Born-Oppenheimer electronic Hamiltonian). Larger basis sets would lead to correspondingly more accurate results. Since both molecules are simple two-electron systems, essentially exact calculations, corresponding to an infinite basis set, will be available for comparison with our very approximate calculations. Before describing these calculations, however, we need to introduce the basis set that we will be using.

3.5.1 The 1s Minimal STO-3G Basis Set

In Section 3.6 we will describe basis sets for the general polyatomic molecule calculation, including s, p, and d-type basis functions. Here we introduce some of the basic ideas involved in the choice of a basis set, by describing basis functions of the $1s$ type, i.e., those that will be used in our simple calculations on H_2 and HeH^+. Better calculations would use many $1s$ functions and/or $2p$, $3d$, etc. functions in the basis set $\{\phi_\mu\}$. The extension of most of the concepts introduced here, for $1s$ functions, to the general case is fairly straightforward.

In a strictly mathematical sense many different kinds of basis set functions ϕ_μ could be used. A variety of choices have been suggested but only two types of basis functions have found common use. The normalized $1s$ *Slater-type function*, centered at $\mathbf{R}_A$, has the form

$$\phi_{1s}^{SF}(\zeta, \mathbf{r} - \mathbf{R}_A) = (\zeta^3/\pi)^{1/2} e^{-\zeta|\mathbf{r} - \mathbf{R}_A|} \tag{3.202}$$

where ζ is the *Slater orbital exponent*. The normalized $1s$ *Gaussian-type function*, centered at $\mathbf{R}_A$, has the form

$$\phi_{1s}^{GF}(\alpha, \mathbf{r} - \mathbf{R}_A) = (2\alpha/\pi)^{3/4} e^{-\alpha|\mathbf{r} - \mathbf{R}_A|^2} \tag{3.203}$$

where α is the *Gaussian orbital exponent*. The $2p$, $3d$, etc. Slater and Gaussian functions are generalizations of (3.202) and (3.203) that have polynomials in the components of $\mathbf{r} - \mathbf{R}_A$ ($x - X_A$, etc.) multiplying the same exponential ($e^{-\zeta r}$) or Gaussian ($e^{-\alpha r^2}$) fall-off. The orbital exponents, which are positive numbers larger than zero, determine the diffuseness or "size" of the basis functions; a large exponent implies a small dense function, a small exponent implies a large diffuse function. The major differences between the two functions $e^{-\zeta r}$ and $e^{-\alpha r^2}$ occur at $r = 0$ and at large r. At $r = 0$, the Slater function has a finite slope and the Gaussian function has a zero slope,

$$[d/dr \; e^{-\zeta r}]_{r=0} \neq 0 \tag{3.204}$$

$$[d/dr \; e^{-\alpha r^2}]_{r=0} = 0 \tag{3.205}$$

At large values of r, the Gaussian function $e^{-\alpha r^2}$ decays much more rapidly than the Slater function $e^{-\zeta r}$.

For electronic wave function calculations one would prefer to use the Slater functions. They more correctly describe the qualitative features of the molecular orbitals ψ_i than do Gaussian functions, and fewer Slater basis functions than Gaussian basis functions would be needed in the basis function expansion of ψ_i, for comparable results. It is possible to show, for example, that at large distances molecular orbitals decay as $\psi_i \sim e^{-a_i r}$, which is of the Slater rather than the Gaussian form. In particular, the exact solution for the $1s$ orbital of the hydrogen atom is the Slater function $(\pi)^{-1/2} e^{-r}$.

The reason why one considers Gaussian functions at all is that, in an SCF calculation, one must calculate of the order of $K^4/8$ two-electron integrals $(\mu\nu|\lambda\sigma)$. These integrals are of the form

$$(\mu_A \nu_B | \lambda_C \sigma_D) = \int d\mathbf{r}_1 \, d\mathbf{r}_2 \; \phi_\mu^{A*}(\mathbf{r}_1)\phi_\nu^B(\mathbf{r}_1)r_{12}^{-1}\phi_\lambda^{C*}(\mathbf{r}_2)\phi_\sigma^D(\mathbf{r}_2) \qquad (3.206)$$

where ϕ_μ^A is a basis function on nucleus A, i.e., centered at $\mathbf{R}_A$. The general integral involves four different centers: $\mathbf{R}_A$, $\mathbf{R}_B$, $\mathbf{R}_C$, and $\mathbf{R}_D$. Evaluation of these four-center integrals is very difficult and time-consuming with Slater basis functions. These integrals are relatively easy to evaluate with Gaussian basis functions, however. The reason is that the product of two $1s$ Gaussian functions, each on different centers, is, apart from a constant, a $1s$ Gaussian function on a third center. Thus

$$\phi_{1s}^{GF}(\alpha, \mathbf{r} - \mathbf{R}_A)\phi_{1s}^{GF}(\beta, \mathbf{r} - \mathbf{R}_B) = K_{AB}\phi_{1s}^{GF}(p, \mathbf{r} - \mathbf{R}_P) \qquad (3.207)$$

where the constant K_{AB} is

$$K_{AB} = (2\alpha\beta/[(\alpha + \beta)\pi])^{3/4} \exp[-\alpha\beta/(\alpha + \beta)|\mathbf{R}_A - \mathbf{R}_B|^2] \qquad (3.208)$$

The exponent of the new Gaussian centered at $\mathbf{R}_P$ is

$$p = \alpha + \beta \qquad (3.209)$$

and the third center P is on a line joining the centers A and B,

$$\mathbf{R}_P = (\alpha\mathbf{R}_A + \beta\mathbf{R}_B)/(\alpha + \beta) \qquad (3.210)$$

This relationship is shown in Fig. 3.1

Exercise 3.19 Derive Eq. (3.207)

As a result of (3.207), the four-center integral in (3.206) immediately reduces, for $1s$ Gaussians, to the two-center integral

$$(\mu_A \nu_B | \lambda_C \sigma_D) = K_{AB}K_{CD}\int d\mathbf{r}_1 \, d\mathbf{r}_2 \; \phi_{1s}^{GF}(p, \mathbf{r}_1 - \mathbf{R}_P)r_{12}^{-1}\phi_{1s}^{GF}(q, \mathbf{r}_2 - \mathbf{R}_Q) \qquad (3.211)$$

These integrals can be readily evaluated as described in Appendix A.

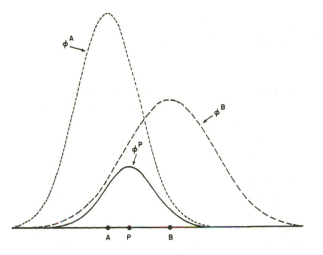

Figure 3.1 The product of two $1s$ Gaussians is a third $1s$ Gaussian.

One is thus faced with somewhat of a dilemma. Two-electron integrals can be calculated rapidly and efficiently with Gaussian functions, but Gaussian functions are not optimum basis functions and have functional behavior different from the known functional behavior of molecular orbitals. One would prefer to use better basis functions. One way around this problem is to use as basis functions fixed linear combinations of the primitive Gaussian functions ϕ_p^{GF}. These linear combinations, called *contractions*, lead to *contracted Gaussian functions* (CGF),

$$\phi_\mu^{\mathrm{CGF}}(\mathbf{r} - \mathbf{R}_A) = \sum_{p=1}^{L} d_{p\mu} \phi_p^{\mathrm{GF}}(\alpha_{p\mu}, \mathbf{r} - \mathbf{R}_A) \tag{3.212}$$

where L is the length of the contraction and $d_{p\mu}$ is a contraction coefficient. The pth normalized primitive Gaussian ϕ_p^{GF} in the basis function ϕ_μ^{CGF} has a functional dependence on the Gaussian orbital exponent (contraction exponent) $\alpha_{p\mu}$. By a proper choice of the contraction length, the contraction coefficients, and the contraction exponents, the contracted Gaussian function can be made to assume any functional form consistent with the primitive functions used. If the primitive functions are all $1s$ Gaussians at the same center, then ϕ_μ^{CGF} can only be of s-symmetry. Although we will not pursue such possibilities, if the primitive functions were allowed to reside on different centers, the expansion (3.212) could in principle describe any basis function. The idea behind the use of contracted Gaussian functions is to choose in advance the contraction length, contraction coefficients, and contraction exponents that fit the right-hand side of (3.212) to a desirable set of basis functions ϕ_μ^{CGF} and then to use these fixed functions in molecular wave function calculations. That is, the contraction coefficients, etc. are not allowed to change in the course of an SCF calculation. The two-electron integrals $(\mu\nu|\lambda\sigma)$ for the contracted basis set functions $\{\phi_\mu^{CGF}\}$ can, from (3.212), be evaluated as sums of rapidly calculated two-electron integrals over the primitive Gaussian functions.

By proper choice of the contraction parameters one can thus use basis functions that are approximate atomic Hartree-Fock functions, Slater-type functions, etc., while still evaluating integrals only with primitive Gaussian functions. A procedure that has come into wide use is to fit a Slater-type orbital (STO) to a linear combination of $L = 1, 2, 3, \ldots$ primitive Gaussian functions. This is the STO-LG procedure (the procedure is commonly referred to as STO-NG, but since N represents the number of electrons everywhere in this book, we prefer an alternative symbol). In particular, STO-3G basis sets are often used in polyatomic calculations, in preference to evaluating integrals with Slater functions. We will use STO-3G basis sets for our model calculations on H_2 and HeH^+. We need to explicitly consider the form that the contraction (3.212) takes if ϕ_{1s}^{CGF} is to approximate a $1s$ Slater-type function.

Let us first consider fitting a Slater function having Slater exponent $\zeta = 1.0$. Later we will return to consider other exponents. We will only consider contractions up to length three so that the three fits we seek to find are

$$\phi_{1s}^{CGF}(\zeta = 1.0, \text{STO-1G}) = \phi_{1s}^{GF}(\alpha_{11}) \tag{3.213}$$

$$\phi_{1s}^{CGF}(\zeta = 1.0, \text{STO-2G}) = d_{12}\phi_{1s}^{GF}(\alpha_{12}) + d_{22}\phi_{1s}^{GF}(\alpha_{22}) \tag{3.214}$$

$$\phi_{1s}^{CGF}(\zeta = 1.0, \text{STO-3G}) = d_{13}\phi_{1s}^{GF}(\alpha_{13}) + d_{23}\phi_{1s}^{GF}(\alpha_{23}) + d_{33}\phi_{1s}^{GF}(\alpha_{33}) \tag{3.215}$$

where the ϕ_{1s}^{CGF} ($\zeta = 1.0$, STO-LG) are the basis functions that approximate as best as possible a Slater-type function with $\zeta = 1.0$. We therefore need to find the coefficients $d_{p\mu}$ and exponents $\alpha_{p\mu}$ in (3.213) to (3.215) that provide the best fit. The fitting criterion is one that fits the contracted Gaussian function to the Slater function in a least-squares sense, i.e., we seek to minimize the integral

$$I = \int d\mathbf{r} \left[\phi_{1s}^{SF}(\zeta = 1.0, \mathbf{r}) - \phi_{1s}^{CGF}(\zeta = 1.0, \text{STO-LG}, \mathbf{r}) \right]^2 \tag{3.216}$$

Equivalently, since the two functions in this equation are normalized, one maximizes the overlap between the two functions, i.e., one maximizes

$$S = \int d\mathbf{r}\, \phi_{1s}^{SF}(\zeta = 1.0, \mathbf{r})\, \phi_{1s}^{CGF}(\zeta = 1.0, \text{STO-LG}, \mathbf{r}) \tag{3.217}$$

For the STO-1G case there are no contraction coefficients, and we only need to find the primitive Gaussian exponent α which maximizes the overlap

$$S = (\pi)^{-1/2}(2\alpha/\pi)^{3/4} \int d\mathbf{r}\, e^{-r} e^{-\alpha r^2} \tag{3.218}$$

Table 3.1 Overlap of a 1s Slater function
($\zeta = 1.0$) and a 1s Gaussian function

$$S = \int dr \, \phi_{1s}^{SF}(\zeta = 1.0) \, \phi_{1s}^{GF}(\alpha)$$

α	S
0.1	0.8641
0.2	0.9673
0.3	0.9772
0.4	0.9606
0.5	0.9355

$$\alpha_{optimum} = 0.270950$$

This overlap is shown in Table 3.1. The optimum fit occurs for $\alpha = 0.270950$ and is shown in Fig. 3.2a. The corresponding radial distribution functions $(4\pi r^2 |\phi_{1s}(r)|^2)$ are compared in Fig. 3.2b. Notice the different behavior near the origin and the more rapid fall-off of the Gaussian function at large r. The overlaps S of (3.217) can be maximized for the STO-2G and STO-3G cases also and, if one does so, the optimum fits are as follows:

$$\phi_{1s}^{CGF}(\zeta = 1.0, \text{STO-1G}) = \phi_{1s}^{GF}(0.270950) \tag{3.219}$$

$$\phi_{1s}^{CGF}(\zeta = 1.0, \text{STO-2G})$$
$$= 0.678914\phi_{1s}^{GF}(0.151623) + 0.430129\phi_{1s}^{GF}(0.851819) \tag{3.220}$$

$$\phi_{1s}^{CGF}(\zeta = 1.0, \text{STO-3G}) = 0.444635\phi_{1s}^{GF}(0.109818) + 0.535328 \, \phi_{1s}^{GF}(0.405771)$$
$$+ 0.154329\phi_{1s}^{GF}(2.22766) \tag{3.221}$$

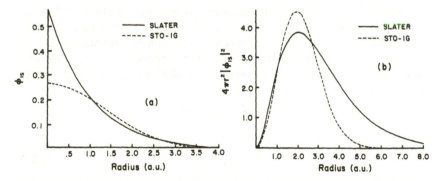

Figure 3.2 Comparison of a Slater function with a Gaussian function: a) least squares fit of a 1s Slater function ($\zeta = 1.0$) by a single STO-1G 1s Gaussian function ($\alpha = 0.270950$); b) comparison of the corresponding radial distribution functions $(4\pi r^2 |\phi_{1s}(r)|^2)$.

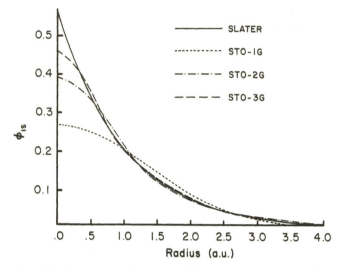

Figure 3.3 Comparison of the quality of the least-squares fit of a 1s Slater function ($\zeta = 1.0$) obtained at the STO-1G, STO-2G, and STO-3G levels.

Figure 3.3 illustrates the improvement of the fit to a Slater 1s function ($\zeta = 1.0$) obtained by increasing the number of Gaussians in the contraction (i.e., upon going from STO-1G to STO-2G to STO-3G).

Exercise 3.20 Calculate the values of $\phi(\mathbf{r})$ at the origin for the three STO-LG contracted functions and compare with the value of $(\pi)^{-1/2}$ for a Slater function ($\zeta = 1.0$).

The STO-LG fits to a Slater function, given in Eqs. (3.219) to (3.221), are for a Slater exponent of $\zeta = 1.0$. How does one obtain a fit to a Slater function with a different orbital exponent? The orbital exponents are scale factors which scale the function in r, i.e., they expand or contract the function, but do not change its functional form. Because the scale factors multiply r as follows,

$$e^{-[\zeta r]} \leftrightarrow e^{-[\sqrt{\alpha} r]^2} \tag{3.222}$$

the proper scaling is

$$\zeta'/\zeta = [\alpha'/\alpha]^{1/2} \tag{3.223}$$

The appropriate contraction exponents α for fitting to a Slater function with orbital exponent ζ are thus

$$\alpha = \alpha(\zeta = 1.0) \times \zeta^2 \tag{3.224}$$

If the Slater exponent doubles, the contraction exponents should be multiplied by a factor of four. This scaling procedure is quite general and contraction parameters need only be determined once for a given type of basis function ϕ_μ^{CGF}. If a different scale factor for ϕ_μ^{CGF} is required, the contraction exponents can be appropriately scaled. The usual description of the STO-3G basis set includes a standard set of Slater orbital exponents ζ, for basis functions centered on particular atoms. For example, the standard exponent for the $1s$ basis function of hydrogen is $\zeta = 1.24$. This is larger than the $\zeta = 1.0$ exponent of the hydrogen atom, since the hydrogen $1s$ orbital in average molecules is known to be "smaller" or "denser" than in the atom. Using the scaling relation (3.224), the standard STO-3G basis function for hydrogen becomes,

$$\phi_{1s}^{CGF}(\zeta = 1.24, \text{STO-3G}) = 0.444635\phi_{1s}^{GF}(0.168856) + 0.535328\phi_{1s}^{GF}(0.623913)$$
$$+ 0.154329\phi_{1s}^{GF}(3.42525) \tag{3.225}$$

This is the basis function we will use for H in our following calculations.

3.5.2 STO-3G H$_2$

In Subsection 2.2.5 we presented our minimal basis H$_2$ model, which has only one occupied molecular orbital and one virtual molecular orbital. With the description of the $1s$ minimal STO-3G basis set given in the last subsection we are now in a position to illustrate *ab initio* Hartree-Fock calculations on H$_2$. The model is simple but extension to larger basis sets is relatively straightforward and most of the aspects of Hartree-Fock theory that we wish to illustrate here are independent of the actual size of the basis set. Unfortunately, however, the model is too simple to be able to illustrate the iterative nature of the SCF procedure. In the next subsection we describe a minimal basis calculation on HeH$^+$, in order to illustrate this aspect of Hartree-Fock theory.

In this subsection, we describe restricted closed-shell calculations on the ground state of H$_2$. As we will see, there is a very basic deficiency in such calculations at long bond lengths. Later in this chapter, when we describe unrestricted open-shell calculations, we will return to minimal basis H$_2$ and partially correct this deficiency. Some of the results obtained here will also be used in later chapters when we use the minimal basis H$_2$ model to illustrate procedures that go beyond the Hartree-Fock approximation.

According to the steps involved in an SCF calculation, as outlined in Subsection 3.4.6, we must first of all choose a geometry for the nuclear framework. We will use the coordinate system of Fig. 2.5 with an internuclear distance $R = |\mathbf{R}_{12}|$ equal to the experimental value of 1.4 atomic units (Bohr). Our basis set is the standard STO-3G basis set, consisting of two functions ϕ_1 and ϕ_2 where each of these functions is a contraction of three primitive Gaussians such that each constitutes a least-squares fit to a Slater function with orbital exponent $\zeta = 1.24$, as previously illustrated in Eq. (3.225). That is,

$$\phi_1(\mathbf{r}) \simeq (\zeta^3/\pi)^{1/2} e^{-\zeta|\mathbf{r} - \mathbf{R}_1|}$$

$$\phi_2(\mathbf{r}) \simeq (\zeta^3/\pi)^{1/2} e^{-\zeta|\mathbf{r} - \mathbf{R}_2|} \tag{3.226}$$

$$\zeta = 1.24$$

It is important to remember, however, that each basis function has the definite form of (3.225) and that, while the functions approximate Slater functions, there is no approximation being made, other than the Hartree-Fock approximation, once the basis functions are chosen. The next step in the SCF calculation involves the evaluation of all integrals over the basis set $\{\phi_\mu\}$, i.e., $S_{\mu\nu}$, $H_{\mu\nu}^{core}$, and the two-electron integrals $(\mu\nu|\lambda\sigma)$. All these integrals can be evaluated using formulas developed in Appendix A. Consider the overlap integral,

$$S_{\mu\nu} = \int d\mathbf{r} \; \phi_\mu^{CGF}(\mathbf{r} - \mathbf{R}_A)\phi_\nu^{CGF}(\mathbf{r} - \mathbf{R}_B) \tag{3.227}$$

Substituting the general contraction (3.212), this integral reduces to the sum of overlap integrals involving primitive Gaussians. That is,

$$
\begin{aligned}
S_{\mu\nu} &= \int d\mathbf{r} \sum_{p=1}^{L} d_{p\mu}^* \phi_p^{GF*}(\alpha_{p\mu}, \mathbf{r} - \mathbf{R}_A) \sum_{q=1}^{L} d_{q\nu} \phi_q^{GF}(\alpha_{q\nu}, \mathbf{r} - \mathbf{R}_B) \\
&= \sum_{p=1}^{L} \sum_{q=1}^{L} d_{p\mu}^* d_{q\nu} \int d\mathbf{r} \; \phi_p^{GF*}(\alpha_{p\mu}, \mathbf{r} - \mathbf{R}_A)\phi_q^{GF}(\alpha_{q\nu}, \mathbf{r} - \mathbf{R}_B) \\
&= \sum_{p=1}^{L} \sum_{q=1}^{L} d_{p\mu}^* d_{q\nu} S_{pq}
\end{aligned}
\tag{3.228}
$$

In a similar manner, if other integrals over primitive Gaussian functions are evaluated using the methods of Appendix A, they can be summed to give integrals over any specific contracted function of interest, such as that of (3.225). One finds that the overlap S_{12} for the functions ϕ_1 and ϕ_2 of minimal basis H_2 at $R = 1.4$ a.u. is 0.6593. The overlap matrix is thus

$$\mathbf{S} = \begin{pmatrix} 1.0 & 0.6593 \\ 0.6593 & 1.0 \end{pmatrix} \tag{3.229}$$

At longer bond lengths the overlap S_{12} would decrease towards zero. At $R = 0$ the overlap S_{12} is, of course, 1.0.

Exercise 3.21 Use definition (3.219) for the STO-1G function and the scaling relation (3.224) to show that the STO-1G overlap for an orbital exponent $\zeta = 1.24$ at $R = 1.4$ a.u., corresponding to result (3.229), is $S_{12} = 0.6648$. Use the formula in Appendix A for overlap integrals. Do not forget normalization.

The elements $H_{\mu\nu}^{\text{core}}$ of the core-Hamiltonian are the sum of elements $T_{\mu\nu}$ describing the kinetic energy and elements describing the coulomb attraction of an electron for the first nucleus $(V_{\mu\nu}^1)$ and the second nucleus $(V_{\mu\nu}^2)$. From Appendix A these integrals can be calculated to be

$$T = \begin{pmatrix} 0.7600 & 0.2365 \\ 0.2365 & 0.7600 \end{pmatrix} \tag{3.230}$$

$$V^1 = \begin{pmatrix} -1.2266 & -0.5974 \\ -0.5974 & -0.6538 \end{pmatrix} \tag{3.231}$$

$$V^2 = \begin{pmatrix} -0.6538 & -0.5974 \\ -0.5974 & -1.2266 \end{pmatrix} \tag{3.232}$$

If the basis functions were the hydrogen atom solutions $(\pi)^{-1/2}e^{-r}$, then T_{11} would be 0.5, the kinetic energy of an electron in the hydrogen atom, and $V_{11}^1 = V_{22}^2$ would be -1.0, the potential energy of an electron in the hydrogen atom. The present values reflect the larger exponent $\zeta = 1.24$, which leads to a "smaller" orbital than in the hydrogen atom. The electron is therefore closer to the nucleus, leading to a more negative value of the potential energy (-1.2266), and it "travels faster to avoid collapsing into the nucleus," leading to a larger kinetic energy (0.7600). The energy of a hydrogen atom in this basis is just $T_{11} + V_{11}^1 = 0.7600 - 1.2266 = -0.4666$ a.u. to be compared with the exact value of -0.5 a.u. If an electron in ϕ_1 were somehow to be localized exactly at the position of nucleus 1, its attraction for nucleus 2 would be $-1/1.4 = -0.7143$. The actual value of this attraction, shown in Fig. 3.4, is $V_{11}^2 = -0.6538$, at $R = 1.4$ a.u. As the internuclear distance R increases, V_{11}^2 will converge asymptotically to $-R^{-1}$. The off-diagonal elements of T and V^{nucl} cannot be given such simple classical interpretations and they constitute the basic quantum mechanical effects of bonding. As the internuclear distance R becomes large, the off-diagonal elements go to zero.

The core-Hamiltonian matrix is the sum of the above three matrices,

$$H^{\text{core}} = T + V^1 + V^2 = \begin{pmatrix} -1.1204 & -0.9584 \\ -0.9584 & -1.1204 \end{pmatrix} \tag{3.233}$$

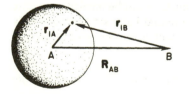

$$\langle V^B \rangle = \langle r_{1B}^{-1} \rangle \longrightarrow R_{AB}^{-1}$$

Figure 3.4 Attraction of an electron for an adjacent nucleus.

This is the Hamiltonian matrix for a single electron in the field of the nuclei, in this case for H_2^+. Solving the matrix eigenvalue problem

$$\mathbf{H}^{\text{core}}\mathbf{C} = \mathbf{SC}\boldsymbol{\varepsilon} \tag{3.234}$$

would lead to the orbital energies and molecular orbitals of H_2^+. For other cases such as H_2O, it would lead to the orbitals energies and molecular orbitals of H_2O^{9+}, which are not of particular interest.

Of the $2^4 = 16$ possible two-electron integrals $(\mu\nu|\lambda\sigma)$ in the minimal basis model, there are only four unique values,

$$
\begin{aligned}
(\phi_1\phi_1|\phi_1\phi_1) &= (\phi_2\phi_2|\phi_2\phi_2) = 0.7746 \text{ a.u.} \\
(\phi_1\phi_1|\phi_2\phi_2) &= 0.5697 \text{ a.u.} \\
(\phi_2\phi_1|\phi_1\phi_1) &= (\phi_2\phi_2|\phi_2\phi_1) = 0.4441 \text{ a.u.} \\
(\phi_2\phi_1|\phi_2\phi_1) &= 0.2970 \text{ a.u.}
\end{aligned}
\tag{3.235}
$$

The other integrals are related to the above by simple interchange of indices, for example, $(\mu\nu|\lambda\sigma) = (\mu\nu|\sigma\lambda) = (\lambda\sigma|\mu\nu)$. The one-center integrals $(\phi_1\phi_1|\phi_1\phi_1)$ and $(\phi_2\phi_2|\phi_2\phi_2)$ just represent the average value of the electron-electron repulsion of two electrons in the same $1s$ orbital. The two-center integral $(\phi_1\phi_1|\phi_2\phi_2)$ is the repulsion between an electron in an orbital on center 1 and an electron in an orbital on center 2. Its value, which is 0.5697 a.u. at $R = 1.4$ a.u., will tend to $1/R$ as the internuclear distance R increases. The other two integrals do not have classical interpretations. They both go to zero at long bond lengths as the overlap S_{12} goes to zero. Having calculated all basic integrals we could proceed to solve Roothaan's equations by the procedure we have previously given, i.e., guess at the density matrix, form the Fock matrix, transform the Fock matrix to a basis of orthonormal orbitals, diagonalize the transformed Fock matrix, etc. Our minimal basis model for H_2 is, however, simple enough that the solutions to Roothaan's equations are determined by simple symmetry arguments. The canonical molecular orbitals will form a representation of the point group of the molecule. That is, for a homonuclear diatomic they can be labeled as having the symmetry σ_g, σ_u, π_g, π_u, etc. With our minimal basis set there are only two molecular orbitals. The lowest energy one will be the occupied molecular orbital, a bonding orbital of σ_g symmetry,

$$\psi_1 = [2(1 + S_{12})]^{-1/2}(\phi_1 + \phi_2) \tag{3.236}$$

The virtual molecular orbital will be the corresponding antibonding combination of σ_u symmetry,

$$\psi_2 = [2(1 - S_{12})]^{-1/2}(\phi_1 - \phi_2) \tag{3.237}$$

The final coefficient matrix for this problem is, therefore,

$$\mathbf{C} = \begin{pmatrix} [2(1 + S_{12})]^{-1/2} & [2(1 - S_{12})]^{-1/2} \\ [2(1 + S_{12})]^{-1/2} & -[2(1 - S_{12})]^{-1/2} \end{pmatrix} \tag{3.238}$$

and the final density matrix is

$$\mathbf{P} = \begin{pmatrix} (1 + S_{12})^{-1} & (1 + S_{12})^{-1} \\ (1 + S_{12})^{-1} & (1 + S_{12})^{-1} \end{pmatrix} = (1 + S_{12})^{-1} \begin{pmatrix} 1 & 1 \\ 1 & 1 \end{pmatrix} \tag{3.239}$$

If a density matrix other than the above were actually used for an initial guess in the SCF procedure and the iterations carried out, the procedure would converge to this symmetry determined solution.

Exercise 3.22 Derive the coefficients $[2(1 + S_{12})]^{-1/2}$ and $[2(1 - S_{12})]^{-1/2}$ in the basis function expansion of ψ_1 and ψ_2 by requiring ψ_1 and ψ_2 to be normalized.

Exercise 3.23 The coefficients of minimal basis H_2^+ are also determined by symmetry and are identical to those of minimal basis H_2. Use the above result for the coefficients to solve Eq. (3.234) for the orbital energies of minimal basis H_2^+ at $R = 1.4$ a.u. and show they are

$$\varepsilon_1 = (H_{11}^{\text{core}} + H_{12}^{\text{core}})/(1 + S_{12}) = -1.2528 \text{ a.u.}$$

$$\varepsilon_2 = (H_{11}^{\text{core}} - H_{12}^{\text{core}})/(1 - S_{12}) = -0.4756 \text{ a.u.}$$

Exercise 3.24 Use the general definition (3.145) of the density matrix to derive (3.239). What is the corresponding density matrix for H_2^+?

Exercise 3.25 Use the general definition (3.154) of the Fock matrix to show that the converged values of its elements for minimal basis H_2 are

$$F_{11} = F_{22} = H_{11}^{\text{core}} + (1 + S_{12})^{-1}[\tfrac{1}{2}(\phi_1\phi_1|\phi_1\phi_1) + (\phi_1\phi_1|\phi_2\phi_2)$$
$$+ (\phi_1\phi_1|\phi_1\phi_2) - \tfrac{1}{2}(\phi_1\phi_2|\phi_1\phi_2)] = -0.3655 \text{ a.u.}$$

$$F_{12} = F_{21} = H_{12}^{\text{core}} + (1 + S_{12})^{-1}[-\tfrac{1}{2}(\phi_1\phi_1|\phi_2\phi_2) + (\phi_1\phi_1|\phi_1\phi_2)$$
$$+ \tfrac{3}{2}(\phi_1\phi_2|\phi_1\phi_2)] = -0.5939 \text{ a.u.}$$

Exercise 3.26 Use the result of Exercise 3.23 to show that the orbital energies of minimal basis H_2, that are a solution to the Roothaan equations $\mathbf{FC} = \mathbf{SC\varepsilon}$, are

$$\varepsilon_1 = (F_{11} + F_{12})/(1 + S_{12}) = -0.5782 \text{ a.u.}$$

$$\varepsilon_2 = (F_{11} - F_{12})/(1 - S_{12}) = +0.6703 \text{ a.u.}$$

Exercise 3.27 Use the general result (3.184) for the total electronic energy to show that the electronic energy of minimal basis H_2 is

$$E_0 = (F_{11} + H_{11}^{core} + F_{12} + H_{12}^{core})/(1 + S_{12}) = -1.8310 \text{ a.u.}$$

and that the total energy including nuclear repulsion is

$$E_{tot} = -1.1167 \text{ a.u.}$$

The results of the last three exercises have described the results of a minimal basis H_2 calculation in terms of integrals and matrices evaluated over members of the basis set $\{\phi_\mu\}$ rather than over members of the set of solutions $\{\psi_i\}$. That is how any actual calculation would be performed; the molecular orbitals ψ_i are not known until the calculation is completed. For a discussion of the SCF results or for use of the SCF results in subsequent treatments of correlation effects, it is convenient to transform the basic integrals in terms of the functions $\{\phi_\mu\}$ to corresponding integrals in terms of the functions $\{\psi_i\}$. Since we know the relation between the two sets of functions, i.e.,

$$\psi_i = \sum_{\mu=1}^{K} C_{\mu i} \phi_\mu \qquad i = 1, 2, \ldots, K \tag{3.240}$$

The transformations proceed as follows

$$h_{ij} = (\psi_i|h|\psi_j) = \sum_\mu \sum_\nu C_{\mu i}^* C_{\nu j} H_{\mu\nu}^{core} \tag{3.241}$$

$$(\psi_i \psi_j | \psi_k \psi_l) = \sum_\mu \sum_\nu \sum_\lambda \sum_\sigma C_{\mu i}^* C_{\nu j} C_{\lambda k}^* C_{\sigma l} (\mu\nu | \lambda\sigma) \tag{3.242}$$

The two-index transformation of (3.241) is relatively easy and requires no more than the multiplication of $K \times K$ matrices. The four-index transformation of the two-electron integrals, however, is a very time consuming process. An optimum algorithm for performing the transformation requires the $O(K^5)$ multiplications. This is an order of K more difficult than any step in the complete SCF calculation. If the transformed two-electron integrals are not required, they certainly should not be calculated. On the other hand, most formulations for proceeding beyond the Hartree-Fock approximation, and all those considered in this book, require integrals over molecular orbitals. For our minimal basis H_2 model, the transformation is, of course, not difficult. The nonzero transformed elements of the core-Hamiltonian and two-electron integral matrix that result are

$h_{11} = (\psi_1|h|\psi_1) = -1.2528$ a.u. $\qquad h_{22} = (\psi_2|h|\psi_2) = -0.4756$ a.u.

$J_{11} = (\psi_1\psi_1|\psi_1\psi_1) = 0.6746$ a.u. $\qquad J_{22} = (\psi_2\psi_2|\psi_2\psi_2) = 0.6975$ a.u.

$J_{12} = (\psi_1\psi_1|\psi_2\psi_2) = 0.6636$ a.u. $\qquad K_{12} = (\psi_1\psi_2|\psi_2\psi_1) = 0.1813$ a.u.

As we've described before, h_{11} is the kinetic energy and nuclear attraction of an electron in ψ_1, h_{22} is the same for an electron in ψ_2, J_{11} is the coulomb interaction of the two electrons in ψ_1, J_{22} is the coulomb interaction of two electrons in ψ_2, J_{12} is the coulomb interaction of an electron in ψ_1 and another electron in ψ_2, and $-K_{12}$ is the exchange interaction between an electron in ψ_1 and an electron with the same spin in ψ_2.

The transformed Fock matrix, $f_{ij} = \langle \psi_i | f | \psi_j \rangle$ is by definition diagonal, with diagonal elements equal to the orbital energies. The closed-shell orbital energies, as derived in Exercise 3.9, are given by

$$\varepsilon_i = h_{ii} + \sum_b 2J_{ib} - K_{ib} \tag{3.243}$$

For our minimal basis model, these are

$$\varepsilon_1 = h_{11} + J_{11} = -0.5782 \text{ a.u.} \tag{3.244}$$

$$\varepsilon_2 = h_{22} + 2J_{12} - K_{12} = +0.6703 \text{ a.u.} \tag{3.245}$$

Note that $\varepsilon_2 \neq h_{22} + J_{22}$, since it is a virtual orbital and describes the energy of an electron in the $(N + 1)$-electron system, as discussed in conjunction with Koopmans' theorem. Appendix D contains the values of these orbital energies, as well as the two-electron integrals J_{11}, etc., as a function of bond length. Using the values of these integrals it will be possible, here and in later chapters, to investigate the behavior of a number of many-electron quantities for H_2, as a function of bond length.

The total electronic energy of the ground state is

$$E_0 = 2h_{11} + J_{11} = -1.8310 \text{ a.u.} \tag{3.246}$$

The total energy, including nuclear repulsion, is

$$E_{\text{tot}} = E_0 + 1/R = -1.1167 \text{ a.u.} \tag{3.247}$$

Since the energy of a hydrogen atom in this basis is -0.4666 a.u., the predicted dissociation energy of H_2 is $2(-0.4666) + 1.1167 = 0.1835$ a.u. $\equiv$ 4.99 eV. This is to be compared with the experimental dissociation energy of 4.75 eV. The agreement is remarkably good; even though the calculated energy of H_2 is much above the exact value, there is a compensating inexact treatment of the hydrogen atom.

The above dissociation energy is in good agreement with experiment, but to explore fully the dissociation question it is necessary to investigate the full potential surface. By repeating the above calculations for different values of the internuclear distance one obtains the potential curve shown in Fig. 3.5, which is to be compared with the essentially exact results of Kolos and Wolniewicz.[5] The minimal basis restricted Hartree-Fock calculation does not go to the limit of two hydrogen atoms as R goes to infinity. This result may at first be surprising. This totally incorrect behavior is not specific

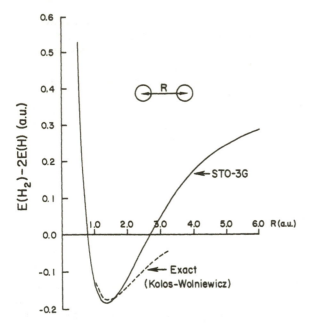

Figure 3.5 Restricted Hartree-Fock potential curve for STO-3G ($\zeta = 1.24$)H_2 compared with the accurate results of Kolos and Wolniewicz.

to H_2. If one stretches any bond for which the correct products of dissociation must be represented by open-shell wave functions, then restricted closed-shell calculations must necessarily give the wrong limit. For H_2 the products of dissociation are two localized hydrogen atoms; that is, one electron is localized near one of the protons and the other electron is localized near the other distant proton. In the restricted calculation, however, both electrons are forced to occupy the same spatial molecular orbital ψ_1. This molecular orbital is symmetry determined, having the form of (3.236). Therefore, independent of the bond length, both electrons are described by exactly the same spatial wave function and have the same probability distribution function in 3-dimensional space. Such a description is inappropriate for two separated hydrogen atoms. A restricted closed-shell Hartree-Fock calculation, which restricts electrons to occupy molecular orbitals in pairs, cannot, therefore, properly describe dissociation unless the products of dissociation are both closed-shells.

We can investigate the dissociation behavior in an analytical way by using the results of Exercises 3.25 and 3.27. As $R \to \infty$, the two-center nuclear attraction of Fig. 3.4 goes to zero and $H_{11}^{\text{core}} \to T_{11} + V_{11}^1$, the energy of a hydrogen atom in the basis (-0.4666). All other integrals go to zero as $R \to \infty$, except the one-center electron-electron repulsion integral ($\phi_1\phi_1|\phi_1\phi_1$).

One therefore obtains

$$\lim_{R \to \infty} E_{tot}(R) = \lim_{R \to \infty} 2H_{11}^{core} + \tfrac{1}{2}(\phi_1\phi_1|\phi_1\phi_1)$$

$$= 2E(H) + \tfrac{1}{2}(\phi_1\phi_1|\phi_1\phi_1)$$

$$= -0.9332 + 0.3873$$

$$= -0.5459 \text{ a.u.}$$

The limit, rather than being twice the energy of a hydrogen atom in the same basis ($2E(H)$), includes the spurious term $\tfrac{1}{2}(\phi_1\phi_1|\phi_1\phi_1)$. This spurious term arises because, since both electrons occupy the same spatial orbital, there remains even at infinity some electron-electron repulsion. Alternatively, the products of the dissociation are not just $2H\cdot$, but also include, incorrectly, H^- and H^+. The energy of H^- includes contributions from the electron-electron repulsion integral $(\phi_1\phi_1|\phi_1\phi_1)$. Another way of looking at this is that a molecular orbital wave function is equivalent to a valence bond wave function in which equal weight is given to covalent terms and ionic terms. Since the wave function is symmetry determined, the ionic terms remain, even on dissociation. We will return to the dissociation question when we consider unrestricted Hartree-Fock calculations.

The poor behavior of restricted closed-shell Hartree-Fock calculations upon dissociation to open-shell products does not detract from their utility in the region of equilibrium. The calculated equilibrium geometry is that at which E_{tot} is a minimum with respect to the coordinates of the nuclei. Table 3.2 shows the value of this energy for internuclear distances in the vicinity of the experimental bond length of 1.4 a.u. The calculated minimum energy occurs at 1.346 a.u. This is in error by 4% and errors of similar magnitude can be expected for equilibrium geometries of other molecules at this level of approximation.

Before leaving minimal basis H_2 (at least, temporarily), we want to use the model to illustrate exponent optimization. We have been using a standard exponent of $\zeta = 1.24$. These orbital exponents are nonlinear parameters upon which our wave function depends. By the variational principle, the best wave

Table 3.2 Energy of minimal basis STO-3G H_2 as a function of bond length

R (a.u.)	E_{tot} (a.u.)
1.32	-1.11731
1.34	-1.11750
1.36	-1.11745
1.38	-1.11719
1.40	-1.11672
$R_{eq} = 1.346$ a.u.	

**Table 3.3 Optimization of the Slater expo-
nent for minimal basis STO-3G H_2 at
$R = 1.4$ a.u.**

ζ	E_{tot} (a.u.)
1.0	-1.08164
1.1	-1.11089
1.2	-1.11912
1.3	-1.10714
$\zeta_{optimum} = 1.19$	

function of a given form is the one in which the energy has been minimized
with respect to all wave function parameters. Because the orbital exponents
are nonlinear parameters, however, there is no computationally easy way of
determining their optimum values. Rather than go to the expense of finding
their optimum values by performing many calculations with different orbital
exponents it is common to choose reasonable "standard" values, such as our
STO-3G value of 1.24 for the $1s$ orbitals of hydrogen. The size of the basis is
increased if greater accuracy is desired. Nevertheless, it is sometimes necessary or desirable to optimize exponents if for no other reason than to determine what a "reasonable" value might be. Table 3.3 shows values of the total
energy for minimal basis H_2 at $R = 1.4$ for a range of values of the $1s$ Slater
exponent ζ. The optimum exponent is 1.19 at $R = 1.4$ a.u. This optimum value
will change with bond length and will be different for hydrogen atoms in
other molecules. One finds, however, that optimum values for a range of
molecules are commonly larger than 1.0 (the exact hydrogen atom value),
and the standard STO-3G value of 1.24 was chosen as representative of
optimum value in a number of small molecules. The optimum value of 1.19
for H_2 means that the hydrogen molecule is in a sense "smaller" than the
sum of 2 hydrogen atoms. This is common in chemical bonding—the
attraction of the bonding electrons for two nuclei, rather than one, contracts
the electron cloud.

3.5.3 An SCF Calculation on STO-3G HeH$^+$

The two-electron molecules H_2 and HeH$^+$ are prototypes for homonuclear
and heteronuclear diatomic molecules. We have just finished describing
restricted Hartree-Fock calculations on minimal basis STO-3G H_2, and
we now do the same for minimal basis STO-3G HeH$^+$. The limitation of
these minimal basis models for describing two-electron systems lies solely
in the basis set. Larger basis sets would lead to correspondingly more
accurate results, but the minimal basis is adequate for illustrating the points
we wish to make. The extension to larger basis sets is straightforward.

Although the minimal basis sets are very small, one can describe results that are "exact" within the one-electron space spanned by these basis functions. In particular, using the minimal basis sets, we can explore any number of different computational approaches and compare the results with the "exact" results for the same basis. In later chapters we will be using the same minimal basis STO-3G H_2 and HeH^+ models to illustrate configuration interaction calculations, perturbation theory calculations, etc. The results obtained there will be compared with the "exact" results for the same basis and the Hartree-Fock calculations of this chapter.

One of the deficiencies of the minimal basis H_2 model was that, with only two basis functions, the molecular orbitals were symmetry determined and the model could not be used to illustrate the iterative nature of the SCF procedure. Because HeH^+ is heteronuclear it has less symmetry than H_2 and the molecular orbitals for the minimal basis model are not determined by symmetry. As such, restricted Hartree-Fock calculations on HeH^+ provide essentially a complete illustration of the principles involved in solving the Roothaan equations.

The singly charged helium hydride molecular ion has been known for many years from mass-spectroscopic studies. It is of interest in astrophysical problems, as the product of the β decay of HT, in scattering of protons off helium, and for a number of other reasons associated with its simplicity. There seems to be, however, little direct experimental evidence for the structure of its various electronic states. Very accurate calculations by Wolniewicz[6] show that its ground state has an equilibrium bond length of 1.4632 a.u. and an electronic binding energy of 0.0749 a.u. (2.039 eV). The ground state dissociates to a helium atom and a proton

$$HeH^+(X^1\Sigma) \rightarrow He(^1S) + H^+ \qquad (3.248)$$

rather than to $He^+ + H$, since the ionization potential of He (24.6 eV) is larger than the electron affinity of a proton (13.6 eV). Alternatively, the electron affinity of He^+ is larger than the electron affinity of H^+. The exact energies of the various species involved are given in Table 3.4. Since

Table 3.4 Exact energy of H and He species

Species	Energy (a.u.)
H^+	0.0
H	−0.5
He^+	−2.0
He	−2.90372[a]
HeH^+ ($R = 1.4632$ a.u.)	−2.97867[b]

[a] C. L. Pekeris, *Phys. Rev.* **115:** 1217 (1959).
[b] L. Wolniewicz, *J. Chem. Phys.* **43:** 1807 (1965).

the products of dissociation are closed-shells, we expect a restricted Hartree-Fock calculation to behave correctly at large bond lengths, unlike the behavior exhibited in the case of H_2.

To perform an SCF calculation, we first choose a geometry for the nuclei. We let the helium nucleus be nucleus 1 with position vector $\mathbf{R}_1$ and the hydrogen nucleus be nucleus 2 with position vector $\mathbf{R}_2$, so that $|\mathbf{R}_1 - \mathbf{R}_2| = R_{12} \equiv R$ is the internuclear distance. We will use the exact internuclear distance so that $R = 1.4632$ a.u. We then need to specify the basis set, which in our case is the STO-3G minimal basis set consisting of a $1s$ basis function on each of the two nuclei,

$$\phi_1 \simeq (\zeta_1^3/\pi)^{1/2} e^{-\zeta_1 |\mathbf{r} - \mathbf{R}_1|} \tag{3.249}$$

$$\phi_2 \simeq (\zeta_2^3/\pi)^{1/2} e^{-\zeta_2 |\mathbf{r} - \mathbf{R}_2|} \tag{3.250}$$

These basis functions are each a contraction of three primitive Gaussians, with the contraction coefficients of (3.221), and contraction exponents which are those of (3.221) scaled by multiplying by the square of either ζ_1 or ζ_2. It remains only to specify the Slater exponents ζ_1 for He and ζ_2 for H. The standard STO-3G exponent for H in molecular environments is $\zeta_2 = 1.24$ which is the value that we used in our minimal basis H_2 calculations. No such standard STO-3G exponent for He has been recommended, however. In choosing exponents for molecular calculations it is common to use exponents obtained by minimizing the energy of a calculation on the isolated atom using the same basis set. These are sometimes called "best atom" exponents. For H the best atom exponent is 1.0. If one performs a restricted Hartree-Fock calculation on the He atom using a basis set consisting of only one Slater orbital, $(\zeta^3/\pi)^{1/2} e^{-\zeta r}$, one finds that the best atom exponent is $27/16 = 1.6875$. The derivation of the value 27/16 is a common textbook example[7] of the variational principle. Since HeH^+ has a net positive charge, we expect, however, that the electron cloud is considerably contracted relative to that for the free atoms He and H. For illustrative purposes we will use the standard STO-3G value of $\zeta_2 = 1.24$ for H and let the standard value for He be $\zeta_1 = 1.6875 \times 1.24 = 2.0925$ which, like the value for H, is also a factor of 1.24 larger than the best atom value. Figure 3.6 shows our coordinate system and basis functions for the HeH^+ calculation. As described previously, each of the STO-3G basis functions of Fig. 3.6 or Eqs. (3.249) and (3.250) is the sum of the three primitive Gaussians.

The next step in our calculation, or most *ab initio* calculations, is the evaluation of the required integrals over the set of basis functions. Appendix A describes how the integrals can be evaluated for $1s$ Gaussian functions and gives explicit formulas for all integrals involving only $1s$ Gaussians. Appendix B gives a FORTRAN listing, and output for our HeH^+ case, of a small program which illustrates the complete steps involved in solving the Roothaan equations and which is capable of performing minimal basis

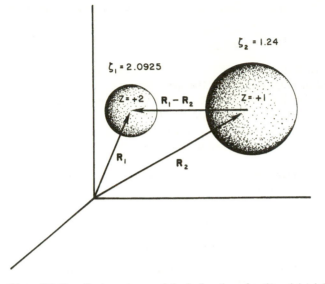

Figure 3.6 Coordinate system and basis functions for the minimal basis STO-3G HeH$^+$ calculation.

Gaussian calculations on any two-electron diatomic molecule. We will only give some of the results here and refer the reader to the output or program in Appendix B, when necessary. From Appendix A, the overlap of the two basis functions decreases exponentially with the internuclear distance. At the internuclear distance $R = 1.4632$ a.u. its value is $S_{12} = S_{21} = 0.4508$. This overlap is smaller than that in H_2 mainly because the He orbital is smaller and more localized than a corresponding H orbital. The overlap matrix is, therefore,

$$\mathbf{S} = \begin{pmatrix} 1.0 & 0.4508 \\ 0.4508 & 1.0 \end{pmatrix} \qquad (3.251)$$

The kinetic energy matrix is

$$\mathbf{T} = \begin{pmatrix} 2.1643 & 0.1670 \\ 0.1670 & 0.7600 \end{pmatrix} \qquad (3.252)$$

The T_{22} value is, of course, identical to that in H_2 since we are using the same exponent. The T_{11} value, which describes the kinetic energy of an electron in the $1s$ orbital around He, is much larger than that for H, reflecting the larger orbital exponent of the He orbital, which in turn reflects the larger nuclear charge of He. The smaller the average distance of electron from the nucleus, the larger is its kinetic energy.

The matrix of nuclear attraction energy to nucleus 1 (the He nucleus) is

$$\mathbf{V}^1 = \begin{pmatrix} -4.1398 & -1.1029 \\ -1.1029 & -1.2652 \end{pmatrix} \tag{3.253}$$

The one-center attraction of an electron in ϕ_1 for its own nucleus (-4.1398) is naturally larger in magnitude that the attraction of an electron in the distant function ϕ_2 for this nucleus (-1.2652). This last two-center integral becomes $-2/R$ at large internuclear distances. The off-diagonal element, the attraction for the helium nucleus of an electron described by the product distribution $\phi_1(1)\phi_2(1)$, is the quantum mechanical term responsible for chemical bonding.

The matrix of nuclear attraction energy to nucleus 2 (the H nucleus) is similar to the above, but smaller in magnitude because of the smaller nuclear charge of the proton.

$$\mathbf{V}^2 = \begin{pmatrix} -0.6772 & -0.4113 \\ -0.4113 & -1.2266 \end{pmatrix} \tag{3.254}$$

The V_{22}^2 element is the same as that for H_2. The function ϕ_1 with its larger exponent is relatively localized about the He nucleus and $V_{11}^2 = $ is close to its asymptotic value of $-1/R = -0.6834$.

Having obtained the kinetic energy and nuclear attraction integrals, we can now form the core-Hamiltonian matrix

$$\mathbf{H}^{core} = \mathbf{T} + \mathbf{V}^1 + \mathbf{V}^2 = \begin{pmatrix} -2.6527 & -1.3472 \\ -1.3472 & -1.7318 \end{pmatrix} \tag{3.255}$$

As we have stated before, this is the correct Hamiltonian matrix for a single electron in the field of the nuclear point charges. The solution of a Roothaan-like equation for the core-Hamiltonian

$$\mathbf{H}^{core}\mathbf{C} = \mathbf{SC}\varepsilon \tag{3.256}$$

would lead to the molecular orbitals and orbital energies (and, in this case, total electronic energies) for the one-electron molecule HeH^{++}. The effect of electron-electron repulsion on the molecular orbitals and orbital energies, within the single determinant approximation, is in the matrix $\mathbf{G}$ which must be added to $\mathbf{H}^{core}$ to obtain the Fock matrix $\mathbf{F}$.

The final remaining integrals to be calculated are the two-electron repulsion integrals. Of the $2^4 = 16$ possible integrals $(\mu\nu|\lambda\sigma)$, there are only six unique integrals,

$$(\phi_1\phi_1|\phi_1\phi_1) = 1.3072 \text{ a.u.} \qquad (\phi_2\phi_2|\phi_1\phi_1) = 0.6057 \text{ a.u.}$$

$$(\phi_2\phi_1|\phi_1\phi_1) = 0.4373 \text{ a.u.} \qquad (\phi_2\phi_2|\phi_2\phi_1) = 0.3118 \text{ a.u.}$$

$$(\phi_2\phi_1|\phi_2\phi_1) = 0.1773 \text{ a.u.} \qquad (\phi_2\phi_2|\phi_2\phi_2) = 0.7746 \text{ a.u.}$$

The one-center integrals $(\phi_1\phi_1|\phi_1\phi_1)$ and $(\phi_2\phi_2|\phi_2\phi_2)$ are the repulsions between an electron in ϕ_1 (or ϕ_2) and another electron in the same orbital ϕ_1 (or ϕ_2). The average distance between two electrons in the "smaller" function ϕ_1 is less than that between two electrons in the "larger" and more diffuse function ϕ_2, and thus $(\phi_1\phi_1|\phi_1\phi_1)$ is larger than $(\phi_2\phi_2|\phi_2\phi_2)$. The two-center integral $(\phi_2\phi_2|\phi_1\phi_1)$ is the repulsion between an electron in ϕ_1 and an electron in ϕ_2. This has the asymptotic value $1/R$ as the inter-nuclear distance becomes large. The other three integrals do not have simple classical interpretations.

We now have all the integrals needed for our SCF calculation on HeH$^+$. Prior to beginning the iterations, however, we need to derive a transformation matrix to orthonormal basis functions,

$$\phi'_\mu = \sum_\nu X_{\nu\mu}\phi_\nu \tag{3.257}$$

There are many transformations $\mathbf{X}$ that we might use to derive an orthonormal set of functions $\{\phi'_\mu\}$. The Schmidt procedure, discussed in Chapter 1, uses the following matrix

$$\mathbf{X}_{\text{Schmidt}} = \begin{pmatrix} 1 & -S_{12}/(1-S_{12}^2)^{1/2} \\ 0 & 1/(1-S_{12}^2)^{1/2} \end{pmatrix} = \begin{pmatrix} 1.0 & -0.5050 \\ 0.0 & 1.1203 \end{pmatrix} \tag{3.258}$$

Exercise 3.28 Show that the above transformation produces orthonormal basis functions.

Two other orthonormalization procedures, that we have previously described, require diagonalizing the overlap matrix. Diagonalizing a 2×2 matrix can be accomplished by the methods outlined in Subsection 1.1.6. For the overlap matrix, the eigenvalues are simply $s_1 = 1 + S_{12} = 1.4508$ and $s_2 = 1 - S_{12} = 0.5492$. The unitary matrix which performs the diagonalization is

$$\mathbf{U} = \begin{pmatrix} [2]^{-1/2} & [2]^{-1/2} \\ [2]^{-1/2} & -[2]^{-1/2} \end{pmatrix} \tag{3.259}$$

To derive the symmetric and canonical orthogonalization transformations we need the matrix

$$\mathbf{s}^{-1/2} = \begin{pmatrix} s_1^{-1/2} & 0 \\ 0 & s_2^{-1/2} \end{pmatrix} = \begin{pmatrix} 0.8302 & 0.0 \\ 0.0 & 1.3493 \end{pmatrix} \tag{3.260}$$

Symmetric orthogonalization then uses the transformation matrix

$$\mathbf{X}_{\text{symmetric}} = \mathbf{S}^{-1/2} = \mathbf{U}\mathbf{s}^{-1/2}\mathbf{U}^\dagger = \begin{pmatrix} 1.0898 & -0.2596 \\ -0.2596 & 1.0898 \end{pmatrix} \tag{3.261}$$

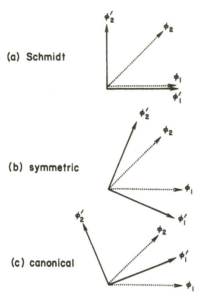

(a) Schmidt

(b) symmetric

(c) canonical

Figure 3.7 Three orthogonalization procedures:
a) Schmidt; b) symmetric; c) canonical.

whereas, canonical orthogonalization uses the transformation matrix

$$\mathbf{X}_{\text{canonical}} = \mathbf{U}\mathbf{s}^{-1/2} = \begin{pmatrix} 0.5871 & 0.9541 \\ 0.5871 & -0.9541 \end{pmatrix} \qquad (3.262)$$

The relationship between these three orthogonalizations is shown in Fig. 3.7. The angle θ between the original basis functions is given by $\cos\theta = S_{12}$. Schmidt orthogonalization leaves the first basis function alone and produces a second orthogonal to it. Symmetric orthogonalization produces two new functions, which most closely resemble the original basis functions. It does this by opening up the angle between the vectors to 90°. Canonical orthogonalization produces one vector which bisects the angle between the original vectors and a second vector orthogonal to the first. We will use canonical orthogonalization so that the transformed basis functions are

$$\phi'_1 = 0.5871\phi_1 + 0.5871\phi_2 \qquad (3.263)$$

$$\phi'_2 = 0.9541\phi_1 - 0.9541\phi_2 \qquad (3.264)$$

We are now ready to begin the SCF iteration procedure. We first need an initial guess at the density matrix. It is convenient to use the null matrix. This is equivalent to neglecting all electron-electron interaction (setting $\mathbf{G}$ equal to the null matrix) and using the core-Hamiltonian as a first guess at the Fock matrix

$$\mathbf{F} \simeq \mathbf{H}^{\text{core}} = \begin{pmatrix} -2.6527 & -1.3472 \\ -1.3472 & -1.7318 \end{pmatrix} \qquad (3.265)$$

This is the easiest initial guess to obtain, but may be a poor one in more complicated situations. The next step is to transform the Fock matrix to the canonically orthonormalized basis set.

$$\mathbf{F'} = \mathbf{X^\dagger F X} = \begin{pmatrix} -2.4397 & -0.5158 \\ -0.5158 & -1.5387 \end{pmatrix} \tag{3.266}$$

Diagonalizing this matrix, i.e., solving

$$\mathbf{F'C'} = \mathbf{C'\varepsilon} \tag{3.267}$$

gives a unitary matrix of coefficients

$$\mathbf{C'} = \begin{pmatrix} 0.9104 & 0.4136 \\ 0.4136 & -0.9104 \end{pmatrix} \tag{3.268}$$

and two eigenvalues

$$\mathbf{\varepsilon} = \begin{pmatrix} -2.6741 & 0.0 \\ 0.0 & -1.3043 \end{pmatrix} \tag{3.269}$$

The coefficients of the original basis functions are then

$$\mathbf{C} = \mathbf{XC'} = \begin{pmatrix} 0.9291 & -0.6259 \\ 0.1398 & 1.1115 \end{pmatrix} \tag{3.270}$$

Equations (3.270) and (3.269) gives the orbitals and orbital energies of HeH^{++}, which we are using as a first guess at the orbitals and orbital energies of HeH^+. Note that the lowest molecular orbital ψ_1 is composed mainly of ϕ_1 (coefficient = 0.9291) with only a little mixing of ϕ_2 (coefficient = 0.1398). With no electron-electron repulsion the electrons tend to congregate near the He nucleus with its higher nuclear charge. The effect of adding electron-electron repulsion, as we iterate further, will be to moderate this effect and "smear" the electrons out a bit, so as to decrease the electron-electron repulsion.

From (3.270) we can now form our first real guess at the density matrix

$$\mathbf{P} = \begin{pmatrix} 1.7266 & 0.2599 \\ 0.2599 & 0.0391 \end{pmatrix} \tag{3.271}$$

The diagonal elements of $\mathbf{P}$ show (only qualitatively) how most of the electron density is in the vicinity of the He rather than the H nucleus. A better appreciation of this would be obtained by a population analysis. The density matrix of (3.271) is not that of HeH^{++} (it differs by a factor of 2) but that of *two* noninteracting electrons in the field of the nuclei.

From $\mathbf{P}$ we can now form a guess at $\mathbf{G}$,

$$\mathbf{G} = \begin{pmatrix} 1.2623 & 0.3740 \\ 0.3740 & 0.9890 \end{pmatrix} \tag{3.272}$$

and a new Fock matrix

$$F = H^{core} + G = \begin{pmatrix} -1.3904 & -0.9732 \\ -0.9732 & -0.7429 \end{pmatrix} \tag{3.273}$$

Because of the positive electron-electron interaction, represented by the positive elements of the matrix G of (3.272), the elements of this new Fock matrix are considerably less negative than our original core-Hamiltonian guess (3.265). We can now solve the eigenvalue problem with this latest Fock matrix to get a new guess at C and P, and repeat the whole procedure until self-consistency is obtained. Appendix B contains a program for doing this, and the program output for our current example of minimal basis STO-3G HeH^+. This output should be followed in conjunction with our description here.

Table 3.5 shows the elements of the density matrix and the corresponding electronic energy as a function of the iteration number. As the iterations proceed, charge builds up around H and decreases around He. To provide a variational value of the energy at each iteration, the formula

$$E_0 = \frac{1}{2} \sum_\mu \sum_\nu P_{\nu\mu}(H_{\mu\nu}^{core} + F_{\mu\nu}) \tag{3.274}$$

must use the same density matrix P as was used to form F. Thus the energy should be calculated immediately after forming a new F, not immediately after forming a new P. Table 3.5 shows the energy converging monotonically from above. Because the energy is a variational quantity, the relative error in the energy is less than that in the wave function or density matrix.

The final wave function and orbital energies are

$$C = \begin{pmatrix} 0.8019 & -0.7823 \\ 0.3368 & 1.0684 \end{pmatrix} \tag{3.275}$$

$$\varepsilon = \begin{pmatrix} -1.5975 & 0.0 \\ 0.0 & -0.0617 \end{pmatrix} \tag{3.276}$$

Table 3.5 Density matrix and electronic energy during the iterative process (STO-3G HeH^+)

Iteration	P_{11}	P_{12}	P_{22}	E_0 (a.u.)
1	1.7266	0.2599	0.0391	−4.141863
2	1.3342	0.5166	0.2000	−4.226492
3	1.2899	0.5384	0.2247	−4.227523
4	1.2864	0.5400	0.2267	−4.227529
5	1.2862	0.5402	0.2269	−4.227529
6	1.2861	0.5402	0.2269	−4.227529

The lowest orbital, the occupied orbital ψ_1, is a bonding orbital as evidenced by the same sign for the two coefficients. It is still composed mainly of the He function ϕ_1. The virtual orbital ψ_2 is an antibonding orbital with opposite signs for the coefficients. It has a heavier weight for the H function ϕ_2 as is necessary if it is to be orthogonal to ψ_1. Koopmans' theorem allows us to predict an ionization potential and electron affinity. The predicted ionization potential is large (1.5975 a.u. = 43.5 eV) as should be the case for a cationic species. The predicted electron affinity (0.0617 a.u. = 1.7 eV) is positive, and the calculation predicts that HeH$^+$ will bind an electron. This does not mean that HeH will be a stable molecule, since the dissociation products of HeH$^+$ (i.e., He + H$^+$) bind an electron much more strongly (the electron affinity of H$^+$ is greater than the sum of the electron affinity and the dissociation energy of HeH$^+$).

A Mulliken population analysis can be obtained from the diagonal elements of **PS**. Such a population analysis associates 1.53 electrons with ϕ_1 and 0.47 electrons with ϕ_2. The net charge is then $+0.47$ on He and $+0.53$ on H. The formal charge of $+1$ is thus predicted to be divided more or less equally between the two atoms. A Löwdin population analysis can be obtained from the primed matrices, that is, the matrices associated with the orthonormal basis $\{\phi'_\mu\}$. The number of electrons associated with the hydrogen orbital is

$$P'_{22} = (S^{1/2}PS^{1/2})_{22} = 2(S^{1/2}C)^2_{12} = 0.5273 \tag{3.277}$$

This predicts a similar separation of charge, with net charges of $+0.53$ and $+0.47$ on the He and H, respectively. Note the reversal.

The total energy of HeH$^+$ is obtained by adding the nuclear repulsion $2/R$ to the electronic energy, to give -2.860662 a.u. From our basic integrals it is also possible to determine the energies of H, He$^+$, and He. The energy of the H atom in this basis is the same as that used in the H$_2$ calculation, i.e., $T_{22} + V^2_{22} = -0.4666$ a.u. The energy of the one-electron atom He$^+$ in the basis is similarly $T_{11} + V^1_{11} = -1.9755$ a.u. The He atom has two electrons in ϕ_1 and, in addition to the kinetic energy and attraction to the helium nucleus, the energy contains a contribution from the electron-electron repulsion of the two electrons. The He atom energy is thus $2(T_{11} + V^1_{11}) + (\phi_1\phi_1|\phi_1\phi_1)$. These energies are shown in Table 3.6 for comparison with exact results in Table 3.4.

From the results in Table 3.6 it is possible to calculate dissociation energies for the processes

$$\text{HeH}^+ \rightarrow \text{He} + \text{H}^+ \qquad \Delta E = 0.2168 \text{ a.u.} \tag{3.278}$$

$$\text{HeH}^+ \rightarrow \text{He}^+ + \text{H} \qquad \Delta E = 0.4168 \text{ a.u.} \tag{3.279}$$

The calculations correctly predict that HeH$^+$ will dissociate to the closed-shells He + H$^+$ rather than the open shells He$^+$ + H. The dissociation energy of 0.2168 a.u. = 5.90 eV is quite a bit larger than the correct value

Table 3.6 Energies for H and He species with the STO-3G basis set ($\zeta_1 = 2.0925$, $\zeta_2 = 1.24$)

Species	Energy (a.u.)
H^+	0.0
H	-0.466582
He^+	-1.975514
He	-2.643876
HeH^+ ($R = 1.4632$ a.u.)	-2.860662

of 2.04 eV. This is mainly because our He exponent of 2.0925, while reasonable for the HeH^+ molecule, is quite a bit larger than the best value of 1.6875 for the dissociation product, the He atom. The He energy is too high, relative to the HeH^+ energy.

Figure 3.8 shows the whole potential curve for our standard exponents, and also shows the essentially exact results of Wolniewicz. The calculated STO-3G equilibrium bond length is 1.3782 a.u., which is in fair agreement with the accurate results, even though the well depth is much too large. Unlike the difficulty encountered in H_2 (Fig. 3.5), the dissociation behavior

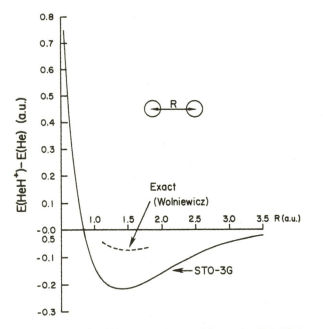

Figure 3.8 Restricted Hartree-Fock potential curve for STO-3G ($\zeta_{He} = 2.0925$, $\zeta_H = 1.24$)HeH^+ compared with the accurate results of Wolniewicz.

is correct since the products of dissociation are closed shells. We can look at this dissociation analytically as we did for H_2. As the bond length is stretched, the coefficient of ϕ_1 in ψ_1 increases and the coefficient of ϕ_2 in ψ_1 decreases. The electrons concentrate more and more around the He nucleus. In the limit, ψ_1 becomes just ϕ_1. The virtual orbital ψ_2, to be orthogonal to ψ_1, correspondingly becomes pure ϕ_2, that is

$$\mathbf{C}_{R\to\infty} = \begin{pmatrix} 1.0 & 0.0 \\ 0.0 & 1.0 \end{pmatrix} \tag{3.280}$$

The corresponding density matrix is

$$\mathbf{P}_{R\to\infty} = \begin{pmatrix} 2.0 & 0.0 \\ 0.0 & 0.0 \end{pmatrix} \tag{3.281}$$

The total electronic energy as $R \to \infty$ can be evaluated by setting all two-center integrals to zero. The only remaining integrals are T_{11}, T_{22}, V_{11}^1, V_{22}^2, $(\phi_1\phi_1|\phi_1\phi_1)$, and $(\phi_2\phi_2|\phi_2\phi_2)$.

Exercise 3.29 Use expression (3.184) for the electronic energy, expression (3.154) for the Fock matrix, and the asymptotic density matrix (3.281) to show that

$$E_0(R \to \infty) = 2T_{11} + 2V_{11}^1 + (\phi_1\phi_1|\phi_1\phi_1)$$

This is just the proper energy of the He atom, for the minimal basis, as discussed previously in the text.

The calculations on HeH^+ described here used a standard set of exponents. A better procedure would have been to optimize exponents at each internuclear distance. If we had done so, the He exponent ζ_1 would have decreased towards the best atom value of 1.6875 as we increased R. The value of the H exponent would have decreased to zero at large R (since in the products of dissociation both electrons reside on the He, the only way the H basis function can contribute at large R is by being extremely diffuse and extending over to the region of the He nucleus). Since there are only two exponents in this problem, it would have been reasonably easy to optimize exponents, for example, just by iteratively optimizing each one in sequence. In the general problem, with many orbital exponents, finding a minimum involves searches on a many-dimensional surface with possibly many local minima. One does not routinely optimize exponents in these larger problems.

Having described restricted closed-shell Hartree-Fock calculations in conjunction with the H_2 and HeH^+ model systems, we now want to illustrate the results of more realistic calculations on polyatomic molecules. We do this not to provide a review of current calculations, but rather to illustrate the main ideas behind all calculations of the closed-shell restricted Hartree-Fock type, and to provide some feeling and intuition for how well (or how badly) such calculations compare with experiment. We will restrict our calculations to H_2, N_2, and CO and molecules in the ten-electron series CH_4, NH_3, H_2O, and FH. For each of these we will use a hierarchy of well-defined basis sets. Prior to describing the results of such calculations, however, we need to discuss the general question of polyatomic basis sets and describe the specific basis sets we will be using.

3.6 POLYATOMIC BASIS SETS

There are probably as many basis sets defined for polyatomic calculations as there are quantum chemists. The choice of a basis set is not nearly the black art, however, that it may first appear. In our sample calculations we will use a reasonably well-defined hierarchy of basis sets starting with a minimal STO-3G basis and proceeding through the 4-31G basis, which effectively doubles the number of functions, the 6-31G* basis, which adds d-type functions to heavy atoms C, N, O, and F and finally, the 6-31G** basis, which, in addition to d-type functions for heavy atoms, adds p-type functions to hydrogen. By performing electronic structure calculations on a small variety of molecules, using this hierarchy of basis sets, it is possible to gain some insight into the size and characteristics of a basis set needed to obtain a given level of calculational accuracy.

The above basis sets have been introduced by Pople and collaborators (see Hehre *et al.* in Further Reading at the end of this chapter) and have been used extensively, by a number of workers for calculations on a large variety of molecules. Apart from a few instances, all the calculations in this book use the STO-3G, 4-31G, 6-31G*, and 6-31G** hierarchy of basis sets. By restricting our example calculations to a very limited set of molecules and the above basis sets, we are attempting to illustrate in a systematic way how specific attributes of a basis set affect calculated quantities. We are not attempting to provide a general review of current calculations. Such a review would be out of date very quickly. While the basis sets we use are not necessarily optimum, and may themselves be out of date shortly, they do have characteristics that can be used to illustrate all basis sets.

Our purpose in this section is to explicitly define the STO-3G, 4-31G, 6-31G*, and 6-31G** basis sets that we will be using in this and subsequent chapters. In the process, however, we will describe attributes that are characteristic of most of the basis sets that are in current use, and we will introduce some of the notation and some of the mechanics of defining and choosing a basis set. In particular, we first present a general treatment of contraction.

3.6.1 Contracted Gaussian Functions

In Subsection 3.5.1, when we defined the $1s$ STO-3G basis set of our model calculations, we indicated some of the ideas pertinent to the concept of contraction. We review these briefly. There are two main considerations in the choice of a basis. The first is that one desires to use the most efficient and accurate functions possible, in the sense that the expansion

$$\psi_i = \sum_{\mu=1}^{K} C_{\mu i} \phi_\mu \tag{3.282}$$

will require the fewest possible terms for an accurate representation of the molecular orbitals ψ_i. From this consideration, Slater functions are better than Gaussian functions. The second consideration in the choice of a basis set is the speed of two-electron integral evaluation. Here Gaussian functions have the advantage. By using a basis set of *contracted Gaussian functions* one can in a sense have one's cake and eat it too. In this procedure, one lets each basis function be a fixed linear combination (contraction) of Gaussian functions (primitives). Prior to a calculation one chooses the exponents of the primitives and the contraction coefficients so as to lead to basis functions with desired qualities. The contracted basis functions might be chosen to approximate Slater functions, Hartree-Fock atomic orbitals, or any other set of functions desired. Integrals involving such basis functions reduce to sums of integrals involving the primitive Gaussian functions. Even though many primitive integrals may need to be calculated for each basis function integral, the basis function integrals will be rapidly calculated provided the method of computing primitive integrals is very fast.

To avoid confusion with a multitude of ϕ's we will use the symbol g here for a normalized Gaussian function. A contraction thus has the form.

$$\phi_\mu^{\text{CGF}}(\mathbf{r} - \mathbf{R}_A) = \sum_{p=1}^{L} d_{p\mu} g_p(\alpha_{p\mu}, \mathbf{r} - \mathbf{R}_P) \tag{3.283}$$

where $\alpha_{p\mu}$ and $d_{p\mu}$ are the contraction exponents and coefficients and L is the length of the contraction. The normalized Gaussian primitive functions are of the $1s$, $2p$, $3d$, ... type,

$$g_{1s}(\alpha, \mathbf{r}) = (8\alpha^3/\pi^3)^{1/4} e^{-\alpha r^2} \tag{3.284}$$

$$g_{2p_x}(\alpha, \mathbf{r}) = (128\alpha^5/\pi^3)^{1/4} x e^{-\alpha r^2} \tag{3.285}$$

$$g_{3d_{xy}}(\alpha, \mathbf{r}) = (2048\alpha^7/\pi^3)^{1/4} xy e^{-\alpha r^2} \tag{3.286}$$

The simplifications that occur for integral evaluation using these functions do not appear for the $2s$, $3p$, ... Gaussians, and so any basis function of s symmetry, for example a $2s$ or $3s$ Slater function, will be expanded in only $1s$ Gaussians, with similar restrictions on the other symmetry types. The origins $\mathbf{R}_P$ of the primitives in (3.283) are almost always equal to $\mathbf{R}_A$. Different origins for the primitives in a contraction are used only with Gaussian lobe basis sets. In these basis sets one approximates s, p, d, ... functions as combinations of spherical $1s$ Gaussians (lobes) placed appropriately in space. For example, a $2p$ Gaussian orbital can be approximated as closely as desired by two $1s$ Gaussian lobes of opposite sign placed an infinitesimal distance apart. We will not be concerned with Gaussian lobes here.

A common way of determining contractions is from the results of atomic SCF calculations. In these atomic calculations one uses a relatively large basis of uncontracted Gaussians, optimizes all exponents, and determines the SCF coefficients of each of the derived atomic orbitals. The optimized exponents and SCF coefficients can then be used to derive suitable contraction exponents and contraction coefficients for a smaller basis set to be used in subsequent molecular calculations. Let us first illustrate this with s-type basis functions for hydrogen. Huzinaga[8] has determined coefficients and exponents of Gaussian expansions that minimize the energy of a hydrogen atom. With four Gaussian functions he obtains

$$\psi_{1s} = 0.50907g_{1s}(0.123317, \mathbf{r}) + 0.47449\dot{g}_{1s}(0.453757, \mathbf{r})$$
$$+ 0.13424g_{1s}(2.01330, \mathbf{r}) + 0.01906g_{1s}(13.3615, \mathbf{r}) \quad (3.287)$$

The basis set is an uncontracted basis consisting of four functions of s-type symmetry, i.e., it is a $(4s)$ basis. A contracted basis set derived from this would use the four Gaussian functions as primitives and contract them to reduce the number of basis functions. There are a number of ways the above four primitives might be contracted. One usually uses disjoint subsets of primitives so that no primitive appears in more than one basis function. From evidence on molecular calculations, it appears that a useful contraction scheme is one which leaves the most diffuse primitive uncontracted and contracts the remaining three primitives into one basis function, with contraction coefficients just equal to the above coefficients (SCF coefficients in the general case). That is,

$$\phi_1(\mathbf{r}) = g_{1s}(0.123317, \mathbf{r}) \quad (3.288)$$

$$\phi_2(\mathbf{r}) = N[0.47449g_{1s}(0.453757, \mathbf{r}) + 0.13424g_{1s}(2.01330, \mathbf{r})$$
$$+ 0.01906g_{1s}(13.3615, \mathbf{r})]$$
$$= 0.817238g_{1s}(0.453757, \mathbf{r}) + 0.231208g_{1s}(2.01330, \mathbf{r})$$
$$+ 0.032828g_{1s}(13.3615, \mathbf{r}) \quad (3.289)$$

In the last equation, the contraction coefficients have been properly renormalized. This scheme leads to a contracted basis set of two s-type functions, i.e., a *[2s] contracted basis set*, coming from a *(4s) uncontracted basis set*. This defines a $(4s)/[2s]$ contraction.

Huzinaga also determined relatively large uncontracted Gaussian $(9s5p)$ basis sets, with optimized exponents, for the first-row atoms Li to Ne. Dunning[9] has suggested useful contractions of these. As an example of the procedure, consider a $[3s2p]$ contracted basis for the oxygen atom. We are going to contract the nine primitives of s type into three basis functions. On inspecting the atomic SCF calculation we find that one of the nine primitives contributes strongly to both the $1s$ and $2s$ orbitals of the oxygen atom; this function is left uncontracted.

$$\phi_1(\mathbf{r}) = g_{1s}(9.5322, \mathbf{r}) \tag{3.290}$$

The two primitives, which are most diffuse, make a negligible contribution to the $1s$ atomic orbital but are the chief contributors to the $2s$ atomic orbital. They are contracted to give the second basis function,

$$\begin{aligned}
\phi_2(\mathbf{r}) &= N[0.59566g_{1s}(0.9398, \mathbf{r}) + 0.52576g_{1s}(0.2846, \mathbf{r})] \\
&= 0.563459g_{1s}(0.9398, \mathbf{r}) + 0.497338g_{1s}(0.2846, \mathbf{r}) \tag{3.291}
\end{aligned}$$

where 0.59566 and 0.52576 are the coefficients of these primitives in the $2s$ atomic orbital of the atomic SCF calculation. The last basis function consist of the remainder of the nine primitives,

$$\begin{aligned}
\phi_3(\mathbf{r}) &= N[0.14017g_{1s}(3.4136, \mathbf{r}) + 0.35555g_{1s}(27.1836, \mathbf{r}) \\
&\quad + 0.14389g_{1s}(81.1696, \mathbf{r}) + 0.04287g_{1s}(273.188, \mathbf{r}) \\
&\quad + 0.00897g_{1s}(1175.82, \mathbf{r}) + 0.00118g_{1s}(7816.54, \mathbf{r})] \\
&= 0.241205g_{1s}(3.4136, \mathbf{r}) + 0.611832g_{1s}(27.1836, \mathbf{r}) \\
&\quad + 0.247606g_{1s}(81.1696, \mathbf{r}) + 0.073771g_{1s}(273.188, \mathbf{r}) \\
&\quad + 0.015436g_{1s}(1175.82, \mathbf{r}) + 0.002031g_{1s}(7816.54, \mathbf{r}) \tag{3.292}
\end{aligned}$$

where 0.14017, 0.35555, etc, are the coefficients of these primitives in the $1s$ atomic orbital of the atomic SCF calculation.

In a similar way, the five primitives of p-type symmetry are contracted to two basis functions. Here the most diffuse p-function is left uncontracted,

$$\phi_1(\mathbf{r}) = g_{2p}(0.2137, \mathbf{r}) \tag{3.293}$$

and the remaining four primitives are contracted using the SCF coefficients of the $2p$ atomic orbital

$$\begin{aligned}
\phi_2(\mathbf{r}) &= N[0.49376g_{2p}(0.7171, \mathbf{r}) + 0.31066g_{2p}(2.3051, \mathbf{r}) \\
&\quad + 0.09774g_{2p}(7.9040, \mathbf{r}) + 0.01541g_{2p}(35.1832, \mathbf{r})] \\
&= 0.627375g_{2p}(0.71706, \mathbf{r}) + 0.394727g_{2p}(2.30512, \mathbf{r}) \\
&\quad + 0.124189g_{2p}(7.90403, \mathbf{r}) + 0.019580g_{2p}(35.1835, \mathbf{r}) \tag{3.294}
\end{aligned}$$

This $(9s5p)/[3s2p]$ contraction reduces the number of basis functions from 24 to 9. Remember the p_x, p_y, and p_z are included for each p orbital exponent. A calculation with either basis set would give almost identical results in a calculation on the oxygen atom, however. The loss of variational flexibility in molecular calculations is not extreme either. For example, a calculation on the water molecule using the fully uncontracted $(9s5p/4s)^{10}$ basis set gives an energy of -76.0133, whereas the $[3s2p/2s]$ contracted basis gives an energy of -76.0080, only 0.007% above the much larger calculation. Because the cost of an SCF calculation increases with the fourth power of the number of basis functions, the reduction from 32 functions to 13 functions is impressive.

3.6.2 Minimal Basis Sets: STO-3G

A minimal basis set is a relatively inexpensive one, which can be used for calculations on quite large molecules. It is minimal in the sense of having the least number of functions per atom required to describe the occupied atomic orbitals of that atom. This is not quite accurate, since one usually considers $1s$, $2s$ and $2p$, i.e., five functions, to constitute a minimal basis set for Li and Be, for example, even though the $2p$ orbital is not occupied in these atoms. The $2sp$ ($2s$ and $2p$), $3sp$, $4sp$, $3d$, ..., etc. shells are considered together. The minimal basis set thus consists of 1 function for H and He, 5 functions for Li to Ne, 9 functions for Na to Ar, 13 functions for K and Ca, 18 functions for Sc to Kr, ..., etc. Because the minimal basis set is so small, it is not one which can lead to quantitatively accurate results. It does, however, contain the essentials of chemical bonding and many useful qualitative results can be obtained with it.

Because of the small number of functions in a minimal basis set, it is particularly important that these functions be of near optimum form. This immediately rules out a single Gaussian function. One would prefer to use Slater functions or functions that closely resemble the known shape of atomic orbitals. A significant advance in minimal basis calculations came with the development of computer programs like "Gaussian 70," which could reproduce the results of minimal basis Slater orbital calculations using contracted Gaussian functions. The STO-LG method uses a contraction of L primitive Gaussians for each basis function, where the contraction coefficients and exponents are chosen so that the basis functions approximate Slater functions. We have already discussed the $1s$ STO-3G basis set in Subsection 3.5.1.

The calculations in this book are restricted to a small number of molecules, all of which include only the first row atoms up to fluorine Although the STO-LG method has been extended to second row atoms, we will only consider its formulation, and the formulation of the other basis sets which follow, for first row atoms and, in particular, for H, C, N, O, and F. We are therefore interested in the expansion of the $1s$, $2s$, and $2p$ Slater functions in a set of primitive Gaussians

$$\phi_{1s}^{CGF}(\zeta = 1.0) = \sum_{i=1}^{L} d_{i,1s} g_{1s}(\alpha_{i,1s}) \tag{3.295}$$

$$\phi_{2s}^{CGF}(\zeta = 1.0) = \sum_{i=1}^{L} d_{i,2s} g_{1s}(\alpha_{i,2sp}) \tag{3.296}$$

$$\phi_{2p}^{CGF}(\zeta = 1.0) = \sum_{i=1}^{L} d_{i,2p} g_{2p}(\alpha_{i,2sp}) \tag{3.297}$$

where the contraction coefficients (d's) and exponents (α's) are to be obtained by a least-squares fit which minimizes the integrals

$$\int d\mathbf{r} \, [\phi_{1s}^{SF}(\mathbf{r}) - \phi_{1s}^{CGF}(\mathbf{r})]^2$$

and

$$\int d\mathbf{r} \, [\phi_{2s}^{SF}(\mathbf{r}) - \phi_{2s}^{CGF}(\mathbf{r})]^2 + \int d\mathbf{r} \, [\phi_{2p}^{SF}(\mathbf{r}) - \phi_{2p}^{CGF}(\mathbf{r})]^2$$

One of the unique aspects of the STO-LG method and the fitting procedure is the sharing of contraction exponents in $2sp, 3sp, \dots$ shells. Thus the exponents in (3.296) and (3.297) are constrained to be identical and the $2s$ and $2p$ fits are performed simultaneously as indicated by the second integral above. The reason for this constraint is that if $2s$ and $2p$ functions have the same exponents, then they have the same radial behavior, and during the radial part of the integral evaluation they can be treated as one function. That is, all integrals involving any sp shell are treated together and one radial integration is sufficient for up to $256 \equiv 4^4$ separate integrals. This grouping of basis functions by shells with shared exponents leads to considerable efficiency in integral evaluation. The general STO-LG procedure uses contraction lengths up to $L = 6$. The longer the length of the contraction, however, the more time is spent in integral evaluation. It has been empirically determined that a contraction of length 3 is sufficient to lead to calculated properties that reproduce essentially all the valence features of a Slater calculation, and STO-3G has become the *de facto* standard for minimal basis calculations. Table 3.7 gives the STO-3G contraction exponents and coefficients of Eqs. (3.295) and (3.297). In the general notation, the STO-3G contraction is $(6s3p/3s)/[2s1p/1s]$.

Once the least-squares fits to Slater functions with orbital exponents $\zeta = 1.0$ (Table 3.7) are available, fits to Slater functions with other orbital exponents can be obtained by simply multiplying the α's in (3.295) to (3.297) by ζ^2. It remains to be determined what Slater orbital exponents ζ to use in electronic structure calculations. Two possibilities might be to use "best atom" exponents ($\zeta = 1.0$ for H, for example) or to optimize exponents in each calculation. The "best atom" exponents might be rather a poor choice

Table 3.7 STO-3G contraction exponents and coefficients for 1s, 2s, and 2p basis functions

α_{1s}	d_{1s}	α_{2sp}	d_{2s}	d_{2p}
0.109818	0.444635	0.0751386	0.700115	0.391957
0.405771	0.535328	0.231031	0.399513	0.607684
2.22766	0.154329	0.994203	−0.0999672	0.155916

Table 3.8 Standard STO-3G exponents

Atom	ζ_{1s}	ζ_{2sp}
H	1.24	——
Li	2.69	0.75
Be	3.68	1.10
B	4.68	1.45
C	5.67	1.72
N	6.67	1.95
O	7.66	2.25
F	8.65	2.55

for molecular environments, and optimization of nonlinear exponents is not practical for large molecules, where the dimension of the space to be searched is very large. A compromise is to use a set of standard exponents which are the average values of exponents optimized for a set of small molecules. The recommended STO-3G exponents are shown in Table 3.8.

The STO-LG basis is not the only possible minimal basis of course. Stewart,[11] for example, has determined fits of contracted Gaussian functions to individual Slater functions, without the constraint of sharing exponents in a shell. Rather than use Slater functions or fits to Slater functions, a reasonable choice is contracted basis functions which closely approximate the individually determined Hartree-Fock atomic orbitals of the atom. Calculations suggest, however, that Slater functions with near optimum exponents are better than these Hartree-Fock atomic orbitals for a minimal basis; orbitals in molecules may be rather different than those in the constituent atoms.

3.6.3 Double Zeta Basis Sets: 4-31G

A minimal basis set has rather limited variational flexibility particularly if exponents are not optimized. The first step in improving upon the minimal basis set involves using two functions for each of the minimal basis functions—a double zeta basis set. The best orbital exponents of the two functions are commonly slightly above and slightly below the optimal exponent of the minimal basis function. This allows effective expansion or "contraction" of the basis functions by variation of linear parameters rather than nonlinear exponents. The SCF procedure will weight either the coefficient of the dense or diffuse component according to whether the molecular environment requires the effective orbital to be expanded or "contracted." In addition, an extra degree of anisotropy is allowed relative to an STO-3G basis since, for example, *p* orbitals in different directions can have effectively different sizes.

The 4-31G basis set is not exactly a double zeta basis since only the valence functions are doubled and a single function is still used for each inner shell orbital. It may be termed a split valence shell basis set. The inner shells contribute little to most chemical properties and usually vary only slightly from molecule to molecule. Not splitting the inner shell functions has some effect on the total energy, but little effect on dipole moments, valence ionization potentials, charge densities, dissociation energies, and most other calculated quantities of chemical interest. The 4-31G basis thus consists of 2 functions for H and He, 9 functions for Li to Ne, 13 functions for Na to Ar, . . . , etc. For hydrogen the contractions are

$$\phi'_{1s}(\mathbf{r}) = \sum_{i=1}^{3} d'_{i,1s} g_{1s}(\alpha'_{i,1s}, \mathbf{r}) \tag{3.298}$$

$$\phi''_{1s}(\mathbf{r}) = g_{1s}(\alpha''_{1s}, \mathbf{r}) \tag{3.299}$$

The outer hydrogen function ϕ''_{1s} is uncontracted and the inner hydrogen function ϕ'_{1s} is a contraction of three primitive Gaussians. Apart from small numerical differences in deriving the contraction coefficients and exponents, the above basis functions are identical to the $(4s)/[2s]$ functions of (3.288) and (3.289). That is, the 4-31G basis is not fit to any particular functional form but is derived by choosing the form of the contraction and then minimizing the energy of an atomic calculation by varying the contraction coefficients and exponents. The 4-31G acronym implies that the valence basis functions are contractions of three primitive Gaussians (the inner function) and one primitive Gaussian (the outer function), whereas the inner shell functions are contractions of four primitive Gaussians. Hydrogen, of course, does not have inner shells.

For the atoms Li to F, the contractions are

$$\phi_{1s}(\mathbf{r}) = \sum_{i=1}^{4} d_{i,1s} g_{1s}(\alpha_{i,1s}, \mathbf{r}) \tag{3.300}$$

$$\phi'_{2s}(\mathbf{r}) = \sum_{i=1}^{3} d'_{i,2s} g_{1s}(\alpha'_{i,2sp}, \mathbf{r}) \tag{3.301}$$

$$\phi''_{2s}(\mathbf{r}) = g_{1s}(\alpha''_{2sp}, \mathbf{r}) \tag{3.302}$$

$$\phi'_{2p}(\mathbf{r}) = \sum_{i=1}^{3} d'_{i,2p} g_{2p}(\alpha'_{i,2sp}, \mathbf{r}) \tag{3.303}$$

$$\phi''_{2p}(\mathbf{r}) = g_{2p}(\alpha''_{2sp}, \mathbf{r}) \tag{3.304}$$

As in the STO-3G basis, the $2s$ and $2p$ functions share exponents for computational efficiency. Given the above functional forms, the contraction coefficients d_{1s}, d'_{2s}, d''_{2s}, d'_{2p}, and d''_{2p} and the contraction exponents α_{1s}, α'_{2sp}, and α''_{2sp} were explicitly varied until the energy of an atomic SCF calculation reached a minimum. Unlike the STO-3G basis, which was

Table 3.9 Standard 4-31G valence shell scale factors

Atom	ζ'	ζ''
H	1.20	1.15
C	1.00	1.04
N	0.99	0.98
O	0.99	0.98
F	1.00	1.00

obtained by a least-squares fit to known functions, or a general contraction scheme based on contraction of previously determined uncontracted atomic calculations, the 4-31G basis sets were determined by choosing the specific form (3.300) to (3.304) for the contractions and then optimizing all contraction parameters. That is, the basis set was obtained by contraction first, then optimization, as opposed to optimization first, then contraction. In our general notation, the 4-31G contraction is written as $(8s4p/4s)/[3s2p/2s]$. The basis consists of inner shell functions, inner valence functions, and outer valence functions. These are contractions of 4, 3, and 1 primitive functions, respectively.

Since the basis set is obtained from atomic calculations, it is still desirable to scale exponents for the molecular environment. This is accomplished by defining an inner valence scale factor ζ' and an outer valence scale factor ζ'' and multiplying the corresponding inner and outer α's by the square of these factors. Only the valence shells are scaled. Table 3.9 gives a set of standard 4-31G scale factors. Only those for H differ significantly from unity, although the outer carbon functions are somewhat denser than in the atom.

Exercise 3.30 A 4-31G basis for He has not been officially defined. Huzinaga,[8] however, in an SCF calculation on the He atom using four uncontracted $1s$ Gaussians, found the coefficients and optimum exponents of the normalized $1s$ orbital of He to be

α_μ	$C_{\mu i}$
0.298073	0.51380
1.242567	0.46954
5.782948	0.15457
38.47497	0.02373

Use the expression for overlaps given in Appendix A to derive the contraction parameters for a 4-31G He basis set.

3.6.4 Polarized Basis Sets: 6-31G* and 6-31G**

The next step in improving a basis set could be to go to triple zeta, quadruple zeta, etc. If one goes in this direction rather than adding functions of higher angular quantum number, the basis set would not be well balanced. In the limit of a large number of only s and p functions, one finds, for example, that the equilibrium geometry of ammonia actually becomes planar. The next step beyond double zeta usually involves adding *polarization functions*, i.e., adding d-type functions to the first row atoms Li-F and p-type functions to H. To see why these are called polarization functions, consider the hydrogen atom. The exact wave function for an isolated hydrogen atom is just the $1s$ orbital. If the hydrogen atom is placed in a uniform electric field, however, the electron cloud is attracted to the direction of the electric field, and the charge distribution about the nucleus becomes asymmetric. It is polarized. The lowest order solution to this problem is a mixture of the original $1s$ orbital and a p-type function, i.e., the solution can be considered to be a hybridized orbital. A hydrogen atom in a molecule experiences a similar, but nonuniform, electric field arising from its nonspherical environment. By adding polarization functions, i.e., p-type functions, to a basis for H we directly accommodate this effect. In a similar way, d-type functions, which are not occupied in first row atoms, play the role of polarization functions for the atoms Li to F. The 6-31G* and 6-31G** basis sets closely resemble the 4-31G basis set with d-type basis functions added to the heavy atoms (*) or d-type functions added to the heavy atoms, *and* p-type functions added to hydrogen (**). It has been empirically determined that adding polarization functions to the heavy atoms is more important than adding polarization functions to hydrogen. The hierarchy of our basis sets is thus STO-3G, 4-31G, 6-31G*, and 6-31G**.

The 6-31G* and 6-31G** basis sets are formed by adding polarization function to a 6-31G basis. The form of the 6-31G contractions are identical to those of the 4-31G basis, except that the inner shell functions ($1s$ only, for Li to F) become a contraction of six primitive Gaussians rather than four. The 6-31G optimization was performed from the beginning and so the valence functions are not identical to those of the 4-31G basis, but are very similar. The 6-31G and 4-31G basis sets give almost identical results for valence properties although the 6-31G basis gives lower energies, because of the improvement in the inner shell.

The d-type functions that are added to a 6-31G basis to form a 6-31G* basis are a single set of uncontracted $3d$ primitive Gaussians. For computational convenience there are "six $3d$ functions" per atom—$3d_{xx}$, $3d_{yy}$, $3d_{zz}$, $3d_{xy}$, $3d_{yz}$, and $3d_{zx}$. These six, the Cartesian Gaussians, are linear combinations of the usual five $3d$ functions—$3d_{xy}$, $3d_{x^2-y^2}$, $3d_{yz}$, $3d_{zx}$, and $3d_{z^2}$ and a $3s$ function $(x^2 + y^2 + z^2)$. The 6-31G* basis, in addition to adding polarization functions to a 6-31G basis, thus includes one more function of

s-type symmetry. The contraction is thus $(11s4p1d/4s)/[4s2p1d/2s]$ and the basis set includes 2 functions for H and 15 functions for Li to F. A standard Gaussian exponent for the six $3d$ functions of $\alpha = 0.8$ has been suggested for C, N, O, and F.

The 6-31G** basis differs from the 6-31G* basis by the addition of one set of uncontracted p-type Gaussian primitives for each H. A standard Gaussian exponent of $\alpha = 1.1$ has been suggested for these functions. The 6-31G** contraction is thus $(11s4p1d/4s1p)[4s2p1d/2s1p]$ and each hydrogen now includes five basis functions.

Exercise 3.31 Determine the total number of basis functions for STO-3G, 4-31G, 6-31G*, and 6-31G** calculations on benzene.

Calculations at the 6-31G* and 6-31G** level provide, in many cases, quantitative results considerably superior to those at the lower STO-3G and 4-31G levels. Even these basis sets, however, have deficiencies that can only be remedied by going to triple zeta or quadruple zeta, adding more than one set of polarization functions, adding f-type functions to heavy atoms and d-type functions to hydrogen, improving the basis function description of the inner shell electrons, etc. As technology improves it will be possible to use more and more accurate basis sets.

3.7 SOME ILLUSTRATIVE CLOSED-SHELL CALCULATIONS

In this section, we illustrate results that are characteristic of Hartree-Fock calculations on the ground state of closed-shell molecules. Now that we have discussed polyatomic basis sets and the closed-shell restricted Hartree-Fock procedure, we are in a position to appreciate the results and the methodology of sample SCF calculations. The results of an extremely large number of SCF calculations are now available in the literature; we make no attempt to review these calculations. Instead, we apply a well-defined hierarchy of basis sets to a small set of "typical" molecules, and use these calculations to illustrate the order of accuracy expected in the general SCF calculation. By restricting our calculations to a few well-defined basis sets and a small set of molecules, we will be able to apply the various methods of later chapters, which go beyond the Hartree-Fock approximation, to the same collection of basis sets and molecules. In this way, we hope to give a more systematic illustration of the results obtained from the many computational methods of quantum chemistry than would be possible by simply reviewing selected results available in the literature. Thus our purpose here, in addition to illustrating SCF results themselves, is to display Hartree-Fock values of a small number of calculated quantities, for comparison with

Table 3.10 Standard geometries used in calculations

Molecule	Bond length (a.u.)	Bond angle
H_2	1.400	
CO	2.132	
N_2	2.074	
CH_4	2.050	109.47°
NH_3	1.913	106.67°
H_2O	1.809	104.52°
FH	1.733	

better values obtained in later chapters. In some cases, the Hartree-Fock results of this section are even qualitatively wrong, but, as we shall see later, these errors are corrected by including the effects of correlation.

The molecules we will use henceforth are H_2, isoelectronic N_2 and CO, and the ten-electron series, CH_4, NH_3, H_2O, and FH. The standard geometries at which all calculations, unless otherwise indicated, have been carried out are shown in Table 3.10. These "experimental" values are close but not always identical to values obtained in the "best" or most recent structure determination. The small set of molecules we have chosen cannot, of course, illustrate the total wealth of chemistry being approached by *ab initio* calculations. They do illustrate, however, some of the interesting quantities that can be derived from an SCF calculation. When we discuss open-shell calculations in the next section, we will introduce a few additional molecules. For the most part, however, the illustrative calculations in this book are performed on the molecules of Table 3.10.

3.7.1 Total Energies

Perhaps the primary quantity available in any *ab initio* calculation is the total energy. The total energy is the electronic energy (the output of the quantum mechanical calculation) plus the classical nuclear repulsion energy. In the SCF approximation, the electronic energy is variational and the "better" the basis set, the lower is the total energy. As the basis set becomes more and more complete the total energy approaches the Hartree-Fock limit. This limit can sometimes be estimated from large basis set calculations. By the variational principle, the Hartree-Fock-limit energy is still above the "exact" energy, which here can be taken as the energy obtained from an exact solution to the nonrelativistic Schrödinger equation in the Born-Oppenheimer approximation. In very accurate calculations on the atoms He, Be, etc., proper account must be taken of relativistic and Born-Oppenheimer corrections when comparing these "exact" energies with experimental ones. For most purposes in quantum chemistry these corrections can be assumed to be negligible and "exact" results equated to experimental results.

Tables 3.11 to 3.13 show the total energies obtained for the molecules of Table 3.10 using the four basis sets, STO-3G, 4-31G, 6-31G*, and 6-31G**. H_2 has no inner shells, or heavy atoms for d-type polarization functions, so the 6-31G* basis set is equivalent to the 4-31G basis set for this molecule. Similarly, N_2 and CO have no hydrogen atoms to add p-type polarization

Table 3.11 SCF total energies (a.u.) of H_2 with the standard basis sets

Basis set	Energy
STO-3G	−1.117
4-31G	−1.127
6-31G**	−1.131
HF-limit[a]	−1.134

[a] J. M. Schulman and D. N. Kaufman, *J. Chem. Phys.* **53**: 477 (1970).

Table 3.12 SCF total energies (a.u.) of N_2 and CO with the standard basis sets

Basis set	N_2	CO
STO-3G	−107.496	−111.225
4-31G	−108.754	−112.552
6-31G*	−108.942	−112.737
HF-limit[a]	−108.997	−112.791

[a] P. C. Hariharan and J. A. Pople, *Theoret. Chim. Acta* **28**: 213 (1973).

Table 3.13 SCF total energies (a.u.) for the ten-electron series with the standard basis sets

Basis set	CH_4	NH_3	H_2O	FH
STO-3G	−39.727	−55.454	−74.963	−98.571
4-31G	−40.140	−56.102	−75.907	−99.887
6-31G*	−40.195	−56.184	−76.011	−100.003
6-31G**	−40.202	−56.195	−76.023	−100.011
HF-limit[a]	−40.225	−56.225	−76.065	−100.071

[a] P. C. Hariharan and J. A. Pople, *Theoret. Chim. Acta* **28**: 213 (1973).

functions to, so the 6-31G** basis set is equivalent to the 6-31G* basis set. These absolute energies in themselves are rather uninteresting; chemical energetics is concerned with energy differences not absolute energies.

Exercise 3.32 Use the results of Tables 3.11 to 3.13 to calculate, for each basis set and at the Hartree-Fock limit, the energy difference for the following two reactions,

$$N_2 + 3H_2 \rightarrow 2NH_3 \qquad \Delta E = ?$$

$$CO + 3H_2 \rightarrow CH_4 + H_2O \qquad \Delta E = ?$$

Are the results consistent for different basis sets? Does Hartree-Fock theory predict these reactions to be exoergic or endoergic? The experimental hydrogenation energies (heats of reaction $\Delta H°$) at zero degrees Kelvin are -18.604 kcal mol^{-1} (N_2) and -45.894 kcal mol^{-1} (CO), with 1 a.u. of energy equivalent to 627.51 kcal mol^{-1}.

Differences in the zero-point vibrational energies of reactants and products also contribute to reaction energies. From the experimental vibrational spectra, the $3N$-6 (or $3N$-5) zero-point energies ($hv_0/2$) for the relevant molecules (with degeneracies in parenthesis) are:

Molecule	$hv_0/2$ (kcal mol^{-1})
H_2	6.18
N_2	3.35
CO	3.08
H_2O	2.28
	5.13
	5.33
NH_3	1.35
	2.32(2)
	4.77
	4.85(2)
CH_4	1.86(3)
	2.17(2)
	4.14
	4.2(3)

Calculate the contribution of zero-point vibrations to the energy of the above two reactions. Is it a reasonable approximation to neglect the effect of zero-point vibrations?

Unfortunately, energy differences satisfy no variational principle, and it is often difficult to estimate the error in an energy difference. Provided equivalent basis sets are used for each species, the error in an energy difference will be, however, much less than the error in the corresponding absolute energies. As the last exercise shows, the SCF approximation often gives valid qualitative results for an energy change, even the energy change involved in a chemical reaction. In the general case, however, some estimate of the changes in correlation energies will be necessary for valid quantitative results.

3.7.2 Ionization Potentials

Koopmans' theorem provides the theoretical justification for interpreting Hartree-Fock orbital energies as ionization potentials and electron affinities. For the series of molecules we are using, the lowest virtual orbital always has a positive orbital energy, and thus Hartree-Fock theory predicts that none of these molecules will bind an electron to form a negative ion. Hartree-Fock almost always provides a very poor description of the electron affinity, and we will not consider the energies of virtual orbitals further.

The occupied orbital energies, on the other hand, commonly provide a reasonable first description of ionization potentials. Except for the interesting case of N_2, Koopmans' ionization potentials for our series of molecules are in reasonable agreement with experiment.

The molecule H_2 has only one occupied orbital. The negative of the energy of this occupied orbital for the various basis sets is shown in Table 3.14, and all the H_2 orbitals are shown in Fig. 3.9. A small change in the occupied orbital energy is observed in going beyond the minimal basis set, but beyond the minimal basis the orbital energy remains fixed at -0.595 Hartrees. The predicted ionization potential of $+0.595$ Hartrees is only in error by $\sim 2\%$. In Table 3.14, as well as in subsequent tables, all ionization potentials are vertical rather than adiabatic. A vertical transition is one in which the final state has the same nuclear geometry as the initial state, rather than its own equilibrium nuclear geometry (adiabatic transition). The excellent agreement between the Koopmans' value and the experimental value arises because of fortuitous cancellation of the correlation and relaxation effects, which are neglected in the Koopmans' approximation. Correlation has no effect on

Table 3.14 Ionization potential (a.u.) of H_2 obtained via Koopmans' theorem

Basic set	Ionization potential
STO-3G	0.578
4-31G	0.596
6-31G**	0.595
Near-HF-limit[a]	0.595
Experiment	0.584

[a] J. M. Schulman and D. M. Kaufman, *J. Chem. Phys.* **53**: 477 (1970).

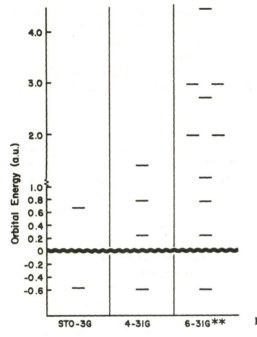

Figure 3.9 Orbital energies of H_2.

the final one-electron H_2^+ but lowers the energy of the initial H_2 state. Relaxation, on the other hand, lowers the energy of the final H_2^+ state. These two effects very nearly cancel in this example.

Table 3.15 shows the first two Koopmans' ionization potentials of CO. The highest occupied molecular orbitals of this molecule are the bonding 5σ and 1π orbitals formed mainly from linear combinations of the $2p$ orbitals

Table 3.15 The first two ionization potentials (a.u.) of CO obtained via Koopmans' theorem

Basis set	Ion symmetry	
	$^2\Sigma$	$^2\Pi$
STO-3G	0.446	0.551
4-31G	0.549	0.640
6-31G*	0.548	0.633
Near-HF-limit[a]	0.550	0.640
Experiment	0.510	0.620

[a] W. M. Huo, *J. Chem. Phys.* **43:** 624 (1965).

of the individual C and O atoms. Ionization of an electron from the 5σ orbital leads to an ion of $^2\Sigma$ symmetry while ionization of an electron from the 1π orbital leads to an ion of $^2\Pi$ symmetry. A primary question in the ionization of CO, as well as in the ionization of isoelectronic N_2, is whether the first ionization removes an electron from the 5σ or the 1π orbital. In the Hartree-Fock approximation, the equivalent question is whether the 5σ or 1π orbital is the highest lying occupied orbital.

For CO, the calculations agree with experiment in predicting that the 5σ orbital lies above the 1π orbital. The usual argument, which rationalizes this result, is that, while $2p\sigma$ orbitals on C and O interact more strongly than the corresponding $2p\pi$ orbitals on C and O and a bonding σ orbital would normally lie below a corresponding bonding π orbital, the 5σ orbital is "pushed" up by interaction with the lower 4σ antibonding orbital formed from $2s$ orbitals on C and O. In any event, the results of *ab initio* SCF calculations are in good agreement with experiment.

The molecule N_2 is isoelectronic with CO and has a similar orbital structure. Unlike CO, however, a fundamental problem arises in using Koopmans' theorem to interpret its ionization spectra. Table 3.16 compares calculated Koopmans' ionization potentials for N_2 with experiment, and Fig. 3.10 shows the calculated orbital energies. The first point to notice is that the STO-3G calculation is not in agreement with calculations using better basis sets and calculations at the Hartree-Fock limit. The STO-3G calculation predicts the $3\sigma_g$ orbital to be higher lying than the $1\pi_u$ orbital, whereas, the "correct" Hartree-Fock result is that the $1\pi_u$ orbital is the highest occupied orbital. As Fig. 3.10 shows, in homonuclear N_2, unlike heteronuclear CO, the $3\sigma_g(5\sigma$ in CO) orbital has a different symmetry than the $2\sigma_u(4\sigma$ in CO) orbital and, hence, the interaction which pushes these two orbitals apart is missing in N_2, but present in CO. This argument can

Table 3.16 The first two ionization potentials (a.u.) of N_2 obtained via Koopmans' theorem

	Ion symmetry	
Basis set	$^2\Sigma$	$^2\Pi$
STO-3G	0.540	0.573
4-31G	0.629	0.621
6-31G*	0.630	0.612
Near-HF-limit[a]	0.635	0.616
Experiment	0.573	0.624

[a] P. E. Cade, K. D. Sales, and A. C. Wahl, *J. Chem. Phys.* **44**: 1973 (1966).

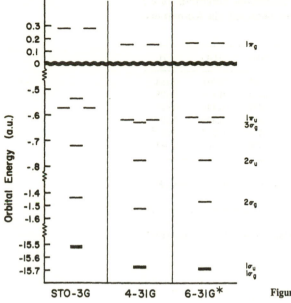

Figure 3.10 Orbital energies of N_2.

be used to rationalize why Hartree-Fock calculations predict the highest occupied orbital to be of π symmetry in N_2, but of σ symmetry in CO.

The second and most important point of Table 3.16 is that the "correct" Hartree-Fock results are in qualitative disagreement with experiment. In the molecular orbital Hartree-Fock model, the $1\pi_u$ orbital is the highest occupied orbital, yet the lowest experimental ionization potential corresponds to the production of an ion with Σ_g symmetry. This implies a breakdown of the simple orbital picture of ionization. The Hartree-Fock picture is an approximation. For the case of N_2 this approximation is not sufficiently accurate for even a qualitative understanding of the ionization phenomena. As we shall see in Chapters 4 and 7, when the single determinant Hartree-Fock model is replaced by a multideterminantal model, with its associated inclusion of correlation effects, theoretical calculations and experiment ultimately agree on the ionization spectra of N_2.

Table 3.17 shows the calculated and experimental results for the first ionization potential of molecules in the ten-electron series, CH_4, NH_3, H_2O, and FH. The largest basis set gives ionization potentials that are all slightly larger than experimental values, with the agreement becoming slightly worse as one moves to the right in the periodic table. The correct ordering FH > CH_4 > H_2O > NH_3 is reproduced, except with the minimal basis set. Figure 3.11 shows all the occupied orbitals and the first virtual orbital for this ten-electron series, using the largest 6-31G** basis set. The lowest

Table 3.17 The lowest ionization potentials (a.u.) of the ten-electron series obtained via Koopmans' theorem

Basis set	CH_4	NH_3	H_2O	FH
STO-3G	0.518	0.353	0.391	0.464
4-31G	0.543	0.414	0.500	0.628
6-31G*	0.545	0.421	0.498	0.628
6-31G**	0.543	0.421	0.497	0.627
Near-HF-limit	0.546[a]	0.428[b]	0.507[c]	0.650[d]
Experiment	0.529	0.400	0.463	0.581

[a] W. Meyer, *J. Chem. Phys.* **58:** 1017 (1973).

[b] A. Rauk, L. C. Allen, and E. Clementi, *J. Chem. Phys.* **52:** 4133 (1970).

[c] B. J. Rosenberg and I. Shavitt, *J. Chem. Phys.* **63:** 2162 (1975).

[d] P. E. Cade and W. M. Huo, *J. Chem. Phys.* **47:** 614 (1967).

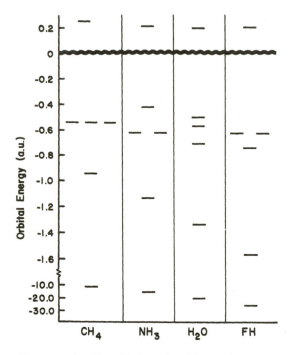

Figure 3.11 6-31G** orbital energies of the ten-electron series.

molecular orbital is essentially the $1s$ inner-shell atomic orbital of the heavy atom. The second molecular orbital is comprised mainly of the $2s$ orbital of the heavy atom, particularly as one moves to the right in the periodic table. In FH this molecular orbital is very atomic in nature and has much of the character of an inner shell. The average energy of the three highest occupied orbitals also decreases slightly as one moves to the right in the periodic table, but the individual energies of these three orbitals are determined by the symmetry of the system. That is, the reason CH_4 does not have the lowest ionization potential, in line with the trend in the rest of the series, is that the tetrahedral symmetry causes these three orbitals to be degenerate. If CH_4 was distorted from its tetrahedral symmetry, its first ionization potential would decrease.

Figure 3.12 shows all the orbital energies of H_2O, using our four standard basis sets. In this as in other examples, the occupied orbital spectrum is nearly invariant to the basis set once the basis set is of double zeta quality or better. The minimal basis set leads to ionization potentials that are significantly different from those given by better basis sets.

The Koopmans' approximation to ionization potentials provides a valuable qualitative tool for interpreting and making assignments to experimental spectra. Like any other Hartree-Fock result, however, it is not

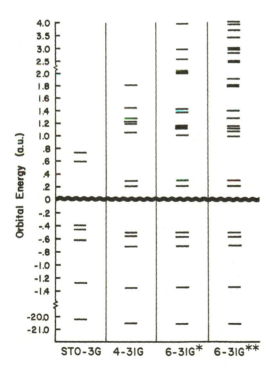

Figure 3.12 Occupied and virtual orbital energies of H_2O.

truly quantitative and, in some cases, it may even give an incorrect qualitative picture.

3.7.3 Equilibrium Geometries

Perhaps the most common use of electronic structure calculations is to predict the equilibrium geometry of molecules. In the Born-Oppenheimer approximation, the total energy, as a function of the coordinates of the nuclei, defines a potential surface. The motion of the nuclei on this potential surface defines the possible chemical reactions, the molecular vibrations, etc. The points on the potential surface of most immediate interest are the stationary points. The saddle points define transition states, and the minima define equilibrium geometries. Although one would like to know the complete details of the potential surface, this quickly becomes an impossible task as the number of nuclei and the 3N-6 degrees of freedom increase. For moderate-sized molecules, however, it may still be possible to find the potential surface minimum. This constitutes a prediction of chemical structure. In cases where the number of degrees of freedom is very large, it may be sufficient to fix certain geometric variables at their standard or expected values and optimize only the geometric variables for which little *a priori* information is available.

Since one cannot, in general, obtain a potential surface which is close to the exact Born-Oppenheimer surface, the best that can be hoped for is a potential surface which is closely parallel to the exact surface. As our discussion of minimal basis H_2 (c.f. Fig. (3.5)) has shown, the potential surface obtained from a closed shell restricted Hartree-Fock calculation will not be parallel to, nor even qualitatively resemble, the exact potential surface in regions of the surface characterizing the stretching and breaking of a bond if, as is the usual case, the dissociation products have open shells. The restricted closed-shell Hartree-Fock procedure is thus inappropriate for the general exploration of a potential surface. In most cases, however, the restricted Hartree-Fock potential surface is "reasonably" parallel to the exact surface near the region of a minimum. That is, restricted Hartree-Fock predictions of equilibrium geometries provide reasonably valid approximations to experimental values.

The problem of finding an equilibrium geometry is equivalent to the mathematical problem of nonlinear unconstrained minimization. Historically, there have been a number of different methods used for such minimizations. An inefficient, but conceptually simple, procedure is the line search. Here one varies only one variable at a time until a minimum for that variable is obtained. One cycles through all of the variables a number of times, varying each one in sequence, until the optimum values no longer change. If there is a large coupling between variables, this procedure may converge

Table 3.18 SCF equilibrium bond lengths (a.u.) of H_2

Basis set	Bond length
STO-3G	1.346
4-31G	1.380
6-31G**	1.385
Experiment	1.401

very slowly. Other procedures depend on knowing the first derivatives and possibly the second derivatives of the energy with respect to the nuclear coordinates. These are much better procedures but require an evaluation of a number of derivatives. In the past these derivatives have been calculated numerically, but there are now a number of programs that can calculate these derivatives by efficient analytical procedures and use them to generate equilibrium geometries automatically. Appendix C discusses the basic ideas behind these important developments.

Table 3.18 gives the calculated equilibrium bond lengths of H_2 with our standard basis sets. The calculated bond length of 1.385 a.u. obtained from using the 6-31G** basis set is close to the Hartree-Fock limit value. The error of 0.016 a.u. ($\sim 1\%$) in the bond length is not uncharacteristic of what can be expected from a good *ab initio* SCF calculation, although the average absolute error is more commonly $\sim 0.02-0.04$ a.u.

Table 3.19 gives the calculated bond lengths for CO and N_2 and Table 3.20 the calculated bond lengths for CH_4, NH_3, H_2O, and FH. The error at the Hartree-Fock limit is somewhat larger for a distance between two heavy atoms than it is for an X-H distance. In these, as in most other molecules, bond lengths predicted by Hartree-Fock limit calculations are too

Table 3.19 SCF equilibrium bond lengths (a.u.) of N_2 and CO

Basis set	N_2	CO
STO-3G	2.143	2.166
4-31G	2.050	2.132
6-31G*	2.039	2.105
Near-HF-limit	2.013[a]	2.081[b]
Experiment	2.074	2.132

[a] P. E. Cade, K. D. Sales, and A. C. Wahl, *J. Chem. Phys.* **44**: 1973 (1966).

[b] W. M. Huo, *J. Chem. Phys.* **43**: 624 (1965).

Table 3.20 SCF equilibrium bond lengths (a.u.) of the ten-electron series

Basis set	CH_4	NH_3	H_2O	FH
STO-3G	2.047	1.952	1.871	1.807
4-31G	2.043	1.873	1.797	1.742
6-31G*	2.048	1.897	1.791	1.722
6-31G**	2.048	1.897	1.782	1.703
Near-HF-limit	2.048[a]	1.890[b]	1.776[c]	1.696[d]
Experiment	2.050	1.912	1.809	1.733

[a] W. Meyer, *J. Chem. Phys.* **58**: 1017 (1973).

[b] A. Rauk, L. C. Allen, and E. Clementi, *J. Chem. Phys.* **52**: 4133 (1970).

[c] B. J. Rosenberg, W. C. Ermler, and I. Shavitt, *J. Chem. Phys.* **65**: 4072 (1976).

[d] P. E. Cade and W. J. Huo, *J. Chem. Phys.* **47**: 614 (1967).

short. The average error in a bond length predicted by poorer basis sets, such as the minimal STO-3G basis set, is larger than for basis sets at the Hartree-Fock limit, and is commonly ~0.05–0.10 a.u.

Values for the calculated bond angles in NH_3 and H_2O are shown in Table 3.21. There is quite reasonable agreement (~1°–2°) at the Hartree-Fock limit. These bond angles are not particularly good at the double zeta (4-31G) level, and it appears that d-type functions in the basis set are necessary for a quantitative description of the angle in these molecules. If the limit is taken of adding only s- and p-type functions to a basis set for NH_3, the predicted geometry becomes planar! This illustrates the necessity of a balanced basis set.

Table 3.21 SCF equilibrium bond angles for NH_3 and H_2O

Basis set	NH_3	H_2O
STO-3G	104.2	100.0
4-31G	115.8	111.2
6-31G*	107.5	105.5
6-31G**	107.6	106.0
Near-HF-limit	107.2[a]	106.1[b]
Experiment	106.7	104.5

[a] A. Rauk, L. C. Allen, and E. Clementi, *J. Chem. Phys.* **52**: 4133 (1970).

[b] B. J. Rosenberg, W. C. Ermler, and I. Shavitt, *J. Chem. Phys.* **65**: 4072 (1976).

As with any other calculated quantity, the predictions of equilibrium geometries obtained from the *ab initio* SCF procedure cannot provide truly quantitative agreement with experiment. Nevertheless, such geometries almost always give the correct trends when comparing a series of related molecules; the *a priori* prediction of chemical structure has been one of the most successful aspects of Hartree-Fock calculations. The book by Hehre *et al.* (see Further Reading) critically evaluates the performance of SCF calculations in predicting equilibrium geometries, with many examples.

3.7.4 Population Analysis and Dipole Moments

An *ab initio* SCF calculation produces a one-electron charge density $\rho(\mathbf{r})$ describing the probability of finding an electron, i.e.,

$$\rho(\mathbf{r}) = \sum_{\mu} \sum_{\nu} P_{\mu\nu} \phi_{\mu}(\mathbf{r}) \phi_{\nu}^{*}(\mathbf{r}) \qquad (3.305)$$

This charge density is commonly plotted as a contour map for visual interpretation of the charge density. Alternatively, one would like to have more quantitative characterizations of the charge density. One way of doing this is to calculate the moments of the charge—dipole moment, quadrupole moment, etc. In addition, chemists would like to ascribe portions of the charge to specific atoms in line with their intuitive notions. As we have described previously, there is no rigorous way of doing this. Nevertheless, a population analysis may sometimes be useful for interpretive purposes. As an illustration of such an analysis, Tables 3.22 and 3.23 contain the net positive charge on each of the hydrogens in the 10-electron series, from either a Mulliken or a Löwdin population analysis. In agreement with standard electronegativity arguments, the hydrogen atom becomes more positively charged as one goes to the right in the periodic table. Very little can be said about the absolute magnitude of these charges, however. It is particularly dangerous to compare numbers for different basis sets. For example, a 6-31G* calculation on CH_4 compared to an STO-3G calculation

Table 3.22 A Mulliken SCF population analysis for the ten-electron series. The entries are the net charges on the hydrogens

Basis set	CH_4	NH_3	H_2O	FH
STO-3G	0.06	0.16	0.18	0.21
4-31G	0.15	0.30	0.39	0.48
6-31G*	0.16	0.33	0.43	0.52
6-31G**	0.12	0.26	0.34	0.40

Table 3.23 A Löwdin SCF population analysis for the ten-electron series. The entries are the net charges on the hydrogens

Basis set	CH_4	NH_3	H_2O	FH
STO-3G	0.03	0.10	0.13	0.15
4-31G	0.10	0.20	0.28	0.36
6-31G*	0.16	0.27	0.36	0.45
6-31G**	0.11	0.18	0.23	0.27

on NH_3 would predict the CH bond in methane to be more polar than the NH bond in ammonia. Since one adds orbitals only to hydrogen atoms in going from the 6-31G* to 6-31G** basis set, the 6-31G** basis set always assigns more electrons (less positive charge) to the hydrogens than does the 6-31G* basis set. In spite of problems, a population analysis can be a useful interpretive device when used properly.

Calculations of the dipole moment of CO are shown in Table 3.24. This particular calculation has had an interesting history, since there has been considerable disagreement as to the proper sign of the dipole moment. The correct experimental result is that the negative end of the molecule is carbon, not oxygen as simple electronegativity arguments would suggest. Although a minimal basis set gives the right sign, all SCF calculations with a basis set of double zeta quality or better predict the wrong sign. The difficulty arises because of the relatively small magnitude of the dipole moment, a result of the cancellation of two large and opposite contributions. One contribution is that of net charge which, in line with electronegativity arguments, has oxygen more negative. In addition, however, there is a lone pair of electrons on carbon, directed away from the bond. This asymmetry of the charge on carbon leads to an additional contribution to the dipole moment, which is

Table 3.24 SCF dipole moment (a.u.) of CO for the standard basis sets. A positive dipole moment corresponds to C^-O^+

Basis set	Dipole moment
STO-3G	0.066
4-31G	−0.237
6-31G*	−0.131
Near-HF-limit[a]	−0.110
Experiment	0.044

[a] A. D. McLean and M. Yoshimine, *Intern. J. Quantum Chem.* **15**: 313 (1967).

Table 3.25 SCF dipole moments (a.u.) for the ten-electron series and the standard basis sets

Basis set	NH_3	H_2O	FH
STO-3G	0.703	0.679	0.507
4-31G	0.905	1.026	0.897
6-31G*	0.768	0.876	0.780
6-31G**	0.744	0.860	0.776
Near-HF-limit	0.653[a]	0.785[b]	0.764[c]
Experiment	0.579	0.728	0.716

[a] A. Rauk, L. C. Allen, and E. Clementi, *J. Chem. Phys.* **52**: 4133 (1970).
[b] B. J. Rosenberg and I. Shavitt, *J. Chem. Phys.* **63**: 2162 (1975).
[c] P. E. Cade and W. M. Huo, *J. Chem. Phys.* **45**: 1063 (1966).

opposite to the first contribution. The cancellation, leading to a small positive (C^-O^+) dipole moment, is not reproduced with sufficient accuracy in the SCF calculations. As we shall see in the next chapter, this disagreement between theory and experiment disappears when proper account is taken of correlation effects.

Table 3.25 contains the calculated dipole moments for NH_3, H_2O, and FH using our standard basis sets. Only at the 6-31G* level and beyond is the proper trend $H_2O > FH > NH_3$ reproduced. At the Hartree-Fock limit the calculated dipole moments are somewhat too large, but the trend is well reproduced. The 6-31G** basis set still appears to be inadequate for accurate calculation of dipole moments, since the values obtained with it are still rather distant from Hartree-Fock-limit values.

3.8 UNRESTRICTED OPEN-SHELL HARTREE-FOCK: THE POPLE-NESBET EQUATIONS

At the beginning of this chapter we derived and discussed formal properties of the Hartree-Fock equations independent of any particular form for the spin orbitals. We then introduced a set of restricted spin orbitals and have since been concerned solely with restricted closed-shell calculations of the type

$$|\Psi_{RHF}\rangle = |\psi_1\bar{\psi}_1\cdots\rangle \tag{3.306}$$

Obviously, not all molecules, nor all states of closed-shell molecules, can be described by pairs of electrons in closed-shell orbitals, and we now need to generalize the previous closed-shell formalism to accommodate situations in which a molecule has one or more open-shell (unpaired) electrons. That is, we need to consider unrestricted wave functions of the type

$$|\Psi_{UHF}\rangle = |\psi_1^\alpha \bar{\psi}_1^\beta \cdots\rangle \tag{3.307}$$

In the previous chapter we gave a preliminary description of open-shell determinants (Section 2.5); we now obtain the SCF equations for unrestricted calculations.

In dealing with open-shell problems, there are two common approaches: the restricted open-shell, and the unrestricted open-shell Hartree-Fock procedures. In the restricted open-shell formalism, all electrons, except those that are explicitly required to occupy open-shell orbitals, occupy closed-shell orbitals. The advantage of this procedure is that the wave functions one obtains are eigenfunctions of the spin operator $\mathscr{S}^2$. The disadvantage is that the constraint of occupying orbitals in pairs raises the variational energy. In addition, the spatial equations defining the closed- and open-shell orbitals of restricted open-shell Hartree-Fock theory are somewhat more involved or at least less straightforward than the spatial equations of unrestricted Hartree-Fock theory. For treating open-shells our emphasis is on unrestricted calculations—mainly for reasons of simplicity and generality.

As we have discussed previously, a restricted Hartree-Fock description is inappropriate at long bond lengths for a molecule like H_2, which dissociates to open-shell species. This problem can be solved to a certain extent by using an unrestricted wave function at long bond lengths. In addition to describing unrestricted wave functions for "true" open shells (doublets, triplets, etc.), we will spend some time in this section analyzing the "singlet" dissociation problem with our minimal basis H_2 model. An unrestricted wave function will allow a closed-shell molecule like H_2 to dissociate to open-shell atoms.

In this section, then, we first introduce a set of unrestricted spin orbitals to derive the spatial eigenvalue equations of unrestricted Hartree-Fock theory. We then introduce a basis set and generate the unrestricted Pople-Nesbet matrix equations, which are analogous to the restricted Roothaan equations. We then perform some sample calculations to illustrate solutions to the unrestricted equations. Finally, we discuss the dissociation problem and its unrestricted solution.

3.8.1 Open-Shell Hartree-Fock: Unrestricted Spin Orbitals

The general Hartree-Fock eigenvalue equation, in terms of spin orbitals, is

$$f(1)\chi_i(1) = \varepsilon_i \chi_i(1) \tag{3.308}$$

What we want to do now is to introduce the specific unrestricted form for the spin orbitals $\{\chi_i\}$ and derive, from the above general Hartree-Fock equation, the spatial equations which determine the unrestricted spatial orbitals. The procedure that we use here is quite analogous to that of Subsection 3.4.1, where we derived the spatial equations determining restricted spatial orbitals. We will not repeat all details of the derivation.

Analogous to Eq. (3.110) for restricted spin orbitals, an unrestricted set of spin orbitals has the following form

$$\chi_i(\mathbf{x}) = \begin{cases} \psi_j^\alpha(\mathbf{r})\alpha(\omega) \\ \psi_j^\beta(\mathbf{r})\beta(\omega) \end{cases} \tag{3.309}$$

That is, electrons of α spin are described by a set of spatial orbitals $\{\psi_j^\alpha | j = 1, 2, \ldots, K\}$, and electrons of β spin are described by a different set of spatial orbitals $\{\psi_j^\beta | j = 1, 2, \ldots, K\}$. In our previous restricted case $\psi_j^\alpha \equiv \psi_j^\beta \equiv \psi_j$. We are now allowing electrons of α and β spin to be described by different spatial functions.

To derive the spatial equations defining $\{\psi_j^\alpha\}$ and $\{\psi_j^\beta\}$, we need to insert Eq. (3.309) for the spin orbitals $\{\chi_i\}$ into the general Hartree-Fock equation (3.308) and integrate out the spin variable ω. For simplicity, we will concentrate on the equation defining ψ_j^α and use the symmetry between α and β spins to write down the corresponding equations defining ψ_j^β. Substituting Eq. (3.309) into Eq. (3.308) leads to

$$f(1)\psi_j^\alpha(\mathbf{r}_1)\alpha(\omega_1) = \varepsilon_i\psi_j^\alpha(\mathbf{r}_1)\alpha(\omega_1) \tag{3.310}$$

Now, ε_i is the energy of the spin orbital $\chi_i \equiv \psi_j^\alpha\alpha$. Since the spin orbitals for electrons of α and β spin have different spatial parts, their energies will also be different. In the above case $\varepsilon_i \equiv \varepsilon_j^\alpha$. There will be a corresponding set of orbital energies $\{\varepsilon_j^\beta | j = 1, 2, \ldots, K\}$ for electrons of β spin. Thus

$$f(1)\psi_j^\alpha(\mathbf{r}_1)\alpha(\omega_1) = \varepsilon_j^\alpha\psi_j^\alpha(\mathbf{r}_1)\alpha(\omega_1) \tag{3.311}$$

If we now multiply this equation by $\alpha^*(\omega_1)$ and integrate over spin we get

$$f^\alpha(1)\psi_j^\alpha(1) = \varepsilon_j^\alpha\psi_j^\alpha(1) \tag{3.312}$$

$$f^\beta(1)\psi_j^\beta(1) = \varepsilon_j^\beta\psi_j^\beta(1) \tag{3.313}$$

as the spatial equations defining the spatial orbitals ψ_j^α and ψ_j^β. The spatial Fock operators $f^\alpha(1)$ and $f^\beta(1)$ are defined by

$$f^\alpha(\mathbf{r}_1) = \int d\omega_1 \, \alpha^*(\omega_1)f(\mathbf{r}_1, \omega_1)\alpha(\omega_1) \tag{3.314}$$

$$f^\beta(\mathbf{r}_1) = \int d\omega_1 \, \beta^*(\omega_1)f(\mathbf{r}_1, \omega_1)\beta(\omega_1) \tag{3.315}$$

We could use the spin orbital definition (3.115) of $f(\mathbf{r}_1, \omega_1)$ to perform these integrations and work out explicit formulas for f^α and f^β. Alternatively, we can just write down expressions for f^α and f^β by considering the possible interactions defined by any unrestricted determinant,

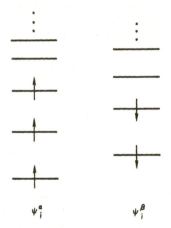

$$\psi_i^\alpha \qquad\qquad \psi_i^\beta$$

The operator $f^\alpha(1)$ is the kinetic energy, nuclear attraction, and effective potential of an electron of α spin. The effective interactions of an electron of α spin include a coulomb and exchange interaction with all other electrons of α spin plus only a coulomb interaction with electrons of β spin. Thus

$$f^\alpha(1) = h(1) + \sum_a^{N^\alpha} \left[J_a^\alpha(1) - K_a^\alpha(1) \right] + \sum_a^{N^\beta} J_a^\beta(1) \qquad (3.316)$$

where the two sums in this equation are over the N^α orbitals ψ_a^α occupied by electrons of α spin and the N^β orbitals ψ_a^β occupied by electrons of β spin. The kinetic energy and nuclear attraction are independent of spin so $h(1)$ is identical to the corresponding operator of the restricted case. The electrons of α spin see a coulomb potential J_a^α and an exchange potential $-K_a^\alpha$ coming from each of the N^α electrons of α spin occupying the orbitals ψ_a^α, plus a coulomb potential J_a^β coming from each of the $N^\beta = N - N^\alpha$ electrons of β spin occupying the orbitals ψ_a^β. The sum over the N^α orbitals ψ_a^α in the above equation formally includes the interaction of an α electron with itself. However, since

$$\left[J_a^\alpha(1) - K_a^\alpha(1) \right] \psi_a^\alpha(1) = 0 \qquad (3.317)$$

this self-interaction is eliminated. The corresponding Fock operator for electrons of β spin is

$$f^\beta(1) = h(1) + \sum_a^{N^\beta} \left[J_a^\beta(1) - K_a^\beta(1) \right] + \sum_a^{N^\alpha} J_a^\alpha(1) \qquad (3.318)$$

The unrestricted coulomb and exchange operators are defined in analogy to our previous definitions (3.124) and (3.125) of the restricted coulomb and exchange operators. That is,

$$J_a^\alpha(1) = \int d\mathbf{r}_2 \, \psi_a^{\alpha*}(2) r_{12}^{-1} \psi_a^\alpha(2) \tag{3.319}$$

$$K_a^\alpha(1)\psi_i^\alpha(1) = \left[\int d\mathbf{r}_2 \, \psi_a^{\alpha*}(2) r_{12}^{-1} \psi_i^\alpha(2) \right] \psi_a^\alpha(1)$$

$$= \left[\int d\mathbf{r}_2 \, \psi_a^{\alpha*}(2) r_{12}^{-1} \mathscr{P}_{12} \psi_a^\alpha(2) \right] \psi_i^\alpha(1) \tag{3.320}$$

The definitions of J_a^β and K_a^β are strictly analogous to the above.

From the definitions (3.316) and (3.318) of the two Fock operators f^α and f^β, we can see that the two integro-differential eigenvalue equations (3.312) and (3.313) are coupled and cannot be solved independently. That is, f^α depends on the occupied β orbitals, ψ_a^β, through J_a^β, and f^β depends on the occupied α orbitals, ψ_a^α, through J_a^α. The two equations must thus be solved by a simultaneous iterative process.

Exercise 3.33 Rather than use the simple technique of writing down $f^\alpha(1)$ by inspection of the possible interactions, as we have done above, use expression (3.314) for $f^\alpha(1)$ and explicitly integrate over spin and carry through the algebra, as was done in Subsection 3.4.1 for the restricted closed-shell case, to derive

$$f^\alpha(1) = h(1) + \sum_a^{N^\alpha} \left[J_a^\alpha(1) - K_a^\alpha(1) \right] + \sum_a^{N^\beta} J_a^\beta(1)$$

Now that we have derived the unrestricted Hartree-Fock equations, we can write down expressions for the unrestricted orbital energies, total unrestricted energy, etc. First, we need to define a few terms. The kinetic energy and nuclear attraction of an electron in one of the unrestricted orbitals ψ_i^α or ψ_i^β is the expectation value

$$h_{ii}^\alpha = \langle \psi_i^\alpha | h | \psi_i^\alpha \rangle \quad \text{or} \quad h_{ii}^\beta = \langle \psi_i^\beta | h | \psi_i^\beta \rangle \tag{3.321}$$

The Coulomb interaction of an electron in ψ_i^α with one in ψ_j^β is

$$J_{ij}^{\alpha\beta} = J_{ji}^{\beta\alpha} = \langle \psi_i^\alpha | J_j^\beta | \psi_i^\alpha \rangle = \langle \psi_j^\beta | J_i^\alpha | \psi_j^\beta \rangle = \langle \psi_i^\alpha \psi_i^\alpha | \psi_j^\beta \psi_j^\beta \rangle \tag{3.322}$$

The corresponding coulomb interactions between electrons of the same spin are

$$J_{ij}^{\alpha\alpha} = \langle \psi_i^\alpha | J_j^\alpha | \psi_i^\alpha \rangle = \langle \psi_j^\alpha | J_i^\alpha | \psi_j^\alpha \rangle = \langle \psi_i^\alpha \psi_i^\alpha | \psi_j^\alpha \psi_j^\alpha \rangle \tag{3.323}$$

and

$$J_{ij}^{\beta\beta} = \langle \psi_i^\beta | J_j^\beta | \psi_i^\beta \rangle = \langle \psi_j^\beta | J_i^\beta | \psi_j^\beta \rangle = \langle \psi_i^\beta \psi_i^\beta | \psi_j^\beta \psi_j^\beta \rangle \tag{3.324}$$

The exchange interactions between electrons of parallel spin are

$$K_{ij}^{\alpha\alpha} = (\psi_i^\alpha | K_j^\alpha | \psi_i^\alpha) = (\psi_j^\alpha | K_i^\alpha | \psi_j^\alpha) = (\psi_i^\alpha \psi_j^\alpha | \psi_j^\alpha \psi_i^\alpha) \qquad (3.325)$$

and

$$K_{ij}^{\beta\beta} = (\psi_i^\beta | K_j^\beta | \psi_i^\beta) = (\psi_j^\beta | K_i^\beta | \psi_j^\beta) = (\psi_i^\beta \psi_j^\beta | \psi_j^\beta \psi_i^\beta) \qquad (3.326)$$

There is, of course, no exchange interaction between electrons of opposite spin.

The total unrestricted electronic energy can now be written down just by considering all the contributing energy terms,

$$E_0 = \sum_a^{N^\alpha} h_{aa}^\alpha + \sum_a^{N^\beta} h_{aa}^\beta + \frac{1}{2} \sum_a^{N^\alpha} \sum_b^{N^\alpha} (J_{ab}^{\alpha\alpha} - K_{ab}^{\alpha\alpha}) + \frac{1}{2} \sum_a^{N^\beta} \sum_b^{N^\beta} (J_{ab}^{\beta\beta} - K_{ab}^{\beta\beta}) + \sum_a^{N^\alpha} \sum_b^{N^\beta} J_{ab}^{\alpha\beta}$$

$$(3.327)$$

The summations with upper limit N^α are summations over the occupied orbitals ψ_a^α or ψ_b^α. A similar convention holds for orbitals occupied by electron of β spin. The factor of $\frac{1}{2}$ in the third and fourth terms eliminates the double counting in the free summation. The self-interaction disappears since $J_{aa}^{\alpha\alpha} - K_{aa}^{\alpha\alpha} = J_{aa}^{\beta\beta} - K_{aa}^{\beta\beta} = 0$ as Eqs. (3.323) to (3.326) verify.

Exercise 3.34 The unrestricted doublet ground state of the Li atom is $|\Psi_0\rangle = |\psi_1^\alpha(1)\bar\psi_1^\beta(2)\psi_2^\alpha(3)\rangle$. Show that the energy of this state is $E_0 = h_{11}^\alpha + h_{11}^\beta + h_{22}^\alpha + J_{12}^{\alpha\alpha} - K_{12}^{\alpha\alpha} + J_{11}^{\alpha\beta} + J_{21}^{\alpha\beta}$.

Exercise 3.35 The unrestricted orbital energies are $\varepsilon_i^\alpha = (\psi_i^\alpha | f^\alpha | \psi_i^\alpha)$ and $\varepsilon_i^\beta = (\psi_i^\beta | f^\beta | \psi_i^\beta)$. Show that these are given by

$$\varepsilon_i^\alpha = h_{ii}^\alpha + \sum_a^{N^\alpha} (J_{ia}^{\alpha\alpha} - K_{ia}^{\alpha\alpha}) + \sum_a^{N^\beta} J_{ia}^{\alpha\beta}$$

$$\varepsilon_i^\beta = h_{ii}^\beta + \sum_a^{N^\beta} (J_{ia}^{\beta\beta} - K_{ia}^{\beta\beta}) + \sum_a^{N^\alpha} J_{ia}^{\beta\alpha}$$

Derive an expression for E_0 in terms of the orbital energies and the coulomb and exchange energies.

3.8.2 Introduction of a Basis: The Pople-Nesbet Equations

To solve the unrestricted Hartree-Fock equations (3.312) and (3.313), we need to introduce a basis set and convert these integro differential equations to matrix equations,[12] just as we did when deriving Roothaan's equations. We thus introduce our set of basis functions $\{\phi_\mu | \mu = 1, 2, \ldots, K\}$ and expand the unrestricted molecular orbitals in this basis,

$$\psi_i^\alpha = \sum_{\mu=1}^{K} C_{\mu i}^\alpha \phi_\mu \qquad i = 1, 2, \ldots, K \qquad (3.328)$$

$$\psi_i^\beta = \sum_{\mu=1}^{K} C_{\mu i}^\beta \phi_\mu \qquad i = 1, 2, \ldots, K \qquad (3.329)$$

The two eigenvalue equations (3.312) and (3.313) guarantee that the sets of eigenfunctions $\{\psi_i^\alpha\}$ and $\{\psi_i^\beta\}$ individually form orthonormal sets. There is no reason, however, that a member of the set $\{\psi_i^\alpha\}$ need be orthogonal to a member of the set $\{\psi_i^\beta\}$. Even though the two sets of spatial orbitals overlap with each other, the set of $2K$ spin orbitals $\{\chi_i\}$ will form an orthonormal set, either from spatial orthogonality ($\alpha\alpha$ and $\beta\beta$ case) or spin orthogonality ($\alpha\beta$ case).

Substituting the expansion (3.328) for the orbitals ψ_j^α into the α Hartree-Fock equation (3.312) gives

$$\sum_\nu C_{\nu j}^\alpha f^\alpha(1)\phi_\nu(1) = \varepsilon_j^\alpha \sum_\nu C_{\nu j}^\alpha \phi_\nu(1) \qquad (3.330)$$

If we multiply this equation by $\phi_\mu^*(1)$ and integrate over the spatial co-ordinates of electron-one, we get

$$\sum_\nu F_{\mu\nu}^\alpha C_{\nu j}^\alpha = \varepsilon_j^\alpha \sum_\nu S_{\mu\nu} C_{\nu j}^\alpha \qquad j = 1, 2, \ldots, K \qquad (3.331)$$

where **S** is the overlap matrix (c.f. Eq. (3.136)) and $\mathbf{F}^\alpha$ is the matrix representation of f^α in the basis $\{\phi_\mu\}$,

$$F_{\mu\nu}^\alpha = \int d\mathbf{r}_1 \, \phi_\mu^*(1) f^\alpha(1)\phi_\nu(1) \qquad (3.332)$$

Identical results can be obtained for β orbitals. The algebraic equations in (3.331) and the corresponding equations for β orbitals can be combined into the two matrix equations,

$$\mathbf{F}^\alpha \mathbf{C}^\alpha = \mathbf{S}\mathbf{C}^\alpha \boldsymbol{\varepsilon}^\alpha \qquad (3.333)$$

$$\mathbf{F}^\beta \mathbf{C}^\beta = \mathbf{S}\mathbf{C}^\beta \boldsymbol{\varepsilon}^\beta \qquad (3.334)$$

These two equations are the unrestricted generalizations of the restricted Roothaan equations (c.f. Eq. (3.139)) and were first given by Pople and Nesbet. The matrices $\boldsymbol{\varepsilon}^\alpha$ and $\boldsymbol{\varepsilon}^\beta$ are diagonal matrices of orbital energies (c.f. Eq. (3.141)). The $K \times K$ square matrices $\mathbf{C}^\alpha$ and $\mathbf{C}^\beta$ have as columns the expansion coefficients for ψ_i^α and ψ_i^β (c.f. Eq. (3.140)). These equations can be solved in a manner similar to the way Roothaan's equations are solved, except that, since $\mathbf{F}^\alpha$ and $\mathbf{F}^\beta$ depend on both $\mathbf{C}^\alpha$ and $\mathbf{C}^\beta$, the two matrix eigenvalue problems must be solved simultaneously. We will return to the solution of these equations after we have described unrestricted density matrices and the explicit form of $F_{\mu\nu}^\alpha$ and $F_{\mu\nu}^\beta$.

3.8.3 Unrestricted Density Matrices

We continue here with the generalization of our previous results for restricted closed-shell wave functions. If an electron is described by the molecular orbital $\psi_a^\alpha(\mathbf{r})$, then the probability of finding that electron in a volume element $d\mathbf{r}$ at $\mathbf{r}$ is $|\psi_a^\alpha(\mathbf{r})|^2 \, d\mathbf{r}$. The probability distribution function (charge density) is $|\psi_a^\alpha(\mathbf{r})|^2$. If we have N^α electrons of α spin, then the total charge density contributed by these electrons is

$$\rho^\alpha(\mathbf{r}) = \sum_a^{N^\alpha} |\psi_a^\alpha(\mathbf{r})|^2 \tag{3.335}$$

The corresponding charge density contributed by electrons of β spin is

$$\rho^\beta(\mathbf{r}) = \sum_a^{N^\beta} |\psi_a^\beta(\mathbf{r})|^2 \tag{3.336}$$

and the total charge density for electrons of either spin is the sum of these

$$\rho^T(\mathbf{r}) = \rho^\alpha(\mathbf{r}) + \rho^\beta(\mathbf{r}) \tag{3.337}$$

Integrating this equation leads, as expected, to

$$\int d\mathbf{r} \, \rho^T(\mathbf{r}) = N = N^\alpha + N^\beta \tag{3.338}$$

In an unrestricted wave function, electrons of α and β spin have different spatial distributions ($\rho^\alpha \neq \rho^\beta$), and it is convenient to define a *spin density* $\rho^S(\mathbf{r})$ by

$$\rho^S(\mathbf{r}) = \rho^\alpha(\mathbf{r}) - \rho^\beta(\mathbf{r}) \tag{3.339}$$

From the above definition of the spin density, it is clear that in regions of space where there is a higher probability of finding an electron of α spin than there is of finding an electron of β spin the spin density is positive. Alternatively, the spin density is negative in regions of space where electrons of β spin are most prevalent. The individual densities ρ^α and ρ^β are, of course positive everywhere. The spin density is a convenient way of describing the distribution of spins in an open-shell system.

Exercise 3.36 Use definitions (3.335) and (3.336) and Eq. (2.254) to show that the integral over all space of the spin density is $2\langle \mathscr{S}_z \rangle$.

By substituting the basis set expansions (3.328) and (3.329) of the α and β molecular orbitals into the expressions (3.335) and (3.336) for the α and β charge densities, one can generate matrix representations (density matrices) of the α and β charge densities,

$$\rho^\alpha(\mathbf{r}) = \sum_a^{N^\alpha} |\psi_a^\alpha(\mathbf{r})|^2 = \sum_\mu \sum_\nu P_{\mu\nu}^\alpha \phi_\mu(\mathbf{r}) \phi_\nu^*(\mathbf{r}) \tag{3.340}$$

$$\rho^\beta(\mathbf{r}) = \sum_a^{N^\beta} |\psi_a^\beta(\mathbf{r})|^2 = \sum_\mu \sum_\nu P_{\mu\nu}^\beta \phi_\mu(\mathbf{r})\phi_\nu^*(\mathbf{r}) \tag{3.341}$$

where the density matrix $\mathbf{P}^\alpha$ for α electrons and the density matrix $\mathbf{P}^\beta$ for β electrons are defined by

$$P_{\mu\nu}^\alpha = \sum_a^{N^\alpha} C_{\mu a}^\alpha (C_{\nu a}^\alpha)^* \tag{3.342}$$

$$P_{\mu\nu}^\beta = \sum_a^{N^\beta} C_{\mu a}^\beta (C_{\nu a}^\beta)^* \tag{3.343}$$

In addition to these two density matrices, one can, of course, define, in analogy to our previous definitions, a total density matrix and a spin density matrix. That is,

$$\mathbf{P}^T = \mathbf{P}^\alpha + \mathbf{P}^\beta \tag{3.344}$$

$$\mathbf{P}^S = \mathbf{P}^\alpha - \mathbf{P}^\beta \tag{3.345}$$

Exercise 3.37 Carry through the missing steps that led to Eqs. (3.340) to (3.343).

Exercise 3.38 Show that expectation values of spin-independent sums of one-electron operators $\sum_{i=1}^N h(i)$ are given by

$$\langle \mathcal{O}_1 \rangle = \sum_\mu \sum_\nu P_{\mu\nu}^T (\nu|h|\mu)$$

for any unrestricted single determinant.

Exercise 3.39 Consider the following spin-dependent operator which is a sum of one-electron operators,

$$\hat{\rho}^S = 2 \sum_{i=1}^N \delta(\mathbf{r}_i - \mathbf{R})s_z(i)$$

Use the rules for evaluating matrix elements, given in Chapter 2, to show that the expectation value of $\hat{\rho}^S$ for any unrestricted single determinant is

$$\langle \hat{\rho}^S \rangle = \rho^S(\mathbf{R}) = \text{tr}(\mathbf{P}^S\mathbf{A})$$

where

$$A_{\mu\nu} = \phi_\mu^*(\mathbf{R})\phi_\nu(\mathbf{R})$$

This matrix element is important in the theory of the Fermi contact contribution to ESR and NMR coupling constants.

Having defined the unrestricted density matrices $\mathbf{P}^\alpha$, $\mathbf{P}^\beta$, $\mathbf{P}^T$, and $\mathbf{P}^S$ we will now use these definitions to give explicit form to the unrestricted Fock matrices $\mathbf{F}^\alpha$ and $\mathbf{F}^\beta$.

3.8.4 Expression for the Fock Matrices

To obtain expressions for the elements of the matrices $\mathbf{F}^\alpha$ and $\mathbf{F}^\beta$, we simply take matrix elements in the basis $\{\phi_\mu\}$ of the two Fock operators f^α (Eq. (3.316)) and f^β (Eq. (3.318)), and use expressions (3.322) to (3.326) for matrix elements of the coulomb and exchange operators. That is,

$$
\begin{aligned}
F_{\mu\nu}^\alpha &= \int d\mathbf{r}_1 \, \phi_\mu^*(1) f^\alpha(1) \phi_\nu(1) \\
&= H_{\mu\nu}^{\text{core}} + \sum_a^{N^\alpha} \left[(\phi_\mu\phi_\nu | \psi_a^\alpha\psi_a^\alpha) - (\phi_\mu\psi_a^\alpha | \psi_a^\alpha\phi_\nu) \right] + \sum_a^{N^\beta} (\phi_\mu\phi_\nu | \psi_a^\beta\psi_a^\beta) \quad (3.346)
\end{aligned}
$$

$$
\begin{aligned}
F_{\mu\nu}^\beta &= \int d\mathbf{r}_1 \, \phi_\mu^*(1) f^\beta(1) \phi_\nu(1) \\
&= H_{\mu\nu}^{\text{core}} + \sum_a^{N^\beta} \left[(\phi_\mu\phi_\nu | \psi_a^\beta\psi_a^\beta) - (\phi_\mu\psi_a^\beta | \psi_a^\beta\phi_\nu) \right] + \sum_a^{N^\alpha} (\phi_\mu\phi_\nu | \psi_a^\alpha\psi_a^\alpha) \quad (3.347)
\end{aligned}
$$

To continue, we substitute the basis set expansions of ψ_a^α and ψ_a^β to get

$$
\begin{aligned}
F_{\mu\nu}^\alpha &= H_{\mu\nu}^{\text{core}} + \sum_\lambda \sum_\sigma \sum_a^{N^\alpha} C_{\lambda a}^\alpha (C_{\sigma a}^\alpha)^* [(\mu\nu|\sigma\lambda) - (\mu\lambda|\sigma\nu)] + \sum_\lambda \sum_\sigma \sum_a^{N^\beta} C_{\lambda a}^\beta (C_{\sigma a}^\beta)^* (\mu\nu|\sigma\lambda) \\
&= H_{\mu\nu}^{\text{core}} + \sum_\lambda \sum_\sigma P_{\lambda\sigma}^\alpha [(\mu\nu|\sigma\lambda) - (\mu\lambda|\sigma\nu)] + \sum_\lambda \sum_\sigma P_{\lambda\sigma}^\beta (\mu\nu|\sigma\lambda) \\
&= H_{\mu\nu}^{\text{core}} + \sum_\lambda \sum_\sigma P_{\lambda\sigma}^T (\mu\nu|\sigma\lambda) - P_{\lambda\sigma}^\alpha (\mu\lambda|\sigma\nu) \quad (3.348)
\end{aligned}
$$

$$
\begin{aligned}
F_{\mu\nu}^\beta &= H_{\mu\nu}^{\text{core}} + \sum_\lambda \sum_\sigma \sum_a^{N^\beta} C_{\lambda a}^\beta (C_{\sigma a}^\beta)^* [(\mu\nu|\sigma\lambda) - (\mu\lambda|\sigma\nu)] + \sum_\lambda \sum_\sigma \sum_a^{N^\alpha} C_{\lambda a}^\alpha (C_{\sigma a}^\alpha)^* (\mu\nu|\sigma\lambda) \\
&= H_{\mu\nu}^{\text{core}} + \sum_\lambda \sum_\sigma P_{\lambda\sigma}^\beta [(\mu\nu|\sigma\lambda) - (\mu\lambda|\sigma\nu)] + \sum_\lambda \sum_\sigma P_{\lambda\sigma}^\alpha (\mu\nu|\sigma\lambda) \\
&= H_{\mu\nu}^{\text{core}} + \sum_\lambda \sum_\sigma P_{\lambda\sigma}^T (\mu\nu|\sigma\lambda) - P_{\lambda\sigma}^\beta (\mu\lambda|\sigma\nu) \quad (3.349)
\end{aligned}
$$

If one compares these expressions with the corresponding restricted closed-shell expression (3.154), one sees that the coulomb term is identical and depends on the total density matrix. The difference is only that here one has separate representations of the α and β density matrices rather than, as in the closed-shell case,

$$
P_{\mu\nu}^\alpha = P_{\mu\nu}^\beta = \tfrac{1}{2} P_{\mu\nu}^T \quad (3.350)
$$

The coupling of the two sets of equations is made explicit in the above expressions, i.e., $\mathbf{F}^\alpha$ depends on $\mathbf{P}^\beta$ (through the total density matrix $\mathbf{P}^T$) and $\mathbf{F}^\beta$ similarly depends on $\mathbf{P}^\alpha$.

3.8.5 Solution of the Unrestricted SCF Equations

The procedure for solving the unrestricted SCF equations is essentially identical to that previously described for solving the Roothaan equations. An initial guess is required for the two density matrices $\mathbf{P}^\alpha$ and $\mathbf{P}^\beta$ and hence $\mathbf{P}^T$. An obvious choice is to set these matrices to zero and use $\mathbf{H}^{core}$ as an initial guess to both $\mathbf{F}^\alpha$ and $\mathbf{F}^\beta$. If this procedure is followed, the first iteration will produce identical orbitals for α and β spin, i.e., a restricted solution. If, however, $N^\alpha \neq N^\beta$, then all subsequent iterations will have $\mathbf{P}^\alpha \neq \mathbf{P}^\beta$ and an unrestricted solution will result.

Given approximations to $\mathbf{P}^\alpha$ and $\mathbf{P}^\beta$, at each step of the iteration, we can form $\mathbf{F}^\alpha$ and $\mathbf{F}^\beta$, solve the two generalized matrix eigenvalue problems

$$\mathbf{F}^\alpha \mathbf{C}^\alpha = \mathbf{S}\mathbf{C}^\alpha \boldsymbol{\varepsilon}^\alpha \tag{3.351}$$

$$\mathbf{F}^\beta \mathbf{C}^\beta = \mathbf{S}\mathbf{C}^\beta \boldsymbol{\varepsilon}^\beta \tag{3.352}$$

for $\mathbf{C}^\alpha$ and $\mathbf{C}^\beta$, and then form new approximations to $\mathbf{P}^\alpha$ and $\mathbf{P}^\beta$. Because of the coupling of the two equations, one cannot obtain a self-consistent solution to the α equations without at the same time obtaining a self-consistent solution to the β equations, although at any one iteration step the two matrix eigenvalue problems (3.351) and (3.352) can be solved independently; the coupling is in the formation of the Fock matrices. Solving the matrix eigenvalue problem will involve knowing a transformation matrix $\mathbf{X}$ to an orthonormal basis set, forming $\mathbf{F}^{\alpha'} = \mathbf{X}^\dagger \mathbf{F}^\alpha \mathbf{X}$, diagonalizing $\mathbf{F}^{\alpha'}$ to get $\mathbf{C}^{\alpha'}$, and then forming $\mathbf{C}^\alpha = \mathbf{X}\mathbf{C}^{\alpha'}$, etc., just as in the restricted closed-shell case.

Exercise 3.40 Substitute the basis set expansion of the unrestricted molecular orbitals into Eq. (3.327) for the electronic energy E_0 to show that

$$E_0 = \frac{1}{2} \sum_\mu \sum_\nu \left[P^T_{\nu\mu} H^{core}_{\mu\nu} + P^\alpha_{\nu\mu} F^\alpha_{\mu\nu} + P^\beta_{\nu\mu} F^\beta_{\mu\nu} \right]$$

Before going on to describe sample unrestricted calculations, an important point should be noted about solutions to the Pople-Nesbet equations for the special case $N^\alpha = N^\beta$, i.e., for the case where a molecule would normally be described by a restricted closed-shell wave function. For this case, there exists the possibility of two independent solutions to the Pople-Nesbet equations. The first solution is a restricted solution. If $\mathbf{P}^\alpha = \mathbf{P}^\beta = \frac{1}{2}\mathbf{P}$, then $\mathbf{F}^\alpha = \mathbf{F}^\beta = \mathbf{F}$ and the Pople-Nesbet equations degenerate to the Roothaan equations. *When $N^\alpha = N^\beta$, a restricted solution to the Roothaan equations is a solution to the unrestricted Pople-Nesbet equations.* This restricted solution always exists and necessarily results if an initial guess $\mathbf{P}^\alpha = \mathbf{P}^\beta$ is used. For $N^\alpha = N^\beta$, however, in addition to the restricted solution there may also exist a second unrestricted solution of lower energy. The restricted solution constrains the density of α electrons to equal the density of β electrons, but under certain conditions (which we shall consider in the last subsection of this chapter) relaxing this constraint will result in an unrestricted solution of lower energy for which $\mathbf{P}^\alpha$ is not equal to $\mathbf{P}^\beta$. *When $N^\alpha = N^\beta$, under certain conditions there exists a second solution, the unrestricted solution to the Pople-Nesbet equations.* In seeking this second solution, it is imperative that an initial guess $\mathbf{P}^\alpha \neq \mathbf{P}^\beta$ be used or the equations will necessarily yield the restricted solution. Even if an unrestricted initial guess is used, there is still the possibility that iteration will lead to the restricted solution. When two solutions exist, the initial guess will strongly determine to which solution the iterations lead.

One normally uses unrestricted wave functions to describe open-shell states of molecules for which $N^\alpha \neq N^\beta$, and the above considerations are not of concern. When, however, one uses unrestricted wave functions as a solution to the dissociation problem, as we shall subsequently do, the possibility of two solutions is of supreme importance.

3.8.6 Illustrative Unrestricted Calculations

An interesting example of the use of unrestricted wave functions occurs for the methyl radical CH_3. This molecule has D_{3h} symmetry, i.e., it is planar with bond angles of 120°. The CH internuclear distance is taken to be 2.039 a.u. The simplest description of the electronic structure of this radical is a restricted Hartree-Fock description, shown in Fig. 3.13. The unpaired

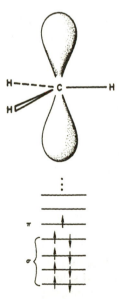

Figure 3.13 Restricted Hartree-Fock description of the planar methyl radical.

electron is in an open-shell π orbital, which in a minimal basis description would be a pure $2p$ orbital on carbon. The remaining electrons are paired in σ orbitals. In this restricted Hartree-Fock description, the spin density $\rho^s(\mathbf{r})$ is everywhere positive, except in the plane of the molecule where it is zero because of the node in the π orbital. Because all σ electrons are paired, the spin density is just

$$\rho^S(\mathbf{r}) = |\psi_\pi(\mathbf{r})|^2 \tag{3.353}$$

where ψ_π is the π molecular orbital containing the unpaired electron.

The above description, while simple, is not in agreement with experimental results. In an electron spin resonance (ESR) experiment on the methyl radical, measurements were made of a^H and a^C, the coupling constants for the hydrogen and carbon nuclei. These ESR coupling constants are a direct measure of the spin densities at the position of the respective nuclei,

$$a^H(\text{Gauss}) = 1592\rho^S(\mathbf{R}_H) \tag{3.354}$$

$$a^C(\text{Gauss}) = 400.3\rho^S(\mathbf{R}_C) \tag{3.355}$$

The experimental measurements of a^H and a^C give not only the magnitude but also the sign of the spin density. It is found that the spin density at the H nucleus is negative, and the spin density at the C nucleus is positive. Unfortunately, the restricted Hartree-Fock description predicts the coupling constants a^H and a^C to be both zero. If the molecule were vibrating so that part of the time the molecule had a bent C_{3v} geometry, then the restricted description would allow nonzero spin densities at the nuclei. But these spin densities and the associated coupling constants would always be positive. Thus the negative spin density at the positions of the hydrogen nuclei cannot be explained by a restricted Hartree-Fock description.

The simplest way of obtaining the correct qualitative result is to use an unrestricted Hartree-Fock description. The electrons of Fig. 3.13 that are paired in a σ orbital have different interactions with the unpaired electron, i.e., the electrons of α spin have a coulomb and exchange interaction with the unpaired electron while the electrons of β spin have only a coulomb interaction. There is thus good reason why the α and β electrons of the sigma system should have different energies and occupy different spatial orbitals. If, indeed, one does relax the constraint of paired electrons, by using the Pople-Nesbet equations, the unrestricted solution shown in Fig. 3.14 is found. This unrestricted wave function does not have the σ electrons paired and as such there will be net nonzero spin density in the sigma system, in particular, at the positions of the carbon and hydrogen nuclei. Unrestricted calculations (Table 3.26), show that the spin density is positive at the carbon nucleus and negative at the hydrogen nuclei, as also shown in the figure. This result is commonly explained by the use of two rules: an "intraatomic Hund's rule", which postulates that electrons tend to have parallel spins on the same atom, and a rule which states that the spins of electrons in orbitals

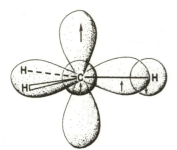

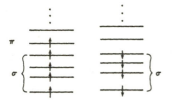

Figure 3.14 Unrestricted Hartree-Fock description of the planar methyl radical.

Table 3.26 Unrestricted SCF spin densities and hyperfine coupling constants for the methyl radical using the standard basis sets. A value of $\langle \mathscr{S}^2 \rangle = 0.75$ corresponds to a pure doublet

Basis set	Spin density (a.u.)		Coupling constant (Gauss)		
	C	H	a^C	a^H	$\langle \mathscr{S}^2 \rangle$
STO-3G	+0.2480	−0.0340	+99.3	−54.2	0.7652
4-31G	+0.2343	−0.0339	+93.8	−54.0	0.7622
6-31G*	+0.1989	−0.0303	+79.6	−48.3	0.7618
6-31G**	+0.1960	−0.0296	+78.5	−47.1	0.7614
Experiment			+38.3	−23.0	0.75

that overlap to form a chemical bond tend to be antiparallel. Negative spin density in the vicinity of the hydrogen nuclei results from application of these two rules.

Table 3.26 shows the results of *ab initio* calculations of the CH_3 hyperfine coupling constants. The correct qualitative results are obtained—a positive spin density at the carbon nucleus and a negative spin density at the hydrogen nuclei. The magnitudes of the spin densities are too large, however. They are in error by about a factor of 2 for the 6-31G** basis set. Without performing more extensive calculations, it is difficult to know whether the source of the error is in the basis sets or in the neglect of correlation. The standard basis sets we are using were derived primarily for the description of valence properties, and they may not be adequate near a nucleus. In particular, Gaussian functions are known to be poor at their origin. Also, the basis sets that we are using contain only a single function for the inner-shell of carbon.

The table also contains expectation values of $\mathscr{S}^2$. One of the deficiencies of an unrestricted calculation is that it does not produce a pure spin state. The ground state of the methyl radical is a doublet with $\langle \mathscr{S}^2 \rangle = S(S + 1) = \frac{3}{4}$. The unrestricted calculations produce a doublet wave function, which is contaminated with small amounts of a quartet, sextet, etc, as discussed in Section 2.5. The expectation values of $\mathscr{S}^2$ are close to the correct value of $\frac{3}{4}$, showing that these contaminants are not large.

Exercise 3.41 Assume the unrestricted Hartree-Fock (UHF) calculations of Table 3.26 contain only the leading quartet contaminant. That is,

$$\Psi_{\text{UHF}} = c_1 \, ^2\Psi + c_2 \, ^4\Psi$$

If the percent contamination is defined as $100c_2^2/(c_1^2 + c_2^2)$, calculate the percent contamination of each of the four calculations from the quoted value of $\langle \mathscr{S}^2 \rangle$.

We have previously used Koopmans' theorem to calculate the first two ionization potentials of N_2. As we saw at that time, calculations at the Hartree-Fock limit, or with our best (6-31G*) basis set, incorrectly predict the $^2\Pi_u$ state of N_2^+ to be lower in energy than $^2\Sigma_g$ state of N_2^+. That is, the highest occupied orbital of N_2 is calculated to be the $1\pi_u$ orbital rather than the $3\sigma_g$ orbital. There are two reasons why Koopmans' theorem might make the wrong prediction: neglect of correlation or neglect of relaxation. We can test the second alternative by explicitly performing Hartree-Fock calculations on the $^2\Pi_u$ and $^2\Sigma_g$ state of N_2^+. Koopmans' theorem calculations assume the orbitals of these two states to be identical to those of ground state N_2. By performing separate unrestricted calculations on these two doublet states of N_2^+, we will be allowing the orbitals to relax to their optimum form. The ionization potentials can then be obtained by subtracting the total restricted energy of the N_2 ground state from the total unrestricted energy of each of the N_2^+ ions.

Table 3.27 shows the results of 6-31G* calculations on the $^1\Sigma_g$ state of N_2 and the $^2\Sigma_g$ and $^2\Pi_u$ states of N_2^+. To compare with experimental vertical ionization potentials, all the calculations were performed at the equilibrium geometry $(R = 2.074 \text{ a.u.})$ of ground state N_2. These calculations still predict that the $^2\Pi_u$ state has a lower energy than the $^2\Sigma_g$ state in disagreement with

Table 3.27 SCF calculations on the ground state of N_2 (restricted) and two states of N_2^+ (unrestricted) with a 6-31G* basis set. Vertical ($R_e = 2.074$ a.u.) ionization potentials are shown, and experimental values are in parenthesis

State	Total Energy (a.u.)	Ionization Potential (a.u.)
$N_2(^1\Sigma_g)$	-108.94235	
$N_2^+(^2\Pi_u)$	-108.37855	0.564 (0.624)
$N_2^+(^2\Sigma_g)$	-108.36597	0.576 (0.573)

experiment. This is therefore an indication that the qualitative disagreement of experiment with Koopmans' theorem ionization potentials for N_2 is a result of the lack of inclusion of correlation effects. Later, inclusion of correlation effects will verify this.

Our final example of *ab initio* unrestricted calculations is O_2. This molecule has unpaired spins and is paramagnetic. The first brilliant success of molecular orbital theory was the explanation of why O_2, with an even

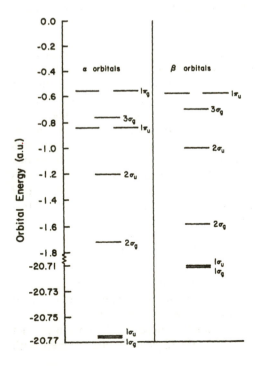

Figure 3.15 Unrestricted occupied molecular orbitals of $O_2(^3\Sigma_g^-$, $R = 2.281$ a.u.) with a 6-31G* basis set.

number of electrons, does not have all its electrons paired. The molecular orbitals of homonuclear diatomics are ordered $1\sigma_g$, $1\sigma_u$, $2\sigma_g$, $2\sigma_u$, $(3\sigma_g, 1\pi_u)$ $1\pi_g$, $3\sigma_u$. The last two electrons of O_2 go into the doubly degenerate anti-bonding $1\pi_g$ orbital. By Hund's rule, these two electrons go into separate $1\pi_g$ orbitals with their spins parallel so as to enjoy the negative exchange interaction. This, therefore, leads to a final $^3\Sigma_g^-$ state. The occupied orbitals of an unrestricted 6-31G* calculation on O_2, for a bond length of 2.281 a.u., are shown in Fig. 3.15. The "open-shell" alpha electrons in the $1\pi_g$ orbital "push" down (stabilize) the α orbitals relative to the β orbitals because of exchange interactions that are present only between electrons of the same spin. In a restricted description, all but the $1\pi_g$ orbitals would be constrained to be paired. Note how the order of the $1\pi_u$ and $3\sigma_g$ orbitals are reversed for electrons of α and β spin.

To complete our discussion of unrestricted Hartree-Fock theory, we will use our minimal basis H_2 model to investigate the description of bond dissociation by unrestricted wave functions.

3.8.7 The Dissociation Problem and Its Unrestricted Solution

The unrestricted wave function is normally used to describe open-shell states–doublets, triplets, etc., as in the examples of the last subsection. Under certain circumstances, however, it may be appropriate to use an unrestricted wave function to describe states that are normally thought of as closed-shell singlets. For the ground state of a molecule like H_2, the restricted formulation, with electrons paired, is the usual description. As we shall shortly see, it is also the only appropriate Hartree-Fock description under certain conditions. At very large bond lengths, however, one is really trying to describe two individual hydrogen atoms. A proper description will have one electron on one H atom and the other electron on the other H atom, i.e., the two electrons will have quite different spatial distributions. They should not have identical spatial distributions as is implied by a restricted wave function, which places both electrons in the same spatial orbital. It would thus appear that at equilibrium distances we want a restricted wave function, but at large bond lengths we want an unrestricted wave function. In a sense, we will be able to have our cake and eat it too. As was discussed in the previous subsection, there may exist two solutions to the unrestricted equations of Pople and Nesbet when $N^\alpha = N^\beta$. The restricted solution of Roothaan's equations is necessarily a solution to the Pople-Nesbet equations. It only remains to discover whether there is a second truly unrestricted solution that is lower in energy than the restricted solution. We shall find that for normal geom-etries there is not always an unrestricted solution. If, however, we stretch a bond which cleaves homolytically, like the bond in H_2 ($H_2 \rightarrow H + H$) but unlike the bond in HeH^+ ($HeH^+ \rightarrow He + H^+$), then an unrestricted solution will always exist at large bond lengths. The unrestricted solution accomodates the unpairing of electrons inherent in the breaking of the bond. To see this explicitly, we will investigate wave functions for our minimal basis model of H_2.

We could numerically solve the Pople-Nesbet equations for minimal basis STO-3G H_2, just as we have solved them for CH_3, N_2^+, and O_2. An appropriate unrestricted initial guess would be required if the iterations were to lead to an unrestricted solution rather than to the restricted solution. The transition from a restricted to an unrestricted wave function will be more transparent, however, if, rather than obtain a numerical solution to the Pople-Nesbet matrix equations, we formulate the problem in an analytical fashion.

The restricted molecular orbitals of minimal basis H_2 are symmetry determined and given by

$$\psi_1 = [2(1 + S_{12})]^{-1/2}(\phi_1 + \phi_2) \tag{3.356}$$

$$\psi_2 = [2(1 - S_{12})]^{-1/2}(\phi_1 - \phi_2) \tag{3.357}$$

Since the minimal basis model has only two basis functions with coefficients that can be varied and since molecular orbitals are constrained to be normalized, the minimal basis model has, in the general case, only one degree of freedom. An unrestricted solution, unlike the restricted solution, is not symmetry determined and a convenient way of incorporating this one degree of freedom into unrestricted calculations is to write the unrestricted occupied molecular orbitals ψ_1^α and ψ_1^β as linear combinations of the restricted symmetry determined orbitals ψ_1 and ψ_2, as follows:

$$\psi_1^\alpha = \cos\theta\psi_1 + \sin\theta\psi_2 \tag{3.358}$$

$$\psi_1^\beta = \cos\theta\psi_1 - \sin\theta\psi_2 \tag{3.359}$$

The single degree of freedom here is in the angle θ. It is sufficient to consider values of θ between $0°$ and $45°$. The value $\theta = 0$ corresponds to the restricted solution $\psi_1^\alpha = \psi_1^\beta = \psi_1$ and nonzero values of θ correspond to unrestricted solutions $\psi_1^\alpha \neq \psi_1^\beta$. The unrestricted virtual orbitals are given by

$$\psi_2^\alpha = -\sin\theta\psi_1 + \cos\theta\psi_2 \tag{3.360}$$

$$\psi_2^\beta = \sin\theta\psi_1 + \cos\theta\psi_2 \tag{3.361}$$

Exercise 3.42 Show that the set of α orbitals $\{\psi_1^\alpha, \psi_2^\alpha\}$ and the set of β orbitals $\{\psi_1^\beta, \psi_2^\beta\}$ form separate orthonormal sets.

If we substitute the basis set expansions (3.356) and (3.357) into the previous four equations, we will obtain basis set expansions for the unrestricted molecular orbitals. The occupied molecular orbitals, which are the only ones we need consider from now on, are given by

$$\psi_1^\alpha = c_1\phi_1 + c_2\phi_2 \tag{3.362}$$

$$\psi_1^\beta = c_2\phi_1 + c_1\phi_2 \tag{3.363}$$

where

$$c_1 = [2(1 + S_{12})]^{-1/2} \cos\theta + [2(1 - S_{12})]^{-1/2} \sin\theta \tag{3.364}$$

$$c_2 = [2(1 + S_{12})]^{-1/2} \cos\theta - [2(1 - S_{12})]^{-1/2} \sin\theta \tag{3.365}$$

By allowing ψ_2 to mix with ψ_1 in the definition of the unrestricted occupied orbitals (Eqs. (3.358) and (3.359)), we allow the weights of ϕ_1 and ϕ_2 in the basis set expansions of ψ_1^α and ψ_1^β to vary as shown by Eqs. (3.362) and (3.363). If $\theta = 0$, the wave function is just the restricted wave function with $c_1 = c_2 = [2(1 + S_{12})]^{-1/2}$. As θ increases from zero, c_1 gets larger and c_2 gets smaller or, equivalently, ψ_1^α acquires a larger admixture of ϕ_1 and ψ_1^β acquires a larger admixture of ϕ_2. If $S_{12} = 0$ as is appropriate for large internuclear distances, then in the limit of $\theta = 45°$ we have $c_1 = 1$, $c_2 = 0$, and

$$\left.\begin{aligned}\psi_1^\alpha \equiv \phi_1\\\psi_1^\beta \equiv \phi_2\end{aligned}\right\} \theta = 45°, \qquad S_{12} = 0 \tag{3.366}$$

This is the result we desire for two separate H atoms—an electron with α spin in ϕ_1 and an electron with β spin in ϕ_2.

We thus characterize molecular orbitals for minimal basis H_2 by the single parameter θ. At one extreme, $\theta = 0$ corresponds to the restricted solution where the occupied molecular orbital is an equal mixture of ϕ_1 and ϕ_2. At the other extreme, $\theta = 45°$ corresponds to an unrestricted solution for isolated hydrogen atoms. Intermediate value of θ correspond to unrestricted solutions where ψ_1^α is mainly ϕ_1 and ψ_1^β is mainly ϕ_2. Figure 3.16 gives a qualitative picture of the unrestricted molecular orbitals of H_2 as a function of θ. While we have derived this picture using the minimal basis, the figure is qualitatively correct for H_2 with any basis set.

We have seen that for the ground state of a closed-shell molecule like H_2 it appears possible to define unrestricted wave functions which have the qualitatively correct behavior that we expect for the dissociation process. It remains to relate these unrestricted wave functions to solutions of the Hartree-Fock equations. If we solve the Pople-Nesbet equations, will a non-zero value of θ be obtained? To investigate this question, we need to determine the energy as a function of θ.

The electronic energy of an unrestricted single determinant wave function for H_2,

$$|\Psi_0\rangle = |\psi_1^\alpha(1)\bar\psi_1^\beta(2)\rangle \tag{3.367}$$

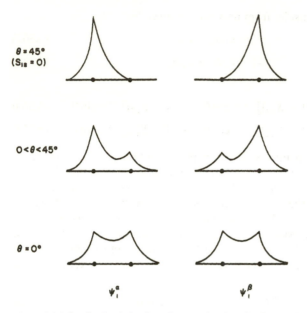

Figure 3.16 Qualitative behavior of unrestricted molecular orbitals ψ_1^α and ψ_1^β for H_2 as a function of θ.

is just the kinetic energy and nuclear attraction of each electron, plus the coulomb repulsion between the two electrons. That is,

$$
\begin{aligned}
E_0 = \langle \Psi_0 | \mathscr{H} | \Psi_0 \rangle &= h_{11}^\alpha + h_{11}^\beta + J_{11}^{\alpha\beta} \\
&= (\psi_1^\alpha | h | \psi_1^\alpha) + (\psi_1^\beta | h | \psi_1^\beta) + (\psi_1^\alpha \psi_1^\alpha | \psi_1^\beta \psi_1^\beta)
\end{aligned} \tag{3.368}
$$

Substituting the expansions (3.358) and (3.359) into this expression, we can write the electronic energy, as a function of θ, in terms of molecular integrals of the restricted problem

$$
\begin{aligned}
E_0(\theta) = 2\cos^2\theta \, h_{11} &+ 2\sin^2\theta \, h_{22} + \cos^4\theta \, J_{11} \\
&+ \sin^4\theta \, J_{22} + 2\sin^2\theta \cos^2\theta (J_{12} - 2K_{12})
\end{aligned} \tag{3.369}
$$

If $\theta = 0$, the unrestricted energy just reduces to the restricted energy

$$
E_0(0) = 2h_{11} + J_{11} \tag{3.370}
$$

The first derivative of the unrestricted energy with respect to θ is

$$
\begin{aligned}
dE_0(\theta)/d\theta = 4\cos\theta\sin\theta [h_{22} - h_{11} &+ \sin^2\theta J_{22} - \cos^2\theta J_{11} \\
&+ (\cos^2\theta - \sin^2\theta)(J_{12} - 2K_{12})]
\end{aligned} \tag{3.371}
$$

To find the values of θ which solve the Pople-Nesbet equations, i.e., to find the values of θ which make the unrestricted energy stationary, we set the first derivative of the unrestricted energy to zero,

$$dE_0(\theta)/d\theta = AB = 0 \qquad (3.372)$$

where

$$A = 4\cos\theta\sin\theta \qquad (3.373)$$

and

$$B = h_{22} - h_{11} + \sin^2\theta J_{22} - \cos^2\theta J_{11} + (\cos^2\theta - \sin^2\theta)(J_{12} - 2K_{12}) \quad (3.374)$$

There are thus two ways the energy could be stationary:

1. $A = 0$. This is the restricted solution. The condition is satisfied if $\theta = 0$.
2. $B = 0$. This is the unrestricted solution. The condition is satisfied and there exists an unrestricted wave function only if there is a solution to:

$$\cos^2\theta = \eta \qquad (3.375)$$

where

$$\eta = (h_{22} - h_{11} + J_{22} - J_{12} + 2K_{12})/(J_{11} + J_{22} - 2J_{12} + 4K_{12}) \quad (3.376)$$

This last equation is obtained by setting B of Equation (3.374) to zero. This equation has a solution only if the internuclear distance and basis functions, and hence the molecular integrals h_{11}, h_{22}, etc., are such that η lies between zero and one, i.e., $0 \leq \eta \leq 1$.

Exercise 3.43 Use the molecular integrals given in Appendix D to show that no unrestricted solution exists for minimal basis STO-3G H_2 at $R = 1.4\,a.u.$ Repeat the calculation for $R = 4.0\,a.u.$ and show that an unrestricted solution exists with $\theta = 39.5°$. Remember that $\varepsilon_1 = h_{11} + J_{11}$ and $\varepsilon_2 = h_{22} + 2J_{12} - K_{12}$.

To proceed with the analysis let us investigate the nature of the restricted solution ($\theta = 0$) by evaluating the second derivative of the energy (at the restricted solution),

$$d^2E_0(\theta)/d\theta^2]_{\theta=0} = E_0''(0) = 4(h_{22} - h_{11} - J_{11} + J_{12} - 2K_{12})$$
$$= 4(\varepsilon_2 - \varepsilon_1 - J_{12} - K_{12}) \qquad (3.377)$$

The nature of the restricted solution is determined by this second derivative. If $E_0''(0) > 0$, it is an energy minimum. If $E_0''(0) < 0$, it is an energy maximum. If $E_0''(0) = 0$, i.e., if

$$h_{22} - h_{11} = J_{11} - J_{12} + 2K_{12} \qquad (3.378)$$

then the restricted solution is a saddle point. Substituting this last saddle point condition into Eq. (3.376), we find that $\eta = 1$ at the saddle point. Using the molecular integrals of Appendix D we can investigate the behavior of

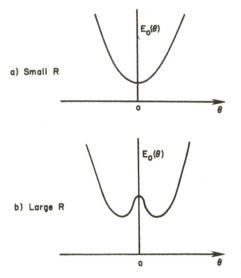

a) Small R

b) Large R

Figure 3.17 Qualitative behavior of the unrestricted energy of H_2 as a function of θ for small and large internuclear distances: a) small R; b) large R.

$E_0''(0)$ and η as a function of bond length. At short bond lengths $E_0''(0) > 0$ and $\eta > 1$. As the bond length increases both $E_0''(0)$ and η decrease monotonically, until they reach a limit at $R = \infty$ of $E_0''(0) = -1/2 \, (\phi_1\phi_1|\phi_1\phi_1)$ and $\eta = 1/2$. At a transition point, which occurs in the vicinity of $R = 2.3$ a.u., the second derivative $E_0''(0)$ becomes negative and simultaneously η becomes less than 1. The behavior of the solutions is therefore as follows: At short bond lengths $\eta > 1$, the restricted solution is a true minimum, and no unrestricted solution exists. On increasing the bond length the value of η decreases until, at a distance of approximately 2.3 a.u., η becomes 1 and a saddle point occurs in the energy. This transition point defines the onset of an unrestricted solution. At a bond length beyond this, the restricted solution ($\theta = 0$) is actually a maximum in the energy as shown in Fig. 3.17. When an unrestricted solution exists ($\eta \leq 1$), the value of η can be equated to $\cos^2\theta$. As the bond length becomes larger and larger, θ gives to the limit of $45°$ appropriate to isolated hydrogen atoms. A potential curve for STO-3G H_2 showing the two solutions is shown in Fig. 3.18. The unrestricted energy goes smoothly to the limit of two hydrogen atoms calculated with the same basis set, i.e., $2(\phi_1|h|\phi_1)$. The restricted energy goes to a limit $1/2(\phi_1\phi_1|\phi_1\phi_1)$ above the right result. Also shown in Fig. 3.18 is the essentially exact result of Kolos and Wolniewicz.[5] The hydrogen atom energies used in the figure $(-0.4666$ and $-0.5)$ are obtained with the basis sets employed in the respective methods. Thus, both curves go to zero at large R. The corresponding curves for a 6-31G** basis set are shown in Fig. 3.19.

The "correct" dissociation of H_2, which we have obtained by using an unrestricted wave function, is not free of faults. The unrestricted wave

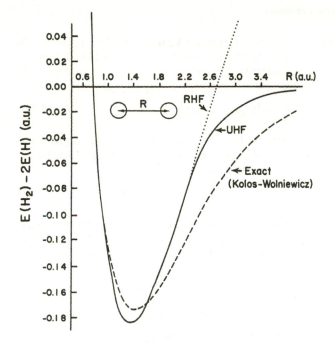

Figure 3.18 STO-3G potential curves for H_2.

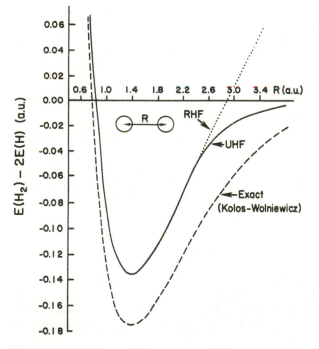

Figure 3.19 6-31G** potential energy curves for H_2.

function is not a pure singlet as one would like it to be. The energy goes to the correct limit but the total wave function does not, as we shall now see. In the limit $R \to \infty$, the molecular orbitals become $\psi_1^\alpha = \phi_1$ and $\psi_1^\beta = \phi_2$ and the unrestricted single determinant $|\Psi_0\rangle$ of Eq. (3.367) becomes

$$\lim_{R \to \infty} |\Psi_0\rangle = |\phi_1(1)\bar{\phi}_2(2)\rangle \qquad (3.379)$$

This, however, is not the correct form for a singlet wave function in which electrons occupy different spatial orbitals ϕ_1 and ϕ_2. In analogy to Eq. (2.260), the singlet wave function should be

$$\lim_{R \to \infty} |\Phi_0\rangle = 2^{-1/2}[|\phi_1(1)\bar{\phi}_2(2)\rangle + |\phi_2(1)\bar{\phi}_1(2)\rangle] \qquad (3.380)$$

The orbitals are correct but the total wave function is not. An alternative way of looking at this problem is obtained by substituting the expansions (3.358) and (3.359) for the unrestricted molecular orbitals into the single determinant $|\Psi_0\rangle$ and expanding the determinant

$$\begin{aligned}
|\Psi_0\rangle = |\psi_1^\alpha \bar{\psi}_1^\beta\rangle &= \cos^2\theta|\psi_1\bar{\psi}_1\rangle - \sin^2\theta|\psi_2\bar{\psi}_2\rangle \\
&\quad - (2)^{1/2}\cos\theta\sin\theta[|\psi_1\bar{\psi}_2\rangle - |\psi_2\bar{\psi}_1\rangle]/(2)^{1/2} \\
&= \cos^2\theta|\psi_1\bar{\psi}_1\rangle - \sin^2\theta|\psi_2\bar{\psi}_2\rangle \\
&\quad - (2)^{1/2}\cos\theta\sin\theta|{}^3\Psi_1^2\rangle
\end{aligned} \qquad (3.381)$$

Here, $|{}^3\Psi_1^2\rangle$ is the singly excited triplet configuration defined in Eq. (2.261). The closed-shell determinants $|\psi_1\bar{\psi}_1\rangle$ and $|\psi_2\bar{\psi}_2\rangle$ are, of course, singlets. An unrestricted single determinant for the ground state of H_2 is thus not a pure singlet but is contaminated by a triplet. The mixing of the doubly excited determinant $|\psi_2\bar{\psi}_2\rangle$ with $|\psi_1\bar{\psi}_1\rangle$ allows the dissociation to go to the correct limit, but the triplet contaminant is required if the final wave function is to be a single determinant. As $R \to \infty$ the triplet contamination increases until it represents 50% of the wave function,

$$\lim_{R \to \infty} |\Psi_0\rangle = 1/2[|\psi_1\bar{\psi}_1\rangle - |\psi_2\bar{\psi}_2\rangle - (2)^{1/2}|{}^3\Psi_1^2\rangle] \qquad (3.382)$$

Although the correct dissociation energy is obtained using an unrestricted wave function, the poor wave function will limit the desirability of using, near the dissociation limit, an unrestricted single determinant as a starting point for configuration interaction or perturbation calculations.

Exercise 3.44 Derive Eq. (3.379) from Eq. (3.382).

We have only discussed the restricted Hartree-Fock dissociation problem for the minimal basis model of H_2. The ideas presented are not limited to H_2, however, and very similar effects will occur for other closed-shell systems when a bond is stretched. In H_2, the onset of unrestricted solutions occurs beyond the equilibrium distance but in the general case there may even be unrestricted solutions at the experimental geometry. By an extension of our analysis, it is possible to derive general conditions under which there exists an unrestricted solution lower in energy than the closed-shell restricted solution (Thouless, 1961).

NOTES

1. J. S. Binkley, R. A. Whiteside, R. Krishnan, R. Seeger, H. B. Schlegel, D. J. Defrees, and J. A. Pople, *Gaussian 80*, program #406, Quantum Chemistry Program Exchange, Indiana University. More recent versions, such as *Gaussian 88*, are available via Professor John Pople of Carnegie-Mellon University.
2. Hartree-Fock theory in special cases, such as for restricted open-shell wave functions, involves a multideterminantal wave function. Since we will be concerned only with unrestricted open-shell wave functions, all Hartree-Fock wave functions will be single determinants.
3. In principle it is possible to use the procedures to be discussed in this section to obtain restricted closed-shell excited states, but there is no general way of keeping such calculations from converging to the ground state $|\Psi_0\rangle$ if the excited state has the same symmetry as the ground state.
4. C. C. J. Roothaan, New developments in molecular orbital theory, *Rev. Mod. Phys.* **23**: 69 (1951).
5. W. Kolos and L. Wolniewicz, Improved theoretical ground-state energy of the hydrogen molecule, *J. Chem. Phys.* **49**: 404 (1968).
6. L. Wolniewicz, Variational treatment of the HeH$^+$ ion and the β-decay in HT, *J. Chem. Phys.* **43**: 1087 (1965).
7. I. N. Levine, *Quantum Chemistry*, Allyn and Bacon, Boston, 1974, p. 190.
8. S. Huzinaga, Gaussian-type functions for polyatomic systems. I, *J. Chem. Phys.* **42**: 1293 (1965).
9. T. H. Dunning, Gaussian basis functions for use in molecular calculations. I. Contraction of $(9s5p)$ atomic basis sets for the first-row atoms, *J. Chem. Phys.* **53**: 2823 (1970).
10. The slash here separates heavy atom (Li-Ne) basis functions from hydrogen atom basis functions.
11. R. F. Stewart, Small Gaussian expansions of Slater-type orbitals, *J. Chem. Phys.* **52**: 431 (1970).
12. J. A. Pople and R. K. Nesbet, Self-consistent orbitals for radicals, *J. Chem. Phys.* **22**: 571 (1954).

FURTHER READING

Cook, D. B., *Ab Initio Valence Calculations in Chemistry*, Wiley, New York, 1974. This book is a good source of information on the practical aspects of performing *ab initio* self-consistent-field calculations.

Davidson, E. R. and Feller, B., Basis set selection in molecular calculations, *Chem. Rev.* **86**: 681 (1986). This article describes and compares the performance of a variety of Gaussian basis sets that are in common use.

England, W., Salmon, L. S., and Ruedenberg, K., Localized MO's: A bridge between chemical intuition and molecular quantum mechanics, in *Topics in Current Chemistry*, Springer-Verlag, 1971; Caldwell, D. and Eyring, H., Localized orbitals in spectroscopy, *Adv. Quantum Chem.* **11**: 93 (1978). These two references discuss the invariance of Hartree-Fock orbitals to a unitary transformation, and the physical interpretation of particular transformations. A particular concern of many quantum chemists is transformations which localize the orbitals.

Hehre, W. J., Radom, L., Schleyer, P. v. R., and Pople, J. A., *Ab Initio Molecular Orbital Theory*, Wiley, New York, 1986. This excellent book contains a simple account of molecular orbital theory focusing on computational techniques and the Gaussian series of software programs. Numerous applications to chemical problems are discussed. Chapter 4 describes the STO-3G, 4-31G, 6-31G*, and 6-31G** basis sets used in many of our illustrative calculations, as well as others (e.g. 3-21G and 6-311G**) that are now in common use.

Hurley, A. C., *Introduction to the Electron Theory of Small Molecules*, Academic Press, New York, 1976. The restricted open-shell Hartree-Fock procedure, which we have not expanded upon is described.

Melius, C. F. and Goddard, W. A. III, *Ab initio* effective potentials for use in molecular quantum mechanics, *Phys. Rev.* **A10:** 1528 (1974); Kahn, L. R., Baybutt, P., and Truhlar, D. G., *Ab initio* effective core potentials: Reduction of all-electron molecular structure calculations to calculations involving only valence electrons, *J. Chem. Phys.* **65:** 3826 (1976). For calculations involving atoms beyond the second long row of the periodic table, it is common to exclude inner-shell electrons from the calculation and to introduce an effective one-electron potential (a pseudo-potential) which accounts for these electrons. These two papers describe an *ab initio* procedure for deriving such a pseudo-potential.

Ohno, K. and Morokuma, K., *Quantum Chemistry Literature Data Base*, Elsevier, New York, 1982. This book contains a bibliography of *ab initio* calculations for 1978–1980, with more than 2500 literature citations. Since 1982, annual updates of this data base are being published in *J. Mol. Structure (THEOCHEM)*.

Pulay, P., Analytical derivative methods in quantum chemistry. *Adv. Chem. Phys.* **69:** 241 (1987). Recent advances in the calculation of first, second, and higher-order derivatives of both SCF and correlated energies with respect to the internuclear coordinates are reviewed.

Richards, W. G., Walker, T. E. H., and Hinkley, R. K., *A Bibliography of ab initio Molecular Wave Functions*, Oxford University Press, 1971; Supplement for 1970–73, Richards, W. G., Walker, T. E. H., Farnell, L., and Scott, P. R. (1974); Supplement for 1974–1977, Richards, W. G., Scott, P. R., Colbourn, E. A., and Marchington, A. F. (1978); Supplement for 1978–1980, Richards, W. G., Scott, P. R., Sackwild, V., and Robins, S. A. (1981). These books provide an essentially complete catalogue of published *ab initio* calculations.

Schlegel, H. B., Optimization of equilibrium geometries and transition structures. *Adv. Chem. Phys.* **67:** 249 (1987). A number of algorithms for finding stationary points of potential surfaces using derivatives of the energy with respect to internuclear coordinates are discussed.

Thouless, D. J., *The Quantum Mechanics of Many-Body Systems*, Academic Press, New York, 1961. The transition from a restricted to an unrestricted wave function is part of the Hartree-Fock stability problem and is closely related to the random phase approximation. The original source of information on this problem is this book.

FOUR

CONFIGURATION INTERACTION

As we have seen in the last chapter, the Hartree-Fock approximation, while it is remarkably successful in many cases, has its limitations. An example of where it predicts *qualitatively* incorrect results is the ordering of the ionization potentials of N_2. Moreover, the restricted HF method cannot describe the dissociation of molecules into open-shell fragments (e.g., $H_2 \rightarrow 2H$). While the unrestricted HF procedure does give a qualitatively correct picture of such dissociations, the resulting potential energy curves (or, in general, surfaces) are not accurate. The rest of this book will focus on a variety of procedures for improving upon the Hartree-Fock approximation. We will be interested in obtaining the correlation energy (E_{corr}), which is defined as the difference between the exact nonrelativistic energy of the system ($\mathscr{E}_0$) and the Hartree-Fock energy (E_0) obtained in the limit that the basis set approaches completeness

$$E_{corr} = \mathscr{E}_0 - E_0 \tag{4.1}$$

Because the Hartree-Fock energy is an upper bound to the exact energy, the correlation energy is negative.

In this chapter we consider the method of *configuration interaction* (CI) for obtaining the correlation energy. Of all the approaches considered in this book, CI is conceptually (but not computationally!) the simplest. The basic idea is to diagonalize the N-electron Hamiltonian in a basis of N-electron functions (Slater determinants). In other words, we represent the exact wave function as a linear combination of N-electron trial functions and use the linear variational method. If the basis were complete, we would obtain the exact energies not only of the ground state but also of all excited states of the system. In order to limit the size of this chapter, we will not consider CI calculations of excited-state energies.

In principle, CI provides an exact solution of the many-electron problem. In practice, however, we can handle only a finite set of N-electron trial functions; consequently, CI provides only upper bounds to the exact energies. How can we obtain a suitable set of N-electron trial functions? As we have seen in Chapter 2, given some arbitrary set of $2K$ one-electron spin orbitals, we can construct $\binom{2K}{N}$ different N-electron Slater determinants. Unfortunately, even for small molecules and moderately sized one-electron basis sets, the number of N-electron determinants is truly enormous. Thus even if one uses a finite one-electron basis, one usually must truncate the trial function in some way and use only a fraction of all possible N-electron functions.

We will begin this chapter by constructing determinantal trial functions from the Hartree-Fock molecular orbitals, obtained by solving Roothaan's equations. It will prove convenient to describe the possible N-electron functions by specifying how they differ from the Hartree-Fock wave function Ψ_0. Wave functions that differ from Ψ_0 by n spin orbitals are called n-tuply excited determinants. We then consider the structure of the full CI matrix, which is simply the Hamiltonian matrix in the basis of all possible N-electron functions formed by replacing none, one, two, ... all the way up to N spin orbitals in Ψ_0. In Section 4.2 we consider various approximations to the full CI matrix obtained by truncating the many-electron trial function at some excitation level. In particular, we discuss, in some detail, a form of truncated CI in which the trial function contains determinants which differ from Ψ_0 by at most two spin orbitals. Such a calculation is referred to as singly and doubly excited CI (SDCI).

In Section 4.3 we present the results of numerical calculations which serve to illustrate various aspects of the general theory. In Section 4.4 we introduce natural orbitals which have the important property that they lead to the most rapidly convergent CI expansion. In Section 4.5 we briefly discuss another approach to limiting the length of the CI expansion called the multiconfiguration self-consistent field (MCSCF) method. The basic idea of this approach is to use the variation principle to determine the orbitals used in the CI expansion. One considers a trial wave function containing relatively few determinants and optimizes not only the expansion coefficients but also the orbitals. Finally, in Section 4.6 we examine a serious deficiency of all forms of truncated CI that makes the approach inapplicable to large systems and highlights the need for the alternate approximation schemes to be discussed in subsequent chapters.

4.1 MULTICONFIGURATIONAL WAVE FUNCTIONS AND THE STRUCTURE OF THE FULL CI MATRIX

For the sake of simplicity, we assume in this chapter that our molecule of interest has an even number of electrons and is adequately represented, to a first approximation, by a closed-shell restricted HF determinant, $|\Psi_0\rangle$. Suppose we have solved Roothaan's equations in a finite basis set and obtained a set of $2K$ spin orbitals $\{\chi_i\}$. The determinant formed from the N lowest energy spin orbitals is $|\Psi_0\rangle$. As we have seen in Chapter 2, we can form, in addition to $|\Psi_0\rangle$, a large number of other N-electron determinants from the $2K$ spin orbitals. It is convenient to describe these other determinants by stating how they differ from $|\Psi_0\rangle$. Thus the set of possible determinants include $|\Psi_0\rangle$, the singly excited determinants $|\Psi_a^r\rangle$ (which differ from $|\Psi_0\rangle$ in having the spin orbital χ_a replaced by χ_r), the doubly excited determinants $|\Psi_{ab}^{rs}\rangle$, etc., up to and including N-tuply excited determinants. We can use these many-electron wave functions as a basis in which to expand the exact many-electron wave function $|\Phi_0\rangle$. If $|\Psi_0\rangle$ is a reasonable approximation to $|\Phi_0\rangle$, then we know from the variation principle that a better approximation (which becomes exact as the basis becomes complete) is

$$|\Phi_0\rangle = c_0|\Psi_0\rangle + \sum_{ar} c_a^r|\Psi_a^r\rangle + \sum_{\substack{a<b \\ r<s}} c_{ab}^{rs}|\Psi_{ab}^{rs}\rangle$$

$$+ \sum_{\substack{a<b<c \\ r<s<t}} c_{abc}^{rst}|\Psi_{abc}^{rst}\rangle + \sum_{\substack{a<b<c<d \\ r<s<t<u}} c_{abcd}^{rstu}|\Psi_{abcd}^{rstu}\rangle + \cdots \quad (4.2a)$$

This is the form of the full CI wave function. The restrictions on the summation indices (e.g., $a < b$, $r < s$, etc.) insure that a given excited determinant is included in the sum only once. When doing formal manipulations it is sometimes convenient to remove this restriction and rewrite Eq. (4.2a) as

$$|\Phi_0\rangle = c_0|\Psi_0\rangle + \left(\frac{1}{1!}\right)^2 \sum_{ar} c_a^r|\Psi_a^r\rangle + \left(\frac{1}{2!}\right)^2 \sum_{abrs} c_{ab}^{rs}|\Psi_{ab}^{rs}\rangle$$

$$+ \left(\frac{1}{3!}\right)^2 \sum_{\substack{abc \\ rst}} c_{abc}^{rst}|\Psi_{abc}^{rst}\rangle + \left(\frac{1}{4!}\right)^2 \sum_{\substack{abcd \\ rstu}} c_{abcd}^{rstu}|\Psi_{abcd}^{rstu}\rangle + \cdots \quad (4.2b)$$

A factor $(1/n!)^2$ is included in front of the summation involving n-tuply excited determinants to insure that a given excitation is really counted but once. For example, the unrestricted summations for double excitations include the following terms

$$c_{ab}^{rs}|\Psi_{ab}^{rs}\rangle, \quad c_{ba}^{rs}|\Psi_{ba}^{rs}\rangle, \quad c_{ab}^{sr}|\Psi_{ab}^{sr}\rangle, \quad \text{and} \quad c_{ba}^{sr}|\Psi_{ba}^{sr}\rangle$$

Now, if we require the coefficient c_{ab}^{rs} to be antisymmetric with respect to the interchange of a and b or r and s just as the wave functions, then all four terms are equal. Thus the factor of $1/4$ insures that each determinant is counted only once.

How many n-tuple excitations are there? If we have $2K$ spin orbitals, N will be occupied in $|\Psi_0\rangle$ and $2K - N$ will be unoccupied. We can choose n spin orbitals from those occupied in $|\Psi_0\rangle$ in $\binom{N}{n}$ ways. Similarly, we can choose n orbitals from the $2K - N$ virtual orbitals in $\binom{2K - N}{n}$ ways. Thus the total number of n-tuply excited determinants is $\binom{N}{n}\binom{2K - N}{n}$. Even for small molecules and one-electron basis sets of only moderate size, the number of n-tuply excited determinants is extremely large for all n except 0 and 1. A significant number of these determinants can be eliminated (although in most cases not enough!) by exploiting the fact that there is no mixing of wave functions with different spin (i.e., $\langle \Psi_i | \mathscr{H} | \Psi_j \rangle = 0$ if $|\Psi_i\rangle$ and $|\Psi_j\rangle$ have different spin). Suppose we are interested in the singlet states of a molecule. Then we can immediately eliminate from the trial function those determinants which do not have the same number of α and β spin orbitals (i.e., keep only those which are eigenfunctions of $\mathscr{S}_z$ with eigenvalue 0). Moreover, as we have seen in Section 2.5, by taking appropriate linear combinations of these remaining determinants we can form *spin-adapted configurations* which are eigenfunctions of $\mathscr{S}^2$. Thus if we are interested in singlet states we need only include singlet spin-adapted configurations in the trial function. Although actual calculations always use spin-adapted configurations, it will be convenient, however, to develop the formalism in terms of determinants, since the resulting expressions have a simpler structure.

Given the trial function of Eq. (4.2) we can find the corresponding energies by using the linear variational method. As we have seen in Chapter 1, this consists of forming the matrix representation of the Hamiltonian in the basis of the N-electron functions of expansion (4.2) and then finding the eigenvalues of this matrix. This is called the full CI matrix, and the method is referred to as full CI. The lowest eigenvalue will be an upper bound to the ground state energy of the system. The higher eigenvalues will be upper bounds to excited states of the system. Here we will focus only on the lowest eigenvalue. The difference between the lowest eigenvalue ($\mathscr{E}_0$) and the Hartree-Fock energy (E_0) obtained within the same one-electron basis is called *the basis set correlation energy*. As the one-electron basis set approaches completeness, this basis set correlation energy approaches the exact correlation energy. The basis set correlation energy obtained by performing a full CI is, however, exact within the subspace spanned by the one-electron basis. Thus it constitutes a benchmark by which all other approaches to the calculation of the correlation energy performed with the same basis set should be judged. For a given one-electron basis set, full CI is the best that one can do.

To examine the structure of the full CI matrix it is convenient to rewrite the expansion of Eq. (4.2) in a symbolic form

$$|\Phi_0\rangle = c_0|\Psi_0\rangle + c_S|S\rangle + c_D|D\rangle + c_T|T\rangle + c_Q|Q\rangle + \cdots \qquad (4.3)$$

where $|S\rangle$ represents the terms involving single excitations, $|D\rangle$ represents the terms involving double excitations, and so on. Using this notation, the full CI matrix is presented in Fig. 4.1. The following observations are important:

1. There is no coupling between the HF ground state and single excitations (i.e., $\langle\Psi_0|\mathscr{H}|S\rangle = 0$). This is a consequence of Brillouin's theorem (see Subsection 3.3.2), which states that all matrix elements of the form $\langle\Psi_0|\mathscr{H}|\Psi_a^r\rangle$ are zero.

2. There is no coupling between $|\Psi_0\rangle$ and triples or quadruples. Similarly, singles do not mix with quadruples. This is a consequence of the fact that all matrix elements of the Hamiltonian between Slater determinants which differ by more than 2 spin orbitals are zero. A corollary of this is that the blocks that are not zero are sparse. For example, the symbol $\langle D|\mathscr{H}|Q\rangle$ represents

$$\langle D|\mathscr{H}|Q\rangle \leftrightarrow \langle\Psi_{ab}^{rs}|\mathscr{H}|\Psi_{cdef}^{tuvw}\rangle$$

For a matrix element of this type to be nonzero, the indices a and b must be included in the set $\{c, d, e, f\}$ and the indices r and s must be included in the set $\{t, u, v, w\}$.

3. Because single excitations do not mix *directly* with $|\Psi_0\rangle$, they can be expected to have a very small effect on the ground state energy. Their effect is not zero because they do mix *indirectly*; that is, they interact with the doubles which in turn interact with $|\Psi_0\rangle$. Although they have

		$|\Psi_a^r\rangle$	$|\Psi_{ab}^{rs}\rangle$	$|\Psi_{abc}^{rst}\rangle$	$|\Psi_{abcd}^{rstu}\rangle$	$\cdots$
	$|\Psi_0\rangle$	$|S\rangle$	$|D\rangle$	$|T\rangle$	$|Q\rangle$	$\cdots$
$\langle\Psi_0|$	$\langle\Psi_0|\mathscr{H}|\Psi_0\rangle$	0	$\langle\Psi_0|\mathscr{H}|D\rangle$	0	0	$\cdots$
$\langle S|$		$\langle S|\mathscr{H}|S\rangle$	$\langle S|\mathscr{H}|D\rangle$	$\langle S|\mathscr{H}|T\rangle$	0	$\cdots$
$\langle D|$			$\langle D|\mathscr{H}|D\rangle$	$\langle D|\mathscr{H}|T\rangle$	$\langle D|\mathscr{H}|Q\rangle$	$\cdots$
$\langle T|$				$\langle T|\mathscr{H}|T\rangle$	$\langle T|\mathscr{H}|Q\rangle$	$\cdots$
$\langle Q|$					$\langle Q|\mathscr{H}|Q\rangle$	$\cdots$
$\vdots$						$\vdots$

$$\langle S|\mathscr{H}|T\rangle \leftrightarrow \langle\Psi_a^r|\mathscr{H}|\Psi_{cde}^{tuv}\rangle$$
$$\langle D|\mathscr{H}|D\rangle \leftrightarrow \langle\Psi_{ab}^{rs}|\mathscr{H}|\Psi_{cd}^{tu}\rangle$$

Figure 4.1 Structure of the full CI matrix. The matrix is Hermitian and only the upper triangle is shown.

almost negligible effect on the *energy* of the ground state, they do influence the charge distribution and, as we shall see later, single excitations are needed for a proper description of one-electron properties such as the dipole moment. The situation is entirely different for excited-electronic states. In the calculation of the electronic spectra of molecules the single excitations play the primary role.

4. Because it is the double excitations that mix directly with $|\Psi_0\rangle$, it is to be expected that these excitations play an important, and, for small systems, a predominant role in determining the correlation energy. Moreover, it turns out that quadruple excitations are more important than triple or single excitations if one is concerned solely with the ground state energy.

5. All the matrix elements required for actual calculations can be found using the rules described in Chapter 2. As mentioned previously, calculations are performed using spin-adapted configurations. These have been discussed in detail in Section 2.5 for singly and doubly excited determinants. Some of the matrix elements in the CI matrix involving these configurations are given in Table 4.1. They will be used several times later in the book. You might like to test your facility with the rules for evaluation of matrix elements and your stamina by checking the entries in the table.

Table 4.1 Some matrix elements between singlet symmetry-adapted configurations constructed from real orbitals

---

SINGLE EXCITATIONS

$$\langle\Psi_0|\mathscr{H}|\Psi_a^r\rangle = 0$$

$$\langle{}^1\Psi_a^r|\mathscr{H} - E_0|{}^1\Psi_b^s\rangle = (\varepsilon_r - \varepsilon_a)\delta_{rs}\delta_{ab} - (rs|ba) + 2(ra|bs)$$

DOUBLE EXCITATIONS

$$\langle\Psi_0|\mathscr{H}|{}^1\Psi_{aa}^{rr}\rangle = K_{ra}$$

$$\langle\Psi_0|\mathscr{H}|{}^1\Psi_{aa}^{rs}\rangle = 2^{1/2}(sa|ra)$$

$$\langle\Psi_0|\mathscr{H}|{}^1\Psi_{ab}^{rr}\rangle = 2^{1/2}(rb|ra)$$

$$\langle\Psi_0|\mathscr{H}|{}^A\Psi_{ab}^{rs}\rangle = 3^{1/2}((ra|sb) - (rb|sa))$$

$$\langle\Psi_0|\mathscr{H}|{}^B\Psi_{ab}^{rs}\rangle = (ra|sb) + (rb|sa)$$

$$\langle{}^1\Psi_{aa}^{rr}|\mathscr{H} - E_0|{}^1\Psi_{aa}^{rr}\rangle = 2(\varepsilon_r - \varepsilon_a) + J_{aa} + J_{rr} - 4J_{ra} + 2K_{ra}$$

$$\langle{}^1\Psi_{aa}^{rs}|\mathscr{H} - E_0|{}^1\Psi_{aa}^{rs}\rangle = \varepsilon_r + \varepsilon_s - 2\varepsilon_a + J_{aa} + J_{rs} + K_{rs} - 2J_{sa} - 2J_{ra} + K_{sa} + K_{ra}$$

$$\langle{}^1\Psi_{ab}^{rr}|\mathscr{H} - E_0|{}^1\Psi_{ab}^{rr}\rangle = 2\varepsilon_r - \varepsilon_a - \varepsilon_b + J_{rr} + J_{ab} + K_{ab} - 2J_{rb} - 2J_{ra} + K_{rb} + K_{ra}$$

$$\langle{}^A\Psi_{ab}^{rs}|\mathscr{H} - E_0|{}^A\Psi_{ab}^{rs}\rangle = \varepsilon_r + \varepsilon_s - \varepsilon_a - \varepsilon_b + J_{ab} + J_{rs} - K_{ab}$$
$$- K_{rs} - J_{sb} - J_{sa} - J_{rb} - J_{ra} + \tfrac{3}{2}(K_{sb} + K_{sa} + K_{rb} + K_{ra})$$

$$\langle{}^B\Psi_{ab}^{rs}|\mathscr{H} - E_0|{}^B\Psi_{ab}^{rs}\rangle = \varepsilon_r + \varepsilon_s - \varepsilon_a - \varepsilon_b + J_{ab} + J_{rs} + K_{ab}$$
$$+ K_{rs} - J_{sb} - J_{sa} - J_{rb} - J_{ra} + \tfrac{1}{2}(K_{sb} + K_{sa} + K_{rb} + K_{ra})$$

$$\langle{}^A\Psi_{ab}^{rs}|\mathscr{H}|{}^B\Psi_{ab}^{rs}\rangle = (3/4)^{1/2}(K_{sb} - K_{sa} - K_{rb} + K_{ra})$$

4.1.1 Intermediate Normalization and an Expression for the Correlation Energy

Now that we have examined the general features of the full CI matrix, we will study the CI formalism in greater detail, as applied to the ground state of the system. When $|\Psi_0\rangle$ is a reasonable approximation to the exact ground state wave function $|\Phi_0\rangle$, the coefficient c_0 in the CI expansion (see Eq. (4.2)) will be much larger than any of the others. It is convenient to write $|\Phi_0\rangle$ in an *intermediate normalized form*

$$|\Phi_0\rangle = |\Psi_0\rangle + \sum_{ct} c_c^t |\Psi_c^t\rangle + \sum_{\substack{c<d \\ t<u}} c_{cd}^{tu} |\Psi_{cd}^{tu}\rangle$$

$$+ \sum_{\substack{c<d<e \\ t<u<v}} c_{cde}^{tuv} |\Psi_{cde}^{tuv}\rangle + \sum_{\substack{c<d<e<f \\ t<u<v<w}} c_{cdef}^{tuvw} |\Psi_{cdef}^{tuvw}\rangle + \cdots \quad (4.4)$$

Because

$$\langle\Phi_0|\Phi_0\rangle = 1 + \sum_{ct} (c_c^t)^2 + \sum_{\substack{c<d \\ t<u}} (c_{cd}^{tu})^2 + \cdots$$

this wave function is not normalized. However, it has the property that

$$\langle\Psi_0|\Phi_0\rangle = 1 \quad (4.5)$$

Given the intermediate normalized $|\Phi_0\rangle$ we can always normalize it, if we so desire, by multiplying each term in the expansion by a constant (i.e., $|\Phi_0'\rangle = c'|\Phi_0\rangle$, chosen so that $\langle\Phi_0'|\Phi_0'\rangle = 1$.)

As discussed in Chapter 1, an equivalent formulation of the linear variation method is simply to write

$$\mathscr{H}|\Phi_0\rangle = \mathscr{E}_0|\Phi_0\rangle \quad (4.6)$$

where $|\Phi_0\rangle$ is given by Eq. (4.4) and then successively multiply this equation by $\langle\Psi_0|$, Ψ_a^r, $\langle\Psi_{ab}^{rs}|$, etc. Before we do this, it is convenient to rewrite Eq. (4.6) by subtracting $E_0|\Phi_0\rangle$ from both sides to obtain

$$(\mathscr{H} - E_0)|\Phi_0\rangle = (\mathscr{E}_0 - E_0)|\Phi_0\rangle = E_{\text{corr}}|\Phi_0\rangle \quad (4.7)$$

where E_{corr} is the correlation energy. If we multiply both sides of this equation by $\langle\Psi_0|$ we obtain

$$\langle\Psi_0|\mathscr{H} - E_0|\Phi_0\rangle = E_{\text{corr}}\langle\Psi_0|\Phi_0\rangle = E_{\text{corr}} \quad (4.8)$$

where we have used the fact that $|\Phi_0\rangle$ is intermediately normalized. Now consider the left-hand side of the equation. Using the expansion in Eq. (4.4), we have

$$\langle\Psi_0|\mathscr{H} - E_0|\Phi_0\rangle = \langle\Psi_0|\mathscr{H} - E_0\left(|\Psi_0\rangle + \sum_{ct} c_c^t|\Psi_c^t\rangle + \sum_{\substack{c<d \\ t<u}} c_{cd}^{tu}|\Psi_{cd}^{tu}\rangle + \cdots\right)$$

$$= \sum_{\substack{c<d \\ t<u}} c_{cd}^{tu}\langle\Psi_0|\mathscr{H}|\Psi_{cd}^{tu}\rangle \quad (4.9)$$

where we have used Brillouin's theorem ($\langle\Psi_0|\mathscr{H}|\Psi_c^t\rangle = 0$) and the fact that triple and higher excitations do not mix with $|\Psi_0\rangle$ because they differ from $|\Psi_0\rangle$ by more than two spin orbitals. Combining Eqs. (4.8) and (4.9) we have the following explicit expression for the correlation energy:

$$E_{\mathrm{corr}} = \sum_{\substack{a<b \\ r<s}} c_{ab}^{rs}\langle\Psi_0|\mathscr{H}|\Psi_{ab}^{rs}\rangle \tag{4.10}$$

Thus the correlation energy is determined solely by the coefficients of the double excitations in the intermediate normalized CI function. This does *not* mean that only double excitations need to be included for an exact CI description of the ground state; the coefficients $\{c_{ab}^{rs}\}$ are affected by the presence of other excitations. To see this, multiply Eq. (4.7) by $\langle\Psi_a^r|$ to obtain

$$\langle\Psi_a^r|\mathscr{H} - E_0|\Phi_0\rangle = E_{\mathrm{corr}}\langle\Psi_a^r|\Phi_0\rangle$$

Using the expansion for $|\Phi_0\rangle$ and Brillouin's theorem this becomes

$$\sum_{ct} c_c^t\langle\Psi_a^r|\mathscr{H} - E_0|\Psi_c^t\rangle + \sum_{\substack{c<d \\ t<u}} c_{cd}^{tu}\langle\Psi_a^r|\mathscr{H}|\Psi_{cd}^{tu}\rangle + \sum_{\substack{c<d<e \\ t<u<v}} c_{cde}^{tuv}\langle\Psi_a^r|\mathscr{H}|\Psi_{cde}^{tuv}\rangle$$

$$= E_{\mathrm{corr}}c_a^r \tag{4.11}$$

This expression can be simplified somewhat by taking into account the fact that there are nonzero matrix elements between singles and triples only when a equals c, d, or e and r equals t, u, or v. This allows us to rewrite Eq. (4.11) as

$$\sum_{ct} c_c^t\langle\Psi_a^r|\mathscr{H} - E_0|\Psi_c^t\rangle + \sum_{\substack{c<d \\ t<u}} c_{cd}^{tu}\langle\Psi_a^r|\mathscr{H}|\Psi_{cd}^{tu}\rangle + \sum_{\substack{c<d \\ t<u}} c_{acd}^{rtu}\langle\Psi_a^r|\mathscr{H}|\Psi_{acd}^{rtu}\rangle$$

$$= E_{\mathrm{corr}}c_a^r \tag{4.12}$$

The important point about this equation is that it contains, and hence couples, the coefficients of the singles, doubles, and triples. If we continue the above procedure by multiplying Eq. (4.7) by $\langle\Psi_{ab}^{rs}|$, $\langle\Psi_{abc}^{rst}|$, etc., we would end up with a hierarchy of equations that must be solved simultaneously to obtain the correlation energy. This set of coupled equations is extremely large if all possible excitations are included. This is just another way of saying that the full CI matrix is extremely large. After illustrating the formalism developed so far, by applying it to minimal basis H_2, we will return to the problem of truncating the CI matrix to a manageable size.

Exercise 4.1 Obtain Eq. (4.12) from (4.11). It will prove convenient to use unrestricted summations.

Let us consider the application of the above formalism to minimal basis H_2. Since this is a two-electron system, full CI involves only single and double excitations. Recall that in this model we have two molecular orbitals: ψ_1 is the bonding orbital with gerade symmetry and ψ_2 is the antibonding orbital with ungerade symmetry. The HF ground state wave function is

$$|\Psi_0\rangle = |\psi_1\bar{\psi}_1\rangle = |1\bar{1}\rangle \tag{4.13}$$

Since we have four spin orbitals ($\chi_1 \equiv 1$, $\chi_2 \equiv \bar{1}$, $\chi_3 \equiv 2$, $\chi_4 \equiv \bar{2}$) we can form in addition to $|\Psi_0\rangle$, five other determinants namely, $|1\bar{2}\rangle$, $|2\bar{1}\rangle$, $|12\rangle$, $|\bar{2}\bar{1}\rangle$, and $|2\bar{2}\rangle$. Using these determinants, the full CI wave function can be written as

$$|\Phi_0\rangle = |\Psi_0\rangle + c_{\bar{1}}^2|2\bar{1}\rangle + c_1^{\bar{2}}|1\bar{2}\rangle + c_1^2|12\rangle + c_{\bar{1}}^{\bar{2}}|\bar{2}\bar{1}\rangle + c_{1\bar{1}}^{2\bar{2}}|2\bar{2}\rangle \tag{4.14}$$

We can rewrite this in terms of spin-adapted configurations as follows. Since the exact ground state is a singlet we know that only configurations of singlet symmetry need be included in the expansion. The doubly excited state is a closed-shell and hence a singlet. Out of the four singly excited determinants $|2\bar{1}\rangle$, $|1\bar{2}\rangle$, $|12\rangle$, $|\bar{2}\bar{1}\rangle$ we can form one singlet state and three triplets. The singlet state is (see Eq. (2.260))

$$|^1\Psi_1^2\rangle = 2^{-1/2}(|1\bar{2}\rangle + |2\bar{1}\rangle)$$

Thus the spin-adapted expansion is

$$|\Phi_0\rangle = |\Psi_0\rangle + c_1^2|^1\Psi_1^2\rangle + c_{1\bar{1}}^{2\bar{2}}|2\bar{2}\rangle \tag{4.15}$$

Finally, we can simplify the expansion further by taking into account the spatial symmetry of the system. Both $|\Psi_0\rangle$ and $|2\bar{2}\rangle$ are of gerade symmetry while $|^1\Psi_1^2\rangle$ is of ungerade symmetry because it contains one orbital with gerade and one with ungerade symmetry. Therefore, this single excitation will not mix with $|\Psi_0\rangle$ or $|2\bar{2}\rangle$. Thus we can write the CI expansion, which is both symmetry and spin-adapted, as

$$|\Phi_0\rangle = |\Psi_0\rangle + c_{1\bar{1}}^{2\bar{2}}|2\bar{2}\rangle = |\Psi_0\rangle + c_{1\bar{1}}^{2\bar{2}}|\Psi_{1\bar{1}}^{2\bar{2}}\rangle \tag{4.16}$$

Given this trial function, the variational method tells us that the corresponding energy ($\mathscr{E}_0$) is the lowest eigenvalue of the CI matrix

$$\mathbf{H} = \begin{pmatrix} \langle\Psi_0|\mathscr{H}|\Psi_0\rangle & \langle\Psi_0|\mathscr{H}|\Psi_{1\bar{1}}^{2\bar{2}}\rangle \\ \langle\Psi_{1\bar{1}}^{2\bar{2}}|\mathscr{H}|\Psi_0\rangle & \langle\Psi_{1\bar{1}}^{2\bar{2}}|\mathscr{H}|\Psi_{1\bar{1}}^{2\bar{2}}\rangle \end{pmatrix}$$

The required matrix elements are readily evaluated using the rules of Chapter 2. Since the molecular orbitals are real, we have

$$\langle\Psi_0|\mathscr{H}|\Psi_0\rangle = E_0 = 2h_{11} + J_{11} \tag{4.17a}$$

$$\langle\Psi_0|\mathscr{H}|\Psi_{1\bar{1}}^{2\bar{2}}\rangle = \langle 1\bar{1}||2\bar{2}\rangle = (12|12) = K_{12} = \langle\Psi_{1\bar{1}}^{2\bar{2}}|\mathscr{H}|\Psi_0\rangle \tag{4.17b}$$

$$\langle\Psi_{1\bar{1}}^{2\bar{2}}|\mathscr{H}|\Psi_{1\bar{1}}^{2\bar{2}}\rangle = 2h_{22} + J_{22} \tag{4.17c}$$

Using the HF orbital energies (see Eqs. (3.130) and (3.131))

$$\varepsilon_1 = h_{11} + J_{11}$$

$$\varepsilon_2 = h_{22} + 2J_{12} - K_{12}$$

the diagonal matrix elements can be rewritten as

$$E_0 = 2\varepsilon_1 - J_{11} \tag{4.17d}$$

$$\langle \Psi_{1\bar{1}}^{2\bar{2}} | \mathcal{H} | \Psi_{1\bar{1}}^{2\bar{2}} \rangle = 2\varepsilon_2 - 4J_{12} + J_{22} + 2K_{12} \tag{4.17e}$$

Having evaluated the matrix elements, it is a straightforward matter to find the lowest eigenvalue of the matrix using the secular determinant or unitary transformation approach discussed in Chapter 1. Here we wish to solve the problem by a somewhat different but completely equivalent way, that we shall use many times in this book. We start by substituting Eq. (4.16) into Eq. (4.7):

$$(\mathcal{H} - E_0)(|\Psi_0\rangle + c|\Psi_{1\bar{1}}^{2\bar{2}}\rangle) = E_{corr}(|\Psi_0\rangle + c|\Psi_{1\bar{1}}^{2\bar{2}}\rangle) \tag{4.18}$$

where we have written c for $c_{1\bar{1}}^{2\bar{2}}$. Multiplying this equation by $\langle \Psi_0|$ we have

$$E_{corr} = c\langle \Psi_0 | \mathcal{H} | \Psi_{1\bar{1}}^{2\bar{2}} \rangle = cK_{12} \tag{4.19a}$$

Similarly, multiplying by $\langle \Psi_{1\bar{1}}^{2\bar{2}}|$ we have

$$\langle \Psi_{1\bar{1}}^{2\bar{2}} | \mathcal{H} | \Psi_0 \rangle + c\langle \Psi_{1\bar{1}}^{2\bar{2}} | \mathcal{H} - E_0 | \Psi_{1\bar{1}}^{2\bar{2}} \rangle = cE_{corr} \tag{4.19b}$$

Defining

$$2\Delta = \langle \Psi_{1\bar{1}}^{2\bar{2}} | \mathcal{H} - E_0 | \Psi_{1\bar{1}}^{2\bar{2}} \rangle = 2(\varepsilon_2 - \varepsilon_1) + J_{11} + J_{22} - 4J_{12} + 2K_{12} \tag{4.20}$$

where we have used the matrix elements in Eq. (4.17), we can rewrite Eq. (4.19b) as

$$K_{12} + 2\Delta c = cE_{corr} \tag{4.21}$$

The two simultaneous equations (4.19a) and (4.21) can be combined into the matrix equation

$$\begin{pmatrix} 0 & K_{12} \\ K_{12} & 2\Delta \end{pmatrix} \begin{pmatrix} 1 \\ c \end{pmatrix} = E_{corr} \begin{pmatrix} 1 \\ c \end{pmatrix} \tag{4.22}$$

We could have obtained this result directly from the CI eigenvalue problem:

$$\begin{pmatrix} E_0 & K_{12} \\ K_{12} & \langle \Psi_{1\bar{1}}^{2\bar{2}} | \mathcal{H} | \Psi_{1\bar{1}}^{2\bar{2}} \rangle \end{pmatrix} \begin{pmatrix} c_0 \\ c_1 \end{pmatrix} = \mathscr{E} \begin{pmatrix} c_0 \\ c_1 \end{pmatrix}$$

by simply subtracting

$$\begin{pmatrix} E_0 & 0 \\ 0 & E_0 \end{pmatrix} \begin{pmatrix} c_0 \\ c_1 \end{pmatrix}$$

from both sides, using the definition of 2Δ (Eq. (4.20)) and setting $c_0 = 1$ (intermediate normalization), $\mathscr{E} - E_0 = E_{corr}$, and $c_1 = c$. To obtain the lowest eigenvalue we solve Eq. (4.21) for c:

$$c = \frac{K_{12}}{E_{corr} - 2\Delta}$$

and substitute this into Eq. (4.19a) to obtain

$$E_{\text{corr}} = \frac{K_{12}^2}{E_{\text{corr}} - 2\Delta}$$

This equation is a quadratic equation for E_{corr}, which can be solved for the lowest root, i.e.,

$$E_{\text{corr}} = \Delta - (\Delta^2 + K_{12}^2)^{1/2} \qquad (4.23)$$

This is the exact correlation energy of H_2 within the minimal basis set of atomic orbitals.

Exercise 4.2 Using the secular determinant approach show that the lowest eigenvalue of the matrix

$$\begin{pmatrix} 0 & K_{12} \\ K_{12} & 2\Delta \end{pmatrix}$$

is given by Eq. (4.23).

The exact energy of minimal basis H_2 is

$$\mathscr{E}_0 = E_0 + E_{\text{corr}} = 2h_{11} + J_{11} + \Delta - (\Delta^2 + K_{12}^2)^{1/2} \qquad (4.24)$$

In contrast to E_0, this full CI energy properly describes the dissociation of H_2, as might be expected, since it is the exact energy in the basis. To see this, recall that as $R \to \infty$, $h_{11} = h_{22} \to E(H)$, where $E(H)$ is the energy of the hydrogen atom in the basis and that *all* molecular orbital two-electron integrals tend to $\frac{1}{2}(\phi_1\phi_1|\phi_1\phi_1)$, where ϕ_1 is a hydrogenic orbital. It then follows that $\Delta \to 0$ as $R \to \infty$ and hence $E_{\text{corr}} \to -K_{12} = -\frac{1}{2}(\phi_1\phi_1|\phi_1\phi_1)$ which exactly cancels the long range limit of J_{11} thus ensuring that $\mathscr{E}_0$ approaches $2E(H)$. The full CI, RHF, and UHF potential energy curves for STO-3G H_2 are compared in Fig. 4.2. Note that although, in contrast to RHF, UHF does describe dissociation properly, the UHF potential curve is significantly different from the full CI one. For comparison the essentially exact nonrelativistic results of Kolos and Wolniewicz are shown. Their calculations, which exploit the simplifications inherent in a two-electron system, use wave functions which explicitly contain the interelectronic distance (i.e., r_{12}). One can see that although full CI is exact in the STO-3G basis, it gives a potential curve which does not agree very well with the exact one. Although the full CI STO-3G well depth is greater than the exact result, this does not imply that the variation principle has been violated. The STO-3G full CI energy of H_2 and the STO-3G energy of the hydrogen atom are both higher than the corresponding exact results. However, the STO-3G potential energy curve is obtained by subtracting the energy of two isolated H atoms from the energy of H_2 and thus need not be an upper bound to the exact curve. The STO-3G UHF and full CI well depths are greater than the exact result because the STO-3G basis is so poor for the hydrogen atom.

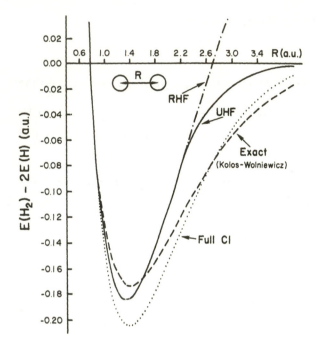

Figure 4.2 STO-3G potential energy curves for H_2.

Exercise 4.3 Calculate the coefficient of the double excitation (c) in the intermediate normalized CI wave function at $R = 1.4$ a.u., using the STO-3G integrals given in Appendix D. Show analytically that as $R \to \infty$, $c \to -1$, and hence that at large distances the Hartree-Fock ground state and the doubly excited configuration have equal weight in the CI ground state. Finally, show that the CI wave function, when normalized to unity, becomes (at $R = \infty$)

$$2^{-1/2}(|\phi_1 \bar{\phi}_2\rangle + |\phi_2 \bar{\phi}_1\rangle)$$

where ϕ_1 and ϕ_2 are atomic orbitals on centers one and two, respectively.

4.2 DOUBLY EXCITED CI

For all but the smallest molecules, even with a minimal basis set, full CI is a computationally impractical procedure. With a one-electron basis of moderate size, there are so many possible spin-adapted configurations that the full CI matrix becomes impossibly large (e.g., its dimensionality is greater than $10^9 \times 10^9$). To obtain a computationally viable scheme one must truncate the full CI matrix or equivalently the CI expansion for the wave function. A systematic procedure for accomplishing this is to consider only those configurations which differ from the Hartree-Fock ground state wave function by no more than a predetermined number, say m, of spin orbitals. For example, if $m = 4$, then one would include single, double, triple, and quadruple excitations in the trial function. For full CI, m would be equal to the number of electrons. The simplest version of this scheme includes only single and double excitations out of $|\Psi_0\rangle$. The resulting ground state energy obtained from such a singly and doubly excited CI (SDCI) calculation is, of course, variational, and for small molecules it gives the major fraction of the correlation energy. As we shall see in Section 4.6, SDCI and, in fact, all forms of truncated CI deteriorate as the number of electrons increases. In spite of this deficiency, SDCI has been a popular approach to the calculation of correlation energies. The effect of single excitations on the correlation energy (but not on the charge distribution) is generally negligible but their number is so much smaller than double excitations that they can be included without complicating the calculations. Here, however, in order to keep the formalism as simple as possible, we ignore single excitations.

We will discuss doubly excited CI (DCI) using a spin orbital basis, but it should be kept in mind that actual calculations are performed using spin-adapted configurations. The intermediate normalized DCI trial function is

$$|\Phi_{\text{DCI}}\rangle = |\Psi_0\rangle + \sum_{\substack{c<d \\ t<u}} c_{cd}^{tu}|\Psi_{cd}^{tu}\rangle \tag{4.25}$$

To obtain the corresponding correlation energy we substitute this into Eq. (4.7)

$$(\mathscr{H} - E_0)\left(|\Psi_0\rangle + \sum_{\substack{c<d \\ t<u}} c_{cd}^{tu}|\Psi_{cd}^{tu}\rangle\right) = E_{\text{corr}}\left(|\Psi_0\rangle + \sum_{\substack{c<d \\ t<u}} c_{cd}^{tu}|\Psi_{cd}^{tu}\rangle\right)$$

and successively multiply by $\langle\Psi_0|$ and $\langle\Psi_{ab}^{rs}|$ to obtain

$$\sum_{\substack{c<d \\ t<u}} c_{cd}^{tu}\langle\Psi_0|\mathscr{H}|\Psi_{cd}^{tu}\rangle = E_{\text{corr}} \tag{4.26a}$$

$$\langle\Psi_{ab}^{rs}|\mathscr{H}|\Psi_0\rangle + \sum_{\substack{c<d \\ t<u}} c_{cd}^{tu}\langle\Psi_{ab}^{rs}|\mathscr{H} - E_0|\Psi_{cd}^{tu}\rangle = c_{ab}^{rs}E_{\text{corr}} \tag{4.26b}$$

These two equations, which are generalizations of Eqs. (4.19a) and (4.19b), determine the correlation energy. By defining the matrices

$$(\mathbf{B})_{rasb} = \langle\Psi_{ab}^{rs}|\mathscr{H}|\Psi_0\rangle \tag{4.27a}$$

$$(\mathbf{D})_{rasb,\,tcud} = \langle\Psi_{ab}^{rs}|\mathscr{H} - E_0|\Psi_{cd}^{tu}\rangle \tag{4.27b}$$

$$(\mathbf{c})_{rasb} = c_{ab}^{rs} \tag{4.27c}$$

Eqs. (4.26a) and (4.26b) can be rewritten as

$$\mathbf{B}^{\dagger}\mathbf{c} = E_{\text{corr}} \tag{4.28a}$$

$$\mathbf{B} + \mathbf{D}\mathbf{c} = \mathbf{c}E_{\text{corr}} \tag{4.28b}$$

which are equivalent to

$$\begin{pmatrix} 0 & \mathbf{B}^{\dagger} \\ \mathbf{B} & \mathbf{D} \end{pmatrix} \begin{pmatrix} 1 \\ \mathbf{c} \end{pmatrix} = E_{\text{corr}} \begin{pmatrix} 1 \\ \mathbf{c} \end{pmatrix} \tag{4.29}$$

This is the matrix form of the DCI equations. The correlation energy is the lowest eigenvalue of the CI matrix

$$\begin{pmatrix} 0 & \mathbf{B}^{\dagger} \\ \mathbf{B} & \mathbf{D} \end{pmatrix}$$

which, as can be seen from the definitions in Eqs. (4.27a) and (4.27b), is simply the matrix representation of the Hamiltonian in the basis $\{|\Psi_0\rangle, \Psi_{ab}^{rs}\rangle\}$ with the Hartree-Fock energy E_0 subtracted from all diagonal elements.

Even for DCI it is not, in general, practical to include all possible double excitations from $|\Psi_0\rangle$. Thus it is desirable to devise procedures for selecting the most important configurations in advance. A simple way of investigating the importance of different doubly excited configurations is to examine a perturbation treatment of this CI eigenvalue problem. If we solve Eq. (4.28b) for $\mathbf{c}$,

$$\mathbf{c} = -(\mathbf{D} - \mathbf{1}E_{\text{corr}})^{-1}\mathbf{B}$$

and substitute the result into Eq. (4.28a) we obtain

$$E_{\text{corr}} = -\mathbf{B}^{\dagger}(\mathbf{D} - \mathbf{1}E_{\text{corr}})^{-1}\mathbf{B} \tag{4.30a}$$

Solving this matrix equation for E_{corr} is completely equivalent to finding the lowest eigenvalue of the CI matrix. Since E_{corr} appears on both sides of the equation an iterative procedure must be used. Since the correlation energy is small compared to the difference between the energy of a doubly excited configuration and E_0 (i.e., the diagonal elements of $\mathbf{D}$), one can set $E_{\text{corr}} = 0$ on the right side of the equation to obtain the approximation

$$E'_{\text{corr}} = -\mathbf{B}^{\dagger}\mathbf{D}^{-1}\mathbf{B} \tag{4.30b}$$

This result can be used in the right-hand side of Eq. (4.30a) to obtain E''_{corr} and so on until convergence is found. It is interesting to note that although the first iteration result E'_{corr} is clearly an approximation to the DCI correlation energy and in fact is no longer an upper bound, as will be discussed subsequently, it has an advantage over the exact DCI result in that it does not deteriorate as the size of the system increases.

When the matrix $\mathbf{D}$ is very large, its inverse cannot be computed and Eq. (4.30b) must be simplified further. In most cases, the largest elements of

D occur on its diagonal. If we assume **D** is diagonal, its inverse is easily obtained namely,

$$(\mathbf{D}^{-1})_{rasb,\,tcud} = \frac{\delta_{ac}\delta_{bd}\delta_{rt}\delta_{su}}{\langle\Psi_{ab}^{rs}|\mathscr{H} - E_0|\Psi_{ab}^{rs}\rangle}$$

So that the correlation energy can be written as

$$E_{\text{corr}} \cong -\sum_{\substack{a<b\\r<s}} \frac{\langle\Psi_0|\mathscr{H}|\Psi_{ab}^{rs}\rangle\langle\Psi_{ab}^{rs}|\mathscr{H}|\Psi_0\rangle}{\langle\Psi_{ab}^{rs}|\mathscr{H} - E_0|\Psi_{ab}^{rs}\rangle} = \sum_{\substack{a<b\\r<s}} E_{\text{corr}}\binom{rs}{ab} \qquad (4.31)$$

where $E_{\text{corr}}\binom{rs}{ab}$ is the contribution of the double excitation $|\Psi_{ab}^{rs}\rangle$ to this approximate correlation energy. Since it can be easily computed, it can be used to identify the configurations that are likely to be most important in the DCI expansion.

4.3 SOME ILLUSTRATIVE CALCULATIONS

In this section we present a variety of results obtained via CI, which serve to illustrate the formalism of the previous sections. We begin by considering the correlation energy of H_2 at $R = 1.4$ a.u., calculated using basis sets of increasing sophistication (See Table 4.2). Because H_2 is a two-electron system, SDCI corresponds to full CI. Of particular interest is that single excitations make an almost negligible contribution to the correlation energy. The contribution of double excitations to the correlation energy is strongly dependent on the quality of the basis. Note that even using a large basis of $10s$, $5p$, and $1d$ Gaussian functions, the full CI correlation energy still does not quite approach the essentially exact value obtained using the highly

Table 4.2 The correlation energy (a.u.) of H_2 at $R = 1.4$ a.u. in a variety of basis sets

Basis set	DCI	SDCI	Contribution of singles
STO-3G	−0.02056	−0.02056	0
4-31G	−0.02487	−0.02494	−0.00007
6-31G**	−0.03373	−0.03387	−0.00014
$(10s, 5p, 1d)^a$	−0.03954	−0.03969	−0.00015
Exact[b]		−0.0409	

[a] C. E. Dykstra (unpublished) using a large Gaussian basis described by J. M. Schulman and D. N. Kaufman, *J. Chem. Phys.* **53**: 477 (1970).

[b] Obtained from the extremely accurate results of W. Kolos and L. Wolniewicz, *J. Chem. Phys.* **49**: 404 (1968).

Table 4.3 Equilibrium bond length (a.u.) of H_2

Basis set	SCF	Full CI
STO-3G	1.346	1.389
4-31G	1.380	1.410
6-31G**	1.385	1.396
Exact[a]		1.401

[a] W. Kolos and L. Wolniewicz, *J. Chem. Phys.* **49**: 404 (1968).

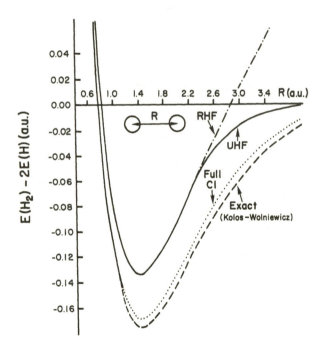

Figure 4.3 6-31G** potential energy curves for H_2.

accurate ground state energy calculated by Kolos and Wolniewicz. This is primarily the result of the absence from the basis set of additional d orbitals and orbitals of higher symmetry.

Table 4.3 presents the equilibrium bond lengths of H_2 obtained by minimizing the energy as a function of the internuclear distance R. For all basis sets the full CI results are better than the SCF ones. At the 6-31G** level, the agreement with the exact results is better than 0.005 a.u.. Figure

Table 4.4 The number of possible symmetry- and spin-adapted configurations for BeH_2 at various excitation levels of CI^a

Basis	Excitation levels			
	SDCI (≤ 2)	SDTCI (≤ 3)	SDTQCI (≤ 4)	Full CI
Minimal	20	36	56	65
Double zeta	146	728	2173	4544

[a] K. Hsu, R. P. Hosteny, and I. Shavitt, unpublished results.

4.3 compares the entire potential energy curves obtained using RHF, UHF, and full CI at the 6-31G** level. As seen from the STO-3G results (Fig. 4.2), full CI correctly describes dissociation and significantly changes the UHF well depth. Moreover, in contrast to Fig. 4.2, the 6-31G** full CI curve now nearly parallels the exact result of Kolos and Wolniewicz.

BeH_2 is among the first systems for which full CI calculations have been performed using nonminimal basis sets. Since it has six electrons, full CI includes all excitations up to and including hextuples. The number of symmetry- and spin-adapted configurations, which are obtained from either a minimal or a double zeta basis set of STO's, when different levels of excitation are included in the trial function, is shown in Table 4.4. A double zeta basis for BeH_2 is still a very small basis in the sense that it can give only about 50% of the exact correlation energy. If this basis were augmented by adding polarization functions, the numbers of configurations in Table 4.4 would increase considerably. For example, if a single set of p orbitals were added to the hydrogens, the number of symmetry- and spin-adapted configurations needed for full CI would be 28,639. Such a basis would still give only about 75% of the exact correlation energy. To obtain a greater fraction of the correlation energy d orbitals on Be would be required.

Table 4.5 presents the correlation energy contributions from different levels of excitation in BeH_2. With either basis set, double excitations are responsible for almost 98% of the full CI correlation energy. This result is not general but is a consequence of the relatively few electrons present in BeH_2. A particularly interesting point is the relative importance of triple and quadruple excitations. It is seen that in the minimal basis they appear equally important, but in the double zeta basis the quadruples are clearly more important. In fact, in calculations performed with a double-zeta (DZ) basis augmented by polarization functions on the hydrogens (SDCI correlation energy $= -0.098468$ a.u.), the contribution to the correlation energy due to triples is essentially the same as in the DZ result, but the contribution due to quadruples almost doubles. Thus quadruple excitations appear to be more important than triples for the correlation energy.

Table 4.5 Correlation energy contributions (a.u.) from different levels of excitation in $BeH_2{}^a$

Excitation level	Minimal basis	Double zeta
1 + 2	− 0.030513	− 0.074033
3	− 0.000329	− 0.000428
4	− 0.000340	− 0.001439
5	− 0.0000001	− 0.000011
6	− 0.00000002	− 0.000006
Total	− 0.031182	− 0.075917
Exact correlation energy	− 0.14	

[a] K. Hsu, R. P. Hosteny, and I. Shavitt, unpublished results.

As we have discussed previously, full CI is impractical for large molecules using extended basis sets, and many CI calculations include only single and double excitations. As an example, we consider the SDCI calculation on H_2O performed by I. Shavitt and his coworkers. They used an extended basis set of Slater-type orbitals (STO's), consisting of $5s$, $4p$, and $2d$ orbitals on the oxygen and $3s$ and $1p$ orbitals on the hydrogens (henceforth referred to as the 39-STO basis). This basis set gives an SCF energy close to the Hartree-Fock limit. In this basis, there are 4120 symmetry- and spin-adapted configurations involving single and double excitations. Shavitt and coworkers examined a variety of properties including the potential energy surface, ionization potentials, and dipole moment of H_2O both at the SCF and SDCI levels. Their results for the correlation energy obtained using only double excitations (DCI) and both single and double excitations (SDCI) are presented in Table 4.6. Notice that the inclusion of single excitations

Table 4.6 Correlation energy (a.u.) of H_2O at the experimental geometry calculated with the 39-STO basis described in the text[a]

	E_{corr}
DCI	− 0.2740
SDCI	− 0.2756
Estimated full CI	− 0.296 ± 0.001
Exact	− 0.37

[a] B. J. Rosenberg and I. Shavitt, *J. Chem. Phys.* **63**: 2162 (1975).

Table 4.7 Equilibrium geometry and some force constants of H_2O using SCF and SDCI calculations with the 39-STO basis described in the text

	SCF[a]	SDCI[a]	Experiment[b]
R_e (a.u.)	1.776	1.800	1.809
θ_e	106.1°	104.9°	104.5°
f_{RR}	9.79	8.88	8.45
$f_{\theta\theta}$	0.88	0.81	0.76

[a] B. J. Rosenberg, W. C. Ermler, and I. Shavitt, *J. Chem. Phys.* **65**: 4072 (1976).

[b] The force constants are in 10^5 dyn/cm, with angles measured in radians. For original references see footnote a of this table.

changes the DCI correlation energy by only a very small amount (i.e., 0.6%). The SDCI result gives almost 75% of the exact correlation energy. However, the exact correlation energy is different from the basis set correlation energy which is the result that would be obtained if full CI were to be performed in the 39-STO basis. Since full CI is out of the question for such a large basis, the resulting energy can only be estimated. An educated guess for the basis set correlation energy is -0.296 ± 0.001 a.u. Thus it seems that SDCI gives about 93% of the basis set correlation energy.

Results for the equilibrium geometry and two force constants are presented in Table 4.7. These were obtained by performing calculations at 36 molecular geometries, finding the equilibrium geometry and then fitting the surface to a Taylor series about the equilibrium geometry, i.e.,

$$E(R_1, R_2, \theta) = E_e + \tfrac{1}{2}f_{RR}[(R_1 - R_e)^2 + (R_2 - R_e)^2] + \tfrac{1}{2}f_{\theta\theta}R_e^2(\theta - \theta_e)^2 + \cdots$$
$$(4.32)$$

where R_1 and R_2 are the two hydrogen-oxygen bond lengths, and θ is the bond angle. A large number of additional force constants can be found in their paper. From Table 4.7 it can be seen that the SDCI results agree much better with experiment than the SCF ones. The fact that the agreement is still far from perfect may be attributed to: 1) the one-electron basis not being complete and 2) the SDCI approximation to full CI. The relative role of these two factors can only be established by performing calculations with larger basis sets and by treating correlation effects beyond the SDCI level.

We now turn to the calculation of ionization potentials (IP's) within the SDCI formalism. The results for H_2O using the 39-STO basis are shown in Table 4.8. The SDCI entry was obtained by performing separate SDCI calculations on H_2O and H_2O^+ and subtracting the resulting energies. The

Table 4.8 Lowest ionization potential (a.u.) of water obtained using Koopmans' theorem and by performing SDCI calculations both on H_2O and H_2O^+ (The 39-STO basis described in the text was used)[a]

	Lowest IP of H_2O
Koopmans'	0.507
SDCI	0.452
Experiment	0.463

[a] B. J. Rosenberg and I. Shavitt, *J. Chem. Phys.* **63**: 2162 (1975).

SDCI result is a considerable improvement over the Koopmans' theorem value but is still not in complete agreement with experiment. A particularly good test of any procedure for calculating ionization potentials is N_2 since for this molecule Koopmans' theorem predicts the incorrect ordering of the first two ionization potentials. The SDCI results, obtained using a large basis set of $6s$, $4p$, $3d$, and $2f$ Slater-type functions on each nitrogen, are shown in Table 4.9. We note that SDCI predicts the correct ordering of the ionization potentials. These results are expected to be close to the SDCI-limit (i.e., SDCI with a complete one-electron basis). Thus the discrepancies with experiment are a reflection of the importance of higher excitations in the CI wave function.

As a final example, we consider the calculation of first-order properties as exemplified by the dipole moment. We have seen that single excitations have an almost negligible effect on the correlation energy. However, their inclusion is essential for a satisfactory description of the charge density and, hence, for properties which are sensitive to this quantity. This point is well

Table 4.9 Ionization potentials (a.u.) of N_2 obtained using Koopmans' theorem and by performing a SDCI calculation on both N_2 and N_2^+ (A large basis consisting of $6s$, $4p$, $3d$, $2f$ Slater-type orbitals was used)[a]

Orbital	Koopmans'	SDCI	Experiment
$3\sigma_g$	0.635	0.580	0.573
$1\pi_u$	0.615	0.610	0.624

[a] W. C. Ermler and A. D. McLean, *J. Chem. Phys.* **73**: 2297 (1980)

Table 4.10 The effect of single excitations on the dipole moment (a.u.) of CO within an extended STO basis[a] **(A positive dipole moment corresponds to C^-O^+)**

Wave function	Energy	Dipole moment
SCF	−112.788	−0.108
SCF + 138 doubles	−113.016	−0.068
SCF + 200 doubles	−113.034	−0.072
SCF + 138 doubles + 62 singles	−113.018	+0.030
Experiment		+0.044

[a] F. Grimaldi, A. Lecourt, and C. Moser, *Int. J. Quantum Chem.* **S1**: 153 (1967).

illustrated by considering the dipole moment of CO (see Table 4.10). Recall that a near-Hartree-Fock-limit calculation predicts the wrong sign of the dipole moment of CO. As Table 4.10 shows, the situation is not rectified even when the 200 most important double excitations are included in the wave function. To obtain the correct sign one must include the single excitations. It is interesting to note that a better variational energy does not necessarily imply an improved description for properties other than the energy. The DCI calculation with 200 doubles has a lower energy than the SDCI, but it still predicts the wrong sign for the dipole moment. Finally, in Table 4.11 we present the dipole moment of H_2O calculated using the 39-STO basis described previously. SDCI reduces the SCF error by about a factor of one-half, but the result is still significantly different from experiment. This is probably a consequence of the absence of diffuse basis functions in the 39-STO basis.

Table 4.11 SCF and SDCI dipole moments (a.u.) of H_2O at the experimental geometry calculated using the 39-STO basis discussed in the text[a]

	Dipole moment
SCF	0.785
SDCI	0.755
Experiment	0.728

[a] B. J. Rosenberg and I. Shavitt, *J. Chem. Phys.* **63**: 2162 (1975).

In summary, for the small molecules considered here, SDCI improves SCF results considerably, but even using extended basis sets the agreement with experiment is not completely satisfactory. It is difficult to ascertain whether this is primarily due to the inadequacy of the basis sets or the SDCI approach itself. As will be discussed in the Section 4.6, for larger systems SDCI becomes increasingly poor due to inherent limitations of this method.

4.4 NATURAL ORBITALS AND THE ONE-PARTICLE REDUCED DENSITY MATRIX

Up to this point we have focused on determinants and configurations formed from a set of canonical Hartree-Fock orbitals. The resulting CI expansion unfortunately turns out to be rather slowly convergent. It is clear, however, that one can perform a CI calculation using N-electron configurations formed from *any* one-electron basis. Therefore, it is of interest to ask whether one can find a one-electron basis for which the CI expansion is more rapidly convergent than it is with the Hartree-Fock basis, and thus be able to obtain equivalent results with a smaller number of configurations. The set of *natural orbitals*, introduced by P.-O. Löwdin,[1] forms such a basis.

In order to define natural orbitals, we now consider the first-order reduced density matrix of an N-electron system. Given a normalized wave function, Φ, then $\Phi(\mathbf{x}_1, \ldots, \mathbf{x}_N) \Phi^*(\mathbf{x}_1, \ldots, \mathbf{x}_N) \, d\mathbf{x}_1 \cdots d\mathbf{x}_N$ is the probability that an electron is in the space-spin volume element $d\mathbf{x}_1$ located at $\mathbf{x}_1$, while simultaneously another electron is in $d\mathbf{x}_2$ at $\mathbf{x}_2$ and so on. If we are interested only in the probability of finding an electron in $d\mathbf{x}_1$ at $\mathbf{x}_1$, independent of where the other electrons are, then we must average over all space-spin coordinates of the other electrons, i.e., integrate over $\mathbf{x}_2, \mathbf{x}_3, \ldots, \mathbf{x}_N$ to obtain

$$\rho(\mathbf{x}_1) = N \int d\mathbf{x}_2 \cdots d\mathbf{x}_N \, \Phi(\mathbf{x}_1, \ldots, \mathbf{x}_N) \Phi^*(\mathbf{x}_1, \ldots, \mathbf{x}_N) \qquad (4.33)$$

$\rho(\mathbf{x}_1)$ is called the *reduced density function* for a single electron in an N-electron system. The normalization factor N is included so that the integral of the density equals the total number of electrons,

$$\int d\mathbf{x}_1 \, \rho(\mathbf{x}_1) = N \qquad (4.34)$$

We now generalize the density function $\rho(\mathbf{x}_1)$ to a density matrix $\gamma(\mathbf{x}_1, \mathbf{x}_1')$ defined as

$$\gamma(\mathbf{x}_1, \mathbf{x}_1') = N \int d\mathbf{x}_2 \cdots d\mathbf{x}_N \, \Phi(\mathbf{x}_1, \mathbf{x}_2, \ldots, \mathbf{x}_N) \Phi^*(\mathbf{x}_1', \mathbf{x}_2, \ldots, \mathbf{x}_N) \qquad (4.35)$$

The matrix $\gamma(\mathbf{x}_1, \mathbf{x}_1')$, which depends on two continuous indices, is called the *first-order reduced density matrix* or alternatively, the *one-electron re-*

duced density matrix or simply the *one-matrix*. Note that the diagonal element of the continuous representation of the one-matrix is the density of electrons

$$\gamma(\mathbf{x}_1, \mathbf{x}_1) = \rho(\mathbf{x}_1) \tag{4.36}$$

Since $\gamma(\mathbf{x}_1, \mathbf{x}_1')$ is a function of two variables, it can be expanded in the orthonormal basis of Hartree-Fock spin orbitals $\{\chi_i\}$ as

$$\gamma(\mathbf{x}_1, \mathbf{x}_1') = \sum_{ij} \chi_i(\mathbf{x}_1)\gamma_{ij}\chi_j^*(\mathbf{x}_1') \tag{4.37}$$

where

$$\gamma_{ij} = \int d\mathbf{x}_1 \, d\mathbf{x}_1' \, \chi_i^*(\mathbf{x}_1)\gamma(\mathbf{x}_1, \mathbf{x}_1')\chi_j(\mathbf{x}_1') \tag{4.38}$$

The matrix γ formed from the elements $\{\gamma_{ij}\}$ is a discrete representation of the one-matrix in the orthonormal basis $\{\chi_i\}$.

Exercise 4.4 Show that γ is a Hermitian matrix.

Exercise 4.5 Show that $\operatorname{tr} \gamma = N$.

Exercise 4.6 Consider the one-electron operator

$$\mathcal{O}_1 = \sum_{i=1}^{N} h(i)$$

a. Show that

$$\langle \Phi | \mathcal{O}_1 | \Phi \rangle = \int d\mathbf{x}_1 \, [h(\mathbf{x}_1)\gamma(\mathbf{x}_1, \mathbf{x}_1')]_{\mathbf{x}_1' = \mathbf{x}_1}$$

where the notation $[\ ]_{\mathbf{x}_1' = \mathbf{x}_1}$ means that $\mathbf{x}_1'$ is set equal to $\mathbf{x}_1$ after $h(\mathbf{x}_1)$ has operated on $\gamma(\mathbf{x}_1, \mathbf{x}_1')$.

b. Show that

$$\langle \Phi | \mathcal{O}_1 | \Phi \rangle = \operatorname{tr} \mathbf{h}\gamma$$

where

$$h_{ij} = \langle i | h | j \rangle = \int d\mathbf{x}_1 \, \chi_i^*(\mathbf{x}_1)h(\mathbf{x}_1)\chi_j(\mathbf{x}_1)$$

Thus the expectation value of any one-electron operator can be expressed in terms of the one-matrix.

In the special case that Φ is the Hartree-Fock ground state wave function Ψ_0, it can be shown from the definition (Eq. (4.35)) that

$$\gamma^{\mathrm{HF}}(\mathbf{x}_1, \mathbf{x}_1') = \sum_{a} \chi_a(\mathbf{x}_1)\chi_a^*(\mathbf{x}_1') \tag{4.39}$$

where the sum runs over only the spin orbitals contained in Ψ_0. Thus the discrete representation of the HF one-matrix is particularly simple—γ^{HF} is diagonal with ones along the diagonal for those elements corresponding to occupied spin orbitals and zeros for unoccupied spin orbitals,

$$
\begin{aligned}
\gamma_{ij}^{HF} &= \delta_{ij} \qquad i, j \in \text{occupied} \\
&= 0 \qquad \text{otherwise}
\end{aligned}
\tag{4.40}
$$

The diagonal elements of γ^{HF} can be regarded as occupation numbers: one for occupied spin orbitals and zero for unoccupied spin orbitals.

Exercise 4.7 Recall that in second quantization a one-electron operator is

$$
\mathcal{C}_1 = \sum_{ij} \langle i|h|j\rangle a_i^\dagger a_j
$$

a. Show that

$$
\gamma_{ij} = \langle \Phi|a_j^\dagger a_i|\Phi\rangle.
$$

b. Show that the matrix elements of γ^{HF} are given by Eq. (4.40).

In general, when Φ is not Ψ_0, the discrete representation of the one-matrix in the basis of HF spin orbitals is *not* diagonal. However, since γ is Hermitian, it is possible to define an orthonormal basis $\{\eta_i\}$, related to $\{\chi_i\}$ by a unitary transformation, in which the matrix representation of the one-matrix is diagonal. The elements of the orthonormal set in which γ is diagonal are called *natural spin orbitals*. To make the above explicit, we start with the relation between two orthonormal bases $\{\eta_i\}$ and $\{\chi_i\}$ (see Eqs. (1.63) and (1.65))

$$
\chi_i = \sum_k \eta_k (\mathbf{U}^\dagger)_{ki} = \sum_k \eta_k U_{ik}^*
\tag{4.41}
$$

$$
\eta_i = \sum_k \chi_k U_{ki}
\tag{4.42}
$$

where $\mathbf{U}$ is a unitary matrix. Substituting Eq. (4.41) into Eq. (4.37), we have

$$
\begin{aligned}
\gamma(\mathbf{x}_1, \mathbf{x}_1') &= \sum_{ijkl} \eta_k(\mathbf{x}_1) U_{ik}^* \gamma_{ij} U_{jl} \eta_l^*(\mathbf{x}_1') \\
&= \sum_{kl} \eta_k(\mathbf{x}_1) \left(\sum_{ij} (\mathbf{U}^\dagger)_{ki} \gamma_{ij} U_{jl} \right) \eta_l^*(\mathbf{x}_1') \\
&= \sum_{kl} \eta_k(\mathbf{x}_1) (\mathbf{U}^\dagger \gamma \mathbf{U})_{kl} \eta_l^*(\mathbf{x}_1') \\
&= \sum_{kl} \eta_k(\mathbf{x}_1) \lambda_{kl} \eta_l^*(\mathbf{x}_1')
\end{aligned}
\tag{4.43}
$$

where we have defined the matrix λ as

$$\lambda = \mathbf{U}^\dagger \gamma \mathbf{U} \qquad (4.44)$$

Now since γ is a Hermitian matrix, it is possible to find a unique unitary matrix $\mathbf{U}$ which diagonalizes γ, i.e.,

$$\lambda_{ij} = \delta_{ij}\lambda_i \qquad (4.45)$$

The corresponding spin orbitals $\{\eta_i\}$ given by Eq. (4.42) are the natural spin orbitals. In terms of the natural spin orbitals, we can write Eq. (4.43) as

$$\gamma(\mathbf{x}_1, \mathbf{x}_1') = \sum_i \lambda_i \eta_i(\mathbf{x}_1)\eta_i^*(\mathbf{x}_1') \qquad (4.46)$$

In analogy to the HF result of Eq. (4.39), λ_i is called the occupation number of the natural spin orbital η_i in the wave function Φ.

The importance of natural orbitals is that, in a certain sense, they give the most rapidly convergent CI expansion. That is, to obtain a given accuracy one requires fewer configurations formed from natural orbitals than configurations formed from any other orthonormal basis. It turns out that only configurations that are constructed from natural orbitals with large occupation numbers make significant contributions to the energy. Thus a natural spin orbital with a negligible occupation number may be omitted from the CI expansion without appreciably affecting the accuracy.

We shall not mathematically show why the use of natural orbitals is expected to improve the convergence of the CI expansion. Rather, we illustrate this point using a numerical example. Shavitt and coworkers performed the following interesting study for H_2O using the 39-STO basis described in Section 4.3. First, they performed a CI calculation containing all 4120 symmetry- and spin-adapted singly and doubly excited configurations constructed from the canonical HF basis. From this wave function they obtained the one-matrix and diagonalized it to determine the natural orbitals within the SDCI approximation. Then they performed a parallel series of truncated SDCI calculations using both the canonical and natural orbitals in order to answer the question, what is the minimum number of configurations needed to recover a given percent of the SDCI correlation energy? The answers are shown in Table 4.12. The faster convergence of the CI expansion based on natural orbitals is apparent. To obtain 60% of the SDCI correlation energy, one needs only 50 configurations formed from natural orbitals, as compared with 140 canonically based configurations. However, it can be also seen that the advantage of natural orbitals over canonical Hartree-Fock orbitals is only for relatively short expansions. It must be emphasized that these results are basis-set dependent, and it is expected that the differences between natural orbitals and Hartree-Fock orbitals are even greater for larger basis sets.

Table 4.12 The number of symmetry- and spin-adapted configurations required to recover given fractions of the SDCI correlation energy of H_2O within the 39-STO basis when canonical SCF (MO) and natural orbitals (NO) are used[a]

	Number of Configurations	
Percent of E_{corr}(SDCI)	MO	NO
20	14	6
40	52	18
60	140	50
80	351	147
90	617	362
99	1760	1652

[a] I. Shavitt, B. J. Rosenberg, and S. Palalikit, *Int. J. Quantum Chem.* **S10**: 33 (1976).

Exercise 4.8 For the special case of a two-electron system, the use of natural orbitals dramatically reduces the size of the full CI expansion. If ψ_1 is the occupied Hartree-Fock spatial orbital and ψ_r, $r = 2, 3, \ldots, K$ are virtual spatial orbitals, the normalized full CI singlet wave function has the form

$$|^1\Phi_0\rangle = c_0|1\bar{1}\rangle + \sum_{r=2}^{K} c_1^r|^1\Psi_1^r\rangle + \frac{1}{2}\sum_{r=2}^{K}\sum_{s=2}^{K} c_{11}^{rs}|^1\Psi_{11}^{rs}\rangle$$

where the singly and doubly excited spin adapted configurations are defined in Subsection 2.5.2.

a. Show that $|^1\Phi_0\rangle$ can be cast into the form

$$|^1\Phi_0\rangle = \sum_{i=1}^{K}\sum_{j=1}^{K} C_{ij}|\psi_i\bar{\psi}_j\rangle$$

where **C** is a symmetric $K \times K$ matrix.

b. Show that

$$\gamma(\mathbf{x}_1, \mathbf{x}_1') = \sum_{ij} (\mathbf{CC}^\dagger)_{ij}(\psi_i(1)\psi_j^*(1') + \bar{\psi}_i(1)\bar{\psi}_j^*(1')).$$

c. Let **U** be the unitary transformation which diagonalizes **C**

$$U^{\dagger}CU = d$$

where $(d)_{ij} = d_i\delta_{ij}$. Show that

$$U^{\dagger}CC^{\dagger}U = d^2.$$

d. Show that

$$\gamma(\mathbf{x}_1, \mathbf{x}_1') = \sum_i d_i^2 \left(\zeta_i(1)\zeta_i^*(1') + \bar{\zeta}_i(1)\bar{\zeta}_i^*(1')\right)$$

where

$$\zeta_i = \sum_k \psi_k U_{ki}$$

Thus **U** diagonalizes the one-matrix, and hence ζ_i are natural spatial orbitals for the two-electron system.

e. Finally, since **C** is symmetric, **U** can be chosen as real. Show that in terms of the natural spatial orbitals, $|{}^1\Phi_0\rangle$ given in part (a) can be re-written as

$$|{}^1\Phi_0\rangle = \sum_{i=1}^{K} d_i \, |\zeta_i\bar{\zeta}_i\rangle$$

and note that this expansion contains only K terms.

Now that we have seen that the use of natural orbitals improves the convergence of the CI expansion, we are faced with the problem of how to exploit this in actual calculations. The difficulty is that the one-matrix and hence the CI wave function is required to calculate natural orbitals. Thus we can obtain the natural orbitals only after the CI calculation is complete. However, we clearly would like to have them before we start the calculation. Fortunately, it turns out that approximate natural orbitals are almost as good as the exact ones. There are several schemes that take advantage of this; here we only mention the iterative natural orbital method of Bender and Davidson.[2] In this approach one performs a series of small CI calculations. The configurations used in a given calculation are constructed from natural orbitals obtained from the wave function of the previous calculation. Thus one starts with a CI calculation involving a small number, say 50, of the most important configurations constructed from canonical Hartree-Fock orbitals. Using the resulting wave function, the one-matrix is calculated and then diagonalized to yield a set of approximate natural orbitals. Using the most important of these natural orbitals (i.e., those with the largest occupation numbers), one constructs a new set of 50 configurations; the procedure is repeated until the natural orbitals and/or the energy has con- ... practice, only a few iterations are performed and, in fact, the ten begins to diverge after several iterations.[3]

4.5 THE MULTICONFIGURATION SELF-CONSISTENT FIELD (MCSCF) AND THE GENERALIZED VALENCE BOND (GVB) METHODS

We have seen that the canonical Hartree-Fock orbitals are not the best choice of orbitals for use in CI calculations. Let us consider a multideterminantal wave function, containing a relatively small number of configurations. What orbitals should we use in constructing these configurations so as to obtain the best possible result? From the variation principle, it is clear that we should vary the orbitals so as to minimize the energy. This is the central idea of the multiconfiguration self-consistent-field (MCSCF) method. Thus the MCSCF wave function is a truncated CI expansion

$$|\Psi_{\text{MCSCF}}\rangle = \sum_I c_I |\Psi_I\rangle \tag{4.47}$$

in which *both* the expansion coefficients (c_I) and the orthonormal orbitals contained in $|\Psi_I\rangle$ are optimized. For a closed-shell system, if only one determinant is included in the expansion (4.47), the MCSCF and Hartree-Fock methods become identical. The general equations, which must be solved to obtain the MCSCF wave function are considerably more complicated than Roothaan's restricted Hartree-Fock equations. Several approaches to the MCSCF problem are discussed in a review by Wahl and Das.[4]

In order to make the MCSCF method more explicit, we consider its application to the ground state of H_2. The simplest possible MCSCF wave function for this molecule contains only two closed-shell configurations

$$|\Psi_{\text{MCSCF}}\rangle = c_A |\psi_A \bar{\psi}_A\rangle + c_B |\psi_B \bar{\psi}_B\rangle \tag{4.48}$$

The orthonormal orbitals ψ_A and ψ_B can be expanded in a basis of atomic orbitals as

$$\psi_i = \sum_\mu C_{\mu i} \phi_\mu \qquad i = A, B \tag{4.49}$$

The MCSCF energy is obtained by minimizing $\langle \Psi_{\text{MCSCF}} | \mathcal{H} | \Psi_{\text{MCSCF}} \rangle$, subject to the constraints

$$\langle \psi_A | \psi_A \rangle = \langle \psi_B | \psi_B \rangle = 1 \qquad \langle \psi_A | \psi_B \rangle = 0 \tag{4.50a}$$

and

$$c_A^2 + c_B^2 = 1 \tag{4.50b}$$

to determine the optimum CI expansion coefficient (i.e., either c_A or c_B) and the optimum form of the orbitals ψ_A and ψ_B (i.e., the optimum set of expansion coefficients $C_{\mu i}$, $i = A, B$). In the minimal basis set description of H_2, ψ_A and ψ_B are determined by symmetry (i.e., they are just ψ_1 and ψ_2) and

Ψ_{MCSCF} is identical to the full CI wave function. However, if an extended basis set is used, the MCSCF energy will be above the full CI energy but below the energy obtained from any two-configuration CI expansion based on canonical Hartree-Fock orbitals.

We conclude this section by noting that the generalized valence bond (GVB) wave function of W. A. Goddard III and coworkers[5] can be regarded as a special form of an MCSCF wave function. The basic idea of the GVB method is most simply illustrated by considering its application to H_2. The valence bond wave function of Heitler and London is

$$|\Psi_{VB}\rangle = (2(1 + S_{12}^2))^{-1/2}[\phi_1(1)\phi_2(2) + \phi_1(2)\phi_2(1)]2^{-1/2}(\alpha(1)\beta(2) - \alpha(2)\beta(1))$$
$$(4.51)$$

where ϕ_1 and ϕ_2 are nonorthogonal *atomic* orbitals ($\langle\phi_1|\phi_2\rangle = S_{12}$) centered on nuclei 1 and 2, respectively. The GVB wave function has the VB *form*

$$|\Psi_{GVB}\rangle = (2(1 + S^2))^{-1/2}[u(1)v(2) + u(2)v(1)]2^{-1/2}(\alpha(1)\beta(2) - \alpha(2)\beta(1))$$
$$(4.52)$$

but the nonorthogonal GVB orbitals u and v ($\langle u|v\rangle = S$) are determined variationally. That is, the nonorthogonal orbitals u and v are expanded in a basis set and the expansion coefficients are varied so as to minimize the energy of $|\Psi_{GVB}\rangle$. Thus the GVB wave function is the self-consistent generalization of the VB wave function. The GVB approach has a number of appealing features including the fact that it properly describes the dissociation of molecules into open shell fragments. If the same set of basis functions are used to expand the orthogonal MCSCF orbitals (ψ_A and ψ_B) and the nonorthogonal GVB orbitals (u and v), it can be shown (see Exercise 4.9) that the simple two configuration MCSCF wave function of Eq. (4.48) and the GVB wave function of Eq. (4.52) are identical.

Exercise 4.9 Consider the transformation

$$u = (a^2 + b^2)^{-1/2}(a\psi_A + b\psi_B)$$
$$v = (a^2 + b^2)^{-1/2}(a\psi_A - b\psi_B)$$

a. Show that

$$\langle u|u\rangle = \langle v|v\rangle = 1 \quad \text{and} \quad \langle u|v\rangle \equiv S = \frac{a^2 - b^2}{a^2 + b^2}.$$

b. Show that $|\Psi_{GVB}\rangle$ in Eq. (4.52) can be rewritten as

$$|\Psi_{GVB}\rangle = (a^4 + b^4)^{-1/2}[a^2\psi_A(1)\psi_A(2)$$
$$- b^2\psi_B(1)\psi_B(2)]2^{-1/2}(\alpha(1)\beta(2) - \alpha(2)\beta(1))$$

and conclude that this is identical to $|\Psi_{MCSCF}\rangle$ in Eq. (4.48) if

$$c_A = (a^4 + b^4)^{-1/2}a^2$$
$$c_B = -(a^4 + b^4)^{-1/2}b^2.$$

As an application of the GVB method (or equivalently the MCSCF method using two configurations) we present numerical results for H_2 using the 6-31G** basis.[6] At $R = 1.4$ a.u., $E_{corr}(GVB) = -0.0183$ a.u. as compared with the full CI result of -0.0339 a.u. (i.e., GVB gives 54% of the exact basis set correlation energy). The GVB, RHF, UHF, and full CI potential energy curves of H_2 are compared in Fig. 4.4. The GVB method properly describes the dissociation of H_2 into hydrogen atoms. Moreover, the GVB potential curve improves with increasing internuclear separation; when R is greater than 3 a.u. the GVB and full CI results are virtually indistinguishable. This is to be contrasted with the potential energy curve obtained from second-order many-body perturbation theory (Chapter 6) using the UHF wave function as the starting point (see Fig. 6.3). Second-order perturbation theory gives a larger fraction of the correlation energy of H_2 at the equilibrium bond length than GVB. However, as R increases the potential curve

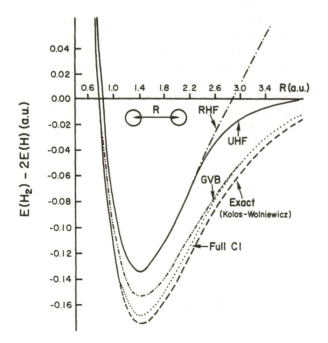

Figure 4.4 6-31G** potential energy curves for H_2.

obtained using perturbation theory deteriorates and, in fact, approaches the UHF result.

4.6 TRUNCATED CI AND THE SIZE-CONSISTENCY PROBLEM

In chemistry one is interested in the relative energies of molecules of different size. For example, suppose we wish to calculate ΔE for the reaction

$$A + B \rightarrow C$$

For our result to be meaningful, it is necessary to use approximation schemes that are equally good, in a certain sense, for molecules with different numbers of electrons. To define in just what sense, let us consider a supermolecule (dimer) composed of two identical but noninteracting molecules (monomers). Two monomers separated by a large distance will serve as an example of such a dimer. Physically, it is clear that the energy of the dimer should be just twice the energy of the monomer, since by assumption the monomers do not interact. An approximation scheme for calculating the energy of such a system that has this property is said to be size consistent. The Hartree-Fock approximation is the simplest example of such a theory: The HF energy of a supermolecule composed of two noninteracting *closed-shell* subsystems is just the sum of the HF energies of the subsystems. A more general definition of size consistency is that the energy of a many-particle system, even in the presence of interactions, becomes proportional to the number of particles (N) in the limit $N \rightarrow \infty$. For example, the HF energy of a crystal is proportional to the number of constituent molecules although the total energy is not simply N times the energy of an isolated molecule.

Full CI, as to be expected from a formally exact theory, is also size consistent. Unfortunately truncated CI does not have this property. At first, this may appear surprising since size consistency seems like a rather modest requirement. This deficiency can however be understood by a simple example. Consider a supermolecule composed of two noninteracting H_2 molecules. Why is it that the DCI energy of the supermolecule is not the sum of the DCI energies of the two monomers? By definition the DCI wave function of each of the monomers contains double excitations within the monomer. If we restrict the supermolecule trial function to double excitations, we exclude the possibility that both monomers are simultaneously doubly excited, since this represents a quadruple excitation in the supermolecule. Thus the supermolecule wave function truncated at the DCI level does not have sufficient flexibility to yield twice the DCI monomer energy.

Let us make this argument more quantitative by using our minimal basis H_2 model. Assume that we have two minimal basis H_2 molecules that are separated by a large (infinite) distance. We can perform calculations on the composite system as if it were one molecule, using the fact that all integrals involving a basis function on one monomer and a basis function on the other monomer will be zero because of the large distance between the basis functions. The results of a Hartree-Fock calculation are shown in Fig. 4.5. Our notation is that 1_1 and 2_1 are the occupied and unoccupied molecular orbitals localized on monomer 1. These are identical to ψ_1 and ψ_2 of our previous monomer calculation. In the same way, the molecular orbitals 1_2 and 2_2 of monomer 2 are the usual σ_g and σ_u bonding and antibonding orbitals formed from linear combinations of the two $1s$ basis functions for that molecule. The Hartree-Fock ground state is the single determinant

$$|\Psi_0\rangle = |1_1\bar{1}_1 1_2\bar{1}_2\rangle \tag{4.53}$$

Any electron-electron repulsion integral involving both molecules, such as $(1_1 1_1 | 1_2 1_2)$, is zero, so that the Hartree-Fock energy of the dimer is just twice the energy of one H_2 molecule

$$^2E_0 = 2(2\varepsilon_1 - J_{11}) \tag{4.54}$$

where

$$J_{11} = (1_1 1_1 | 1_1 1_1) = (1_2 1_2 | 1_2 1_2)$$

Now let us consider a doubly excited CI calculation. We can neglect single excitations again, because of symmetry. There are two doubly excited

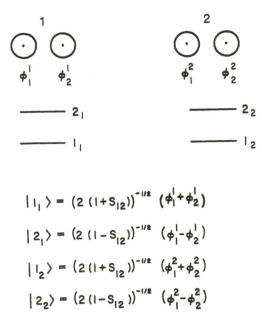

$$|1_1\rangle = (2(1+S_{12}))^{-1/2}(\phi_1^1 + \phi_2^1)$$

$$|2_1\rangle = (2(1-S_{12}))^{-1/2}(\phi_1^1 - \phi_2^1)$$

$$|1_2\rangle = (2(1+S_{12}))^{-1/2}(\phi_1^2 + \phi_2^2)$$

$$|2_2\rangle = (2(1-S_{12}))^{-1/2}(\phi_1^2 - \phi_2^2)$$

Figure 4.5 Two minimal basis H_2 molecules separated by infinity.

configurations that will mix with $|\Psi_0\rangle$, the configuration obtained by exciting two electrons from the occupied orbital of monomer 1 to the unoccupied orbital of the same monomer and the corresponding configuration for monomer 2

$$|\Phi_0\rangle = |\Psi_0\rangle + c_1|2_1\bar{2}_11_2\bar{1}_2\rangle + c_2|1_1\bar{1}_12_2\bar{2}_2\rangle$$

$$= |\Psi_0\rangle + \sum_{i=1}^{2} c_i|\Psi_{1_i\bar{1}_i}^{2_i\bar{2}_i}\rangle \tag{4.55}$$

The doubly excited configuration

and any of the other doubly excited configurations will not mix because the matrix elements involve only zero-valued integrals.

Exercise 4.10 Show that $|1_1\bar{1}_12_1\bar{2}_1\rangle$ has a zero matrix element with any of the configurations in Eq. (4.55).

The matrix element between $|2_1\bar{2}_11_2\bar{1}_2\rangle$ and the Hartree-Fock ground state is

$$\langle\Psi_0|\mathcal{H}|2_1\bar{2}_11_2\bar{1}_2\rangle = \langle\Psi_0|\mathcal{H}|\Psi_{1_1\bar{1}_1}^{2_1\bar{2}_1}\rangle$$

$$= \langle 1_1\bar{1}_1||2_1\bar{2}_1\rangle = [1_12_1|\bar{1}_1\bar{2}_1] - [1_1\bar{2}_1|\bar{1}_12_1]$$

$$= (1_12_1|1_12_1) = (1_12_1|2_11_1)$$

$$= K_{12}$$

which is the same as the matrix element between $|\Psi_0\rangle$ and the other doubly excited state. The two doubly excited states will not mix with each other as they differ by four spin orbitals and, therefore, the CI matrix equation is

$$\begin{pmatrix} 0 & K_{12} & K_{12} \\ K_{12} & 2\Delta & 0 \\ K_{12} & 0 & 2\Delta \end{pmatrix}\begin{pmatrix} 1 \\ c_1 \\ c_2 \end{pmatrix} = {}^2E_{\text{corr}}(\text{DCI})\begin{pmatrix} 1 \\ c_1 \\ c_2 \end{pmatrix} \tag{4.56}$$

where the excitation energy to the doubly excited configurations is 2Δ, as in the monomer calculation. This CI equation for the dimer is to be compared with Eq. (4.22) for the monomer. The matrix Eq. (4.56) contains the following three simultaneous equations

$$K_{12}(c_1 + c_2) = {}^2E_{\text{corr}}(\text{DCI}) \tag{4.57a}$$

$$K_{12} + 2\Delta c_1 = {}^2E_{\text{corr}}(\text{DCI})c_1 \tag{4.57b}$$

$$K_{12} + 2\Delta c_2 = {}^2E_{\text{corr}}(\text{DCI})c_2 \tag{4.57c}$$

From the last two of these it is evident that $c_1 = c_2$, as to be expected from symmetry. Solving for c_1 gives

$$c_1 = c_2 = \frac{K_{12}}{{}^2E_{\text{corr}}(\text{DCI}) - 2\Delta} \tag{4.58}$$

and substituting this result into Eq. (4.57a) we have

$$^2E_{\text{corr}}(\text{DCI}) = \frac{2K_{12}^2}{{}^2E_{\text{corr}}(\text{DCI}) - 2\Delta} \tag{4.59}$$

If we solve the quadratic equation for $^2E_{\text{corr}}(\text{DCI})$, we obtain

$$^2E_{\text{corr}}(\text{DCI}) = \Delta - (\Delta^2 + 2K_{12}^2)^{1/2} \tag{4.60}$$

The corresponding monomer result (Eq. (4.23)) is

$$^1E_{\text{corr}}(\text{DCI}) = {}^1E_{\text{corr}}(\text{exact}) = \Delta - (\Delta^2 + K_{12}^2)^{1/2} \tag{4.61}$$

The deficiency of doubly excited CI, which we discussed at the beginning of this section, is illustrated in these last two equations. *The energy of two noninteracting molecules is not twice the energy of one of them, calculated in the same approximation.* Thus doubly excited CI is not size consistent.

Moreover, DCI deteriorates as the size of the system increases; i.e., it is a much worse approximation for large molecules than it is for small molecules. To see this, let us generalize the above analysis to N, instead of just two, noninteracting H_2 molecules. The DCI wave function is

$$|\Phi_0\rangle = |\Psi_0\rangle + \sum_{i=1}^{N} c_i |\Psi_{1_i \bar{1}_i}^{2_i \bar{2}_i}\rangle \tag{4.62}$$

which is the appropriate generalization of Eq. (4.55). The corresponding CI eigenvalue problem is

$$\begin{pmatrix} 0 & K_{12} & K_{12} & \cdots & K_{12} \\ K_{12} & 2\Delta & 0 & \cdots & 0 \\ K_{12} & 0 & 2\Delta & \cdots & 0 \\ \vdots & \vdots & \vdots & \ddots & 0 \\ K_{12} & 0 & 0 & 0 & 2\Delta \end{pmatrix} \begin{pmatrix} 1 \\ c_1 \\ c_2 \\ \vdots \\ c_N \end{pmatrix} = {}^NE_{\text{corr}}(\text{DCI}) \begin{pmatrix} 1 \\ c_1 \\ c_2 \\ \vdots \\ c_N \end{pmatrix} \tag{4.63}$$

where $^NE_{\text{corr}}(\text{DCI})$ is the DCI correlation energy for the N molecule system. Again, the coefficients of the double excitations localized on the individual molecules are identical:

$$c_i = c_1 = \frac{K_{12}}{{}^NE_{\text{corr}}(\text{DCI}) - 2\Delta} \tag{4.64}$$

and the correlation energy, from the first row of the CI matrix, is

$$^NE_{\text{corr}}(\text{DCI}) = K_{12} \sum_{i=1}^{N} c_i = NK_{12}c_1 \tag{4.65}$$

so that combining these equations gives

$$^{N}E_{corr}(DCI) = \frac{NK_{12}^{2}}{^{N}E_{corr}(DCI) - 2\Delta} \tag{4.66}$$

Finally, solving for $^{N}E_{corr}(DCI)$, we have

$$^{N}E_{corr}(DCI) = \Delta - (\Delta^{2} + NK_{12}^{2})^{1/2} \tag{4.67}$$

which is to be compared to the exact energy of N noninteracting minimal basis H_2 molecules, viz.

$$^{N}E_{corr}(exact) = N\,^{1}E_{corr}(exact) = N(\Delta - (\Delta^{2} + K_{12}^{2})^{1/2}) \tag{4.68}$$

For small N the difference between the two energies is fairly small. For example, for two STO-3G H_2 molecules both at $R = 1.4$ a.u., $^{2}E_{corr}(DCI) = -0.0406$ a.u. while the exact basis set correlation energy is -0.0411 a.u. However, for large N the discrepancy between the two results increases. In particular the limiting behavior of $^{N}E_{corr}(DCI)$ as $N \to \infty$ is

$$^{N}E_{corr}(DCI) \sim -N^{1/2}K_{12} \tag{4.69}$$

which means that the correlation energy per monomer vanishes in the limit of large N, i.e.,

$$\lim_{N \to \infty} \frac{^{N}E_{corr}(DCI)}{N} = 0 \tag{4.70}$$

Thus DCI is a completely useless approximation for the correlation energy for a macroscopic system such as a crystal. The lack of size consistency demonstrated above is not only a property of DCI but of all forms of truncated CI. For example, if we included both double and quadruple excitations in the supermolecule trial function (DQCI), we would obtain the exact energy *only* when $N = 2$ since DQCI corresponds to full CI in this case. As N increases the discrepancy between $^{N}E_{corr}(DQCI)$ and $^{N}E_{corr}(exact)$ grows, but of course more slowly than the difference between $^{N}E_{corr}(DCI)$ and $^{N}E_{corr}(exact)$. Nevertheless, as $N \to \infty$ the DQCI correlation energy per particle still vanishes.

In principle, no form of truncated CI is size consistent. In practice, however, CI including quadruple excitations is effectively size consistent for relatively small molecules containing less than, say, 50 electrons. Thus much current research is directed towards finding efficient ways of performing CI calculations in which the most important quadruple excitations are included. Are there approximation methods for the correlation energy which are rigorously size consistent? In fact, there are a number of such schemes and they constitute the subject matter of much of the rest of this book.

Exercise 4.11 Use the integrals for STO-3G H_2 at $R = 1.4$ a.u., given in Appendix D, to calculate $^{N}E_{corr}(DCI)/N$ for $N = 1$, 10, and 100.

Exercise 4.12 Show that full CI is size consistent for a dimer of non-interacting minimal basis H_2 molecules. A full CI calculation includes, in addition to the excitations in Eq. (4.55), the quadruply excited state $|2_1\bar{2}_12_2\bar{2}_2\rangle = |\Psi_{1_1\bar{1}_11_2\bar{1}_2}^{2_1\bar{2}_12_2\bar{2}_2}\rangle$

$$|\Phi_0\rangle = |\Psi_0\rangle + c_1|2_1\bar{2}_11_2\bar{1}_2\rangle + c_2|1_1\bar{1}_12_2\bar{2}_2\rangle + c_3|2_1\bar{2}_12_2\bar{2}_2\rangle$$

a. Show that the full CI matrix equation is

$$\begin{pmatrix} 0 & K_{12} & K_{12} & 0 \\ K_{12} & 2\Delta & 0 & K_{12} \\ K_{12} & 0 & 2\Delta & K_{12} \\ 0 & K_{12} & K_{12} & 4\Delta \end{pmatrix} \begin{pmatrix} 1 \\ c_1 \\ c_2 \\ c_3 \end{pmatrix} = {}^2E_{\text{corr}} \begin{pmatrix} 1 \\ c_1 \\ c_2 \\ c_3 \end{pmatrix}.$$

Go directly to part (e). If you need help return to part (b).

b. Show that $c_1 = c_2$ and hence ${}^2E_{\text{corr}} = 2K_{12}c_1$.
c. Show that

$$c_3 = \frac{{}^2E_{\text{corr}}}{{}^2E_{\text{corr}} - 4\Delta}.$$

d. Show that

$$c_1 = \frac{2K_{12}}{{}^2E_{\text{corr}} - 4\Delta}.$$

e. Finally, show that

$$^2E_{\text{corr}} = 2(\Delta - (\Delta^2 + K_{12}^2)^{1/2})$$

which is indeed exact for the model.

It is interesting to note that we can express the coefficient of the quadruple excitation (c_3) as the product of the coefficients of the double excitations:

$$c_3 = \frac{{}^2E_{\text{corr}}}{{}^2E_{\text{corr}} - 4\Delta} = \frac{2K_{12}c_1}{{}^2E_{\text{corr}} - 4\Delta} = (c_1)^2$$

where we have used results in parts (b), (c), and (d). This result is not true in general but is a consequence of the fact that the two monomers are independent. However, it suggests that it might be reasonable to *approximate* the coefficient of a quadruply excited configuration as a product of the coefficients of the double excitations that combine to give the quadruply excited configuration. This idea plays a central role in the next chapter.

Exercise 4.13 Consider the exact basis set correlation energy of minimal basis H_2 given in Eq. (4.23). Assuming that $K_{12}^2/\Delta^2 \ll 1$ show that

$$^1E_{\text{corr}}(\text{exact}) \cong -\frac{K_{12}^2}{2\Delta}$$

Hint: $(1 + x)^{1/2} \cong 1 + \frac{1}{2}x$ when $x \ll 1$. This approximate result is the same as the simplest expression for the correlation energy obtained via a form of perturbation theory. Show that by expanding $^{N}E_{corr}(DCI)$ in the same way, assuming that $NK_{12}^{2}/\Delta^{2} \ll 1$, one obtains simply N times the above result. This approximation is equivalent to a perturbation result for the correlation energy of a supermolecule, and the form of the result is a reflection of the fact that, in contrast to truncated CI, perturbation theory is size consistent (see Chapter 6).

Exercise 4.14 DCI calculations have become relatively routine; therefore, it is of interest to ask whether one can correct the DCI correlation energy so that it becomes approximately size consistent. A simple prescription for doing this, which is approximately valid *only* for relatively small systems, is to write

$$E_{corr} = E_{corr}(DCI) + \Delta E_{Davidson}$$

where the Davidson correction is given by

$$\Delta E_{Davidson} = (1 - c_{0}^{2})E_{corr}(DCI)$$

where c_{0} is the coefficient of the Hartree-Fock wave function in the *normalized* DCI wave function. The Davidson correction can be computed without additional labor since c_{0} is available in a DCI calculation. Moreover, there is numerical evidence that it leads to an improvement over DCI for relatively small molecules. For example, for $H_{2}O$, using the 39-STO basis and the Davidson correction, $\theta_{e} = 104.6°$, $R_{e} = 1.809$ a.u., $f_{RR} = 8.54$, and $f_{\theta\theta} = 0.80$ (see Table 4.7). Also, the ionization potentials of N_{2}, using the same basis as in Table 4.9 and the Davidson correction, are 0.575 a.u. $(3\sigma_{g})$ and 0.617 a.u. $(1\pi_{u})$. The purpose of this exercise is to explore the nature of the Davidson correction.

a. For the model of N independent minimal basis H_{2} molecules, assume that N is large, yet small enough that NK_{12}^{2}/Δ^{2} is still less than unity. In addition, remember that $\Delta \gg K_{12}$. Using the identity $(1 + x)^{1/2} \cong 1 + \frac{1}{2}x - \frac{1}{8}x^{2} + \cdots$ show that

$$^{N}E_{corr}(DCI) = \frac{-NK_{12}^{2}}{2\Delta} + \frac{N^{2}K_{12}^{4}}{8\Delta^{3}} + \cdots$$

The term proportional to N^{2} is spurious and is not present in the similar expansion of $^{N}E_{corr}(exact)$.

b. Show that

$$1 - c_{0}^{2} = \frac{Nc_{1}^{2}}{1 + Nc_{1}^{2}}.$$

c. Show that

$$c_1 = \frac{-K_{12}}{2\Delta} + \cdots.$$

d. Show that

$$\Delta E_{\text{Davidson}} = \frac{-N^2 K_{12}^4}{8\Delta^3} + \cdots.$$

Finally, note that the Davidson correction exactly cancels the term proportional to N^2 in the expansion of $^N E_{\text{corr}}(\text{DCI})$. However, spurious terms containing higher powers of N still remain. For large N, the whole analysis breaks down since $N K_{12}^2 / \Delta^2$ eventually becomes greater than unity. For $N = 1$, $E_{\text{corr}}(\text{DCI})$ is exact within the model, yet $\Delta E_{\text{Davidson}}$ is not zero.

e. Numerically investigate the range of validity of the Davidson correction for N independent H_2 molecules. Calculate $^N E_{\text{corr}}(\text{DCI})/^N E_{\text{corr}}(\text{exact})$ and $(^N E_{\text{corr}}(\text{DCI}) + \Delta E_{\text{Davidson}})/^N E_{\text{corr}}(\text{exact})$ for $N = 1, \ldots, 20$ using the values of the two-electron integrals for $R = 1.4$ a.u. given in Appendix D. You will find that the correlation energy calculated using the Davidson correction is within 1% of the exact value for $3 < N < 11$. The DCI correlation energy, on the other hand, errs by 2.5% for $N = 3$ and by 10% for $N = 11$. For $N = 100$, the correlation energy with and without the Davidson correction errs by 25% and 42%, respectively. Hint: show that

$$\Delta E_{\text{Davidson}} = (^N E_{\text{corr}}(\text{DCI}))^3 / (N K_{12}^2 + (^N E_{\text{corr}}(\text{DCI}))^2).$$

f. Calculate the Davidson-corrected correlation energy of H_2O using $E_{\text{corr}}(\text{DCI})$ and c_0 given by Saxe et al. (see Further Reading). Compare your result with $E_{\text{corr}}(\text{DQCI})$ and the exact basis set correlation energy.

Exercise 4.15 The normalized full CI wave function for a minimal basis H_2 molecule is

$$|\Phi_0\rangle = (1 + c^2)^{-1/2}|1\bar{1}\rangle + c(1 + c^2)^{-1/2}|2\bar{2}\rangle$$

where $c = {}^1 E_{\text{corr}}/K_{12}$. Show that for N independent minimal basis H_2 molecules, the overlap between the Hartree-Fock wave functions, $|\Psi_0\rangle$, and the exact *normalized* ground state wave function is

$$\langle \Psi_0 | \Phi_0 \rangle = (1 + c^2)^{-N/2}$$

Using the values of the two-electron integrals for $R = 1.4$ a.u., given in Appendix D, calculate $\langle \Psi_0 | \Phi_0 \rangle$ for $N = 1, 10,$ and 100. Note that this overlap decreases quickly (in fact, exponentially) as N increases. Thus the overlap between the Hartree-Fock and the exact wave functions of the system exponentially approaches zero as the size of the system increases, even though the Hartree-Fock energy is size consistent. *Hint:* Because the N independent H_2 molecules are infinitely separated we can, for all intents and purposes, ignore the requirement that the wave function of this system be antisymmetric with respect to the interchange of electrons which belong to different H_2 molecules. Thus we can write

$$|\Phi_0\rangle \sim \prod_{i=1}^{N} ((1 + c^2)^{-1/2}|1_i\bar{1}_i\rangle + c(1 + c^2)^{-1/2}|2_i\bar{2}_i\rangle)$$

and

$$|\Psi_0\rangle \sim \prod_{i=1}^{N} |1_i\bar{1}_i\rangle$$

NOTES

1. P.-O. Löwdin, Quantum theory of many-particle systems. I. Physical interpretation by means of density matrices, natural spin orbitals, and convergence problems in the method of configuration interaction, *Phys. Rev.* **97**: 1474 (1955).
2. C. F. Bender and E. R. Davidson, A natural orbital based energy calculation for helium hydride and lithium hydride, *J. Phys. Chem.* **70**: 2675 (1966).
3. K. H. Thunemann, J. Romelt, S. D. Peyerimhoff, and R. J. Buenker, A study of the convergence in iterative natural orbital procedures, *Int. J. Quantum Chem.* **11**: 743 (1977).
4. A. C. Wahl and G. Das, The multiconfiguration self-consistent field method, in *Methods of Electronic Structure Theory*, H. F. Schaefer III (Ed.), Plenum, New York, 1977, p. 51. In addition to a survey of the formalism, this article contains a selection of numerical results.
5. W. A. Goddard III, T. H. Dunning, Jr., W. J. Hunt, and P. J. Hay, Generalized valence bond description of bonding in low-lying states of molecules, *Acc. Chem. Res.* **6**: 368 (1973); W. A. Goddard III and L. B. Harding. The description of chemical bonding from *ab initio* calculations, *Ann. Rev. Phys. Chem.* **29**: 363 (1978).
6. A. Voter and W. A. Goddard III, unpublished results.

FURTHER READING

Bauschlicher, C. W. Jr. and Taylor, P. R., Benchmark full configuration-interaction calculations on H_2O, F, and F^-. *J. Chem. Phys.* **85**: 2779 (1986). The full valence CI calculation on water used a "double-zeta plus polarization function" basis set (O basis: (9s5p)/[4s2p] plus a set of d functions; H basis: (4s)/[2s] plus a set of p functions) and involved 6,740,280 configurations expanded into 28,233,466 determinants. Compare this with the 1981 calculations of Saxe *et al.* (see below). The full CI energies are compared with those obtained using various levels of the coupled cluster approximation, many body perturbation theory, and quadratic CI in S. J. Cole and R. J. Bartlett, *J. Chem. Phys.* **86**: 873 (1987) and J. A. Pople, M. Head-Gordon and K. Raghavachari, *J. Chem. Phys.* **87**: 5968 (1987).
Bruna, P. J., and Peyerimhoff, S. D., Excited state potentials, *Adv. Chem. Phys.* **67**: 1 (1987). Discusses recent advances in the calculation of excited states, focusing on potential surfaces.
Knowles, P. J., and Handy, N. C., A new determinant-based full configuration interaction

method, *Chem. Phys. Lett.* **111:** 315 (1984). Describes the theoretical advances that made the full CI calculations of Bauschlicher and Taylor (see above) possible.

Löwdin, P.-O., Correlation problem in many-electron quantum mechanics. *Adv. Chem. Phys.* **2:** 207 (1959). A classic review article on the correlation problem.

Paunz, R., *Spin Eigenfunctions*. Plenum, New York, 1979. The last chapter of this monograph contains an introduction to the unitary group approach to CI. This approach, which has recently attracted much attention, provides a formalism for the efficient computation of the Hamiltonian matrix elements between spin-adapted configurations.

Peyerimhoff, S. D. and Buenker, R. J., CI Calculations of vertical excitation energies and oscillator strengths for Rydberg and valence states of molecules, in *Excited States in Chemistry* C. A. Nikolaides and D. R. Beck (Eds.), Reidel, Dordrecht, 1978, p. 79. A survey of calculations of the electronic spectra of molecules using multireference double-excitation CI (MRD-CI). MRD-CI includes single and double excitations relative not only to the Hartree-Fock wave function (Ψ_0) but also relative to other important configurations (which may themselves be doubly excited with respect to Ψ_0). Thus MRD-CI effectively includes higher than double (e.g., quadruple) excitations from Ψ_0 and hence can be essentially size consistent for moderate sized molecules. Other articles found in this book are also recommended.

Roos, B. O., The complete active space self-consistent-field method and its applications in electronic structure calculations. *Adv. Chem. Phys.* **69:** 399 (1987). One of the most successful methods for finding highly accurate wave functions. CASSCF is a variant of MCSCF and uses all the tricks of the trade (direct CI, unitary group approach, etc.) in its implementation.

Roos, B. O. and Siegbahn, P. E. M., The direct configuration interaction method from molecular integrals, in *Methods of Electronic Structure Theory* H. F. Schaefer III (Ed.), Plenum, New York, 1977, p. 277. A discussion by the originators of an important method in which the CI matrix is never explicitly constructed.

Saxe, P., Schaefer III, H. F. and Handy, N. C., Exact solution (within a double-zeta basis set) of the Schrödinger electronic equation for water, *Chem. Phys. Lett.* **79:** 202 (1981). This landmark paper compares the correlation energy of H_2O obtained using DCI, SDCI, SDTCI, DQCI, SDTQCI, and *full* CI with results obtained using the coupled-cluster approximation (see Chapter 5) and many-body perturbation theory (see Chapter 6). The full CI calculation included 256,473 spin- and symmetry-adapted configurations!

Shavitt, I., The method of configuration interaction, in *Methods of Electronic Structure Theory* H. F. Schaefer III (Eds), Plenum, New York, 1977 p. 189. An authoritative review by a leader in the field discussing many aspects of CI (symmetry- and spin-adapted configurations, construction and diagonalization of the CI matrix, selection of configurations, etc.).

Shepard, R., The multiconfiguration self-consistent-field method, *Adv. Chem. Phys.* **69:** 63 (1987). An authoritative discussion of MCSCF.

CHAPTER FIVE

PAIR AND COUPLED-PAIR THEORIES

We saw in Chapter 4 that configuration interaction (CI) using only doubly excited states (DCI) predicts that the correlation energy of N noninteracting minimal basis H_2 molecules is proportional to $N^{1/2}$ as N becomes large. Because the energy of a macroscopic system is an extensive thermodynamic property, it must be proportional to the number of particles; thus DCI is not a satisfactory procedure for treating large systems. For example, the correlation energy per atom of a crystal obtained using DCI is zero! It is clear that to describe correlation in infinite systems one must use methods that yield energies that are proportional to the number of particles. Even for finite systems, it is desirable to use approximations which give results that can be meaningfully compared for molecules of different size. For example, when studying the dissociation of a molecule one should use a method which is equally good, in a certain sense, for the intact molecule and the resulting fragments. Approximations, which have the property that the calculated energies vary linearly with the number of particles as the size of the system increases, are said to be *size consistent*. In the special case of a supermolecule formed from N closed-shell noninteracting "monomers," a size-consistent scheme yields a supermolecule energy which is just N times the monomer energy.

Although size consistency seems like a modest requirement, all forms of CI except full CI, which is of course exact, do not have this property. In this chapter we consider pair and coupled-pair theories, which are size consistent, and in the next chapter we discuss a form of perturbation theory, which also has this property. The price one pays for size consistency is that unlike DCI, pair and perturbation theories are not variational and the total electronic energy obtained using them can be lower than the true energy. For example, pair theory can give, in certain cases, as much as 120% of the correlation energy.

In Section 5.1 we describe the independent electron pair approximation (IEPA). We use an approach that leads quickly to the computational formalism but which may give the misleading impression that IEPA is an approximation to DCI. After showing what is involved in performing pair calculations, we will return to the physical basis of the formalism and show that in fact both IEPA and DCI are different approximations to *full* CI. In Subsection 5.1.1 we describe a deficiency of the IEPA, not shared by DCI or the perturbation theory of Chapter 6 namely, that the IEPA is not invariant under unitary transformations of degenerate molecular orbitals. In Subsection 5.1.2 we present some numerical results which show that while the IEPA is reasonably accurate for small atoms, it has serious deficiencies when applied to larger molecules.

In Section 5.2 we consider how to go beyond the IEPA by incorporating coupling between different pairs of electrons. We will discuss the coupled pair many-electron theory (CPMET) which is also called the coupled-cluster approximation (CCA). We then describe a number of simplifications to this rather sophisticated approach; in particular, we consider the coupled electron pair approximation (CEPA). Finally, in Subsection 5.2.4 we present some numerical applications of coupled-pair theories.

Since coupled pair theories are important but somewhat complicated, in Section 5.3 we discuss, as a pedagogical device, the use of these methods to calculate the energy of an N-electron system described by a Hamiltonian containing only single particle interactions. This problem can easily be solved exactly in an elementary way. However, by seeing how "high-powered" approaches work in a simple context, we will be able to gain insight into the nature of these approximations. In particular, the relationship between different many-electron theories will become clear. As a concrete application of many-electron formalisms (CI, IEPA, CEPA, etc.) to a system described by a one-electron Hamiltonian, we consider the resonance energy of cyclic polyenes within the framework of Hückel theory in Subsection 5.3.2. Our purpose here is not to advocate the use of Hückel theory or the use of many-electron methods to obtain the resonance energy. Rather, we wish to exploit the analogy between the resonance energy and the correlation energy and provide an analytically tractable model which can be used to illustrate some of the computational aspects of various many-electron approaches.

5.1 THE INDEPENDENT ELECTRON PAIR APPROXIMATION (IEPA)

We have seen in the last chapter that the correlation energy obtained from the intermediate normalized (i.e., $\langle \Phi_0 | \Psi_0 \rangle = 1$) full CI wave function formed by making all possible spin orbital excitations from the HF determinant is

$$E_{\text{corr}} = \sum_{a<b} \sum_{r<s} c_{ab}^{rs} \langle \Psi_0 | \mathcal{H} | \Psi_{ab}^{rs} \rangle = \frac{1}{4} \sum_{ab} \sum_{rs} c_{ab}^{rs} \langle \Psi_0 | \mathcal{H} | \Psi_{ab}^{rs} \rangle \qquad (5.1)$$

where c_{ab}^{rs} are the variationally determined coefficients of the doubly excited determinants in the full CI wave function. Recall that the coefficients of single excitations are absent because of Brillouin's theorem, while triple and higher excitations do not appear because the Hamiltonian contains at most two-particle interactions. This expression suggests that we write the total correlation energy as a sum of contributions from each occupied pair of spin orbitals,

$$E_{\text{corr}} = \sum_{a<b} e_{ab} \qquad (5.2)$$

with

$$e_{ab} = \sum_{r<s} c_{ab}^{rs} \langle \Psi_0 | \mathcal{H} | \Psi_{ab}^{rs} \rangle \qquad (5.3)$$

where e_{ab} is the correlation energy resulting from the interaction of the pair of electrons which, in the Hartree-Fock approximation, occupy spin orbitals χ_a and χ_b. Although the above decomposition is exact, it is also deceptive since we need the full CI wave function to obtain the coefficients c_{ab}^{rs}. Thus for an N-electron system, we must consider all the electrons to calculate exactly e_{ab} for the ab pair. Can we devise a scheme for approximating the pair energy of each pair of electrons independently of other pairs? If so, then in effect we could approximately reduce an N-electron problem to $N(N-1)/2$ two-electron problems. The independent electron pair approximation (IEPA) is such a scheme. The IEPA was introduced in quantum chemistry, independently, by O. Sinanoğlu[1] and R. K. Nesbet.[2] These authors used different terminology and formulations, but their final results are essentially equivalent. Sinanoğlu called his theory the Many-Electron Theory (MET) while Nesbet used the name Bethe-Goldstone Theory. In addition, for reasons which will become apparent shortly, the IEPA has been referred to as "pair-at-a-time" CI.

How can we calculate the correlation energy associated with a pair of electrons in spin orbitals a and b (i.e., the pair energy e_{ab})? The simplest approach is to forget about the remaining $N-2$ electrons (i.e., leave them in their HF spin orbitals) and let only the electrons in spin orbitals a and b correlate by exciting them into the virtual orbitals. We construct a correlated wave function for the pair ab, denoted by $|\Psi_{ab}\rangle$, by allowing the HF wave function to interact with determinants formed by exciting only this pair. If for the sake of simplicity we ignore single excitations (which as we have seen in the last chapter have little effect on the correlation energy), the *pair function* $|\Psi_{ab}\rangle$ is

$$|\Psi_{ab}\rangle = |\Psi_0\rangle + \sum_{r<s} c_{ab}^{rs} |\Psi_{ab}^{rs}\rangle \qquad (5.4)$$

where, as usual, we use intermediate normalization. The energy of this wave function, E_{ab}, is just the sum of the HF energy and the pair correlation energy,

$$E_{ab} = \langle\Psi_0|\mathscr{H}|\Psi_0\rangle + e_{ab} = E_0 + e_{ab} \qquad (5.5)$$

Thus the energy of this wave function is below the HF energy by the pair correlation energy. To obtain the best possible energy for the above pair-function, we use the linear variation method. Thus we construct the matrix representation of the Hamiltonian in the subspace spanned by $|\Psi_0\rangle$ and all double excitations involving χ_a and χ_b and then find the lowest eigenvalue of this matrix. Equivalently, we substitute the expansion for $|\Psi_{ab}\rangle$ into

$$\mathscr{H}|\Psi_{ab}\rangle = E_{ab}|\Psi_{ab}\rangle \qquad (5.6)$$

to obtain

$$\mathscr{H}\left(|\Psi_0\rangle + \sum_{t<u} c_{ab}^{tu}|\Psi_{ab}^{tu}\rangle\right) = E_{ab}\left(|\Psi_0\rangle + \sum_{t<u} c_{ab}^{tu}|\Psi_{ab}^{tu}\rangle\right) \qquad (5.7)$$

We then multiply this equation successively by $\langle\Psi_0|$ and $\langle\Psi_{ab}^{rs}|$ to obtain

$$E_0 + \sum_{t<u} c_{ab}^{tu}\langle\Psi_0|\mathscr{H}|\Psi_{ab}^{tu}\rangle = E_{ab} \qquad (5.8a)$$

and

$$\langle\Psi_{ab}^{rs}|\mathscr{H}|\Psi_0\rangle + \sum_{t<u} \langle\Psi_{ab}^{rs}|\mathscr{H}|\Psi_{ab}^{tu}\rangle c_{ab}^{tu} = E_{ab}c_{ab}^{rs} \qquad (5.8b)$$

Using Eq. (5.5), these equations become

$$\sum_{t<u} c_{ab}^{tu}\langle\Psi_0|\mathscr{H}|\Psi_{ab}^{tu}\rangle = e_{ab} \qquad (5.9a)$$

and

$$\langle\Psi_{ab}^{rs}|\mathscr{H}|\Psi_0\rangle + \sum_{t<u} \langle\Psi_{ab}^{rs}|\mathscr{H} - E_0|\Psi_{ab}^{tu}\rangle c_{ab}^{tu} = e_{ab}c_{ab}^{rs} \qquad (5.9b)$$

Finally, one can rewrite these equations in matrix form by defining

$$(\mathbf{D}_{ab})_{rs,tu} = \langle\Psi_{ab}^{rs}|\mathscr{H} - E_0|\Psi_{ab}^{tu}\rangle \qquad (5.10a)$$

$$(\mathbf{B}_{ab})_{rs} = \langle\Psi_{ab}^{rs}|\mathscr{H}|\Psi_0\rangle \qquad (5.10b)$$

$$(\mathbf{c}_{ab})_{rs} = c_{ab}^{rs} \qquad (5.10c)$$

so that Eqs. (5.9a) and (5.9b) are equivalent to

$$\begin{pmatrix} 0 & \mathbf{B}_{ab}^\dagger \\ \mathbf{B}_{ab} & \mathbf{D}_{ab} \end{pmatrix}\begin{pmatrix} 1 \\ \mathbf{c}_{ab} \end{pmatrix} = e_{ab}\begin{pmatrix} 1 \\ \mathbf{c}_{ab} \end{pmatrix} \qquad (5.11)$$

This matrix eigenvalue problem is solved for each of the $N(N-1)/2$ pairs of occupied electrons. The total correlation energy is then obtained by adding up the respective pair energies,

$$E_{corr}(IEPA) = \sum_{a<b} e_{ab} \tag{5.12}$$

Although each pair energy e_{ab} is obtained by doing a variational, CI-type calculation, the sum of the pair energies is not necessarily above the exact correlation energy. The individual pair CI matrices are much smaller than the DCI matrix and one never has to calculate matrix elements of the type $\langle \Psi_{ab}^{rs} | \mathcal{H} | \Psi_{cd}^{tu} \rangle$ which couple the pairs ab and cd. The IEPA is computationally equivalent to doing DCI for each pair separately; thus, it is sometimes called "pair-at-a-time" CI. This terminology and the fact that the IEPA is computationally simpler than DCI would appear to suggest that the IEPA is an approximation to DCI. However, this is misleading and as we will see in the next section, where we consider how to incorporate coupling between different pairs, it is more appropriate to view the IEPA as an approximation to full CI which is different from DCI.

Before we give a simple illustration of pair theory, we consider some approximations to the pair equations in order to make contact with perturbation theory. If in Eq. (5.9b) we neglect coupling between excited determinants (i.e., we keep only the term $t = r$ and $u = s$ in the summation), we have

$$\langle \Psi_{ab}^{rs} | \mathcal{H} | \Psi_0 \rangle + \langle \Psi_{ab}^{rs} | \mathcal{H} - E_0 | \Psi_{ab}^{rs} \rangle c_{ab}^{rs} = e_{ab} c_{ab}^{rs} \tag{5.13}$$

If we solve this equation for c_{ab}^{rs} and substitute it into Eq. (5.9a), we obtain

$$e_{ab} = -\sum_{r<s} \frac{|\langle \Psi_0 | \mathcal{H} | \Psi_{ab}^{rs} \rangle|^2}{\langle \Psi_{ab}^{rs} | \mathcal{H} - E_0 | \Psi_{ab}^{rs} \rangle - e_{ab}} \tag{5.14}$$

Since e_{ab} is small compared with the difference between the energies of the HF and doubly excited states, it can be set equal to zero in the denominator, so that we have

$$e_{ab}^{EN} = -\sum_{r<s} \frac{|\langle \Psi_0 | \mathcal{H} | \Psi_{ab}^{rs} \rangle|^2}{\langle \Psi_{ab}^{rs} | \mathcal{H} - E_0 | \Psi_{ab}^{rs} \rangle} \tag{5.15}$$

Since the correlation energy obtained using this approximation to the pair energies,

$$E_{corr}(EN) = \sum_{a<b} e_{ab}^{EN} \tag{5.16}$$

was derived by Epstein and Nesbet, e_{ab}^{EN} is called the Epstein-Nesbet pair correlation energy. Since the total correlation energy using Epstein-Nesbet pairs, with large basis sets, overestimates the correlation energy even more than pair theory, it is only occasionally discussed in this book. Finally, if we further approximate the energy difference between $|\Psi_0\rangle$ and $|\Psi_{ab}^{rs}\rangle$ by differences in HF orbital energies,

$$\langle\Psi_{ab}^{rs}|\mathcal{H}-E_0|\Psi_{ab}^{rs}\rangle \cong \varepsilon_r + \varepsilon_s - \varepsilon_a - \varepsilon_b \qquad (5.17)$$

then we obtain

$$e_{ab}^{FO} = \sum_{r<s} \frac{|\langle\Psi_0|\mathcal{H}|\Psi_{ab}^{rs}\rangle|^2}{\varepsilon_a + \varepsilon_b - \varepsilon_r - \varepsilon_s} \qquad (5.18)$$

which is called the first-order pair energy and is the simplest possible approximation to pair theory. The correlation energy obtained using first-order pairs is

$$E_{corr}(FO) = \sum_{a<b}\sum_{r<s} \frac{|\langle\Psi_0|\mathcal{H}|\Psi_{ab}^{rs}\rangle|^2}{\varepsilon_a + \varepsilon_b - \varepsilon_r - \varepsilon_s} = \sum_{a<b}\sum_{r<s} \frac{|\langle ab||rs\rangle|^2}{\varepsilon_a + \varepsilon_b - \varepsilon_r - \varepsilon_s} \qquad (5.19)$$

As we will see in the next chapter, this expression is identical to the first correction to the HF energy obtained within the framework of many-body perturbation theory (i.e., it is the second-order energy). Thus the simplest form of perturbation theory immediately leads to a form of pair theory.

Exercise 5.1 The application of pair theory to minimal basis H_2 is trivial since we are dealing with a two-electron system for which the IEPA is exact, i.e., it gives the full CI result obtained in the last chapter, viz.

$$^1E_{corr} = \Delta - (\Delta^2 + K_{12}^2)^{1/2}$$

where (see Eq. (4.20))

$$\Delta = (\varepsilon_2 - \varepsilon_1) + \tfrac{1}{2}(J_{11} + J_{22} - 4J_{12} + 2K_{12})$$

a. Calculate the correlation energy using first-order pairs. Remember that the summations in Eq. (5.19) go over spin orbitals (i.e., $a = 1, \bar{1}$ and $r = 2, \bar{2}$). Show that

$$^1E_{corr}(FO) = \frac{K_{12}^2}{2(\varepsilon_1 - \varepsilon_2)}.$$

b. Approximate Δ in the exact correlation energy by $\varepsilon_2 - \varepsilon_1$ and recover the first-order pair correlation energy by expanding the exact answer to first order using the relation $(1 + x)^{1/2} \simeq 1 + x/2$.

As a simple illustration of pair theory we calculate the correlation energy of a dimer consisting of two noninteracting minimal basis H_2 molecules. Since H_2 contains only two electrons, pair theory is identical to full CI and hence is exact for the monomer. Recall that we label the occupied and virtual orbitals of the ith monomer as 1_i and 2_i, respectively, and that, since the monomers are noninteracting, all two-electron integrals containing orbitals from different subunits are zero. Since the dimer has the configuration

$(1_1)^2(1_2)^2$, we need to find six pair correlation energies, i.e., $e_{1_1\bar{1}_1}$, $e_{1_1 1_2}$, $e_{1_1 \bar{1}_2}$, $e_{\bar{1}_1 1_2}$, $e_{\bar{1}_1 \bar{1}_2}$, and $e_{1_2 \bar{1}_2}$. However, only two of these six pair energies are non-zero namely, $e_{1_1 \bar{1}_1}$ and $e_{1_2 \bar{1}_2}$. To see this, consider the mixing of a doubly excited state in which a pair, say $(1_i, \bar{1}_j)$ is excited into the virtual orbitals 2_k and $\bar{2}_l$, with the HF ground state

$$\langle \Psi_0 | \mathscr{H} | \Psi_{1_i \bar{1}_j}^{2_k \bar{2}_l} \rangle = \langle 1_i \bar{1}_j | | 2_k \bar{2}_l \rangle$$

$$= \begin{cases} \langle 11 | 22 \rangle = K_{12} & \text{if } i = j = k = l \\ 0 & \text{otherwise} \end{cases} \tag{5.20}$$

Since this matrix element is zero unless both electrons are located within a given subunit and unless they are excited to the virtual orbitals of the same subunits, the pair function for $(1_i, \bar{1}_i)$, is

$$|\Psi_{1_i \bar{1}_i}\rangle = |\Psi_0\rangle + c_{1_i \bar{1}_i}^{2_i \bar{2}_i} | \Psi_{1_i \bar{1}_i}^{2_i \bar{2}_i} \rangle \tag{5.21}$$

The corresponding pair equations obtained from Eqs. (5.9a) and (5.9b) are

$$K_{12} c_{1_i \bar{1}_i}^{2_i \bar{2}_i} = e_{1_i \bar{1}_i} \tag{5.22a}$$

$$K_{12} + 2\Delta c_{1_i \bar{1}_i}^{2_i \bar{2}_i} = e_{1_i \bar{1}_i} c_{1_i \bar{1}_i}^{2_i \bar{2}_i} \tag{5.22b}$$

where $2\Delta = 2(\varepsilon_2 - \varepsilon_1) + J_{11} + J_{22} - 4J_{12} + 2K_{12}$. These equations are identical to the full CI equations of a single minimal basis H_2 molecule (see Eqs. (4.19a) and (4.21)). Thus $e_{1_1 \bar{1}_1} = e_{1_2 \bar{1}_2} = {}^1 E_{corr}$ where ${}^1 E_{corr}$ is the exact correlation energy of a single H_2 molecule. Then the total correlation energy in the IEPA for the dimer is

$$^2 E_{corr}(\text{IEPA}) = e_{1_1 \bar{1}_1} + e_{1_2 \bar{1}_2} = 2 {}^1 E_{corr} \tag{5.23}$$

so that it is twice the exact correlation energy of the monomer. The above discussion can immediately be extended to show that the IEPA correlation energy of N independent H_2 molecules is N times the correlation energy of the monomer. Thus we see that the IEPA is size consistent as stated previously. Since IEPA is exact while DCI fails badly for this model, one cannot really consider the IEPA to be an approximation to DCI.

Exercise 5.2 Derive Eqs. (5.22a) and (5.22b).

Exercise 5.3 Calculate the total first-order pair correlation energy for the dimer using Eq. (5.19) and show that it is twice the result obtained in Exercise 5.1.

5.1.1 Invariance under Unitary Transformations: An Example

In spite of the fact that the IEPA is not variational, it has an advantage over DCI in that it is size consistent. Unlike DCI and many-body perturbation theory (discussed in Chapter 6) the IEPA has the distinct disadvantage of not being invariant to unitary transformations of the occupied spin orbitals in the HF determinant. As we have seen in Chapter 3, if we transform the occupied spin orbitals $\{\chi_a\}$ among themselves, $|\Psi_0\rangle$ will not change; however, the total correlation energy obtained using the IEPA will be different. Since the canonical HF orbitals are unique, apart from degeneracies, the IEPA using nondegenerate orbitals does give a unique correlation energy. In the case of degeneracies, however, an arbitrary mixing of degenerate orbitals can occur in the SCF calculation. The particular linear combination obtained can be a function, for example, of the iteration step at which it was decided to terminate the procedure. The fact that the sum of pair correlation energies is not invariant to this very arbitrary mixing is particularly distressing.

We now investigate this invariance problem with our minimal basis model of two noninteracting H_2 molecules. The following example also provides us with an opportunity to introduce the concept of *spin-adapted pair correlation* energies. Up to this point we have considered *spin-orbital* pair energies. For example, in a 4-electron system with ψ_1 and ψ_2 doubly occupied, we must calculate six spin-orbital pair energies, i.e., $e_{1\bar{1}}, e_{\bar{1}2}, e_{12},$ $e_{\bar{1}\bar{2}}, e_{1\bar{2}},$ and $e_{2\bar{2}}$. However, not all of the corresponding spin-orbital pair functions are eigenfunctions of $\mathscr{S}^2$. In particular, $\Psi_{12}, \Psi_{1\bar{2}}, \Psi_{\bar{1}2},$ and $\Psi_{\bar{1}\bar{2}}$ are not pure spin states. The idea behind spin-adapted pair theory is simply to use only pair functions which are eigenfunctions of $\mathscr{S}^2$. In general, the correlation energy calculated from spin-orbital or spin-adapted pair energies differs, as will be shown by our example.

For two independent H_2 molecules using molecular orbitals *localized* on the two monomers, (i.e., $1_1, 2_1, 1_2, 2_2$) we have seen that the IEPA gives the exact result for the correlation energy of the dimer. We now repeat our analysis using a set of equivalent *delocalized* molecular orbitals. Since orbitals 1_1 and 1_2 as well as 2_1 and 2_2 are degenerate, we can take an arbitrary linear combination of them and retain the same HF description. In particular we consider the completely delocalized orbitals,

$$a = 2^{-1/2}(1_1 + 1_2) \qquad \overset{+}{.}\ \overset{+}{.} \qquad \overset{+}{.}\ \overset{+}{.} \quad (g) \qquad (5.24a)$$

$$b = 2^{-1/2}(1_1 - 1_2) \qquad \overset{+}{.}\ \overset{+}{.} \qquad \overset{-}{.}\ \overset{-}{.} \quad (u) \qquad (5.24b)$$

$$r = 2^{-1/2}(2_1 - 2_2) \qquad \overset{+}{.}\ \overset{-}{.} \qquad \overset{-}{.}\ \overset{+}{.} \quad (g) \qquad (5.24c)$$

$$s = 2^{-1/2}(2_1 + 2_2) \qquad \overset{+}{.}\ \overset{-}{.} \qquad \overset{+}{.}\ \overset{-}{.} \quad (u) \qquad (5.24d)$$

The pluses and minuses indicate the sign of the atomic basis functions in the LCAO description of the four orbitals. Note that a and r are of gerade symmetry while b and s are of ungerade symmetry. The HF wave function in terms of these delocalized orbitals is

$$|\Psi_0\rangle = |a\bar{a}b\bar{b}\rangle \tag{5.25}$$

This wave function is identical to $|1_1\bar{1}_11_2\bar{1}_2\rangle$, which is a reflection of the fact that the HF wave function is invariant to a unitary transformation of the occupied orbitals.

Exercise 5.4 Show that $|a\bar{a}b\bar{b}\rangle = |1_1\bar{1}_11_2\bar{1}_2\rangle$. *Hint:* use Eq. (1.40) repeatedly. Eq. (1.40) for Slater determinants is

$$|\chi_1\chi_2\cdots\left(\sum_k c_k\chi'_k\right)\cdots\chi_N\rangle = \sum_k c_k|\chi_1\chi_2\cdots\chi'_k\cdots\chi_N\rangle$$

Using the expansions in Eqs. (5.24) one can show that the two-electron integrals over the delocalized orbitals are

$$(aa|aa) = (aa|bb) = (bb|bb) = (ab|ab) = \tfrac{1}{2}J_{11} \tag{5.26a}$$

$$(rr|rr) = (rr|ss) = (ss|ss) = (rs|rs) = \tfrac{1}{2}J_{22} \tag{5.26b}$$

$$(rr|aa) = (ss|aa) = (rr|bb) = (ss|bb) = (ab|rs) = \tfrac{1}{2}J_{12} \tag{5.26c}$$

$$(ra|ra) = (sb|sb) = (rb|rb) = (sa|sa) = (ra|sb) = (rb|sa) = \tfrac{1}{2}K_{12} \tag{5.26d}$$

and the orbital energies are $\varepsilon_a = \varepsilon_b = \varepsilon_1$ and $\varepsilon_r = \varepsilon_s = \varepsilon_2$.

We now find the six spin-orbital correlation energies: $e_{a\bar{a}}$, e_{ab}, $e_{a\bar{b}}$, $e_{\bar{a}b}$, $e_{\bar{a}\bar{b}}$, $e_{b\bar{b}}$. First consider the pair $a\bar{a}$. Since $|\Psi_0\rangle$ is of gerade symmetry and is an eigenfunction of $\mathscr{S}_z$ with eigenvalue zero (i.e. $M_S = 0$), only doubly excited determinants of gerade symmetry having equal numbers of spin up and spin down electrons, interact with $|\Psi_0\rangle$. Therefore the required pair function is

$$|\Psi_{a\bar{a}}\rangle = |\Psi_0\rangle + c_1|\Psi_{a\bar{a}}^{r\bar{r}}\rangle + c_2|\Psi_{a\bar{a}}^{s\bar{s}}\rangle \tag{5.27}$$

Since

$$\langle\Psi_0|\mathscr{H}|\Psi_{a\bar{a}}^{r\bar{r}}\rangle = \langle a\bar{a}|r\bar{r}\rangle = (ar|ar) = (ra|ra) = \tfrac{1}{2}K_{12} \tag{5.28a}$$

$$\langle\Psi_0|\mathscr{H}|\Psi_{a\bar{a}}^{s\bar{s}}\rangle = \langle a\bar{a}|s\bar{s}\rangle = (as|as) = (sa|sa) = \tfrac{1}{2}K_{12} \tag{5.28b}$$

it is convenient to use symmetric and antisymmetric combinations of these functions (i.e., $2^{-1/2}(|\Psi_{a\bar{a}}^{r\bar{r}}\rangle \pm |\Psi_{a\bar{a}}^{s\bar{s}}\rangle)$) so that only the plus combination mixes with the ground state. Introducing a new notation for this plus combination

$$|\Psi_{a\bar{a}}^{**}\rangle = 2^{-1/2}(|\Psi_{a\bar{a}}^{r\bar{r}}\rangle + |\Psi_{a\bar{a}}^{s\bar{s}}\rangle) \tag{5.29}$$

we can write the $a\bar{a}$ pair function as

$$|\Psi_{a\bar{a}}\rangle = |\Psi_0\rangle + c|\Psi_{a\bar{a}}^{**}\rangle \tag{5.30}$$

The required matrix elements are

$$\langle \Psi_0 | \mathcal{H} | \Psi_{a\bar{a}}^{**} \rangle = 2^{-1/2} K_{12} \tag{5.31a}$$

$$\langle \Psi_{a\bar{a}}^{**} | \mathcal{H} - E_0 | \Psi_{a\bar{a}}^{**} \rangle = 2(\varepsilon_2 - \varepsilon_1) + J_{22} + \tfrac{1}{2}(J_{11} - 4J_{12} + 2K_{12})$$

$$\equiv 2\Delta' \tag{5.31b}$$

Exercise 5.5 Derive Eqs. (5.31a) and (5.31b).

The equations which determine the pair energy are (see Eqs. (5.9a,b)).

$$e_{a\bar{a}} = 2^{-1/2} K_{12} c \tag{5.32a}$$

$$2^{-1/2} K_{12} + 2\Delta' c = e_{a\bar{a}} c \tag{5.32b}$$

By eliminating c from these equations and solving the resulting quadratic equation for $e_{a\bar{a}}$ we find

$$e_{a\bar{a}} = \Delta' - ((\Delta')^2 + K_{12}^2/2)^{1/2} \tag{5.33}$$

Because of the high symmetry of this problem, as manifested by the relationship between the two-electron integrals in Eqs. (5.26), it follows that $e_{b\bar{b}} = e_{a\bar{a}}$. We now show that $e_{ab} = e_{a\bar{b}} = 0$. To see this, consider the pair function for the $a\bar{b}$ pair. Since only double excitations of gerade symmetry and $M_S = 0$ need be considered, we have

$$|\Psi_{a\bar{b}}\rangle = |\Psi_0\rangle + c_1 |\Psi_{a\bar{b}}^{r\bar{s}}\rangle \tag{5.34}$$

However, this double excitation does not mix with $|\Psi_0\rangle$,

$$\langle \Psi_0 | \mathcal{H} | \Psi_{a\bar{b}}^{r\bar{s}} \rangle = \langle a\bar{b} | | r\bar{s} \rangle = (ar|bs) - (as|br) = 0 \tag{5.35}$$

and thus the pair correlation energy of the $a\bar{b}$ pair is zero. The argument for the ab pair is the same. This result has an interesting physical interpretation. Since correlation between electrons with the same spin is included in the HF approximation (Fermi hole), pair theory does not give any additional correlation. While it is not true in general that pair energies of electrons with the same spin are zero, they are smaller than the other pair energies.

Finally, we must find $e_{a\bar{b}}$ and $e_{\bar{a}b}$. Proceeding in the same way as for the $a\bar{a}$ pair (i.e., eliminating double excitations of the wrong symmetry and then constructing $\pm$ linear combinations) the $a\bar{b}$ pair function becomes

$$|\Psi_{a\bar{b}}\rangle = |\Psi_0\rangle + c|\Psi_{a\bar{b}}^{**}\rangle \tag{5.36}$$

where

$$|\Psi_{a\bar{b}}^{**}\rangle = 2^{-1/2}(|\Psi_{a\bar{b}}^{r\bar{s}}\rangle + |\Psi_{a\bar{b}}^{s\bar{r}}\rangle) \tag{5.37}$$

The $\bar{a}b$ function is obtained by interchanging a and b. Now it can be shown that the equations which determine the $a\bar{b}$ pair are identical to Eqs. (5.32) so that the pair correlation energies for the $a\bar{b}$ and $a\bar{a}$ pair are the same. Thus the total pair correlation energy for the dimer using delocalized orbitals is

$$^2E_{corr}(\text{IEPA}(D)) = e_{a\bar{a}} + e_{b\bar{b}} + e_{ab} + e_{a\bar{b}} + e_{\bar{a}b} + e_{\bar{a}\bar{b}}$$

$$= 4e_{a\bar{a}}$$

$$= 4(\Delta' - ((\Delta')^2 + K_{12}^2/2)^{1/2}) \tag{5.38}$$

This is to be compared to the result obtained using *localized* orbitals

$$^2E_{corr}(\text{IEPA}(L)) = 2(\Delta - (\Delta^2 + K_{12}^2)^{1/2}) \tag{5.39}$$

which is exact for the model. It is clear that the two expressions are different, and using the minimal STO-3G basis set two-electron integrals in Appendix D for $R = 1.4$ a.u., we find

$$^2E_{corr}(\text{exact}) = {}^2E_{corr}(\text{IEPA}(L)) = -0.0411 \text{ a.u.}$$

$$^2E_{corr}(\text{IEPA}(D)) = -0.0275 \text{ a.u.}$$

so that there is almost a factor of 2 difference between the two results! In real systems, the situation is not quite so bad, and the results obtained using localized and delocalized molecular orbitals, although different, are closer. For example, Bender and Davidson[3] have found for boron hydride, using a large basis set of Slater orbitals, that with canonical SCF orbitals the IEPA correlation energy is -0.141, a.u., while with localized orbitals it is -0.139 a.u..

Exercise 5.6 Show that $e_{a\bar{b}} = e_{\bar{a}b} = e_{a\bar{a}}$.

Exercise 5.7 Show that DCI is invariant to unitary transformations for the above model.

a. The DCI wave function is

$$|\Psi_{DCI}\rangle = |\Psi_0\rangle + c_1|\Psi_{a\bar{a}}^{**}\rangle + c_2|\Psi_{b\bar{b}}^{**}\rangle + c_3|\Psi_{a\bar{b}}^{**}\rangle + c_4|\Psi_{\bar{a}b}^{**}\rangle$$

Show that the corresponding eigenvalue problem which determines the DCI correlation energy of the dimer ($^2E_{corr}(\text{DCI})$) is

$$
\begin{vmatrix}
0 & 2^{-1/2}K_{12} & 2^{-1/2}K_{12} & 2^{-1/2}K_{12} & 2^{-1/2}K_{12} \\
2^{-1/2}K_{12} & 2\Delta' & \tfrac{1}{2}J_{11} & \tfrac{1}{2}K_{12} - J_{12} & \tfrac{1}{2}K_{12} - J_{12} \\
2^{-1/2}K_{12} & \tfrac{1}{2}J_{11} & 2\Delta' & \tfrac{1}{2}K_{12} - J_{12} & \tfrac{1}{2}K_{12} - J_{12} \\
2^{-1/2}K_{12} & \tfrac{1}{2}K_{12} - J_{12} & \tfrac{1}{2}K_{12} - J_{12} & 2\Delta' & \tfrac{1}{2}J_{11} \\
2^{-1/2}K_{12} & \tfrac{1}{2}K_{12} - J_{12} & \tfrac{1}{2}K_{12} - J_{12} & \tfrac{1}{2}J_{11} & 2\Delta'
\end{vmatrix}
\begin{pmatrix}
1 \\ c_1 \\ c_2 \\ c_3 \\ c_4
\end{pmatrix}
$$

$$
= {}^2E_{corr}(\text{DCI})
\begin{pmatrix}
1 \\ c_1 \\ c_2 \\ c_3 \\ c_4
\end{pmatrix}
$$

b. Show that $c_1 = c_2 = c_3 = c_4 = c$ and then solve the equations to show

$$^2E_{\text{corr}}(\text{DCI}) = \Delta - (\Delta^2 + 2K_{12}^2)^{1/2}$$

which is the same result as found in the last chapter (see Eq. (4.60)).

The pair functions $|\Psi_{a\bar{b}}\rangle$ and $|\Psi_{\bar{a}b}\rangle$ are not pure singlet wave functions. For example, as can be seen from Eq. (5.37), the doubly excited states in the pair function Ψ_{ab}^{**} are $|r\bar{a}b\bar{s}\rangle$ and $|s\bar{a}b\bar{r}\rangle$, which correspond to

$$\begin{array}{ccc} \underline{\uparrow}\ r & \underline{\downarrow}\ s & \\ \underline{\downarrow}\ a & \underline{\uparrow}\ b & \end{array} \quad \text{and} \quad \begin{array}{cc} \underline{\downarrow}\ r & \underline{\uparrow}\ s \\ \underline{\downarrow}\ a & \underline{\uparrow}\ b \end{array}$$

respectively. Recall (see Table 2.7) that there are two linearly independent singlet spin functions arising from the configuration $(a)^1(b)^1(r)^1(s)^1$ namely, $|^A\Psi_{ab}^{rs}\rangle$ and $|^B\Psi_{ab}^{rs}\rangle$. The matrix elements we require for our discussion are in Table 4.1. Using these results, along with the integrals in Eqs. (5.26), it follows that

$$\langle\Psi_0|\mathscr{H}|^A\Psi_{ab}^{rs}\rangle = \langle^B\Psi_{ab}^{rs}|\mathscr{H}|^A\Psi_{ab}^{rs}\rangle = 0 \qquad (5.40)$$

and hence we only need consider $|^B\Psi_{ab}^{rs}\rangle$ in constructing the spin-adapted pair function for the ab pair,

$$|^1\Psi_{ab}\rangle = |\Psi_0\rangle + c|^B\Psi_{ab}^{rs}\rangle \qquad (5.41)$$

The required matrix elements are

$$\langle\Psi_0|\mathscr{H}|^B\Psi_{ab}^{rs}\rangle = K_{12} \qquad (5.42a)$$

$$\langle^B\Psi_{ab}^{rs}|\mathscr{H} - E_0|^B\Psi_{ab}^{rs}\rangle = 2(\varepsilon_2 - \varepsilon_1) + J_{11} + J_{22} - 2J_{12} + K_{12} = 2\Delta'' \qquad (5.42b)$$

the corresponding pair equations are

$$e_{ab}^{\text{singlet}} = K_{12}c \qquad (5.43a)$$

$$K_{12} + 2\Delta''c = e_{ab}^{\text{singlet}}c \qquad (5.43b)$$

Solving these in the standard way, we find

$$e_{ab}^{\text{singlet}} = \Delta'' - ((\Delta'')^2 + K_{12}^2)^{1/2} \qquad (5.44)$$

Since the pair functions for the $a\bar{a}$ and $b\bar{b}$ pairs are pure singlets the total correlation energy for the dimer using spin-adapted pairs is

$$\begin{aligned} ^2E_{\text{corr}}^{\text{singlet}}(\text{IEPA}(D)) &= e_{a\bar{a}} + e_{b\bar{b}} + e_{ab}^{\text{singlet}} \\ &= 2(\Delta' - ((\Delta')^2 + K_{12}^2/2)^{1/2}) + (\Delta'' - ((\Delta'')^2 + K_{12}^2)^{1/2}) \end{aligned}$$

$$(5.45)$$

With the STO-3G minimal basis set, the above correlation energy is -0.0258 a.u. as compared to -0.0275 a.u. obtained using spin-orbital pairs. Thus pair theory is not only variant to unitary transformations of degenerate HF orbitals, but also gives different answers depending on whether one uses spin-orbital or spin-adapted pair functions.

It is interesting to note, however, that the simplest form of pair theory, namely, first-order pairs (see Eq. (5.19)), *is* invariant to unitary transformations of degenerate orbitals. To see this, we approximate $\Delta' = \Delta'' = \varepsilon_2 - \varepsilon_1$ in Eqs. (5.38) and (5.45),

$$^2E_{\text{corr}}(\text{IEPA}(D)) = 4\left((\varepsilon_2 - \varepsilon_1) - (\varepsilon_2 - \varepsilon_1)\left(1 + \frac{K_{12}^2}{2(\varepsilon_2 - \varepsilon_1)^2}\right)^{1/2}\right)$$

$$^2E_{\text{corr}}^{\text{singlet}}(\text{IEPA}(D)) = 2\left((\varepsilon_2 - \varepsilon_1) - (\varepsilon_2 - \varepsilon_1)\left(1 + \frac{K_{12}^2}{2(\varepsilon_2 - \varepsilon_1)^2}\right)^{1/2}\right)$$

$$+ \left((\varepsilon_2 - \varepsilon_1) - (\varepsilon_2 - \varepsilon_1)\left(1 + \frac{K_{12}^2}{(\varepsilon_2 - \varepsilon_1)^2}\right)^{1/2}\right)$$

Notice we have factored $(\varepsilon_2 - \varepsilon_1)$ out of the square roots. Expanding the square roots using $(1 + x)^{1/2} = 1 + x/2$, we find

$$^2E_{\text{corr}}(\text{FO}(D)) = 4\left(\frac{-K_{12}^2}{4(\varepsilon_2 - \varepsilon_1)}\right) = 2\left(\frac{K_{12}^2}{2(\varepsilon_1 - \varepsilon_2)}\right) \qquad (5.46a)$$

and

$$^2E_{\text{corr}}^{\text{singlet}}(\text{FO}(D)) = 2\left(\frac{-K_{12}^2}{4(\varepsilon_2 - \varepsilon_1)}\right) + \left(\frac{-K_{12}^2}{2(\varepsilon_2 - \varepsilon_1)}\right) = 2\left(\frac{K_{12}^2}{2(\varepsilon_1 - \varepsilon_2)}\right) \qquad (5.46b)$$

These results are equal to the total first-order pair correlation energy, obtained in Exercise 5.3, for the dimer using localized orbitals. The total first-order pair correlation energy is identical to the second-order many-body perturbation result for the correlation energy (see Chapter 6). The above results are a reflection of the fact that many-body perturbation theory is invariant to unitary transformations of degenerate orbitals.

Finally, we have seen that using localized orbitals, the IEPA applied to a dimer of noninteracting minimal basis H_2 molecules gives twice the exact energy of the monomer. However, using delocalized orbitals, the energy of the dimer is no longer twice the monomer energy because of the invariance problem. This appears to contradict our statement that IEPA is size consistent; however, the essential requirement for size consistency is the linear variation of the correlation energy with the number of particles. Thus while the energy of N H_2 molecules calculated within the IEPA may not be N times the exact energy of H_2, it will be proportional to N as N becomes large rather than $N^{1/2}$ as with DCI.

Exercise 5.8 Show directly from Eq. (5.19) using delocalized orbitals and the two-electron integrals in Eq. (5.26) that the total first-order pair correlation energy (which is the same as the many-body second-order perturbation energy) of the dimer is given by Eq. (5.46).

Exercise 5.9 Show that the total correlation energy obtained using Epstein-Nesbet pairs is not invariant to unitary transformations.

a. Show, using localized orbitals, that

$$^2E_{corr}(EN(L)) = -\frac{K_{12}^2}{\Delta}.$$

b. Show, using delocalized spin-orbital pairs, that

$$^2E_{corr}(EN(D)) = -\frac{K_{12}^2}{\Delta'}.$$

c. Show, using delocalized spin-adapted pairs, that

$$^2E_{corr}^{singlet}(EN(D)) = -\frac{K_{12}^2}{2\Delta'} - \frac{K_{12}^2}{2\Delta''}.$$

d. Using the STO-3G integrals for H_2 in Appendix D compare the numerical values of the above expressions at $R = 1.4$ a.u.

Exercise 5.10 The DCI wave function for the H_2 dimer using spin-adapted configurations is

$$|\Psi_{DCI}\rangle = |\Psi_0\rangle + c_1|\Psi_{a\bar{a}}^{**}\rangle + c_2|\Psi_{b\bar{b}}^{**}\rangle + c_3|{}^B\Psi_{ab}^{rs}\rangle$$

Show that the corresponding DCI eigenvalue problem is

$$\begin{pmatrix} 0 & 2^{-1/2}K_{12} & 2^{-1/2}K_{12} & K_{12} \\ 2^{-1/2}K_{12} & 2\Delta' & \frac{1}{2}J_{11} & 2^{-1/2}(K_{12} - 2J_{12}) \\ 2^{-1/2}K_{12} & \frac{1}{2}J_{11} & 2\Delta' & 2^{-1/2}(K_{12} - 2J_{12}) \\ K_{12} & 2^{-1/2}(K_{12} - 2J_{12}) & 2^{-1/2}(K_{12} - 2J_{12}) & 2\Delta'' \end{pmatrix}$$

$$\times \begin{pmatrix} 1 \\ c_1 \\ c_2 \\ c_3 \end{pmatrix} = {}^2E_{corr}(DCI) \begin{pmatrix} 1 \\ c_1 \\ c_2 \\ c_3 \end{pmatrix}$$

and then solve the resulting equations to show that

$$^2E_{corr}(DCI) = \Delta - (\Delta^2 + 2K_{12}^2)^{1/2}$$

5.1.2 Some Illustrative Calculations

An impressive success of the IEPA is its prediction of the ground state energy of the beryllium atom. The spin-orbital pair correlation energies obtained by

Table 5.1 Spin-orbital pair correlation energies (a.u.) of Be[a]

Pair	Correlation energies
$(1s, \overline{1s})$	-0.0418
$(1s, 2s)$	-0.0008
$(1s, \overline{2s})$	-0.0021
$(\overline{1s}, 2s)$	-0.0021
$(\overline{1s}, \overline{2s})$	-0.0008
$(2s, \overline{2s})$	-0.0454
Total pair energy	-0.0930
Exact correlation energy	-0.094

[a] R. K. Nesbet, *Phys. Rev.* **155**: 51 (1967).

Nesbet using a large Slater basis set containing s, p, d, and f orbitals are shown in Table 5.1. The total IEPA correlation energy is 98.9% of the exact correlation energy. It is estimated that if $g, h, i, \ldots$ orbitals were to be included in the basis, the IEPA would give very close to 100% of the exact correlation energy. Of course, because IEPA is not variational, it is possible to obtain over 100%. The individual pair correlation energies listed in Table 5.1 show some interesting trends. The largest values are for pairs with opposite spin in the same orbital (e.g., $(1s, \overline{1s})$) and the smallest are for pairs with parallel spins (e.g., $(1s, 2s)$ or $(\overline{1s}, \overline{2s})$). This is a reflection of the fact that the HF approximation does in fact correlate electrons with the same spin (the Fermi hole); its major defect is that it allows electrons of different spin to be in the same place.

For larger atoms, the IEPA works less well. Using a very large basis containing $s, p, d, f, g, h,$ and i orbitals, Nesbet, Barr, and Davidson[4] obtained a correlation energy for neon of 107% of the exact result. Since their basis was still not complete, they estimate that the IEPA gives about 110% of the exact correlation energy. The IEPA appears to be even poorer for molecules. For example, Langhoff and Davidson[5] studied N_2 using a moderately large Gaussian basis set. It was estimated that if full CI were to be performed in this basis one would obtain a correlation energy of -0.35 a.u. They performed an essentially complete DCI calculation, which gave -0.324 a.u. for the correlation energy. This result is necessarily less than the basis set correlation energy. The IEPA, on the other hand, gave a correlation energy of -0.412 a.u., which is 18% larger than the estimated exact result. With larger basis sets, giving 80 or 90% of the exact correlation energy, it is likely that the IEPA is even worse. Thus to obtain a really accurate approximation for the correlation energy one has to improve upon the IEPA. This can be done by incorporating coupling between pairs as will be discussed in the next section.

5.2 COUPLED-PAIR THEORIES

In this section we will extend the IEPA to incorporate coupling between different pairs ab and cd. Although DCI does include such coupling, since one needs matrix elements of the type $\langle \Psi_{ab}^{rs}|\mathcal{H}|\Psi_{cd}^{tu}\rangle$, DCI is not size consistent. Because full CI is the exact, but computationally prohibitive, solution to the many-electron problem, it seems reasonable to begin our search for a size-consistent extension of the IEPA by considering the full CI equations and seeing if we can find a novel approximation to them. For the sake of simplicity, we ignore the presence of singles, triple, etc. excitations; thus, the intermediate normalized full CI wave function is

$$|\Phi_0\rangle = |\Psi_0\rangle + \sum_{\substack{a<b \\ r<s}} c_{ab}^{rs}|\Psi_{ab}^{rs}\rangle + \sum_{\substack{a<b<c<d \\ r<s<t<u}} c_{abcd}^{rstu}|\Psi_{abcd}^{rstu}\rangle + \cdots \qquad (5.47)$$

where the dots represent hextuple etc. excitations. As we have seen in the last chapter, to determine the variational energy of this wave function, we substitute it into

$$(\mathcal{H} - E_0)|\Phi_0\rangle = E_{\text{corr}}|\Phi_0\rangle$$

and then successively multiply by $\langle\Psi_0|$, $\langle\Psi_{ab}^{rs}|$, $\langle\Psi_{abcd}^{rstu}|$, etc. to obtain the following set of coupled equations

$$\sum_{\substack{c<d \\ t<u}} \langle\Psi_0|\mathcal{H}|\Psi_{cd}^{tu}\rangle c_{cd}^{tu} = E_{\text{corr}} \qquad (5.48a)$$

$$\langle\Psi_{ab}^{rs}|\mathcal{H}|\Psi_0\rangle + \sum_{\substack{c<d \\ t<u}} \langle\Psi_{ab}^{rs}|\mathcal{H} - E_0|\Psi_{cd}^{tu}\rangle c_{cd}^{tu}$$

$$+ \sum_{\substack{c<d \\ t<u}} \langle\Psi_{ab}^{rs}|\mathcal{H}|\Psi_{abcd}^{rstu}\rangle c_{abcd}^{rstu} = E_{\text{corr}}c_{ab}^{rs} \qquad (5.48b)$$

and so on. For example, the next equation involves the coefficients of the quadruples and the hextuples. Note that in writing the matrix element between doubles and quadruples we incorporated the fact that matrix elements between determinants which differ by more than 2 spin orbitals are zero. The above equations form a hierarchy in which the correlation energy depends on the coefficients of the doubles, but the equation for these coefficients involves the coefficients of the quadruples and so on. Clearly, to make progress, we must terminate or decouple this hierarchy. The simplest procedure is to set c_{abcd}^{rstu} equal to zero, obtaining a closed set of equations involving only the coefficients of the doubles. The resulting equations are identical to the DCI equations, which can be derived by starting with a CI wave function containing only double excitations and then using the linear variation method. Is there an alternate and perhaps less drastic way to decouple the hierarchy?

5.2.1 The Coupled-Cluster Approximation (CCA)

If we could express the coefficients of the quadruples as some function of the doubles coefficients, then we would obtain a closed set of equations. In Chapter 4 we found an indication of how this might be done. The full CI wave function for two noninteracting minimal basis H_2 molecules is

$$|\Phi_0\rangle = |1_1\bar{1}_1 1_2 \bar{1}_2\rangle + c_{1_1\bar{1}_1}^{2_1\bar{2}_1}|2_1\bar{2}_1 1_2 \bar{1}_2\rangle + c_{1_2\bar{1}_2}^{2_2\bar{2}_2}|1_1\bar{1}_1 2_2 \bar{2}_2\rangle$$
$$+ c_{1_1\bar{1}_1 1_2\bar{1}_2}^{2_1\bar{2}_1 2_2\bar{2}_2}|2_1\bar{2}_1 2_2 \bar{2}_2\rangle \tag{5.49}$$

In Exercise 4.12 we found that

$$c_{1_1\bar{1}_1 1_2\bar{1}_2}^{2_1\bar{2}_1 2_2\bar{2}_2} = (c_{1_1\bar{1}_1}^{2_1\bar{2}_1})(c_{1_2\bar{1}_2}^{2_2\bar{2}_2}) \tag{5.50}$$

That is, the coefficient of the quadruple excitation is just the product of the coefficients of the two double excitations. This result can be readily understood without our previous algebraic manipulations. The two H_2 molecules are separated by infinity so that, for all intents and purposes, we can ignore the requirement that the total wave function be antisymmetric with respect to the interchange of electrons which belong to different H_2 molecules. Thus since two H_2 molecules are independent, we can write the exact wave function of the dimer as a product of the exact wave functions of the monomers,

$$|\Phi_0\rangle = [|1_1\bar{1}_1\rangle + c_{1_1\bar{1}_1}^{2_1\bar{2}_1}|2_1\bar{2}_1\rangle][|1_2\bar{1}_2\rangle + c_{1_2\bar{1}_2}^{2_2\bar{2}_2}|2_2\bar{2}_2\rangle]$$
$$= |1_1\bar{1}_1\rangle|1_2\bar{1}_2\rangle + c_{1_1\bar{1}_1}^{2_1\bar{2}_1}|2_1\bar{2}_1\rangle|1_2\bar{1}_2\rangle + c_{1_2\bar{1}_2}^{2_2\bar{2}_2}|1_1\bar{1}_1\rangle|2_2\bar{2}_2\rangle$$
$$+ c_{1_1\bar{1}_1}^{2_1\bar{2}_1}c_{1_2\bar{1}_2}^{2_2\bar{2}_2}|2_1\bar{2}_1\rangle|2_2\bar{2}_2\rangle \tag{5.51}$$

Comparing this equation with Eq. (5.49), note that (aside from antisymmetry) for the two functions to be identical the coefficients of the doubles and quadruples must be related by Eq. (5.50). In this simple four-electron model system, the two pairs of electrons are independent and the coefficient of the quadruply excited configuration in the full CI wave function is exactly equal to the product of the coefficients of the double excitations.

In a real many-electron system two pairs of electrons ab and cd are, of course, not independent. However, since the IEPA works fairly well, it seems reasonable to *approximate* the coefficients of the quadruples as products of the coefficients of the doubles. Thus we write symbolically

$$c_{abcd}^{rstu} \cong c_{ab}^{rs} * c_{cd}^{tu} \tag{5.52}$$

The reason why c_{abcd}^{rstu} is not simply the product of c_{ab}^{rs} and c_{cd}^{tu} can be understood as follows. We can obtain a quadruply excited configuration in which electrons in spin orbitals $abcd$ are excited to $rstu$ in many ways. For example, not only can we excite $\begin{smallmatrix}a \to r\\b \to s\end{smallmatrix}$ and $\begin{smallmatrix}d \to u\\c \to t\end{smallmatrix}$ but we can excite $\begin{smallmatrix}a \to r\\b \to t\end{smallmatrix}$ and $\begin{smallmatrix}d \to u\\c \to s\end{smallmatrix}$. In the first case, if $|\Psi_0\rangle$ is $|\cdots abcd \cdots\rangle$, we would obtain a quadruply excited determinant $|\cdots rstu \cdots\rangle$ while in the second case we would get $|\cdots rtsu \cdots\rangle$. These determinants represent the same quadruply excited state but have different signs (i.e., $|\cdots rtsu \cdots\rangle = -|\cdots rstu \cdots\rangle$). Thus we could represent c_{abcd}^{rstu} either as $c_{ab}^{rs} c_{cd}^{tu}$ or $-c_{ab}^{rt} c_{cd}^{su}$. Unfortunately, there are 18 distinct ways we can get a particular quadruple excitation from independent double excitations and c_{abcd}^{rstu} is the sum of all possible products of such double excitation coefficients. The rather formidable result is

$$
\begin{aligned}
c_{abcd}^{rstu} \cong c_{ab}^{rs} * c_{cd}^{tu} &= c_{ab}^{rs} c_{cd}^{tu} - \langle c_{ab}^{rs} * c_{cd}^{tu} \rangle \\
&= c_{ab}^{rs} c_{cd}^{tu} - c_{ac}^{rs} c_{bd}^{tu} + c_{ad}^{rs} c_{bc}^{tu} - c_{ab}^{rt} c_{cd}^{su} + c_{ac}^{rt} c_{bd}^{su} - c_{ad}^{rt} c_{bc}^{su} \\
&\quad + c_{ab}^{ru} c_{cd}^{st} - c_{ac}^{ru} c_{bd}^{st} + c_{ad}^{ru} c_{bc}^{st} + c_{ab}^{tu} c_{cd}^{rs} - c_{ac}^{tu} c_{bd}^{rs} + c_{ad}^{tu} c_{bc}^{rs} \\
&\quad - c_{ab}^{su} c_{cd}^{rt} + c_{ac}^{su} c_{bd}^{rt} - c_{ad}^{su} c_{bc}^{rt} + c_{ab}^{st} c_{cd}^{ru} - c_{ac}^{st} c_{bd}^{ru} + c_{ad}^{st} c_{bc}^{ru}
\end{aligned}
\qquad (5.53)
$$

The signs in front of the various terms are a result of the antisymmetry property of Slater determinants as explained above. Substituting this expression into Eq. (5.48b), we have

$$
\langle \Psi_{ab}^{rs} | \mathscr{H} | \Psi_0 \rangle + \sum_{\substack{c<d \\ t<u}} \langle \Psi_{ab}^{rs} | \mathscr{H} - E_0 | \Psi_{cd}^{tu} \rangle c_{cd}^{tu}
$$

$$
+ \sum_{\substack{c<d \\ t<u}} \langle \Psi_{ab}^{rs} | \mathscr{H} | \Psi_{abcd}^{rstu} \rangle (c_{ab}^{rs} * c_{cd}^{tu}) = \left(\sum_{\substack{c<d \\ t<u}} \langle \Psi_0 | \mathscr{H} | \Psi_{cd}^{tu} \rangle c_{cd}^{tu} \right) c_{ab}^{rs} \qquad (5.54)
$$

where we explicitly used the result in Eq. (5.48a) for E_{corr}. Now

$$
\langle \Psi_{ab}^{rs} | \mathscr{H} | \Psi_{abcd}^{rstu} \rangle = \langle \Psi_0 | \mathscr{H} | \Psi_{cd}^{tu} \rangle
$$

when $ab \neq cd$ and $rs \neq tu$. Since $c_{ab}^{rs} * c_{cd}^{tu}$ vanishes when $ab = cd$ and $rs = tu$, Eq. (5.54) becomes

$$
\langle \Psi_{ab}^{rs} | \mathscr{H} | \Psi_0 \rangle + \sum_{\substack{c<d \\ t<u}} \langle \Psi_{ab}^{rs} | \mathscr{H} - E_0 | \Psi_{cd}^{tu} \rangle c_{cd}^{tu} - \sum_{\substack{c<d \\ t<u}} \langle \Psi_0 | \mathscr{H} | \Psi_{cd}^{tu} \rangle \langle c_{ab}^{rs} * c_{cd}^{tu} \rangle
$$

$$
+ \sum_{\substack{c<d \\ t<u}} \langle \Psi_0 | \mathscr{H} | \Psi_{cd}^{tu} \rangle c_{ab}^{rs} c_{cd}^{tu} = \left(\sum_{\substack{c<d \\ t<u}} \langle \Psi_0 | \mathscr{H} | \Psi_{cd}^{tu} \rangle c_{cd}^{tu} \right) c_{ab}^{rs}
$$

where we have used the definition $c_{ab}^{rs} * c_{cd}^{tu} = c_{ab}^{rs} c_{cd}^{tu} - \langle c_{ab}^{rs} * c_{cd}^{tu} \rangle$ (see Eq. (5.53)). Note that the expression on the right-hand side cancels with a term on the left-hand side, to give

$$
\langle \Psi_{ab}^{rs} | \mathscr{H} | \Psi_0 \rangle + \sum_{\substack{c<d \\ t<u}} \langle \Psi_{ab}^{rs} | \mathscr{H} - E_0 | \Psi_{cd}^{tu} \rangle c_{cd}^{tu} - \sum_{\substack{c<d \\ t<u}} \langle \Psi_0 | \mathscr{H} | \Psi_{cd}^{tu} \rangle \langle c_{ab}^{rs} * c_{cd}^{tu} \rangle = 0
$$

$$
(5.55)
$$

This equation, along with the definition of the symbol $\langle c_{ab}^{rs} * c_{cd}^{tu} \rangle$ given in Eq. (5.53), and the expression for the correlation energy in Eq. (5.48a) are the equations of Coupled-Pair Many-Electron Theory (CPMET), which is also called the Coupled-Cluster Approximation (CCA). We shall use the latter name throughout this book. In quantum chemistry the CCA is associated with the names of J. Čížek and J. Paldus[6] who first derived these equations and studied their properties. The casting of Eq. (5.55) into a computationally convenient form involves fairly laborious manipulations. It should be remembered that we have neglected single excitations in obtaining the CCA equations. The CCA including only double excitations has been called CCD (coupled-clusters doubles) to distinguish it from more general versions of the theory that also incorporate single (CCSD) and higher excitations. Since such extensions are not considered here we will continue to use the acronym CCA.

We now briefly discuss various aspects of CCA. One of the interesting features of this formalism is that Eq. (5.55) does not explicitly contain the correlation energy. Moreover, since it contains products of the doubles coefficients, it is *nonlinear*. Thus, unlike CI, the correlation energy within the CCA cannot be obtained by simple matrix diagonalization. The CCA, although quite complicated, has many attractive properties. Although it incorporates coupling between different pairs just as DCI (i.e., it contains matrix elements of the form $\langle \Psi_{ab}^{rs} | \mathscr{H} | \Psi_{cd}^{tu} \rangle$) it is, in contrast to DCI, size consistent. Moreover, it does not suffer from the invariance problems of the IEPA (i.e., it is invariant to unitary transformations of degenerate orbitals). However, it is still not a variational scheme; that is, it is possible to obtain more than 100% of the exact correlation energy using the CCA.

After applying the CCA to the problem of N independent minimal basis H_2 molecules, we will briefly reconsider the ideas behind this formalism from a more fundamental point of view.

Consider a supermolecule consisting of N noninteracting minimal basis H_2 molecules. In this system, there are N double excitations of the form $|\Psi_{1_i \bar{1}_i}^{2_i \bar{2}_i}\rangle$ $i = 1, \ldots, N$. Since all the H_2 monomers are identical, all the coefficients of these doubles are equal. Denoting these coefficients by c, and since $\langle \Psi_0 | \mathscr{H} | \Psi_{1_i \bar{1}_i}^{2_i \bar{2}_i} \rangle = K_{12}$ for all i, Eq. (5.48a) becomes

$$^N E_{\text{corr}} = NcK_{12} \tag{5.56}$$

A given double excitation $|\Psi_{1_i \bar{1}_i}^{2_i \bar{2}_i}\rangle$ will mix with $N - 1$ quadruple excitations of the type $|\Psi_{1_i \bar{1}_i 1_j \bar{1}_j}^{2_i \bar{2}_i 2_j \bar{2}_j}\rangle j \neq i$. The matrix element between the double and quadruple excitations is again just K_{12}

$$\langle \Psi_{1_i \bar{1}_i}^{2_i \bar{2}_i} | \mathscr{H} | \Psi_{1_i \bar{1}_i 1_j \bar{1}_j}^{2_i \bar{2}_i 2_j \bar{2}_j} \rangle = \langle \Psi_0 | \mathscr{H} | \Psi_{1_j \bar{1}_j}^{2_j \bar{2}_j} \rangle = K_{12} \qquad i \neq j$$

If we write the coefficient of this quadruple excitation as $c_{1_i \bar{1}_i}^{2_i \bar{2}_i} c_{1_j \bar{1}_j}^{2_j \bar{2}_j} = c^2$, Eq. (5.54) becomes

$$K_{12} + 2\Delta c + (N - 1)K_{12}c^2 = {}^N E_{\text{corr}} c = (NcK_{12})c \tag{5.57}$$

The factor $(N - 1)$ appears in front of $K_{12}c^2$ since there are $N - 1$ quadruple excitations that mix with a given double excitation. Note that $NK_{12}c^2$ cancels, and we can rewrite Eq. (5.57) as

$$K_{12} + 2\Delta c - K_{12}c^2 = 0 \tag{5.58}$$

Solving this equation for c, we find

$$c = \frac{\Delta - (\Delta^2 + K_{12}^2)^{1/2}}{K_{12}} \tag{5.59}$$

so that the correlation energy in Eq. (5.56) becomes

$${}^N E_{\text{corr}}(\text{CCA}) = N(\Delta - (\Delta^2 + K_{12}^2)^{1/2}) \tag{5.60}$$

which is just N times the exact correlation energy of a single H_2 molecule. Thus the CCA, unlike DCI, is exact for our model problem. Recall that DQCI is not exact for $N > 2$. Thus by approximating the coefficients of the quadruples by the square of the coefficient of the doubles we did much more that just approximate DQCI. We have in fact implicitly approximated the coefficient of the hextuples as the cube of the coefficients of the doubles and so on. The reason CCA gave the exact answer for our idealized model is because in our model the coefficient of *all* higher (hextuple, octuple, etc) excitations are *exactly* equal to products of the coefficients of the double excitations. This aspect of CCA is brought out clearly using a more fundamental point of view (which is how historically CCA arose) discussed in the next subsection. This section uses some second quantization notation and may be skipped without loss of continuity.

5.2.2 The Cluster Expansion of the Wave Function

Recall that, using second quantization, a doubly excited determinant $|\Psi_{ab}^{rs}\rangle$ can be written as

$$|\Psi_{ab}^{rs}\rangle = a_r^\dagger a_s^\dagger a_b a_a |\Psi_0\rangle$$

where a_a, a_b remove occupied spin orbital from the HF determinant and $a_r^\dagger$, $a_s^\dagger$ replace these by unoccupied spin orbitals. Thus the doubly excited CI wave function can be written as

$$|\Psi_{\text{DCI}}\rangle = \left(1 + \frac{1}{4} \sum_{abrs} c_{ab}^{rs} a_r^\dagger a_s^\dagger a_b a_a\right)|\Psi_0\rangle$$

We now introduce a wave function, which not only contains double excitations but also quadruples, hextuples, etc. excitations in such a way that the coefficients of the $2n$th-tuple excitations are approximated by products of n doubly excited coefficients. Such a wave function $|\Phi_{\text{CCA}}\rangle$, can be written as

$$|\Phi_{\text{CCA}}\rangle = \exp(\mathcal{T}_2)|\Psi_0\rangle \tag{5.61a}$$

where

$$\mathscr{T}_2 = \frac{1}{4} \sum_{abrs} c_{ab}^{rs} a_r^\dagger a_s^\dagger a_b a_a \qquad (5.61b)$$

This is called the cluster form of the wave function. To get some feeling for it, we expand the exponential as $\exp(x) = 1 + x + \frac{1}{2}x^2 + \cdots$ to obtain

$$|\Phi_{CCA}\rangle = \left(1 + \frac{1}{4}\sum_{abrs} c_{ab}^{rs} a_r^\dagger a_s^\dagger a_b a_a + \frac{1}{32}\sum_{\substack{abcd \\ rstu}} c_{ab}^{rs} c_{cd}^{tu} a_r^\dagger a_s^\dagger a_b a_a a_t^\dagger a_u^\dagger a_d a_c + \cdots\right)|\Psi_0\rangle$$

$$= |\Psi_0\rangle + \frac{1}{4}\sum_{abrs} c_{ab}^{rs}|\Psi_{ab}^{rs}\rangle + \frac{1}{32}\sum_{\substack{abcd \\ rstu}} c_{ab}^{rs} c_{cd}^{tu}|\Psi_{abcd}^{rstu}\rangle + \cdots$$

which, after somewhat lengthy manipulations, can be written as

$$|\Phi_{CCA}\rangle = |\Psi_0\rangle + \sum_{\substack{a<b \\ r<s}} c_{ab}^{rs}|\Psi_{ab}^{rs}\rangle + \sum_{\substack{a<b<c<d \\ r<s<t<u}} c_{ab}^{rs} * c_{cd}^{tu}|\Psi_{abcd}^{rstu}\rangle + \cdots \quad (5.62)$$

where $c_{ab}^{rs} * c_{cd}^{tu}$ is our shorthand notation for the sum of the 18 products of doubly excited coefficients in Eq. (5.53). Thus this form of the wave function has the feature that higher excitations are products of double excitations.

An alternate, but equivalent, derivation of CCA can be given using the wave function $|\Phi_{CCA}\rangle$ in Eq. (5.62). By substituting $|\Phi_{CCA}\rangle$ into the Schrödinger equation

$$(\mathscr{H} - E_0)|\Phi_{CCA}\rangle = E_{corr}|\Phi_{CCA}\rangle$$

and then successively multiplying by $\langle\Psi_0|$, and $\langle\Psi_{ab}^{rs}|$ one can show that the resulting equations are identical to our previous Eqs. (5.48a) and (5.55), since the matrix element between $\langle\Psi_{ab}^{rs}|$ and any hextuple excitation is zero. The above theory can be generalized to incorporate the effect of single and/or triple and higher excitations. For example, single excitations can be included by replacing $\mathscr{T}_2$ in Eq. (5.61a) by $\mathscr{T}_1 + \mathscr{T}_2$, where

$$\mathscr{T}_1 = \sum_{ra} c_a^r a_r^\dagger a_a$$

To distinguish various extensions, the theory using only double excitations (i.e., $\mathscr{T}_2$) is commonly called CCD, while the acronym CCSD is used when both single and double excitations are included (i.e., $\mathscr{T}_1 + \mathscr{T}_2$).

Exercise 5.11 Show that the wave function two independent H_2 molecules in Eqs. (5.49) and (5.50) can be written as

$$|\Phi\rangle = \exp(c_{1_1\bar{1}_1}^{2_1\bar{2}_1} a_{2_1}^\dagger a_{\bar{2}_1}^\dagger a_{\bar{1}_1} a_{1_1} + c_{1_2\bar{1}_2}^{2_2\bar{2}_2} a_{2_2}^\dagger a_{\bar{2}_2}^\dagger a_{\bar{1}_2} a_{1_2})|1_1\bar{1}_1 1_2\bar{1}_2\rangle$$

5.2.3 Linear CCA and the Coupled-Electron Pair Approximation (CEPA)

Although CCA is an excellent approximation, it is rather demanding from the computational standpoint. The final equations of CCA have a complicated algebraic structure and can be approximated in numerous ways. In this subsection we consider two possibilities that are most often encountered in the literature. The simplest possible approximation is to set $\langle c_{ab}^{rs} * c_{cd}^{tu} \rangle$ equal to zero in Eq. (5.55) to obtain

$$\langle \Psi_{ab}^{rs} | \mathscr{H} | \Psi_0 \rangle + \sum_{\substack{c<d \\ t<u}} \langle \Psi_{ab}^{rs} | \mathscr{H} - E_0 | \Psi_{cd}^{tu} \rangle c_{cd}^{tu} = 0 \qquad (5.63a)$$

This along with the usual expression for the correlation energy

$$E_{\text{corr}} = \sum_{\substack{c<d \\ t<u}} \langle \Psi_0 | \mathscr{H} | \Psi_{cd}^{tu} \rangle c_{cd}^{tu} \qquad (5.63b)$$

constitute what is called *linear* CCA (L-CCA). Note that we have eliminated the nonlinear terms *after* the cancellation involving the correlation energy was performed. If we would have simply set $c_{ab}^{rs} * c_{cd}^{tu}$ equal to zero, we would have obtained the usual DCI equations.

Introducing the matrix notation

$$(\mathbf{B}^\dagger)_{rasb} = \langle \Psi_0 | \mathscr{H} | \Psi_{ab}^{rs} \rangle$$

$$(\mathbf{D})_{rasb,\, tcud} = \langle \Psi_{ab}^{rs} | \mathscr{H} - E_0 | \Psi_{cd}^{tu} \rangle$$

$$(\mathbf{c})_{rasb} = c_{ab}^{rs}$$

Eqs. (5.63a, b) become

$$\mathbf{B} + \mathbf{Dc} = 0 \qquad (5.64a)$$

$$E_{\text{corr}} = \mathbf{B}^\dagger \mathbf{c} \qquad (5.64b)$$

Solving Eq. (5.64a) for $\mathbf{c}$ and substituting the result into Eq. (5.64b), we have

$$E_{\text{corr}} = -\mathbf{B}^\dagger (\mathbf{D})^{-1} \mathbf{B} \qquad (5.65)$$

which is a convenient form for the correlation energy in linear CCA. It is interesting to compare this with an expression for the DCI correlation energy obtained in the last chapter (see Eq. (4.30a)),

$$E_{\text{corr}} = -\mathbf{B}^\dagger (\mathbf{D} - E_{\text{corr}} \mathbf{1})^{-1} \mathbf{B} \qquad (5.66)$$

Recall that one way of finding the DCI correlation energy is to solve Eq. (5.66) iteratively (i.e., begin by setting $E_{\text{corr}} = 0$ on the right-hand side, find E'_{corr} and use this to find E''_{corr} and so on until convergence is found). Thus Eq. (5.65) appears to be an approximation to DCI. This is, however, misleading because the linear CCA, although no longer variational, is size consistent. This approximation is identical to the infinite-order doubly excited many-body perturbation theory (D-MBPT(∞)) of Bartlett and coworkers[7] which was originally derived by diagrammatic summation techniques (see Subsection 6.7.3).

Exercise 5.12

a. Show that if the matrix **D** is approximated by its diagonal elements, the L-CCA correlation energy is identical to the result obtained using Epstein-Nesbet pairs (i.e., Eqs. (5.15) and (5.16)).

b. Show that linear CCA is invariant under unitary transformations for the problem of two independent H_2 molecules. First show that for this model the correlation energy of the dimer using localized orbitals is the same as that obtained in Exercise 5.9a. Then show using delocalized spin orbitals that, in contrast to the results of Exercise 5.9, one gets the same correlation energy. You will find the DCI matrix given in Exercise 5.7 useful.

We now consider a different approximation to CCA, originally proposed and implemented by W. Meyer[8] called the Coupled Electron Pair Approximation (CEPA). We begin again with the CCA equation (5.55). Instead of ignoring all the terms involving $\langle c_{ab}^{rs} * c_{cd}^{tu} \rangle$ we retain those where $c = a$ and $d = b$. Thus we have

$$\langle \Psi_{ab}^{rs} | \mathcal{H} | \Psi_0 \rangle + \sum_{\substack{c<d \\ t<u}} \langle \Psi_{ab}^{rs} | \mathcal{H} - E_0 | \Psi_{cd}^{tu} \rangle c_{cd}^{tu} = \sum_{t<u} \langle \Psi_0 | \mathcal{H} | \Psi_{ab}^{tu} \rangle \langle c_{ab}^{rs} * c_{ab}^{tu} \rangle$$

$$(5.67a)$$

From Eq. (5.53), which defines $\langle c_{ab}^{rs} * c_{cd}^{tu} \rangle$, it follows that

$$\langle c_{ab}^{rs} * c_{ab}^{tu} \rangle = c_{ab}^{rs} c_{ab}^{tu}$$

where we have used the fact that a coefficient like c_{ab}^{rs} is antisymmetric in the occupied or unoccupied indices (e.g., $c_{ab}^{rs} = -c_{ba}^{rs} = -c_{ab}^{sr}$). Substituting this into Eq. (5.67a) we have

$$\langle \Psi_{ab}^{rs} | \mathcal{H} | \Psi_0 \rangle + \sum_{\substack{c<d \\ t<u}} \langle \Psi_{ab}^{rs} | \mathcal{H} - E_0 | \Psi_{cd}^{tu} \rangle c_{cd}^{tu} = \left(\sum_{t<u} \langle \Psi_0 | \mathcal{H} | \Psi_{ab}^{tu} \rangle c_{ab}^{tu} \right) c_{ab}^{rs} \quad (5.67b)$$

Recognizing that the sum in parentheses is the expression for the pair energy e_{ab} (see Eq. (5.9a))

$$e_{ab} = \sum_{t<u} \langle \Psi_0 | \mathcal{H} | \Psi_{ab}^{tu} \rangle c_{ab}^{tu} \quad (5.68a)$$

Eq. (5.67b) becomes

$$\langle \Psi_{ab}^{rs} | \mathcal{H} | \Psi_0 \rangle + \sum_{\substack{c<d \\ t<u}} \langle \Psi_{ab}^{rs} | \mathcal{H} - E_0 | \Psi_{cd}^{tu} \rangle c_{cd}^{tu} = e_{ab} c_{ab}^{rs} \quad (5.68b)$$

The CEPA correlation energy is given by

$$E_{\text{corr}} = \sum_{a<b} e_{ab} \tag{5.68c}$$

These are the equations of CEPA. Note that they are very similar to both the IEPA and DCI equations. If e_{ab} in Eq. (5.68b) were replaced by the total correlation energy, we would obtain the DCI equation (See Eq. (4.26b)). On the other hand, if the summation over $c < d$ were approximated by a single term (i.e., $c = a$, $d = b$) we would recover the IEPA result given in Eq. (5.9b). CEPA includes coupling between pairs ab and cd, unlike the IEPA, yet it remains size consistent, unlike DCI. Note that computationally, because the equation which determines e_{ab} contains the coefficients of other pairs (c_{cd}^{tu}), the equations in CEPA must be solved iteratively. The major advantage of CEPA over CCA is that it is much simpler computationally. The price one has to pay is (among others) that CEPA is no longer invariant to unitary transformations as is CCA. However, it appears to be more nearly invariant than the IEPA.

As an illustration of the above formalism we calculate the CEPA correlation energy of our dimer of two noninteracting H_2 molecules using both localized and delocalized orbitals. Since with localized orbitals there is no coupling between different pairs in this model (i.e., $\langle \Psi_{1_i 1_i}^{2_i 2_i} | \mathscr{H} | \Psi_{1_j 1_j}^{2_j 2_j} \rangle = 0$ when $i \neq j$) CEPA and IEPA are the same. Using delocalized orbitals, the situation is different. The CCA is exact in this case since it is invariant to unitary transformations, while CEPA is not. However, as we will see CEPA works much better than the IEPA. Since CEPA can formally be obtained from the DCI equations by merely replacing E_{corr} by the appropriate pair correlation energies, we immediately get from the result given in Exercise 5.7

$$^2E_{\text{corr}} = 2^{-1/2}K_{12}(c_1 + c_2 + c_3 + c_4) = e_{a\bar{a}} + e_{b\bar{b}} + e_{a\bar{b}} + e_{\bar{a}b}$$

$$2^{-1/2}K_{12} + 2\Delta'c_1 + \tfrac{1}{2}J_{11}c_2 + (\tfrac{1}{2}K_{12} - J_{12})c_3 + (\tfrac{1}{2}K_{12} - J_{12})c_4 = e_{a\bar{a}}c_1$$

and similar equations for $e_{b\bar{b}}$, $e_{a\bar{b}}$, $e_{\bar{a}b}$. From the symmetry of the equations, it follows that $c_1 = c_2 = c_3 = c_4 = c$ and thus all four pair energies are equal. Hence

$$^2E_{\text{corr}} = 4e = 4(2^{-1/2}K_{12}c) \tag{5.69a}$$

$$2^{-1/2}K_{12} + (2\Delta' + \tfrac{1}{2}J_{11} + K_{12} - 2J_{12})c = ec \tag{5.69b}$$

Recognizing the quantity in parentheses in Eq. (5.69b) as 2Δ (see Eqs. (4.20) and (5.31b)) we have

$$2^{-1/2}K_{12} + 2\Delta c = ec$$

Multiplying this equation by $2^{-1/2}K_{12}$ and using Eq. (5.69a), we obtain

$$K_{12}^2/2 + 2\Delta e = e^2 \tag{5.70}$$

Solving the quadratic equation for e and then using Eq. (5.69a), we finally have

$$^2E_{corr}(CEPA(D)) = 4(\Delta - (\Delta^2 + K_{12}^2/2)^{1/2}) \tag{5.71}$$

Note that this is not equal to the exact correlation energy (i.e., $2(\Delta - (\Delta^2 + K_{12}^2)^{1/2}))$ of the dimer. However, using the STO-3G minimal basis H_2 integrals, in Appendix D, $^2E_{corr}(CEPA(D)) = -0.0414$ a.u. as compared to the exact value of -0.0411 a.u. Thus CEPA is nearly invariant for this problem.

Before considering some applications of coupled-pair theories, it is appropriate to summarize the formal relationship of these theories among themselves and with DCI (see Table 5.2). It can be seen that DCI, L-CCA, and CEPA are rather similar from the computational point of view. Formally, L-CCA can be obtained from DCI by setting $E_{corr} = 0$, while by setting $E_{corr} = e_{ab}$ one obtains CEPA. One must be careful not to be mislead by this formal similarity. It might appear that L-CCA is in fact an approximation to DCI. This is not the case; L-CCA is size consistent in contrast to DCI. A better point of view is that all these schemes are different approximations to full CI. DCI is variational but is not size consistent. L-CCA, CEPA, and CCA are not variational but are size consistent. Among the three coupled-pair theories, the CCA is expected to be the best. As we have seen, the L-CCA can be obtained from the CCA by setting $\langle c_{ab}^{rs} * c_{cd}^{tu} \rangle = 0$ while to obtain CEPA, one makes an apparently less drastic approximation. However, it turns out that while L-CCA is invariant to unitary transformations of degenerate orbitals, CEPA is not. On the other hand, it is difficult to argue, on first principles without numerical applications, that L-CCA should work better than CEPA when applied to a variety of molecules.

Table 5.2 The formal relationship between DCI and various coupled-pair theories

$$E_{corr} = \sum_{a<b} e_{ab} \qquad e_{ab} = \sum_{r<s} \langle \Psi_0 | \mathcal{H} | \Psi_{ab}^{rs} \rangle c_{ab}^{rs}$$

$$\langle \Psi_{ab}^{rs} | \mathcal{H} | \Psi_0 \rangle + \sum_{\substack{c<d \\ t<u}} \langle \Psi_{ab}^{rs} | \mathcal{H} - E_0 | \Psi_{cd}^{tu} \rangle c_{cd}^{tu} = X$$

Method	X
DCI	$E_{corr} c_{ab}^{rs}$
L-CCA	0
CEPA	$e_{ab} c_{ab}^{rs}$
CCA	$\sum_{\substack{c<d \\ t<u}} \langle \Psi_0 | \mathcal{H} | \Psi_{cd}^{tu} \rangle \langle c_{ab}^{rs} * c_{cd}^{tu} \rangle$

These comments emphasize that one must be careful in evaluating various approximation schemes. The consequence of an approximation is often more subtle than it appears at first glance.

5.2.4 Some Illustrative Calculations

In this section we present results for the correlation energy and equilibrium geometry of H_2O obtained with a near-HF one-electron basis (the 39-STO basis described in Chapter 4) via DCI, IEPA, L-CCA, and CCA. Meyer and his coworkers have studied potential energy surfaces, ionization potentials, dipole moments, and polarizabilities of a variety of molecules using CEPA. Some of their impressive results are contained in a review by Meyer.[9]

The correlation energies of H_2O obtained via a variety of many-electron theories within the same one-electron basis are shown in Table 5.3. These results must be compared with the exact basis set correlation energy (-0.296 a.u.), which is different from the exact correlation energy (-0.37 a.u.) because the one-electron basis is incomplete. Otherwise one would obtain the erroneous conclusion that of all the methods, the IEPA works the best. It is, in fact, the worst; it overestimates the correlation energy by 13%. We note that the correlation energies obtained via the L-CCA and CCA are close to each other and to the exact result. Both are superior to SDCI. The closeness of the L-CCA and CCA is somewhat surprizing because the L-CCA involved an apparently drastic approximation to CCA.

In Table 5.4 we present the calculated equilibrium geometries and two force constants obtained using the above methods. It can be seen that

Table 5.3 Correlation energies (a.u.) of H_2O at the experimental geometry calculated with the 39-STO basis described in Chapter 4

	E_{corr}
SDCI[a]	-0.2756
IEPA[a]	-0.3274
L-CCA[b]	-0.2908
CCA[b]	-0.2862
Estimated full CI	-0.296 ± 0.001
Exact	-0.37

[a] B. J. Rosenberg and I. Shavitt, *J. Chem. Phys.* **63**: 2162 (1975).

[b] R. J. Bartlett, I. Shavitt, and G. D. Purvis, *J. Chem. Phys.* **71**: 281 (1979).

Table 5.4 Equilibrium geometry and some force constants of H_2O calculated with the 39-STO basis described in Chapter 4

	SCF[a]	SDCI[a]	L-CCA[b]	CCA[b]	Experiment
R_e(a.u.)	1.776	1.800	1.810	1.806	1.809
θ_e	106.1°	104.9°	104.6°	104.7°	104.5°
f_{RR}	9.79	8.88	8.51	8.67	8.45
$f_{\theta\theta}$	0.88	0.81	0.80	0.80	0.76

[a] B. J. Rosenberg, W. C. Ermler, and I. Shavitt, *J. Chem. Phys.* **65**: 4072 (1976).

[b] R. J. Bartlett, I. Shavitt, and G. D. Purvis, *J. Chem. Phys.* **71**: 281 (1979). These references contain a large number of additional force constants.

L-CCA and CCA represent a significant improvement over SDCI, which is in turn significantly better than SCF. When compared to experiment, the L-CCA appears to be slightly better than the CCA! However, one should not jump to the conclusion that the L-CCA is a better approximation. After all, the one-electron basis is clearly not complete and we do not know the equilibrium geometry of H_2O that would be obtained if a full CI were to be performed in this basis. On the other hand, it is possible that the L-CCA accidently accounts for the effect of single and triple excitations absent in the CCA so that the L-CCA results are really closer to the exact values in the basis. These considerations highlight the need to apply approximation schemes to a variety of molecules, using basis sets of increasing sophistication, before drawing conclusions as to the superiority of one method over another.

5.3 MANY-ELECTRON THEORIES WITH SINGLE PARTICLE HAMILTONIANS

We have encountered a variety of techniques (CI, IEPA, CCA, CEPA) for calculating the correlation energy of a many-electron system, and in Chapter 6 we will discuss still another approach based on perturbation theory. The complexity of these formalisms and of the many-electron problem itself is the result of the two-particle nature of the coulomb repulsion between electrons. If the Hamiltonian contained only single particle interactions, there would be no need for sophisticated many-electron theories since we could solve the problem exactly simply by diagonalizing the Hamiltonian in a basis of one-electron functions (i.e., the orbital picture would be exact). Nevertheless, it is instructive to apply the formalism of many-electron theories to an N-electron problem described by a Hamiltonian that contains only single particle interactions. By seeing how these sophisticated techniques work in such a simple context, insight into the nature of these approaches can be gained. As we shall see here and in the next chapter, when one approaches a many-electron system with only single particle interactions, using the formalism of many-electron theories, one obtains equations that are almost completely analogous, yet much simpler, than those encountered previously. In particular, the nature of the approximations and the relationship between different approaches becomes especially transparent. Moreover, the calculations can be done analytically but still contain many of the features of *ab initio* calculations. This section is a rather long diversion and may be skipped without loss of continuity. Only the second half of Section 6.3 depends on some of the results obtained here.

We begin by posing the problem and solving it exactly in an elementary way. Suppose we have a N-electron system described by the Hamiltonian

$$\mathscr{H} = \mathscr{H}_0 + \mathscr{V} = \sum_i h_0(i) + \sum_i v(i) \tag{5.72}$$

To obtain a zeroth-order description we assume that the perturbation $\mathscr{V}$ is negligible, and find the eigenfunctions and eigenvalues of the N-electron Hamiltonian $\mathscr{H}_0$. Since $\mathscr{H}_0$ contains only single-particle interactions, we proceed as follows. First, we find the complete orthonormal set of spin orbitals $\{\chi_i^{(0)}\}$, which are eigenfunctions of h_0

$$h_0\chi_i^{(0)} = \varepsilon_i^{(0)}\chi_i^{(0)} \tag{5.73}$$

The ground state wave function, $|\Psi_0\rangle$, is then a Slater determinant constructed from the N spin orbitals with the lowest energies. As usual, we label occupied spin orbitals by $a, b, \ldots$ and the unoccupied spin orbitals by $r, s, \ldots$. Thus we have

$$|\Psi_0\rangle = |\chi_1^{(0)} \cdots \chi_a^{(0)} \cdots \chi_N^{(0)}\rangle \tag{5.74}$$

This wave function is an eigenfunction of $\mathscr{H}_0$

$$\mathscr{H}_0|\Psi_0\rangle = \sum_a \varepsilon_a^{(0)}|\Psi_0\rangle \tag{5.75}$$

with an eigenvalue equal to the sum of the occupied spin orbital energies. The approximate ground state energy of the system, in the presence of $\mathscr{V}$ is

$$E_0 = \langle\Psi_0|\mathscr{H}|\Psi_0\rangle = \sum_a \varepsilon_a^{(0)} + \sum_a \langle a|v|a\rangle = \sum_a \varepsilon_a^{(0)} + \sum_a v_{aa} \tag{5.76}$$

so that the total energy is not simply the sum of orbital energies. Note the analogy to the HF energy of a real system having two-particle interactions.

The excited determinants formed from the spin orbitals $\{\chi_i^{(0)}\}$, i.e., $|\Psi_a^r\rangle$, $|\Psi_{ab}^{rs}\rangle$, etc., form a complete set of N-electron basis functions. Thus the exact wave function, $|\Phi_0\rangle$, of the N-electron system can be expressed as a linear combination of these as

$$|\Phi_0\rangle = |\Psi_0\rangle + \sum_{ra} c_a^r |\Psi_a^r\rangle + \sum_{\substack{a<b \\ r<s}} c_{ab}^{rs} |\Psi_{ab}^{rs}\rangle + \cdots \qquad (5.77)$$

where we have used, as usual, intermediate normalization (i.e., $\langle\Psi_0|\Phi_0\rangle = 1$). The exact ground state energy, $\mathscr{E}_0$, can be obtained by diagonalizing $\mathscr{H}$ in the basis of N-electron functions (i.e., using (5.77) and the linear variation method).

However, there is a much simpler way of finding $\mathscr{E}_0$. Since the total Hamiltonian $\mathscr{H}$ is, just like $\mathscr{H}_0$, a sum of single-particle interactions, we can find a set of spin orbitals $\{\chi_i\}$, which are eigenfunctions of $h_0 + v$

$$(h_0 + v)\chi_i = \varepsilon_i\chi_i \qquad (5.78)$$

Then the exact ground state wave function is just a Slater determinant constructed from the N exact spin orbitals with the lowest energies

$$|\Phi_0\rangle = |\chi_1 \cdots \chi_a \cdots \chi_N\rangle \qquad (5.79)$$

Since this wave function is an eigenfunction of the total Hamiltonian $\mathscr{H}$

$$\mathscr{H}|\Phi_0\rangle = \sum_a \varepsilon_a |\Phi_0\rangle \qquad (5.80)$$

with an eigenvalue equal to the sum of the N lowest spin orbital energies, the exact ground state energy is

$$\mathscr{E}_0 = \langle\Phi_0|\mathscr{H}|\Phi_0\rangle = \sum_a \varepsilon_a \qquad (5.81)$$

Clearly, the two procedures 1) diagonalizing $\mathscr{H}$ in the basis of N-electron functions formed from the eigenfunctions of h_0 and 2) diagonalizing $h_0 + v$ and adding up the N lowest spin orbital energies must give identical ground state energies. It is equally clear that method (2) is much easier. As a pedagogical device, we will approach the problem the hard way using N-electron functions. In analogy to the correlation energy, we define the *relaxation energy* of our system as the difference between the exact energy and the energy of $|\Psi_0\rangle$,

$$E_R = \mathscr{E}_0 - E_0 \qquad (5.82)$$

Our objective will be to calculate E_R using many-electron formalisms.

First, let us calculate the relaxation energy simply by solving the eigenvalue problem for $h_0 + v$,

$$(h_0 + v)|\chi\rangle = \varepsilon|\chi\rangle \qquad (5.83)$$

in the basis $\{|\chi_i^{(0)}\rangle\}$. We expand $|\chi\rangle$ as

$$|\chi\rangle = \sum_i |\chi_i^{(0)}\rangle c_i = \sum_b |\chi_b^{(0)}\rangle c_b + \sum_s |\chi_s^{(0)}\rangle c_s \qquad (5.84)$$

Note that we have divided the sum over all zeroth-order spin orbitals into the sum of the first N occupied ones and the rest. Substituting this expansion into Eq. (5.83) and then multiplying by $\langle\chi_a^{(0)}|$ and $\langle\chi_r^{(0)}|$ we find

$$\sum_b \langle\chi_a^{(0)}|h_0 + v|\chi_b^{(0)}\rangle c_b + \sum_s \langle\chi_a^{(0)}|v|\chi_s^{(0)}\rangle c_s = \varepsilon c_a \tag{5.85a}$$

$$\sum_b \langle\chi_r^{(0)}|v|\chi_b^{(0)}\rangle c_b + \sum_s \langle\chi_r^{(0)}|h_0 + v|\chi_s^{(0)}\rangle c_s = \varepsilon c_r \tag{5.85b}$$

or equivalently

$$\sum_b (\varepsilon_a^{(0)}\delta_{ab} + v_{ab})c_b + \sum_s v_{as}c_s = \varepsilon c_a \tag{5.86a}$$

$$\sum_b v_{rb}c_b + \sum_s (\varepsilon_r^{(0)}\delta_{rs} + v_{rs})c_s = \varepsilon c_r \tag{5.86b}$$

To rewrite these equations in matrix notation, we introduce an $N \times N$ matrix $\mathbf{H}_{AA}$ with elements

$$(\mathbf{H}_{AA})_{ab} = \varepsilon_a^{(0)}\delta_{ab} + v_{ab} \tag{5.87a}$$

Similarly, if we define

$$(\mathbf{H}_{BB})_{rs} = \varepsilon_r^{(0)}\delta_{rs} + v_{rs} \tag{5.87b}$$

$$(\mathbf{H}_{AB})_{as} = v_{as} \tag{5.87c}$$

$$(\mathbf{H}_{BA})_{rb} = v_{rb} \tag{5.87d}$$

Eqs. (5.86a,b) become

$$\begin{pmatrix} \mathbf{H}_{AA} & \mathbf{H}_{AB} \\ \mathbf{H}_{BA} & \mathbf{H}_{BB} \end{pmatrix} \begin{pmatrix} \mathbf{c}_A \\ \mathbf{c}_B \end{pmatrix} = \varepsilon \begin{pmatrix} \mathbf{c}_A \\ \mathbf{c}_B \end{pmatrix} \tag{5.88}$$

Note that $\mathbf{H}_{AA}$ is a square matrix with dimensionality equal to the number of occupied orbitals while the dimensionality of $\mathbf{H}_{BB}$ is equal to the number of unoccupied spin orbitals.

To obtain the exact energy, $\mathscr{E}_0$, we must solve this eigenvalue problem and add up the N eigenvalues with the lowest energies. To make this more explicit we let

$$\mathbf{U} = \begin{pmatrix} \mathbf{U}_{AA} & \mathbf{U}_{AB} \\ \mathbf{U}_{BA} & \mathbf{U}_{BB} \end{pmatrix}$$

be the unitary transformation, which diagonalizes the Hamiltonian matrix in Eq. (5.88),

$$\begin{pmatrix} \mathbf{H}_{AA} & \mathbf{H}_{AB} \\ \mathbf{H}_{BA} & \mathbf{H}_{BB} \end{pmatrix} \begin{pmatrix} \mathbf{U}_{AA} & \mathbf{U}_{AB} \\ \mathbf{U}_{BA} & \mathbf{U}_{BB} \end{pmatrix} = \begin{pmatrix} \mathbf{U}_{AA} & \mathbf{U}_{AB} \\ \mathbf{U}_{BA} & \mathbf{U}_{BB} \end{pmatrix} \begin{pmatrix} \boldsymbol{\varepsilon}_A & 0 \\ 0 & \boldsymbol{\varepsilon}_B \end{pmatrix} \tag{5.89}$$

where $\boldsymbol{\varepsilon}_A$ is a diagonal matrix containing the N lowest eigenvalues. Then the exact energy is

$$\mathscr{E}_0 = \operatorname{tr}\boldsymbol{\varepsilon}_A \tag{5.90}$$

Note that we have defined the matrix $\mathbf{H}_{AA}$ in such a way that the sum of its diagonal elements is equal to E_0. Thus the relaxation energy can be written in the compact form

$$E_R = \mathscr{E}_0 - E_0 = \mathscr{E}_0 - \left(\sum_a \varepsilon_a^{(0)} + v_{aa}\right) = \text{tr}(\mathbf{\varepsilon}_A - \mathbf{H}_{AA}) \qquad (5.91)$$

Finally, using the unitary matrix $\mathbf{U}$ one can express the exact orbitals $|\chi_i\rangle$ in terms of the zeroth-order orbitals $|\chi_i^{(0)}\rangle$. Since the ith column of $\mathbf{U}$ contains the coefficients of the ith eigenvector, the exact occupied orbitals can be written as

$$|\chi_a\rangle = \sum_b |\chi_b^{(0)}\rangle(\mathbf{U}_{AA})_{ba} + \sum_r |\chi_r^{(0)}\rangle(\mathbf{U}_{BA})_{ra} \qquad a = 1, 2, \ldots, N \quad (5.92)$$

A similar relation holds for the exact unoccupied orbitals.

For future reference we now reformulate the above theory in a way which might appear unfamiliar at first glance but on closer inspection will turn out to be a generalization of the procedure we have used many times to find the lowest eigenvalue of a matrix. We are now interested in finding the sum of the N lowest eigenvalues. The matrix eigenvalue problem in Eq. (5.89) is equivalent to four equations, two of which are

$$\mathbf{H}_{AA}\mathbf{U}_{AA} + \mathbf{H}_{AB}\mathbf{U}_{BA} = \mathbf{U}_{AA}\mathbf{\varepsilon}_A \qquad (5.93a)$$

$$\mathbf{H}_{BA}\mathbf{U}_{AA} + \mathbf{H}_{BB}\mathbf{U}_{BA} = \mathbf{U}_{BA}\mathbf{\varepsilon}_A \qquad (5.93b)$$

Multiplying the first of these by $\mathbf{U}_{AA}^{-1}$ on the right we have

$$\mathbf{H}_{AA} + \mathbf{H}_{AB}\mathbf{U}_{BA}\mathbf{U}_{AA}^{-1} = \mathbf{U}_{AA}\mathbf{\varepsilon}_A\mathbf{U}_{AA}^{-1} \qquad (5.94)$$

Taking the trace of both sides of this equation and using the fact that $\text{tr}\,\mathbf{AB} = \text{tr}\,\mathbf{BA}$ we have

$$\text{tr}\,\mathbf{H}_{AA} + \text{tr}\,\mathbf{H}_{AB}\mathbf{U}_{BA}\mathbf{U}_{AA}^{-1} = \text{tr}\,\mathbf{U}_{AA}\mathbf{\varepsilon}_A\mathbf{U}_{AA}^{-1} = \text{tr}\,\mathbf{\varepsilon}_A\mathbf{U}_{AA}^{-1}\mathbf{U}_{AA} = \text{tr}\,\mathbf{\varepsilon}_A$$

Using this identity and defining

$$\mathbf{C}_{BA} = \mathbf{U}_{BA}\mathbf{U}_{AA}^{-1} \qquad (5.95)$$

we can rewrite the relaxation energy as

$$E_R = \text{tr}(\mathbf{\varepsilon}_A - \mathbf{H}_{AA}) = \text{tr}\,\mathbf{H}_{AB}\mathbf{C}_{BA} \qquad (5.96)$$

Now let us find the equation which determines $\mathbf{C}_{BA}$. To do this, multiply Eq. (5.94) by $\mathbf{U}_{BA}\mathbf{U}_{AA}^{-1}$ on the left to obtain

$$\mathbf{U}_{BA}\mathbf{U}_{AA}^{-1}\mathbf{H}_{AA} + \mathbf{U}_{BA}\mathbf{U}_{AA}^{-1}\mathbf{H}_{AB}\mathbf{U}_{BA}\mathbf{U}_{AA}^{-1} = \mathbf{U}_{BA}\mathbf{\varepsilon}_A\mathbf{U}_{AA}^{-1} \qquad (5.97a)$$

Then multiply Eq. (5.93b) by $\mathbf{U}_{AA}^{-1}$ on the right to obtain

$$\mathbf{H}_{BA} + \mathbf{H}_{BB}\mathbf{U}_{BA}\mathbf{U}_{AA}^{-1} = \mathbf{U}_{BA}\mathbf{\varepsilon}_A\mathbf{U}_{AA}^{-1} \qquad (5.97b)$$

Substracting these two equations and using the definition of $\mathbf{C}_{BA}$ in Eq. (5.95) we finally have

$$\mathbf{H}_{BA} + \mathbf{H}_{BB}\mathbf{C}_{BA} - \mathbf{C}_{BA}\mathbf{H}_{AA} - \mathbf{C}_{BA}\mathbf{H}_{AB}\mathbf{C}_{BA} = 0 \tag{5.98}$$

This relation together with Eq. (5.96) completely determine the relaxation energy. Note that Eq. (5.98) is nonlinear (i.e., it has a quadratic dependence on the matrix $\mathbf{C}_{BA}$) and therefore must be solved iteratively. Finally, we translate these equations back into a form containing the zeroth-order orbital energies and matrix elements of the perturbation. If we define

$$(\mathbf{C}_{BA})_{ra} = C_{ra} \tag{5.99}$$

then we can rewrite Eq. (5.96) as

$$E_R = \text{tr } \mathbf{H}_{AB}\mathbf{C}_{BA} = \sum_{bs} (\mathbf{H}_{AB})_{bs}(\mathbf{C}_{BA})_{sb} = \sum_{bs} v_{bs}C_{sb} \tag{5.100}$$

where we have used Eq. (5.87). Similarly, the ra element of the matrix equation (5.98) is

$$(\mathbf{H}_{BA})_{ra} + \sum_s (\mathbf{H}_{BB})_{rs}(\mathbf{C}_{BA})_{sa} - \sum_b (\mathbf{C}_{BA})_{rb}(\mathbf{H}_{AA})_{ba} - \sum_{bs} (\mathbf{C}_{BA})_{rb}(\mathbf{H}_{AB})_{bs}(\mathbf{C}_{BA})_{sa} = 0$$

which upon using the definitions in Eq. (5.87) becomes

$$v_{ra} + \sum_s (\varepsilon_r^{(0)}\delta_{rs} + v_{rs})C_{sa} - \sum_b C_{rb}(\varepsilon_a^{(0)}\delta_{ab} + v_{ba}) - \sum_{bs} C_{rb}v_{bs}C_{sa} = 0 \tag{5.101}$$

If this equation is solved for the C_{ra}'s, then the exact relaxation energy can be found from Eq. (5.100).

In the past when we considered eigenvalue problems we were usually interested in only the lowest eigenvalue. For the problem at hand we need the N lowest eigenvalues. The above procedure is just a generalization of our standard approach to finding the lowest eigenvalue. To see this, suppose $\mathbf{H}_{AA}$ is a 1×1 matrix and that the entire matrix is $M \times M$. In this case Eq. (5.96) simplifies to

$$E_R = \varepsilon_1 - H_{11} = \sum_{i=2}^{M} H_{1i}C_{i1} \tag{5.102}$$

and Eq. (5.98) becomes

$$H_{i1} + \sum_{j=2}^{M} H_{ij}C_{j1} - C_{i1}H_{11} - C_{i1}\sum_{j=2}^{M} H_{1j}C_{j1} = 0 \qquad i = 2, 3, \ldots, M \tag{5.103}$$

Setting $C_{i1} = c_i$ and using Eq. (5.102) in Eq. (5.103) we have

$$H_{i1} + \sum_{j=2}^{M} (H_{ij} - H_{11}\delta_{ij})c_j = E_R c_i \tag{5.104}$$

Equations (5.102) and (5.104) are equivalent to the matrix equation

$$\begin{pmatrix} 0 & H_{12} & \cdots & H_{1M} \\ H_{21} & H_{22}-H_{11} & \cdots & H_{2M} \\ \vdots & \vdots & & \vdots \\ H_{M1} & H_{M2} & \cdots & H_{MM}-H_{11} \end{pmatrix} \begin{pmatrix} 1 \\ c_2 \\ \vdots \\ c_M \end{pmatrix} = E_R \begin{pmatrix} 1 \\ c_2 \\ \vdots \\ c_M \end{pmatrix} = (\varepsilon_1 - H_{11}) \begin{pmatrix} 1 \\ c_2 \\ \vdots \\ c_M \end{pmatrix}$$

which is precisely the equation we would start with if we were after the lowest eigenvalue.

Exercise 5.13 For the 2×2 matrix

$$\begin{pmatrix} \mathbf{H}_{AA} & \mathbf{H}_{AB} \\ \mathbf{H}_{BA} & \mathbf{H}_{BB} \end{pmatrix} \equiv \begin{pmatrix} H_{11} & H_{12} \\ H_{21} & H_{22} \end{pmatrix} \qquad \begin{matrix} H_{11} < H_{22} \\ H_{12} > 0 \end{matrix}$$

Equation (5.96) simplifies to

$$E_R = \varepsilon_1 - H_{11} = H_{12}C$$

and Eq. (5.98) is

$$H_{21} + H_{22}C - CH_{11} - C^2 H_{12} = 0$$

Solve this quadratic equation for the lowest C and then show that ε_1 thus obtained is the lowest eigenvalue of the matrix.

5.3.1 The Relaxation Energy via CI, IEPA, CEPA, and CCA

Many-electron methods use N-electron wave functions (i.e., Slater determinants). For example, to perform a full CI calculation one must diagonalize the N-electron Hamiltonian in a basis of N-electron functions formed by making all possible excitations from a reference function, $|\Psi_0\rangle$. Formally, one can proceed in the identical manner irrespective of whether the Hamiltonian contains one or two-particle interactions or both. Of course there are considerable simplifications in the case that the N-electron Hamiltonian has only single particle interactions. In particular, the evaluation of the required matrix elements is easy, and the final equations have a much simpler structure. Recall that in a real many-electron system the correlation energy can be written in terms of the coefficients of the doubly excited determinants in the full CI wave function. As we shall see, because of the one-particle nature of the Hamiltonian, the relaxation energy can be expressed in terms of the coefficients of single excitations. Thus, in this section, single excitations play the same role as double excitations do in real many-body systems. Moreover, double excitations here are analogous to quadruple excitations in previous sections. In particular, when we consider the analogue of CCA for our model problem, we will need to express the coefficient of the doubles in terms of products of the coefficients of the singles. This is much simpler; instead of a messy 18-term result (see Eq. (5.53)) we will have a simple two term expression. Finally, when the Hamiltonian only contains single particle interactions, the analogues of pair theories are just "particle" theories. For example, instead of writing the correlation energy as a sum of independently calculated pair energies, e_{ab}, we will express the relaxation energy as a sum of "particle" energies, e_a. We can retain the various acronyms used for pair and coupled-pair theories if we simply change the meaning of the P's.

We now calculate the relaxation energy using the various many-electron approaches.

1. Full CI. The intermediate normalized full CI wave function is given in Eq. (5.77)

$$|\Phi_0\rangle = |\Psi_0\rangle + \sum_{sb} c_b^s |\Psi_b^s\rangle + \sum_{\substack{b<c \\ s<t}} c_{bc}^{st} |\Psi_{bc}^{st}\rangle + \cdots \tag{5.105}$$

To obtain the full CI equations, we substitute the above expansion for $|\Phi_0\rangle$ into

$$(\mathscr{H} - E_0)|\Phi_0\rangle = (\mathscr{H}_0 + \mathscr{V} - E_0)|\Phi_0\rangle = (\mathscr{E}_0 - E_0)|\Phi_0\rangle = E_R|\Phi_0\rangle \tag{5.106}$$

and then successively multiply by $\langle \Psi_0|$, $\langle \Psi_a^r|$, $\langle \Psi_{ab}^{rs}|$, and so on, remembering that the matrix element of a Hamiltonian containing single particle interactions vanishes when the determinants differ by more than *one* spin orbital. In this way we find

$$\sum_{bs} \langle \Psi_0|\mathscr{H}|\Psi_b^s\rangle c_b^s = E_R \tag{5.107a}$$

$$\langle \Psi_a^r|\mathscr{H}|\Psi_0\rangle + \sum_{bs} \langle \Psi_a^r|\mathscr{H} - E_0|\Psi_b^s\rangle c_b^s + \sum_{bs} \langle \Psi_a^r|\mathscr{H}|\Psi_{ab}^{rs}\rangle c_{ab}^{rs} = E_R c_a^r \tag{5.107b}$$

and so on (i.e., the next equation in this hierarchy involves the coefficients of the doubles and triples).

Exercise 5.14 Show that

a. $\langle \Psi_0|\mathscr{H}|\Psi_b^s\rangle = v_{bs}$.

b. $\langle \Psi_a^r|\mathscr{H}|\Psi_0\rangle = v_{ra}$.

c. $\langle \Psi_a^r|\mathscr{H} - E_0|\Psi_b^s\rangle = 0$ if $a \neq b$ $r \neq s$

                 $= v_{rs}$ if $a = b$ $r \neq s$

                 $= -v_{ba}$ if $a \neq b$ $r = s$

                 $= \varepsilon_r^{(0)} + v_{rr} - \varepsilon_a^{(0)} - v_{aa}$ if $a = b$ $r = s$.

d. $\langle \Psi_a^r|\mathscr{H}|\Psi_{ab}^{rs}\rangle = v_{bs}$ if $a \neq b$ $r \neq s$

                 $= 0$ otherwise.

Using the matrix elements found in Exercise 5.14, Eqs. (5.107a, b) become

$$\sum_{bs} v_{bs} c_b^s = E_R \tag{5.108a}$$

$$v_{ra} + \sum_s (\varepsilon_r^{(0)} \delta_{rs} + v_{rs}) c_a^s - \sum_b (\varepsilon_a^{(0)} \delta_{ab} + v_{ba}) c_b^r + \sum_{\substack{b \neq a \\ s \neq r}} v_{bs} c_{ab}^{rs} = E_R c_a^r \tag{5.108b}$$

and so on. We now consider various approximations to full CI.

2. SCI. The analogue of truncating the CI expansion for the correlation energy at double excitations is to use only single excitations for the relaxation energy. This amounts to setting $c_{ab}^{rs} = 0$ in Eq. (5.108b),

$$v_{ra} + \sum_s (\varepsilon_r^{(0)} \delta_{rs} + v_{rs}) c_a^s - \sum_b (\varepsilon_a^{(0)} \delta_{ab} + v_{ba}) c_b^r = E_R c_a^r \tag{5.109}$$

This along with Eq. (5.108a) for the relaxation energy completely specify the simplest form of truncated CI.

3. IEPA. Since our Hamiltonian contains only single particle interactions, we must devise an independent "particle" theory. In analogy to pair theory we define a "particle" function

$$|\Psi_a\rangle = |\Psi_0\rangle + \sum_s c_a^s |\Psi_a^s\rangle \tag{5.110}$$

with a variational energy equal to $E_0 + e_a$. The relaxation energy then is approximated by a sum of "particle" energies as

$$E_R = \sum_a e_a \tag{5.111}$$

To find e_a, we substitute the expansion for the "particle" function into

$$(\mathscr{H} - E_0)|\Psi_a\rangle = e_a|\Psi_a\rangle \tag{5.112}$$

and then multiply successively by $\langle\Psi_0|$ and $\langle\Psi_a^r|$. Using the matrix elements in Exercise 5.14, we obtain

$$\sum_r v_{ar} c_a^r = e_a \tag{5.113a}$$

$$v_{ra} + \sum_s (\varepsilon_r^{(0)} \delta_{rs} + v_{rs}) c_a^s - (\varepsilon_a^{(0)} + v_{aa}) c_a^r = e_a c_a^r \tag{5.113b}$$

which are the required equations. Note that they can be formally obtained from the SCI equations by restricting the summation over b to one term (i.e., $b = a$, this corresponds to neglecting "particle-particle" interactions) and replacing E_R by e_a.

4. CCA. As we have seen in Section 5.2, the idea of this method, as applied to a real many-particle system, is to truncate the full CI hierarchy by approximating the coefficients of the quadruples as sums of products of the coefficients of the doubles. Here we express the coefficients of the doubles as sums of products of the coefficients of the singles. There are two independent ways in which we can obtain the double excitation Ψ_{ab}^{rs}: 1) we can excite $a \rightarrow r$ and then $b \rightarrow s$ to get $|\cdots rs \cdots\rangle$ and 2) we can excite $a \rightarrow s$ and then $b \rightarrow r$ to get $|\cdots sr \cdots\rangle$. Because the second determinant is the negative of the first, we can represent c_{ab}^{rs} either as $c_a^r c_b^s$ or $-c_a^s c_b^r$. In analogy to Eq. (5.53), we can write

$$c_{ab}^{rs} = c_a^r c_b^s - c_a^s c_b^r \tag{5.114}$$

Substituting this into Eq. (5.108b) with the expression for E_R in Eq. (5.108a) we have

$$v_{ra} + \sum_s (\varepsilon_r^{(0)}\delta_{rs} + v_{rs})c_a^s - \sum_b (\varepsilon_a^{(0)}\delta_{ab} + v_{ba})c_b^r + \sum_{bs} v_{bs}c_a^r c_b^s - \sum_{bs} v_{bs}c_a^s c_b^r = \left(\sum_{bs} v_{bs}c_b^s\right)c_a^r$$

Note that we need not worry about the restriction $b \neq a$ and $s \neq r$ since the expression for c_{ab}^{rs} in Eq. (5.114) vanishes when $b = a$ or $r = s$. Cancelling the common term on the right and left-hand sides, we finally have

$$v_{ra} + \sum_s (\varepsilon_r^{(0)}\delta_{rs} + v_{rs})c_a^s - \sum_b (\varepsilon_a^{(0)}\delta_{ab} + v_{ba})c_b^r - \sum_{bs} c_b^r v_{bs} c_a^s = 0 \tag{5.115}$$

Note that this equation is nonlinear and does not explicitly contain the relaxation energy. Before we discuss its properties, we consider two approximations to CCA.

5. L-CCA. The simplest approximation is to linearize Eq. (5.115) by setting the term which is a quadratic function of the coefficients equal to zero, i.e.,

$$v_{ra} + \sum_s (\varepsilon_r^{(0)}\delta_{rs} + v_{rs})c_a^s - \sum_b (\varepsilon_a^{(0)}\delta_{ab} + v_{ba})c_b^r = 0 \tag{5.116}$$

This is called the linear approximation to CCA.

6. CEPA. Finally, we consider the analogue of the coupled-electron pair approximation. Here instead of throwing away the quadratic term, we approximate it as follows. First note that

$$\sum_{bs} c_b^r v_{bs}c_a^s = \sum_s c_a^r v_{as}c_a^s + \sum_{\substack{b \neq a \\ s}} c_b^r v_{bs}c_a^s$$

and then ignoring the $b \neq a$ terms, we obtain

$$\sum_{bs} c_b^r v_{bs}c_a^s \cong c_a^r \sum_s v_{as}c_a^s = e_a c_a^r$$

Finally, using this in Eq. (5.115), we have

$$v_{ra} + \sum_s (\varepsilon_r^{(0)}\delta_{rs} + v_{rs})c_a^s - \sum_b (\varepsilon_a^{(0)}\delta_{ab} + v_{ba})c_b^r = e_a c_a^r \tag{5.117}$$

which is the required equation.

Table 5.5 Equations for the relaxation energy in many-electron theories

$$E_R = \sum_a e_a \qquad e_a = \sum_r v_{ar}c_a^r$$

SCI:	$v_{ra} + \sum_s (\varepsilon_r^{(0)}\delta_{rs} + v_{rs})c_a^s - \sum_b (\varepsilon_a^{(0)}\delta_{ab} + v_{ba})c_b^r = E_R c_a^r$
IEPA:	$v_{ra} + \sum_s (\varepsilon_r^{(0)}\delta_{rs} + v_{rs})c_a^s - (\varepsilon_a^{(0)} + v_{aa})c_a^r = e_a c_a^r$
L-CCA:	$v_{ra} + \sum_s (\varepsilon_r^{(0)}\delta_{rs} + v_{rs})c_a^s - \sum_b (\varepsilon_a^{(0)}\delta_{ab} + v_{ba})c_b^r = 0$
CEPA:	$v_{ra} + \sum_s (\varepsilon_r^{(0)}\delta_{rs} + v_{rs})c_a^s - \sum_b (\varepsilon_a^{(0)}\delta_{ab} + v_{ba})c_b^r = e_a c_a^r$
CCA:	$v_{ra} + \sum_s (\varepsilon_r^{(0)}\delta_{rs} + v_{rs})c_a^s - \sum_b (\varepsilon_a^{(0)}\delta_{ab} + v_{ba})c_b^r = \sum_{bs} c_b^r v_{bs} c_a^s$

In Table 5.5 we summarize the various results obtained so far. Although the structure of these equations is considerably simpler than the corresponding ones for a real many-body system, the insights concerning the interrelations among the various approaches gained from a perusal of this table are generally valid. In particular, note that 1) CEPA can formally be obtained from SCI by replacing the total relaxation energy E_R by e_a; 2) IEPA can be obtained from CEPA by ignoring the coupling between different "particles," i.e., setting $v_{ab} \to v_{aa}\delta_{ab}$; 3) L-CCA can be obtained from SCI by setting $E_R = 0$ or from CEPA by putting $e_a = 0$. It is important to remember that just because a method can be obtained from another by setting something to zero does not mean that it is inferior. Table 5.5 also shows that all methods except CCA are closely related from the computational point of view. In the next section we will apply the various formalisms to a simple model problem where the exact answer for the relaxation energy is known.

First, let us consider the CCA equations more closely. If we set

$$c_a^r = C_{ra}$$

in Eqs. (5.108a) and (5.115) we see that they are identical to Eqs. (5.100) and (5.101), which determine the exact relaxation energy. Thus CCA is exact for an N-electron system with only single particle interactions! It is gratifying that among all the many-electron theories we have considered at least one is exact when applied to a problem where the orbital picture is exact. This must mean that the approximation used in deriving the CCA equations from the exact full CI ones, i.e.,

$$c_{ab}^{rs} = c_a^r c_b^s - c_a^s c_b^r$$

is really not an approximation in the present context, but an exact relation.

This relation can be mathematically proved using the property of determinants given in Eq. (1.40) and the fact that the exact wave function $|\Phi_0\rangle$ is a single determinant. Recall that (see Eq. (5.79))

$$|\Phi_0\rangle = |\chi_1 \cdots \chi_a \cdots \chi_N\rangle \tag{5.118}$$

and that the exact orbitals $|\chi_a\rangle$ can be expressed as linear combinations of the zeroth-order orbitals $|\chi_i^{(0)}\rangle$ as (see Eq. (5.92))

$$|\chi_a\rangle = \sum_b |\chi_b^{(0)}\rangle (\mathbf{U}_{AA})_{ba} + \sum_r |\chi_r^{(0)}\rangle (\mathbf{U}_{BA})_{ra} \tag{5.119}$$

If we substitute the above expansion for the occupied orbitals into Eq. (5.118) and make repeated use of Eq. (1.40), it can be shown that the exact wave function is

$$|\Phi_0\rangle = |\mathbf{U}_{AA}| \left(|\Psi_0\rangle + \sum_{ar} c_a^r |\Psi_a^r\rangle + \sum_{\substack{a<b \\ r<s}} c_{ab}^{rs} |\Psi_{ab}^{rs}\rangle + \cdots \right) \tag{5.120a}$$

where the coefficients c_a^r are given by

$$c_a^r = (\mathbf{U}_{BA}\mathbf{U}_{AA}^{-1})_{ra} \tag{5.120b}$$

and

$$c_{ab}^{rs} = c_a^r c_b^s - c_a^s c_b^r \tag{5.120c}$$

In an analogous way, coefficients of the higher excitations can be expressed as sums of products of the singles coefficients. Note that for relation (5.120c) to hold, we must use the intermediate normalized form of $|\Phi_0\rangle$. Let us illustrate the above with an example. Suppose we have a two-electron problem (i.e., $|\Phi_0\rangle = |\chi_1\chi_2\rangle$) and

$$|\chi_1\rangle = |\chi_1^{(0)}\rangle + a_3|\chi_3^{(0)}\rangle + a_4|\chi_4^{(0)}\rangle \tag{5.121a}$$

$$|\chi_2\rangle = |\chi_2^{(0)}\rangle + b_3|\chi_3^{(0)}\rangle + b_4|\chi_4^{(0)}\rangle \tag{5.121b}$$

Then by repeated applications of Eq. (1.40) one can show that

$$\begin{aligned}
|\Phi_0\rangle &= |\chi_1^{(0)}\chi_2^{(0)}\rangle + b_3|\chi_1^{(0)}\chi_3^{(0)}\rangle + b_4|\chi_1^{(0)}\chi_4^{(0)}\rangle \\
&\quad + a_3|\chi_3^{(0)}\chi_2^{(0)}\rangle + a_4|\chi_4^{(0)}\chi_2^{(0)}\rangle + (a_3b_4 - a_4b_3)|\chi_3^{(0)}\chi_4^{(0)}\rangle \\
&= |\Psi_0\rangle + b_3|\Psi_2^3\rangle + b_4|\Psi_2^4\rangle + a_3|\Psi_1^3\rangle + a_4|\Psi_1^4\rangle \\
&\quad + (a_3b_4 - a_4b_3)|\Psi_{12}^{34}\rangle
\end{aligned} \tag{5.122}$$

It is then clear that

$$c_{12}^{34} = (a_3b_4 - a_4b_3) = c_1^3 c_2^4 - c_1^4 c_2^3 \tag{5.123}$$

as we claimed.

Exercise 5.15 Repeat the above analysis using

$$|\chi_1\rangle = a_1|\chi_1^{(0)}\rangle + a_2|\chi_2^{(0)}\rangle + a_3|\chi_3^{(0)}\rangle + a_4|\chi_4^{(0)}\rangle$$

$$|\chi_2\rangle = b_1|\chi_1^{(0)}\rangle + b_2|\chi_2^{(0)}\rangle + b_3|\chi_3^{(0)}\rangle + b_4|\chi_4^{(0)}\rangle$$

instead of Eqs. (5.121a, b).

a. By repeated use of Eq. (1.40) show that

$$\begin{aligned}|\Phi_0\rangle = {}&(a_1b_2 - b_1a_2)|\chi_1^{(0)}\chi_2^{(0)}\rangle + (a_1b_3 - b_1a_3)|\chi_1^{(0)}\chi_3^{(0)}\rangle \\ &+ (a_1b_4 - b_1a_4)|\chi_1^{(0)}\chi_4^{(0)}\rangle + (a_3b_2 - b_3a_2)|\chi_3^{(0)}\chi_2^{(0)}\rangle \\ &+ (a_4b_2 - b_4a_2)|\chi_4^{(0)}\chi_2^{(0)}\rangle + (a_3b_4 - b_3a_4)|\chi_3^{(0)}\chi_4^{(0)}\rangle\end{aligned}$$

Intermediate normalize this wave function by dividing the right-hand side by $a_1b_2 - b_1a_2$ and then explicitly verify that Eq. (5.123) is satisfied.

b. To make contact with the general formalism, note that

$$\mathbf{U}_{AA} = \begin{pmatrix} a_1 & b_1 \\ a_2 & b_2 \end{pmatrix} \quad \text{and} \quad \mathbf{U}_{BA} = \begin{pmatrix} a_3 & b_3 \\ a_4 & b_4 \end{pmatrix}$$

Note that $|\mathbf{U}_{AA}| = a_1b_2 - b_1a_2$ as required to make Eq. (5.120a) consistent with the result obtained in part (a). Use the result of Exercise 1.4(f) to evaluate $\mathbf{U}_{AA}^{-1}$ and then verify the general result given in Eq. (5.120b) by calculating

$$(\mathbf{U}_{BA}\mathbf{U}_{AA}^{-1})_{11} = c_1^3$$

and showing that it is identical to the coefficient of $|\chi_3^{(0)}\chi_2^{(0)}\rangle$ obtained in part (a).

5.3.2 The Resonance Energy of Polyenes in Hückel Theory

We now consider an interesting application of the various many-electron approaches just discussed. We will use them to calculate the resonance energy of a cyclic polyene with an even number of carbon atoms ($N = 2n$) within the framework of Hückel theory, and then compare the predicted results with the exact value of this quantity. The *resonance energy* is defined as the difference between the exact energy of the polyene (as obtained by diagonalizing the Hückel matrix and adding up to the occupied orbital energies) and the energy of n localized and noninteracting double bonds or ethylenic units. Since Hückel theory is a one-particle theory (i.e., the effective Hamiltonian does not contain two-particle interactions), the resonance energy is analogous to the relaxation energy previously defined.

To introduce notation, we briefly review the application of Hückel theory to a cyclic polyene with $N = 2n$ carbon atoms. Each carbon atom contributes one electron to the π system of the molecule. The total Hamiltonian is approximated as a sum of one-particle terms as

$$\mathscr{H} = \sum_{i=1}^{N} h_{\text{eff}}(i) \tag{5.124}$$

where h_{eff} is completely specified by its matrix elements between an orthonormal set of atomic orbitals ($\langle \phi_\mu | \phi_\nu \rangle = \delta_{\mu\nu}$, $\mu, \nu = 1, \ldots, N$) as

$$\begin{aligned}
(\mathbf{H})_{\mu\nu} = H_{\mu\nu} = \langle \phi_\mu | h_{\text{eff}} | \phi_\nu \rangle &= \alpha \quad \text{if } \nu = \mu \\
&= \beta \quad \text{if } \nu = \mu \pm 1 \\
&= 0 \quad \text{otherwise} \tag{5.125}
\end{aligned}$$

Both the parameters α and β are negative. Note that in this model only nearest-neighbor (or "bonded") carbon atoms interact. We seek a set of molecular orbitals, $|\psi_i\rangle$, $i = 1, \ldots, N$, which are eigenfunctions of h_{eff}

$$h_{\text{eff}} |\psi_i\rangle = \varepsilon_i |\psi_i\rangle \tag{5.126}$$

If we expand $|\psi_i\rangle$ as a linear combination of atomic orbitals as

$$|\psi_i\rangle = \sum_\mu C_{\mu i} |\phi_\mu\rangle \qquad i = 1, 2, \ldots, N \tag{5.127}$$

and use the linear variational principle, we are lead to the matrix eigenvalue problem

$$\mathbf{HC} = \mathbf{C\varepsilon} \tag{5.128}$$

where the matrix elements of the Hückel matrix, $\mathbf{H}$, are given in Eq. (5.125). Since the n molecular orbitals with smallest orbital energies are doubly occupied, the total Hückel energy is

$$\mathscr{E}_0 = 2 \sum_{i=1}^{n} \varepsilon_i = 2 \sum_{a}^{N/2} \varepsilon_a \tag{5.129}$$

The exact wave function of the system is

$$|\Phi_0\rangle = |\psi_1 \bar{\psi}_1 \cdots \psi_a \bar{\psi}_a \cdots \psi_n \bar{\psi}_n\rangle \tag{5.130}$$

where the spin orbitals are obtained by multiplying the spatial orbitals by a spin function corresponding to spin up (no bar) or spin down (bar).

For the case of cyclic polyenes $(CH)_N$, the Hückel problem is analytically soluble.[10] The orbital energies are

$$\varepsilon_i = \alpha + 2\beta \cos(\pi i/n) \qquad i = 0, \pm 1, \pm 2, \ldots, \pm(n-1), n \tag{5.131}$$

Polyenes with $N = 4\nu + 2$, $\nu = 1, 2, \ldots$ carbon atoms have closed-shell electronic configurations. In this case, the exact ground state energy is

$$\mathscr{E}_0 = 2 \sum_{i=-v}^{v} \varepsilon_i = 2 \sum_{i=-v}^{v} \alpha + 4\beta \sum_{i=-v}^{v} \cos\left(\frac{\pi i}{2v+1}\right)$$
$$= N\alpha + 4\beta/\sin(\pi/N) \tag{5.132}$$

For example, for benzene ($N = 6$, n $= 3$, $v = 1$) the ground state energy is $6\alpha + 8\beta$.

Exercise 5.16 Set up the Hückel matrix for benzene and find its eigenvalues. Remember that if the carbon atoms are labeled clockwise from 1 to 6, then atoms 1 and 6 are nearest neighbors. Show that the six eigenvalues are identical to those given by Eq. (5.131). Find the total energy and compare it with the result given by Eq. (5.132).

We now turn to the localized description of polyenes. As the zeroth-order approximation, we assume that the polyene consists of n noninteracting double bonds or ethylenic units. To obtain the energy corresponding to this description we note that the Hückel matrix for ethylene is

$$\mathbf{H} = \begin{pmatrix} \alpha & \beta \\ \beta & \alpha \end{pmatrix}$$

with eigenvalues and eigenfunctions

$$\varepsilon_1^{(0)} = \alpha + \beta \qquad |1\rangle = 2^{-1/2}(|\phi_1\rangle + |\phi_2\rangle) \tag{5.133a}$$

$$\varepsilon_{1^*}^{(0)} = \alpha - \beta \qquad |1^*\rangle = 2^{-1/2}(|\phi_1\rangle - |\phi_2\rangle) \tag{5.133b}$$

where $|1\rangle$ is the bonding orbital, while $|1^*\rangle$ is the antibonding orbital. Since an ethylenic unit has two electrons, its energy is $2\alpha + 2\beta$. Therefore, the energy of n noninteracting units is

$$E_0 = n(2\alpha + 2\beta) = N\alpha + N\beta \tag{5.134}$$

The difference between the exact energy of the polyene and this approximate localized energy is the resonance energy. Using Eqs. (5.132) and (5.134), we have

$$E_R = \mathscr{E}_0 - E_0 = \beta(4/\sin(\pi/N) - N) \tag{5.135}$$

For benzene, this expression gives $E_R = 2\beta$. In the limit that the number of carbons becomes large ($N \to \infty$) (since $\sin(x) \approx x$ for small x) the resonance energy approaches

$$\lim_{N \to \infty} E_R = N\beta(4/\pi - 1) = 0.2732N\beta \tag{5.136}$$

Since β is negative, the resonance energy, just like the correlation energy of a real many-electron system, is negative. Moreover, it is proportional to the number of particles as these become large. Therefore, only a size consistent approach is suitable for the calculation of the resonance energy of large polyenes.

To make contact with the various many-electron approaches to the relaxation energy, we define the set of occupied ($|i\rangle$) and unoccupied ($|i^*\rangle$) ethylenic orbitals as

$$|i\rangle = 2^{-1/2}(|\phi_{2i-1}\rangle + |\phi_{2i}\rangle) \qquad i = 1, 2, \ldots, n \qquad (5.137a)$$
$$|i^*\rangle = 2^{-1/2}(|\phi_{2i-1}\rangle - |\phi_{2i}\rangle) \qquad (5.137b)$$

Using these localized orbitals, the wave function corresponding to our zeroth-order description is

$$|\Psi_0\rangle = |1\bar{1}2\bar{2}\cdots n\bar{n}\rangle \qquad (5.138)$$

Since the atomic orbitals are orthonormal (i.e., $\langle\phi_\mu|\phi_\nu\rangle = \delta_{\mu\nu}$), it follows that

$$\langle i|j\rangle = \langle i^*|j^*\rangle = \delta_{ij} \qquad (5.139a)$$
$$\langle i|j^*\rangle = 0 \qquad (5.139b)$$

Using the matrix elements given in Eq. (5.125), it can be shown that the nonzero matrix elements of h_{eff} in the localized basis are

$$\langle i|h_{\text{eff}}|i\rangle = \alpha + \beta = \varepsilon_i^{(0)} \qquad (5.140a)$$
$$\langle i^*|h_{\text{eff}}|i^*\rangle = \alpha - \beta = \varepsilon_{i^*}^{(0)} \qquad (5.140b)$$
$$\langle i|h_{\text{eff}}|i \pm 1\rangle = \beta/2 \qquad (5.140c)$$
$$\langle i^*|h_{\text{eff}}|(i \pm 1)^*\rangle = -\beta/2 \qquad (5.140d)$$
$$\langle i|h_{\text{eff}}|(i \pm 1)^*\rangle = \langle(i \mp 1)|h_{\text{eff}}|i^*\rangle = \pm\beta/2 \qquad (5.140e)$$

Exercise 5.17 Verify Eqs. (5.139) and (5.140).

Finally to make contact with the general formalism previously given, we partition h_{eff} as

$$h_{\text{eff}} = h_0 + v \qquad (5.141)$$

where h_0 is defined so that the localized bonding and antibonding orbitals are its eigenfunctions

$$h_0|i\rangle = (\alpha + \beta)|i\rangle = \varepsilon_i^{(0)}|i\rangle \qquad (5.142a)$$
$$h_0|i^*\rangle = (\alpha - \beta)|i^*\rangle = \varepsilon_{i^*}^{(0)}|i^*\rangle \qquad (5.142b)$$

Since the localized orbitals are orthonormal (see Eq. (5.139)), it follows from Eq. (5.140) that the nonzero matrix elements of the perturbation v are

$$\langle i|v|i \pm 1\rangle = \langle i|v|(i+1)^*\rangle = \langle i^*|v|i-1\rangle = \beta/2 \qquad (5.143a)$$
$$\langle i^*|v|(i \pm 1)^*\rangle = \langle i|v|(i-1)^*\rangle = \langle i^*|v|i+1\rangle = -\beta/2 \qquad i = 1, 2, \ldots, n \qquad (5.143b)$$

In using the above matrix elements, it is important to remember that because of the cyclic nature of the problem $|n + 1\rangle = |1\rangle$ and $|0\rangle = |n\rangle$. As a consistency check, let us calculate $E_0 = \langle\Psi_0|\mathscr{H}|\Psi_0\rangle$ using the definition of $|\Psi_0\rangle$ in Eq. (5.138) and the above matrix elements.

$$E_0 = \sum_{i=1}^{n} [\langle i|h_0|i\rangle + \langle\bar{i}|h_0|\bar{i}\rangle] + [\langle i|v|i\rangle + \langle\bar{i}|v|\bar{i}\rangle]$$

$$= \sum_{i=1}^{n} [(\alpha + \beta) + (\alpha + \beta)] = N\alpha + N\beta$$

where we have used the fact that $\langle i|v|i\rangle$ and $\langle\bar{i}|v|\bar{i}\rangle$ are zero. Note that this result for E_0 agrees with that obtained in Eq. (5.134).

We are now ready to use the various many-electron theories to calculate the resonance energy. We restrict ourselves to benzene and leave the extension to larger systems to the exercises. We begin with the IEPA because it is the easiest. Since benzene has six occupied spin orbitals, we need to calculate six "particle" energies e_1, $e_{\bar{1}}$, e_2, $e_{\bar{2}}$, e_3, and $e_{\bar{3}}$. Since all localized bonds are equivalent, all these pair energies are equal to, say, e_1 so that

$$E_R = 6e_1 \tag{5.144}$$

To calculate e_1 we consider the "particle" function $|\Psi_1\rangle$ obtained by mixing the ground state wave function $|\Psi_0\rangle$ with single excitations obtained by promoting the electron in orbital $|1\rangle$ to all possible virtual orbitals,

$$|\Psi_1\rangle = |\Psi_0\rangle + c_1|\Psi_1^{1*}\rangle + c_2|\Psi_1^{2*}\rangle + c_3|\Psi_1^{3*}\rangle \tag{5.145}$$

We did not include excitations involving spin-flips (e.g., $|\Psi_1^{\bar{2}*}\rangle$) because these do not mix with $|\Psi_0\rangle$. Now consider the matrix elements

$$\langle\Psi_0|\mathscr{H}|\Psi_1^{1*}\rangle = \langle 1|h_0|1^*\rangle + \langle 1|v|1^*\rangle = 0$$

$$\langle\Psi_0|\mathscr{H}|\Psi_1^{2*}\rangle = \langle 1|h_0|2^*\rangle + \langle 1|v|2^*\rangle = \beta/2$$

$$\langle\Psi_0|\mathscr{H}|\Psi_1^{3*}\rangle = \langle 1|h_0|3^*\rangle + \langle 1|v|3^*\rangle = -\beta/2$$

where we have used Eqs. (5.142) and (5.143). Note that $|\Psi_1^{1*}\rangle$ does not interact with $|\Psi_0\rangle$, so we need not consider it further. Moreover, since the other two excitations have matrix elements that are negatives of each other, we consider the normalized linear combinations $2^{-1/2}(|\Psi_1^{2*}\rangle \pm |\Psi_1^{3*}\rangle)$. The plus combination does not mix with $|\Psi_0\rangle$. Introducing a new notation for the minus combination

$$|_1^*\rangle = 2^{-1/2}(|\Psi_1^{2*}\rangle - |\Psi_1^{3*}\rangle)$$

or in general

$$|_i^*\rangle = 2^{-1/2}(|\Psi_i^{(i+1)*}\rangle - |\Psi_i^{(i-1)*}\rangle) \tag{5.146}$$

we can write the "particle" function $|\Psi_1\rangle$ as

$$|\Psi_1\rangle = |\Psi_0\rangle + c|^*_1\rangle \tag{5.147}$$

The corresponding "particle" equations are

$$c\langle\Psi_0|\mathscr{H}|^*_1\rangle = e_1 \tag{5.148a}$$

$$\langle^*_1|\mathscr{H}|\Psi_0\rangle + \langle^*_1|\mathscr{H} - E_0|^*_1\rangle c = e_1 c \tag{5.148b}$$

The required matrix elements are

$$\langle\Psi_0|\mathscr{H}|^*_1\rangle = 2^{-1/2}\beta \tag{5.149a}$$

$$\langle^*_1|\mathscr{H} - E_0|^*_1\rangle = -\tfrac{3}{2}\beta \tag{5.149b}$$

Substituting these into Eq. (5.148) and solving these in the usual way, we find

$$e_1 = \beta\left(\frac{(17)^{1/2} - 3}{4}\right)$$

so that

$$E_R(\text{IEPA}) = 6e_1 = 1.685\beta \tag{5.150}$$

which is to be compared with the exact value of 2β (i.e., the IEPA gives 84% of the exact result).

Exercise 5.18 Evaluate the matrix elements given in Eq. (5.149) and fill in the remaining steps leading to Eq. (5.150).

Exercise 5.19 a) Extend the above analysis to calculate the IEPA resonance energy for a cyclic polyene with $N = 2n$ ($N > 6$) carbon atoms. As before, argue that all "particle" energies are the same so that

$$E_R = Ne_1$$

Consider only single excitations that mix with $|\Psi_0\rangle$. Show that the "particle" function $|\Psi_1\rangle$ is

$$|\Psi_1\rangle = |\Psi_0\rangle + c|^*_1\rangle$$

where $|^*_1\rangle$ is obtained from Eq. (5.146),

$$|^*_1\rangle = 2^{-1/2}(|\Psi_1^{2*}\rangle - |\Psi_1^{n*}\rangle)$$

Now show that

$$\langle\Psi_0|\mathscr{H}|^*_1\rangle = 2^{-1/2}\beta$$

as before, but that here

$$\langle^*_1|\mathscr{H} - E_0|^*_1\rangle = -2\beta.$$

instead of the result in Eq. (5.149b). Why the difference? Finally, solve the resulting "particle" equations to show that

$$E_R(\text{IEPA}) = N((3/2)^{1/2} - 1)\beta = 0.2247N\beta.$$

Note that the IEPA is indeed size consistent and that in the limit of large N it gives 82% of the exact resonance energy.

b) The above result is not really exact within the IEPA. The reason for this is that there exist single excitations involving orbital $|1\rangle$ that do not mix with $|\Psi_0\rangle$ but do mix with $|{}^*_1\rangle$ and thus have some effect on the "particle" energy e_1. These excitations are analogous to single excitations in CI for a real many-particle system in the sense that although single excitations do not mix with the HF wave function because of Brillouin's theorem, they do mix indirectly through the double excitations. Investigate the effect of such excitations for the case $N = 10$. Show that the exact "particle" function $|\Psi_1\rangle$ is

$$|\Psi_1\rangle = |\Psi_0\rangle + c_1|{}^*_1\rangle + c_3|\Psi_1^{3*}\rangle + c_4|\Psi_1^{4*}\rangle$$

Now show that

$$\langle\Psi_0|\mathcal{H}|\Psi_1^{3*}\rangle = \langle\Psi_0|\mathcal{H}|\Psi_1^{4*}\rangle = 0$$

$$\langle\Psi_1^{3*}|\mathcal{H} - E_0|\Psi_1^{3*}\rangle = \langle\Psi_1^{4*}|\mathcal{H} - E_0|\Psi_1^{4*}\rangle = -2\beta$$

$$\langle\Psi_1^{3*}|\mathcal{H}|\Psi_1^{4*}\rangle = -\beta/2$$

$$\langle{}^*_1|\mathcal{H}|\Psi_1^{3*}\rangle = -\beta/2^{3/2}$$

$$\langle{}^*_1|\mathcal{H}|\Psi_1^{4*}\rangle = \beta/2^{3/2}$$

Finally, show from the resulting "particle" equations that e_1 is the solution of

$$4e_1^3 + 14\beta e_1^2 + 9\beta^2 e_1 - 3\beta^3 = 0$$

This cubic equation can be solved to yield $e_1 = 0.2387\beta$ so that the exact IEPA resonance energy for $N = 10$ is 2.387β, which is to be compared with the approximate result of 2.247β obtained in part (a), so that there is a 6% difference. The exact resonance energy found from Eq. (5.135) is 2.944β for this case.

We now consider the use of singly excited CI to calculate the resonance energy of benzene. The SCI wave function is

$$|\Psi_{\text{SCI}}\rangle = |\Psi_0\rangle + \sum_{i=1}^{3} c_i|{}^*_i\rangle + \sum_{i=1}^{3} \bar{c}_i|{}^{\bar{*}}_{\bar{i}}\rangle \tag{5.151}$$

where $|{}^*_i\rangle$ is defined in Eq. (5.146); $|{}^{\bar{*}}_{\bar{i}}\rangle$ is simply the corresponding wave function involving a spin orbital with spin down. Because of the symmetry of benzene, it follows that $c_1 = c_2 = c_3 = \bar{c}_1 = \bar{c}_2 = \bar{c}_3 = c$. Furthermore, the required matrix elements, when $i = 1, 2, 3$

$$\langle {}^*_i | \mathscr{H} | \Psi_0 \rangle = 2^{-1/2} \beta \qquad (5.152a)$$

$$\langle {}^*_i | \mathscr{H} - E_0 | {}^*_i \rangle = -\tfrac{3}{2} \beta \qquad (5.152b)$$

$$\langle {}^*_1 | \mathscr{H} | {}^*_2 \rangle = \langle {}^*_1 | \mathscr{H} | {}^*_3 \rangle = \langle {}^*_2 | \mathscr{H} | {}^*_3 \rangle = \beta/4 \qquad (5.152c)$$

are readily evaluated. The matrix elements for the "barred" states are identical and $\langle {}^*_i | \mathscr{H} | {}^{\bar{*}}_j \rangle = 0$ for all i and j. Proceeding in the standard way, the SCI equations can be shown to be

$$E_R(\text{SCI}) = 6(2^{-1/2} c \beta) \qquad (5.153a)$$

$$2^{-1/2} \beta - \beta c = E_R(\text{SCI}) c \qquad (5.153b)$$

Solving these for $E_R(\text{SCI})$ we find

$$E_R(\text{SCI}) = ((13)^{1/2} - 1)\beta/2 = 1.303\beta \qquad (5.154)$$

which is to be compared with the exact value of 2β for the resonance energy. Thus SCI gives only 65% of the correct answer. Moreover, as will be seen in Exercise 5.21, this theory is not size consistent, and it predicts that the resonance energy of a cyclic polyene is proportional to $N^{1/2}$ as the number of carbons increase. This result is quite analogous to the case of N non-interacting H_2 molecules, where the correlation energy obtained using DCI also has the same incorrect N-dependence.

Exercise 5.20 Verify Eq. (5.152c), derive Eq. (5.153a,b), and solve them to obtain the result shown in Eq. (5.154).

Exercise 5.21 Extend the above analysis to calculate the SCI resonance energy for a cyclic polyene with $N = 2n$ ($N > 6$) carbon atoms. If we restrict ourselves to only those configurations which interact with $|\Psi_0\rangle$, then the appropriate generalization of Eq. (5.151) is

$$|\Psi_{\text{SCI}}\rangle = |\Psi_0\rangle + \sum_{i=1}^{n} c_i |{}^*_i\rangle + \sum_{i=1}^{n} \bar{c}_i |{}^{\bar{*}}_i\rangle$$

As discussed in Exercise 5.19b, this is not the complete SCI wave function because there exist additional singly excited configurations which, although they do not mix with $|\Psi_0\rangle$, they do mix with $|{}^*_i\rangle$. The omission of these does not affect our qualitative conclusions. Show that the required matrix elements are

$$\langle \Psi_0 | \mathscr{H} | {}^*_i \rangle = 2^{-1/2} \beta$$

$$\langle {}^*_i | \mathscr{H} - E_0 | {}^*_j \rangle = (-2\beta) \delta_{ij}$$

Why are Eqs. (5.152b) and (5.152c) different? Using these matrix elements, show that the SCI equations are

$$E_R(\text{SCI}) = Nc\beta 2^{-1/2}$$

$$2^{-1/2}\beta - 2c\beta = E_R(\text{SCI})c$$

Finally, solve them to obtain

$$E_R(\text{SCI}) = ((1 + N/2)^{1/2} - 1)\beta$$

which is proportional to $N^{1/2}$ as N becomes large.

Finally, we consider the various coupled-"particle" theories. There is no need to do a CCA calculation because we have already proved in general that it gives the exact relaxation energy. To obtain the L-CCA and the CEPA equations, we use the formal relationship of these methods to SCI which was brought out by Table 5.5. Thus we can obtain the L-CCA equations by formally setting $E_R(\text{SCI})$ equal to zero in Eq. (5.153b). Hence

$$E_R(\text{L-CCA}) = 6c\beta 2^{-1/2} \tag{5.155a}$$

$$2^{-1/2}\beta - \beta c = 0 \tag{5.155b}$$

Solving the second equation for c and substituting the result into the first, we find

$$E_R(\text{L-CCA}) = 3\beta \tag{5.156}$$

which grossly overestimates the exact resonance energy.

The CEPA equations are formally obtained from the SCI equations by replacing $E_R(\text{SCI})$ in Eq. (5.153b) by the particle energy e. Thus we have

$$E_R(\text{CEPA}) = 6e \tag{5.157a}$$

$$e = c\beta 2^{-1/2} \tag{5.157b}$$

$$2^{-1/2}\beta - \beta c = ec \tag{5.157c}$$

Eliminating c from Eqs. (5.157b, c), we obtain the following quadratic equation for e

$$2e^2 - 2e\beta - \beta^2 = 0 \tag{5.158}$$

Solving this for e and substituting it into Eq. (5.157a), we have

$$E_R(\text{CEPA}) = 3(3^{1/2} - 1)\beta = 2.196\beta \tag{5.159}$$

Thus CEPA gives a respectable approximation to the resonance energy of benzene.

The results obtained by the various many-electron approaches are summarized in Table 5.6. It can be seen that, aside from CCA, which is

Table 5.6 Resonance energy of benzene

Exact	2β	Percent of exact
SCI	1.303β	65
IEPA	1.685β	84
CCA	2β	100
L-CCA	3β	150
CEPA	2.196β	110

exact, CEPA works the best. It should be emphasized that one should not draw any conclusions about the validity or accuracy of these methods when applied to a real many-particle problem from this table. For example, for a six-electron system, DCI in a large basis set certainly gives more than 65% of the correlation energy. In some sense the calculation of the resonance energy in this model is more demanding than the calculation of the correlation energy. We certainly do not mean that it is computationally more difficult but merely that our zeroth-order description here is worse than is the HF description for a real many-electron system. In benzene all the six nearest-neighbor atoms interact equivalently (i.e., the matrix elements between adjacent carbons are all equal to β). In the localized description we assume that the resonance integral is β between atoms 1 and 2, 3 and 4, and 5 and 6, but is zero between 2 and 3, 4 and 5, and 6 and 1. Thus our zeroth-order picture is poor, and a general approach like CI does not work very well then truncated at the lowest nontrivial level. Of course, the CCA takes full advantage of the fact that the Hamiltonian contains only single particle interactions and, hence, is exact no matter how poor the starting point.

NOTES

1. O. Sinanoğlu, Many-electron theory of atoms, molecules, and their interactions, *Adv. Chem. Phys.* **6**: 315 (1964).
2. R. K. Nesbet, Electronic correlation in atoms and molecules, *Adv. Chem. Phys.* **9**: 321 (1965).
3. C. F. Bender and E. R. Davidson, Unitary transformations and pair energies, *Chem. Phys. Lett.* **3**: 33 (1969).
4. R. K. Nesbet, T. L. Barr, and E. R. Davidson, Correlation energy of the neon atom, *Chem. Phys. Lett.* **4**: 203 (1969).
5. S. R. Langhoff and E. R. Davidson, Configuration interaction calculations on the nitrogen molecule, *Int. J. Quantum Chem.* **8**: 61 (1974).
6. J. Čížek and J. Paldus, Coupled-cluster approach, *Physica Scripta* **21**: 251 (1980). A non-mathematical review with a historical perspective.
7. R. J. Bartlett, I. Shavitt, and G. D. Purvis, The quartic force field of H_2O determined by many-body methods that include quadruple excitation effects, *J. Chem. Phys.* **71**: 281 (1979), and references therein.

8. W. Meyer, PNO-CI Studies of electron correlation effects. I. Configuration expansion by means of nonorthogonal orbitals, and application of the ground state and ionized states of methane, *J. Chem. Phys.* **58**: 1017 (1973).
9. W. Meyer, Configuration expansion by means of pseudonatural orbitals, in *Methods of Electronic Structure Theory*, H. F. Schaefer III (Ed.), Plenum, New York, 1977, p. 413.
10. See for example, L. Salem, *Molecular Orbital Theory of Conjugated Systems*, Benjamin, New York, 1966, pp. 110–115.

FURTHER READING

Ahlrichs, R. and Scharf, P., The coupled pair approximation, *Adv. Chem. Phys.* **67**: 501 (1987). Emphasis is on the ideas behind CEPA-like methods and on their application. (CEPA discussed in this book corresponds to CEPA-2 in this article).

Bartlett, R. J., Many-body perturbation theory and coupled cluster theory for electron correlation in molecules, *Ann. Rev. Phys. Chem.* **32**: 359 (1981). An excellent review of both theory and applications with two hundred literature references.

Bartlett, R. J., The coupled-cluster approach to molecular structure and spectra: A step towards predictive quantum chemistry. *J. Phys. Chem.* **93**: xxx (1989). Summarizes some of the important advances made in the field since the publication of Bartlett's 1981 review (see above).

Condon, E. U., On pair correlation in the theory of atomic structure, *Rev. Mod. Phys.* **40**: 872 (1968); Nesbet, R. K., Electronic pair correlation in atoms and molecules, *Int. J. Quantum Chem.* **S4**: 117 (1971). In the 1960's there was a disagreement between Sinanoğlu and Nesbet about the origins of the IEPA. These two articles present complementary views of the history of the subject and make fascinating reading.

Hurley, A. C., *Electron Correlation in Small Molecules*, Academic Press, New York, 1976. An advanced monograph containing a detailed discussion of CI, IEPA, and various forms of coupled-pair theories with numerous literature references. The CCA equations are given in an explicit form.

Kutzelnigg, W., Pair correlation theories, in *The Methods of Electronic Structure Theory*, H. F. Schaefer III (Ed.), Plenum, New York, 1977, p. 129. An excellent place to learn more about the topics of this chapter.

Pople, J. A., Head-Gordon, M., and Raghavachari, K., Quadratic configuration interaction. A general technique for determining electron correlation energies. *J. Chem. Phys.* **87**: 5968 (1987). Quadratic CI, which is implemented in the *Gaussian* 88 system of programs, is a method closely related to the coupled cluster approximation differing only in the way single and triple excitations are handled. If only doubly excited configurations are considered, it is identical to the CCA discussed in this chapter (recall that we use CCA to denote CCD (coupled cluster doubles) since we do not explicitly consider extensions involving singles (CCSD) and triples (CCSDT)).

Sinanoğlu, O. and Brueckner, K. A., *Three Approaches to Electron Correlation in Atoms*, Yale, New Haven, 1970. A good discussion of Sinanoğlu's approach to pair theory. Contains reprints of many of the original papers.

CHAPTER SIX

MANY-BODY PERTURBATION THEORY

Configuration interaction (CI) is a systematic procedure for going beyond the Hartree-Fock approximation. It has the important advantage that it is variational (i.e., at each level it gives an upper bound to the exact energy), but it has the disadvantage that it is only size consistent when all possible excitations are incorporated into the trial function (i.e., full CI). A different systematic procedure for finding the correlation energy, which is not variational but is size consistent at each level, is perturbation theory (PT). In this approach, the total Hamiltonian of the system is divided or partitioned into two pieces: a zeroth-order part, $\mathscr{H}_0$, which has known eigenfunctions and eigenvalues, and a perturbation, $\mathscr{V}$. The exact energy is then expressed as an infinite sum of contributions of increasing complexity. The expressions for these contributions contain the eigenvalues of $\mathscr{H}_0$ and matrix elements of the perturbation between the eigenfunctions of $\mathscr{H}_0$. Terms that involve products of n such matrix elements are grouped together and constitute the nth-order perturbation energy. If we have chosen $\mathscr{H}_0$ wisely, then $\mathscr{V}$ is small and the perturbation expansion (i.e., the sum of the 1st, 2nd, ..., nth-order energies) converges quickly.

In this chapter, we consider a form of perturbation theory which is associated with the names of Rayleigh and Schrödinger (RSPT). Since we are interested in obtaining a perturbation expansion for the correlation energy, we choose the Hartree-Fock Hamiltonian as our zeroth-order Hamiltonian. RSPT with this choice of $\mathscr{H}_0$ was applied to N-electron systems in the early days of quantum mechanics by C. Møller and M. S. Plesset and hence is sometimes called Møller-Plesset perturbation theory (MPPT).

Alternatively, this approach is called many-body perturbation theory (MBPT) because of its popularity with physicists interested in treating infinite systems. Clearly, to treat such large systems one needs a theory that is size consistent. It was K. A. Brueckner, while worrying about infinite nuclear matter, who first conjectured that MBPT was size consistent in each order. When one casually looks at the standard expressions for higher MBPT energies, as applied to large systems, one finds that they contain terms that are proportional to the *square* of the number of particles. Brueckner was able to show, for the first few orders, that these ill-behaved terms actually cancel and thus each order of MBPT was size consistent. He could not prove, however, that this was the case for all orders. In order to prove such a theorem, J. Goldstone, inspired by R. P. Feynman's work in quantum electrodynamics, devised a pictorial or diagrammatic representation of the algebraic expressions which occur in the RS perturbation expansion of the energy of a many-particle system. He found that in each order the ill-behaved terms, which were proportional to the square of the number of particles, were represented by pictures made up of disconnected pieces; i.e., these diagrams were *unlinked*. He then proved that such unlinked terms always cancel in every order. This is the famous *linked-cluster theorem*, which says that the perturbation expansion of the energy of a many-body system can be represented solely by linked diagrams. It assures us that the Møller-Plesset perturbation expansion is size consistent no matter in which order it is terminated.

The diagrammatic techniques introduced by Goldstone were first applied to atoms by H. P. Kelly, while he was still a graduate student. Instead of calculating the sum of the perturbation energies up to, say, second-order and ignoring all of the higher order energies, he was able to approximately incorporate certain contributions made in all orders of perturbation theory by summing the values of certain diagrams to infinite order. In other words, he obtained an approximate expression for the correlation energy which contained not only all the contributions made, say, up to second-order but also certain parts of the higher order terms. Using this kind of approach, he was the first to perform a calculation which was later shown to be essentially equivalent to the independent electron pair approximation (IEPA).

In this chapter, we give a simple introduction to diagrammatic perturbation theory. However, this chapter is so organized that it can be read without learning about diagrams. The sections that discuss diagrams are starred and can be skipped without loss of continuity. Any result that can be obtained diagrammatically can always be obtained algebraically. In quantum chemistry, it is basically a matter of taste whether one finds that diagrams are useful and give valuable insight (e.g., suggest certain approximations). We hope that our simplified treatment will stimulate some to learn more about these techniques.

We begin this chapter by deriving the standard expressions for the first-, second-, and third-order energies of RSPT, which are applicable to any quantum mechanical system. In Section 6.2 we introduce a diagrammatic representation of the various algebraic terms that occur in each order. We begin by considering a simple system having only two eigenstates and then generalize our results to handle a system with an arbitrary number of states. We then consider the perturbation expansion of the energy of an N-electron system using the Hartree-Fock description as our starting point. In Section 6.3 we treat the simple case where the perturbation contains only one-particle interactions; Section 6.4 discusses the corresponding diagrammatic representation. In Section 6.5 we consider perturbations containing two-particle interactions and develop the perturbation expansion for the correlation energy. In Section 6.6 we consider the size-consistency problem and apply RSPT to our standard model of N independent minimal basis H_2 molecules. In Section 6.7 we discuss the diagrammatic representation of MBPT, and finally, in Section 6.8 we present some illustrative calculations.

6.1 RAYLEIGH-SCHRÖDINGER (RS) PERTURBATION THEORY

In this section we will derive the standard expressions of Rayleigh-Schrödinger (RS) perturbation theory. Our formulas will be general and apply equally to one-particle or N-particle systems. Suppose we wish to solve the eigenvalue problem

$$\mathscr{H}|\Phi_i\rangle = (\mathscr{H}_0 + \mathscr{V})|\Phi_i\rangle = \mathscr{E}_i|\Phi_i\rangle \tag{6.1}$$

where we know the eigenfunctions and eigenvalues of $\mathscr{H}_0$,

$$\mathscr{H}_0|\Psi_i^{(0)}\rangle = E_i^{(0)}|\Psi_i^{(0)}\rangle \quad \text{or} \quad \mathscr{H}_0|i\rangle = E_i^{(0)}|i\rangle \tag{6.2}$$

where we have written $|\Psi_i^{(0)}\rangle = |i\rangle$ for compactness. If the perturbation, $\mathscr{V}$, is small in some sense, we expect $|\Phi_i\rangle$ and $\mathscr{E}_i$ to be reasonably close to $|i\rangle$ and $E_i^{(0)}$, respectively. We wish to devise a procedure by which we can systematically improve the eigenfunctions and eigenvalues of $\mathscr{H}_0$ so that they become closer and closer to the eigenvalues and eigenfunctions of the total Hamiltonian, $\mathscr{H}$. We can do this by introducing an ordering parameter λ, which will later be set equal to unity, and writing

$$\mathscr{H} = \mathscr{H}_0 + \lambda\mathscr{V} \tag{6.3}$$

We now expand the exact eigenfunctions and eigenvalues in a Taylor series in λ,

$$\mathscr{E}_i = E_i^{(0)} + \lambda E_i^{(1)} + \lambda^2 E_i^{(2)} + \cdots \tag{6.4a}$$

$$|\Phi_i\rangle = |i\rangle + \lambda|\Psi_i^{(1)}\rangle + \lambda^2|\Psi_i^{(2)}\rangle + \cdots \tag{6.4b}$$

We call $E_i^{(n)}$ the nth-order energy. The problem at hand is to express these quantities in terms of the zeroth-order energies and matrix elements of the perturbation $\mathscr{V}$ between the unperturbed wave functions, $\langle i|\mathscr{V}|j\rangle$.

Let us take the wave functions of $\mathscr{H}_0$ to be normalized ($\langle i|i\rangle = 1$) and then *choose* the normalization of $|\Phi_i\rangle$ such that $\langle i|\Phi_i\rangle = 1$. This choice is called intermediate normalization and can always be made unless $|i\rangle$ and $|\Phi_i\rangle$ are orthogonal. Therefore, by multiplying Eq. (6.4b) by $\langle i|$, we have

$$\langle i|\Phi_i\rangle = \langle i|i\rangle + \lambda\langle i|\Psi_i^{(1)}\rangle + \lambda^2\langle i|\Psi_i^{(2)}\rangle + \cdots = 1 \qquad (6.5)$$

The above equation holds for all values of λ. Therefore, the coefficients of λ^n on both sides must be equal, and hence

$$\langle i|\Psi_i^{(n)}\rangle = 0 \qquad n = 1, 2, 3, \ldots \qquad (6.6)$$

Substituting Eqs. (6.4a, b) into Eq. (6.1),

$$(\mathscr{H}_0 + \lambda\mathscr{V})(|i\rangle + \lambda|\Psi_i^{(1)}\rangle + \lambda^2|\Psi_i^{(2)}\rangle + \cdots)$$
$$= (E_i^{(0)} + \lambda E_i^{(1)} + \lambda^2 E_i^{(2)} + \cdots)(|i\rangle + \lambda|\Psi_i^{(1)}\rangle + \cdots)$$

and equating coefficients of λ^n, we find

$$\mathscr{H}_0|i\rangle = E_i^{(0)}|i\rangle \qquad\qquad\qquad n = 0 \quad (6.7a)$$

$$\mathscr{H}_0|\Psi_i^{(1)}\rangle + \mathscr{V}|i\rangle = E_i^{(0)}|\Psi_i^{(1)}\rangle + E_i^{(1)}|i\rangle \qquad\qquad n = 1 \quad (6.7b)$$

$$\mathscr{H}_0|\Psi_i^{(2)}\rangle + \mathscr{V}|\Psi_i^{(1)}\rangle = E_i^{(0)}|\Psi_i^{(2)}\rangle + E_i^{(1)}|\Psi_i^{(1)}\rangle + E_i^{(2)}|i\rangle \qquad n = 2 \quad (6.7c)$$

$$\mathscr{H}_0|\Psi_i^{(3)}\rangle + \mathscr{V}|\Psi_i^{(2)}\rangle = E_i^{(0)}|\Psi_i^{(3)}\rangle + E_i^{(1)}|\Psi_i^{(2)}\rangle$$
$$+ E_i^{(2)}|\Psi_i^{(1)}\rangle + E_i^{(3)}|i\rangle \qquad n = 3 \quad (6.7d)$$

and so on. Multiplying each of these equations by $\langle i|$ and using the orthogonality relation (6.6), we obtain the following expressions for the nth-order energies

$$E_i^{(0)} = \langle i|\mathscr{H}_0|i\rangle \qquad\qquad (6.8a)$$

$$E_i^{(1)} = \langle i|\mathscr{V}|i\rangle \qquad\qquad (6.8b)$$

$$E_i^{(2)} = \langle i|\mathscr{V}|\Psi_i^{(1)}\rangle \qquad\qquad (6.8c)$$

$$E_i^{(3)} = \langle i|\mathscr{V}|\Psi_i^{(2)}\rangle \qquad\qquad (6.8d)$$

All that remains is to solve the set of equations (6.7) for $|\Psi_i^{(n)}\rangle$ and then determine the nth-order energy using (6.8).

First consider Eq. (6.7b), which determines the first-order wave function $|\Psi_i^{(1)}\rangle$. This can be rewritten as

$$(E_i^{(0)} - \mathscr{H}_0)|\Psi_i^{(1)}\rangle = (\mathscr{V} - E_i^{(1)})|i\rangle = (\mathscr{V} - \langle i|\mathscr{V}|i\rangle)|i\rangle \qquad (6.9)$$

This is no longer an eigenvalue equation but an inhomogeneous differential (or in general integro-differential) equation. One way of solving such equations is to expand $|\Psi_i^{(1)}\rangle$ in terms of the eigenfunctions of $\mathscr{H}_0$, which are taken to be complete,

$$|\Psi_i^{(1)}\rangle = \sum_n c_n^{(1)}|n\rangle$$

Since the eigenfunctions of $\mathscr{H}_0$ are orthonormal, multiplying this equation by $\langle n|$, we find

$$\langle n|\Psi_i^{(1)}\rangle = c_n^{(1)}$$

Moreover, from Eq. (6.6) it is clear that $c_i^{(1)} = 0$, so we can write

$$|\Psi_i^{(1)}\rangle = \sum_n{}' |n\rangle\langle n|\Psi_i^{(1)}\rangle \tag{6.10}$$

where the prime on the summation serves as a reminder that the term $n = i$ is excluded. Multiplying Eq. (6.9) by $\langle n|$ and using the fact that the zeroth-order wave functions are orthogonal, we have

$$(E_i^{(0)} - E_n^{(0)})\langle n|\Psi_i^{(1)}\rangle = \langle n|\mathscr{V}|i\rangle \tag{6.11}$$

Using the expansion (6.10) in expression (6.8c) for the second-order energy, we obtain

$$E_i^{(2)} = \langle i|\mathscr{V}|\Psi_i^{(1)}\rangle = \sum_n{}' \langle i|\mathscr{V}|n\rangle\langle n|\Psi_i^{(1)}\rangle$$

and hence, using (6.11), we finally have

$$E_i^{(2)} = \sum_n{}' \frac{\langle i|\mathscr{V}|n\rangle\langle n|\mathscr{V}|i\rangle}{E_i^{(0)} - E_n^{(0)}} = \sum_n{}' \frac{|\langle i|\mathscr{V}|n\rangle|^2}{E_i^{(0)} - E_n^{(0)}} \tag{6.12}$$

which is the desired expression for the second-order energy.

To obtain the third-order energy, $E_i^{(3)}$, we proceed similarly. We first expand the second-order wave function as

$$|\Psi_i^{(2)}\rangle = \sum_n{}' |n\rangle\langle n|\Psi_i^{(2)}\rangle \tag{6.13}$$

Then we multiply Eq. (6.7c) by $\langle n|$ to obtain

$$(E_i^{(0)} - E_n^{(0)})\langle n|\Psi_i^{(2)}\rangle = \langle n|\mathscr{V}|\Psi_i^{(1)}\rangle - E_i^{(1)}\langle n|\Psi_i^{(1)}\rangle \tag{6.14}$$

Next we combine Eqs. (6.8d), (6.13), and (6.14) as follows:

$$
\begin{aligned}
E_i^{(3)} &= \langle i|\mathscr{V}|\Psi_i^{(2)}\rangle \\
&= \sum_n{}' \langle i|\mathscr{V}|n\rangle\langle n|\Psi_i^{(2)}\rangle \\
&= \sum_n{}' \frac{\langle i|\mathscr{V}|n\rangle\langle n|\mathscr{V}|\Psi_i^{(1)}\rangle}{E_i^{(0)} - E_n^{(0)}} - E_i^{(1)} \sum_n{}' \frac{\langle i|\mathscr{V}|n\rangle\langle n|\Psi_i^{(1)}\rangle}{E_i^{(0)} - E_n^{(0)}}
\end{aligned}
$$

and finally, using (6.10) and (6.11) we have

$$E_i^{(3)} = \sum_{nm}' \frac{\langle i|\mathscr{V}|n\rangle\langle n|\mathscr{V}|m\rangle\langle m|\mathscr{V}|i\rangle}{(E_i^{(0)} - E_n^{(0)})(E_i^{(0)} - E_m^{(0)})} - E_i^{(1)} \sum_n' \frac{|\langle i|\mathscr{V}|n\rangle|^2}{(E_i^{(0)} - E_n^{(0)})^2}$$

$$= A_i^{(3)} + B_i^{(3)} \tag{6.15}$$

which is the desired third-order energy.

We now present a simple application of the above formalism. Let us consider a simple quantum mechanical system containing only two states $|\text{I}\rangle$ and $|\text{II}\rangle$, which are eigenfunctions of the Hamiltonian $\mathscr{H}$

$$\mathscr{H}|\text{I}\rangle = \mathscr{E}_1|\text{I}\rangle \quad \text{and} \quad \mathscr{H}|\text{II}\rangle = \mathscr{E}_2|\text{II}\rangle$$

Suppose we write

$$\mathscr{H} = \mathscr{H}_0 + \mathscr{V}$$

where we know the eigenfunctions and eigenvalues of $\mathscr{H}_0$,

$$\mathscr{H}_0|1\rangle = E_1^{(0)}|1\rangle \quad \text{and} \quad \mathscr{H}_0|2\rangle = E_2^{(0)}|2\rangle$$

and we wish to determine the exact ground state energy $\mathscr{E}_1$. We assume that the zeroth-order states are nondegenerate and that $E_1^{(0)}$ is the lower eigenvalue (i.e., $E_1^{(0)} < E_2^{(0)}$). We shall first solve the problem exactly and then compare our results with those obtained using perturbation theory. We solve

$$\mathscr{H}|\Phi\rangle = \mathscr{E}|\Phi\rangle \tag{6.16}$$

by expanding

$$|\Phi\rangle = c_1|1\rangle + c_2|2\rangle \tag{6.17}$$

and thus obtaining

$$\begin{bmatrix} E_1^{(0)} + \langle 1|\mathscr{V}|1\rangle & \langle 1|\mathscr{V}|2\rangle \\ \langle 2|\mathscr{V}|1\rangle & E_2^{(0)} + \langle 2|\mathscr{V}|2\rangle \end{bmatrix} \begin{bmatrix} c_1 \\ c_2 \end{bmatrix} = \mathscr{E}\begin{bmatrix} c_1 \\ c_2 \end{bmatrix} \tag{6.18}$$

To simplify the notation we set $\langle 1|\mathscr{V}|1\rangle = V_{11}$, $\langle 1|\mathscr{V}|2\rangle = V_{12}$, $\langle 2|\mathscr{V}|1\rangle = V_{21}$, $\langle 2|\mathscr{V}|2\rangle = V_{22}$. The lower eigenvalue of the above matrix is easily found to be

$$\mathscr{E}_1 = \tfrac{1}{2}\{E_1^{(0)} + V_{11} + E_2^{(0)} + V_{22} - [(E_1^{(0)} - E_2^{(0)} + V_{11} - V_{22})^2 + 4V_{12}V_{21}]^{1/2}\} \tag{6.19a}$$

We wish to expand $\mathscr{E}_1$ in a Taylor series in the matrix elements of the perturbation. To do this, it is convenient to multiply each matrix element by λ,

$$\mathscr{E}_1 = \tfrac{1}{2}\{E_1^{(0)} + \lambda V_{11} + E_2^{(0)} + \lambda V_{22}$$
$$- [(E_1^{(0)} - E_2^{(0)} + \lambda(V_{11} - V_{22}))^2 + 4\lambda^2 V_{12}V_{21}]^{1/2}\} \tag{6.19b}$$

and then expand $\mathscr{E}_1$ in a Taylor series in λ.

We will need two identities, valid when $|x| < 1$,

$$(1 + x)^{1/2} = 1 + \tfrac{1}{2}x - \tfrac{1}{8}x^2 + \cdots \tag{6.20}$$

$$(1 - x)^{-1} = 1 + x + x^2 + x^3 + \cdots \tag{6.21}$$

We begin by rewriting $\mathscr{E}_1$ as

$$\mathscr{E}_1 = \frac{1}{2}\left\{ E_1^{(0)} + \lambda V_{11} + E_2^{(0)} + \lambda V_{22} + (E_1^{(0)} - E_2^{(0)} + \lambda(V_{11} - V_{22})) \right.$$

$$\left. \times \left[1 + \frac{4\lambda^2 V_{12} V_{21}}{(E_1^{(0)} - E_2^{(0)} + \lambda(V_{11} - V_{22}))^2} \right]^{1/2} \right\}$$

Note that when we factored out $(E_1^{(0)} - E_2^{(0)} + \lambda(V_{11} - V_{22}))$ from the square root, this quantity was taken to be *negative*. This follows from the assumptions that $E_1^{(0)} < E_2^{(0)}$ and that the perturbation is small.

By first expanding the square root and then expanding $(E_1^{(0)} - E_2^{(0)} + \lambda(V_{11} - V_{22}))^{-1}$ as

$$\frac{1}{E_1^{(0)} - E_2^{(0)} + \lambda(V_{11} - V_{22})} = \frac{1}{E_1^{(0)} - E_2^{(0)}} \cdot \frac{1}{1 + \dfrac{\lambda(V_{11} - V_{22})}{E_1^{(0)} - E_2^{(0)}}}$$

$$= \frac{1}{E_1^{(0)} - E_2^{(0)}} - \frac{\lambda(V_{11} - V_{22})}{(E_1^{(0)} - E_2^{(0)})^2} + \cdots$$

it can be shown that

$$\mathscr{E}_1 = E_1^{(0)} + \lambda E_1^{(1)} + \lambda^2 E_1^{(2)} + \lambda^3 E_1^{(3)} + \lambda^4 E_1^{(4)} + \cdots \tag{6.22}$$

where

$$E_1^{(1)} = V_{11} \tag{6.23a}$$

$$E_1^{(2)} = \frac{V_{12} V_{21}}{E_1^{(0)} - E_2^{(0)}} \tag{6.23b}$$

$$E_1^{(3)} = \frac{V_{12} V_{22} V_{21}}{(E_1^{(0)} - E_2^{(0)})^2} - \frac{V_{12} V_{21} V_{11}}{(E_1^{(0)} - E_2^{(0)})^2} \tag{6.23c}$$

$$E_1^{(4)} = \frac{V_{12} V_{21} V_{11}^2}{(E_1^{(0)} - E_2^{(0)})^3} - \frac{2 V_{12} V_{22} V_{21} V_{11}}{(E_1^{(0)} - E_2^{(0)})^3} + \frac{V_{12} V_{22}^2 V_{21}}{(E_1^{(0)} - E_2^{(0)})^3} - \frac{(V_{12} V_{21})^2}{(E_1^{(0)} - E_2^{(0)})^3} \tag{6.23d}$$

Finally, it can be seen that the general formulas of RS perturbation theory for the first-, second-, and third-order energies (Eqs. (6.8b), (6.12), and (6.15)) reproduce the above results for $E_1^{(n)}$, $n = 1, 2, 3$. This follows immediately once it is recognized that there is only one intermediate state (i.e., $|n\rangle = |2\rangle$) in the two-state problem. Such a correspondence is to be expected since the general formulas were derived using a formal Taylor's series expansion in the parameter λ.

*6.2 DIAGRAMMATIC REPRESENTATION OF RS PERTURBATION THEORY

At the end of the last section, we saw that even in the simple case of a two-state system the formulas for higher order energies become quite complicated. Our primary interest in this chapter is to obtain the perturbation expansion of the correlation energy with the Hartree-Fock taken as the zeroth-order description. For this case, the formulas for the higher order energies are much more complicated. Therefore, it would be convenient if we could find a relatively simple way of representing and classifying the various terms that appear in higher orders of perturbation theory. In this section, we introduce a diagrammatic representation of the RS perturbation expansion. We will not attempt to derive diagrammatic perturbation theory; rather the approach we have adopted starts by considering a simple example of perturbation theory and *conjecturing* a one-to-one correspondence between the mathematical expressions and a set of pictures drawn in a well-defined way. We then "guess" the rules for translating these pictures into formulas and show that everything works. We then proceed to generalize the diagrams and rules obtained. At each stage, we check consistency by comparing diagrammatic results with those obtained independently by algebraic methods. In this way we hope to make the reader feel comfortable with diagrams.

The diagrammatic representation of a perturbation expansion was first used in quantum electrodynamics. It was invented by Feynman who, in his Nobel lecture,[1] discussed the evolution of his ideas. We quote, "The rest of my work was simply to improve the techniques then available for calculations, making diagrams to help analyze perturbation theory quicker. Most of this was first worked out by guessing. . ." and again, "Often, even in the physicist's sense, I did not have a demonstration of how to get all these rules and equations from conventional electrodynamics. But, I did know, from fooling around, that everything was, in fact, equivalent to regular electrodynamics."

6.2.1 Diagrammatic Perturbation Theory for Two States

We now introduce a pictorial representation of the various terms that occur in the perturbation expansion of the lowest eigenvalue of the two-state problem, which we considered in the previous section. Let us represent the perturbation $\mathscr{V}$ by a dot,

$$\mathscr{V} \longleftrightarrow \bullet$$

The two zeroth-order states $|1\rangle$ and $|2\rangle$ are represented by lines with arrows pointing down and up, respectively.

A line with an arrow pointing down is called a hole line, and a line with an arrow pointing up is called a particle line. These terms have no significance at this stage; however, they are introduced now so that one does not have to relearn names later on. These definitions suggest simple pictorial representations of the matrix elements of $\mathscr{V}$:

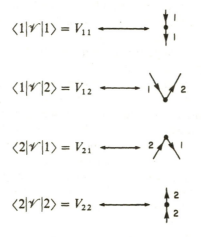

$$\langle 1|\mathscr{V}|1\rangle = V_{11}$$

$$\langle 1|\mathscr{V}|2\rangle = V_{12}$$

$$\langle 2|\mathscr{V}|1\rangle = V_{21}$$

$$\langle 2|\mathscr{V}|2\rangle = V_{22}$$

In summary, each dot has a line going into it and a line coming out of it. The corresponding matrix element is $\langle$label of line *in* $|\mathscr{V}|$ label of line *out*$\rangle$.

Recall that the mathematical expressions for each term in the nth-order energy, $n = 1, 2, 3, 4$ contain a product of n matrix elements of $\mathscr{V}$ in the numerator. Since each matrix element is represented by a dot with a line going into it and a line coming out of it, it might be possible to represent the algebraic expressions by pictures or diagrams containing n dots connected in some way. We now give a well-defined prescription for drawing such diagrams containing n dots:

1. Draw n dots vertically ordered as follows:

-     1
-     2
-     3
      ⋮
-     n.

2. Connect *all n* dots together with a continuous line, so that each dot has one line passing through it.
3. Do this in all possible distinct ways. Two diagrams are equivalent if each and every dot is connected to an identical pair of dots in both diagrams.

The second rule forces us to draw only diagrams that are completely connected or *linked*; i.e., they are all in one piece. Unlinked diagrams ($n = 4$) like

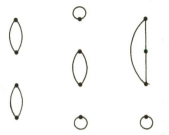

are not allowed. The third rule tells us that the diagrams

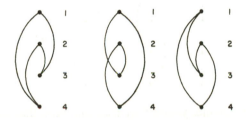

are really the same because they are connected in the same way: 1 is linked to 3 and 4, 2 is linked to 3 and 4, 3 is linked with 1 and 2, and 4 is linked with 1 and 2. To quickly establish the equivalence of diagrams, it is useful to think of the connecting lines as rubber bands glued to the dots, which can be distorted in any way we please. Two diagrams are then distinct if there is no way to distort the connecting lines, so as to make the pictures look identical. Thus

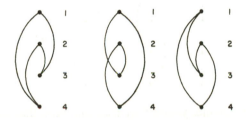

Using these rules, we can draw the following diagrams up to fourth order

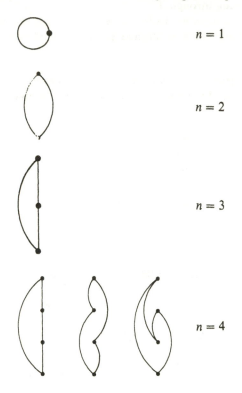

$n = 1$

$n = 2$

$n = 3$

$n = 4$

You should convince yourself, by trial and error, that the above are the only distinct diagrams that one can draw. Now let us draw arrows on each of the connecting lines in all possible ways so that each dot has a line going into it and a line coming out of it. Finally, we label lines going down with 1 and up with 2. For example, if $n = 3$, there are two possible labelings,

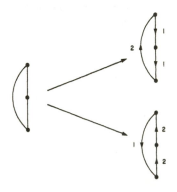

For $n = 1$, an ambiguity arises since the line is circular and an arrow on it would point up in one part of the diagram and down in another,

We resolve this by defining circular lines to be hole lines. It is easy to show that the diagrams we can generate in this way are

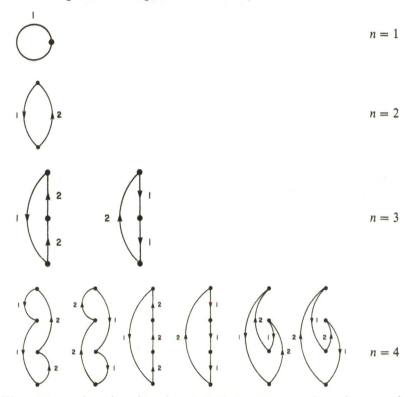

We now *postulate* that there is a one-to-one correspondence between the above pictures with n dots and the terms which contribute to the nth-order energy, and we give the rules allowing one to translate a diagram into an algebraic expression. It is rather remarkable that there exists such a simple correspondence. The translation rules are:

1. *Each dot contributes a factor* $\langle$ *label of line in* $|\mathscr{V}|$ *label of line out* $\rangle$ *to the numerator.*
2. *Each pair of adjacent dots contributes the denominator factor*

$$\sum E^{(0)}_{\text{hole}} - \sum E^{(0)}_{\text{particle}}$$

where the sums run over the labels of all hole and particle lines crossing an imaginary horizontal line separating the two adjacent dots.

3. *The overall sign of the expression is* $(-)^{h+l}$, *where h is the number of hole lines and l is the number of closed loops in the diagram. For the diagrams discussed in this section l is always one. We give this rule in its more general form for future reference.*

These rules are very easy to apply. Consider the fourth-order diagram:

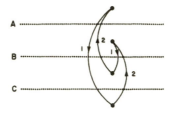

The upper two dots each contribute a factor $\langle 2|\mathscr{V}|1\rangle = V_{21}$, and the remaining dots each contribute a factor $\langle 1|\mathscr{V}|2\rangle = V_{12}$ to the numerator. Imaginary lines A and C contribute a factor $E_1^{(0)} - E_2^{(0)}$ each to the denominator while imaginary line B contributes the factor $(2E_1^{(0)} - 2E_2^{(0)})$. Since we have two hole lines $(h = 2)$ and one closed loop $(l = 1)$ we multiply the expression by $(-1)^3 = -1$. So we have

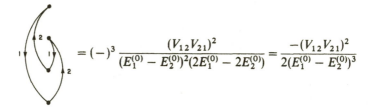

$$= (-)^3 \frac{(V_{12}V_{21})^2}{(E_1^{(0)} - E_2^{(0)})^2(2E_1^{(0)} - 2E_2^{(0)})} = \frac{-(V_{12}V_{21})^2}{2(E_1^{(0)} - E_2^{(0)})^3}$$

Notice now that our translation rules give the same algebraic expression for the following pairs of diagrams

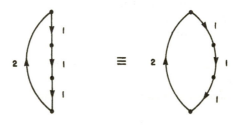

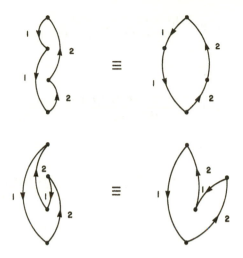

This observation allows us to remove the restriction that all the dots in a given diagram must be on top of one another. The rules are such that we can distort a given diagram in any way *as long as we do not change the vertical ordering of dots*.

In Table 6.1 we have evaluated all second-, third-, and fourth-order diagrams using our rules. By convention, the first-order bubble diagram is

We can see by inspection of Table 6.1 that in each order the sum of the values of all possible diagrams, constructed and evaluated according to the rules, gives precisely the results we obtained for the two-state system in the last section (c.f. Eq. (6.23)).

Exercise 6.1 Write down and evaluate all fifth-order diagrams that have the property that an imaginary horizontal line crosses only one hole and one particle line. Show that the sum of such diagrams is

$$\frac{V_{12}V_{21}(V_{22} - V_{11})^3}{(E_1^{(0)} - E_2^{(0)})^4}$$

Hint: There are eight such diagrams, and they can be generated by adding three dots to the second-order diagram in all possible ways.

Table 6.1 Translation of diagrams into formulas

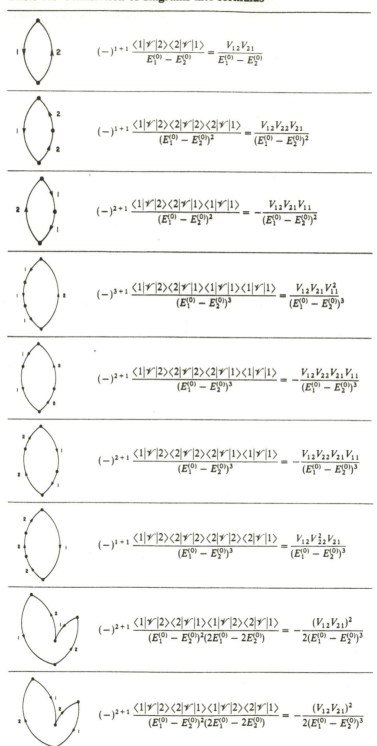

$$(-)^{1+1} \frac{\langle 1|\mathscr{V}|2\rangle\langle 2|\mathscr{V}|1\rangle}{E_1^{(0)} - E_2^{(0)}} = \frac{V_{12}V_{21}}{E_1^{(0)} - E_2^{(0)}}$$

$$(-)^{1+1} \frac{\langle 1|\mathscr{V}|2\rangle\langle 2|\mathscr{V}|2\rangle\langle 2|\mathscr{V}|1\rangle}{(E_1^{(0)} - E_2^{(0)})^2} = \frac{V_{12}V_{22}V_{21}}{(E_1^{(0)} - E_2^{(0)})^2}$$

$$(-)^{2+1} \frac{\langle 1|\mathscr{V}|2\rangle\langle 2|\mathscr{V}|1\rangle\langle 1|\mathscr{V}|1\rangle}{(E_1^{(0)} - E_2^{(0)})^2} = -\frac{V_{12}V_{21}V_{11}}{(E_1^{(0)} - E_2^{(0)})^2}$$

$$(-)^{3+1} \frac{\langle 1|\mathscr{V}|2\rangle\langle 2|\mathscr{V}|1\rangle\langle 1|\mathscr{V}|1\rangle\langle 1|\mathscr{V}|1\rangle}{(E_1^{(0)} - E_2^{(0)})^3} = \frac{V_{12}V_{21}V_{11}^2}{(E_1^{(0)} - E_2^{(0)})^3}$$

$$(-)^{2+1} \frac{\langle 1|\mathscr{V}|2\rangle\langle 2|\mathscr{V}|2\rangle\langle 2|\mathscr{V}|1\rangle\langle 1|\mathscr{V}|1\rangle}{(E_1^{(0)} - E_2^{(0)})^3} = -\frac{V_{12}V_{22}V_{21}V_{11}}{(E_1^{(0)} - E_2^{(0)})^3}$$

$$(-)^{2+1} \frac{\langle 1|\mathscr{V}|2\rangle\langle 2|\mathscr{V}|2\rangle\langle 2|\mathscr{V}|1\rangle\langle 1|\mathscr{V}|1\rangle}{(E_1^{(0)} - E_2^{(0)})^3} = -\frac{V_{12}V_{22}V_{21}V_{11}}{(E_1^{(0)} - E_2^{(0)})^3}$$

$$(-)^{1+1} \frac{\langle 1|\mathscr{V}|2\rangle\langle 2|\mathscr{V}|2\rangle\langle 2|\mathscr{V}|2\rangle\langle 2|\mathscr{V}|1\rangle}{(E_1^{(0)} - E_2^{(0)})^3} = \frac{V_{12}V_{22}^2V_{21}}{(E_1^{(0)} - E_2^{(0)})^3}$$

$$(-)^{2+1} \frac{\langle 1|\mathscr{V}|2\rangle\langle 2|\mathscr{V}|1\rangle\langle 1|\mathscr{V}|2\rangle\langle 2|\mathscr{V}|1\rangle}{(E_1^{(0)} - E_2^{(0)})^2(2E_1^{(0)} - 2E_2^{(0)})} = -\frac{(V_{12}V_{21})^2}{2(E_1^{(0)} - E_2^{(0)})^3}$$

$$(-)^{2+1} \frac{\langle 1|\mathscr{V}|2\rangle\langle 2|\mathscr{V}|1\rangle\langle 1|\mathscr{V}|2\rangle\langle 2|\mathscr{V}|1\rangle}{(E_1^{(0)} - E_2^{(0)})^2(2E_1^{(0)} - 2E_2^{(0)})} = -\frac{(V_{12}V_{21})^2}{2(E_1^{(0)} - E_2^{(0)})^3}$$

6.2.2 Diagrammatic Perturbation Theory for N States

We now generalize our previous development to obtain a diagrammatic representation of RS perturbation theory as applied to an N-state system. Consider the problem of finding the perturbation expansion for the lowest eigenvalue of such a system. Here we still have only one hole state, $|1\rangle$, but there are now $N - 1$ particle states $|n\rangle$, $n = 2, 3, \ldots, N$. We draw the same set of diagrams as before. However, now we can label the particle lines with any index n. For example, the diagram

can now be labeled as

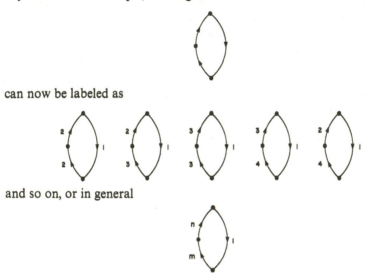

and so on, or in general

where the labels m and n can take on values between 2 and N independently. The subset of diagrams

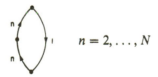

$$n = 2, \ldots, N$$

are called *diagonal*. To obtain mathematical expressions corresponding to these generalized diagrams, we supplement our previous rules with:

4. *Sum the expression over all particle indices.*

As an example, we evaluate the second- and third-order energies:

$$\sum_{n \neq 1} \frac{V_{1n}V_{n1}}{E_1^{(0)} - E_n^{(0)}} = {\sum_{n}}' \frac{V_{1n}V_{n1}}{E_1^{(0)} - E_n^{(0)}} \tag{6.24a}$$

$$\text{(diagrams)} = \sum_{nm}' \frac{V_{1m}V_{mn}V_{n1}}{(E_1^{(0)} - E_m^{(0)})(E_1^{(0)} - E_n^{(0)})} - \sum_n' \frac{V_{11}V_{1n}V_{n1}}{(E_1^{(0)} - E_n^{(0)})^2}$$

(6.24b)

These expressions are identical to our previous results for the second- and third-order energies (Eqs. (6.12) and (6.15)) when $i = 1$. What if we want the perturbation expansion for some state i, which is not necessarily the lowest? What do the diagrams look like? One can easily verify that we get the same answers as before if we label our hole lines by the index i and the particle lines by the indices $m, n, k, \ldots$, which can take on the values $1, 2, \ldots, i - 1$, $i + 1, \ldots, N$. Thus we now have a complete diagrammatic representation of RS perturbation theory, which is applicable to any perturbation and any zeroth-order state.

Exercise 6.2 Use diagrammatic techniques to obtain the fourth-order perturbation energy of a particular state (say, i) of an N-state system. That is, evaluate the diagrams

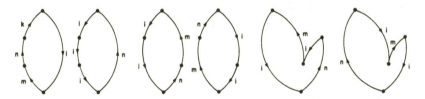

where the indices $m, n, k, \ldots$ exclude i. Using the approach of Section 6.1, obtain an algebraic expression for the fourth-order energy and compare it to the diagrammatic result.

6.2.3 Summation of Diagrams

An important technique in diagrammatic perturbation theory is the summation of certain classes of diagrams to infinite order. To illustrate the technique consider our two-state system. The only diagram that contributes in second-order is

Now consider the sum of all diagrams, which are generated from this diagram, by adding dots in all possible ways,

$$(6.25)$$

Using Table 6.1 and the result of Exercise 6.1, where such fifth-order diagrams are evaluated, we find

$$\Delta = \left[\frac{V_{12}V_{21}}{E_1^{(0)} - E_2^{(0)}} \right] + \left[\frac{V_{12}V_{21}(V_{22} - V_{11})}{(E_1^{(0)} - E_2^{(0)})^2} \right] + \left[\frac{V_{12}V_{21}(V_{22} - V_{11})^2}{(E_1^{(0)} - E_2^{(0)})^3} \right]$$

$$+ \left[\frac{V_{12}V_{21}(V_{22} - V_{11})^3}{(E_1^{(0)} - E_2^{(0)})^4} \right] + \cdots$$

$$\Delta = \frac{V_{12}V_{21}}{E_1^{(0)} - E_2^{(0)}} \left[1 + \left(\frac{V_{22} - V_{11}}{E_1^{(0)} - E_2^{(0)}} \right) + \left(\frac{V_{22} - V_{11}}{E_1^{(0)} - E_2^{(0)}} \right)^2 \right.$$

$$\left. + \left(\frac{V_{22} - V_{11}}{E_1^{(0)} - E_2^{(0)}} \right)^3 + \cdots \right]$$

$$(6.26)$$

Notice that the quantity in brackets is a geometric series. Using the formula

$$(1 - x)^{-1} = 1 + x + x^2 + x^3 + \cdots$$

the series (6.26) can be summed to give

$$\Delta = \frac{V_{12}V_{21}}{E_1^{(0)} - E_2^{(0)}} \left[\frac{1}{1 - \dfrac{V_{22} - V_{11}}{E_1^{(0)} - E_2^{(0)}}} \right] = \frac{V_{12}V_{21}}{(E_1^{(0)} + V_{11}) - (E_2^{(0)} + V_{22})} \qquad (6.27)$$

Thus summing this class of diagrams to infinite order yields an expression which looks like the second-order perturbation result but with a *shifted energy denominator*. This can be restated as follows: summing a geometric series of diagrams to infinite order gives results which can be obtained using perturbation theory to a finite order, when a different *partitioning* of the Hamiltonian into perturbed and unperturbed parts is used. Specifically, in our two-state problem, we originally partitioned our Hamiltonian as

$$\mathbf{H}_0 = \begin{pmatrix} E_1^{(0)} & 0 \\ 0 & E_2^{(0)} \end{pmatrix}$$

$$\mathbf{V} = \begin{pmatrix} V_{11} & V_{12} \\ V_{21} & V_{22} \end{pmatrix}$$

If we repartition our Hamiltonian as

$$\mathbf{H}_0' = \begin{pmatrix} E_1^{(0)} + V_{11} & 0 \\ 0 & E_2^{(0)} + V_{22} \end{pmatrix}$$

$$\mathbf{V}' = \begin{pmatrix} 0 & V_{12} \\ V_{21} & 0 \end{pmatrix}$$

then the geometric sum of diagrams with the original partitioning is simply the second-order perturbation energy with this partitioning.

Let us generalize these results to the N-state system. The diagrams generated by adding dots to the second-order diagram are

$$\Delta = \left[\begin{array}{c} \vphantom{x} \end{array} \right] + \left[\begin{array}{c} \vphantom{x} \end{array} \right] + \cdots \qquad (6.28)$$

Note that the class of diagrams generated in this way are all diagonal. These diagrams can be summed to give

$$\Delta = \sum_n{}' \frac{V_{1n}V_{n1}}{(E_1^{(0)} + V_{11}) - (E_n^{(0)} + V_{nn})} \qquad (6.29)$$

where the prime on the summation serves as a reminder that the $n = 1$ term is excluded. Notice again that the result obtained is simply the second-order expression with shifted energy denominators. We will see later that the geometric series summation technique can be used to obtain perturbation corrections to the Hartree-Fock energy.

6.3 ORBITAL PERTURBATION THEORY: ONE-PARTICLE PERTURBATIONS

The theory we have developed so far is applicable to any quantum mechanical system. In this section we consider the important special case where the unperturbed Hamiltonian is a sum of one-particle Hamiltonians

$$\mathscr{H}_0 = \sum_i h_0(i) \qquad (6.30)$$

In particular, the Hartree-Fock Hamiltonian is of this form. We are interested in improving such an independent-particle description of a many-electron system by means of perturbation theory. For the sake of simplicity, we will first consider the case where the perturbation is also a sum of one-particle interactions,

$$\mathscr{V} = \sum_i v(i) \qquad (6.31)$$

and in the next section we generalize our results to perturbations containing two-particle interactions. As an example of a physical situation where the perturbation is a sum of one-particle terms, consider a molecule in the presence of an electric field $\vec{F}$. In this case the perturbation is $\vec{F} \cdot \sum_i \mathbf{r}(i)$, where $\mathbf{r}(i)$ is the position vector of the ith electron.

Suppose we obtain a set of spin orbitals and orbital energies that are the eigenfunctions and eigenvalues of h_0,

$$h_0 \chi_i^{(0)} = \varepsilon_i^{(0)} \chi_i^{(0)} \tag{6.32}$$

The ground state wave function ($|\Psi_0\rangle$) of an N-electron system with Hamiltonian $\mathscr{H}_0$ is a determinant formed from the N spin orbitals with lowest energies,

$$|\Psi_0\rangle = |\chi_1^{(0)} \cdots \chi_a^{(0)} \cdots \chi_N^{(0)}\rangle \tag{6.33}$$

We label the occupied (hole) spin orbitals by $a, b, c, \ldots$ and the unoccupied (particle) spin orbitals by $r, s, t, \ldots$. The wave function (6.33) is an eigenfunction of $\mathscr{H}_0$ with an eigenvalue equal to the sum of occupied orbital energies,

$$\mathscr{H}_0 |\Psi_0\rangle = \left(\sum_a \varepsilon_a^{(0)} \right) |\Psi_0\rangle$$

In the presence of the perturbation $\mathscr{V}$, $|\Psi_0\rangle$ is an approximation to the exact wave function, $|\Phi_0\rangle$. Within this approximation, the ground state energy of the N-electron system in the presence of the perturbation is

$$E_0 = \langle \Psi_0 | \mathscr{H} | \Psi_0 \rangle = \langle \Psi_0 | \mathscr{H}_0 + \mathscr{V} | \Psi_0 \rangle$$
$$= \sum_a \varepsilon_a^{(0)} + \sum_a \langle a|v|a \rangle = \sum_a \varepsilon_a^{(0)} + \sum_a v_{aa} \tag{6.34}$$

In the special case that the perturbation $\mathscr{V}$ is a sum of one-particle interactions, the total Hamiltonian of the system

$$\mathscr{H} = \mathscr{H}_0 + \mathscr{V} = \sum_i (h_0(i) + v(i)) = \sum_i h(i) \tag{6.35}$$

is also a sum of one-particle interactions. Therefore, we can find a set of exact spin orbitals and corresponding orbital energies,

$$h\chi_i = (h_0 + v)\chi_i = \varepsilon_i \chi_i \tag{6.36}$$

and construct the exact ground state wave function of the system from the N exact spin orbitals with the lowest energies,

$$|\Phi_0\rangle = |\chi_1 \cdots \chi_a \cdots \chi_N\rangle \tag{6.37}$$

This wave function is an eigenfunction of $\mathscr{H}$ with an eigenvalue equal to the sum of occupied orbital energies,

$$\mathscr{H}|\Phi_0\rangle = \left(\sum_a \varepsilon_a \right) |\Phi_0\rangle = \mathscr{E}_0 |\Phi_0\rangle$$

so that the exact energy of the system in the presence of the perturbation is

$$\mathscr{E}_0 = \sum_a \varepsilon_a \tag{6.38}$$

Suppose we wish to obtain a perturbation expansion for the exact energy $\mathscr{E}_0$,

$$\mathscr{E}_0 = E_0^{(0)} + E_0^{(1)} + E_0^{(2)} + \cdots$$

Since the exact energy can be written as a sum of occupied orbital energies, ε_a, we can simply find the perturbation expansion of each ε_a and then sum the result over all occupied spin orbitals. Since the general theory of Section 6.1 is applicable to one-particle zeroth-order Hamiltonians (i.e., h_0) and one-particle perturbations (i.e., v), we can immediately write

$$\varepsilon_a = \varepsilon_a^{(0)} + \langle a|v|a \rangle + \sum_i{}' \frac{\langle a|v|i \rangle \langle i|v|a \rangle}{\varepsilon_a^{(0)} - \varepsilon_i^{(0)}} + \cdots$$

$$= \varepsilon_a^{(0)} + v_{aa} + \sum_i{}' \frac{v_{ai}v_{ia}}{\varepsilon_a^{(0)} - \varepsilon_i^{(0)}} + \cdots \tag{6.39}$$

The sum over i can be divided into two parts, one involving the summation over all particle orbitals, and the other over all hole orbitals except a,

$$\varepsilon_a = \varepsilon_a^{(0)} + v_{aa} + \sum_r \frac{v_{ar}v_{ra}}{\varepsilon_a^{(0)} - \varepsilon_r^{(0)}} + \sum_{b \neq a} \frac{v_{ab}v_{ba}}{\varepsilon_a^{(0)} - \varepsilon_b^{(0)}} + \cdots \tag{6.40}$$

To obtain the perturbation expansion of the exact energy, we substitute (6.40) into (6.38) and find

$$\mathscr{E}_0 = \sum_a \varepsilon_a = \sum_a \varepsilon_a^{(0)} + \sum_a v_{aa} + \sum_{ar} \frac{v_{ar}v_{ra}}{\varepsilon_a^{(0)} - \varepsilon_r^{(0)}} + \sum_{\substack{ab \\ a \neq b}} \frac{v_{ab}v_{ba}}{\varepsilon_a^{(0)} - \varepsilon_b^{(0)}} + \cdots \tag{6.41}$$

The term

$$X = \sum_{\substack{ab \\ a \neq b}} \frac{v_{ab}v_{ba}}{\varepsilon_a^{(0)} - \varepsilon_b^{(0)}}$$

is equal to zero. To show this, we interchange the dummy indices a and b to obtain

$$X = \sum_{\substack{ab \\ a \neq b}} \frac{v_{ba}v_{ab}}{\varepsilon_b^{(0)} - \varepsilon_a^{(0)}} = -\sum_{\substack{ab \\ a \neq b}} \frac{v_{ab}v_{ba}}{\varepsilon_a^{(0)} - \varepsilon_b^{(0)}} = -X$$

and hence $X = 0$. Thus to second-order, we have

$$\mathscr{E}_0 = \sum_a \varepsilon_a^{(0)} + \sum_a v_{aa} + \sum_{ar} \frac{v_{ar}v_{ra}}{\varepsilon_a^{(0)} - \varepsilon_r^{(0)}} + \cdots \tag{6.42}$$

So that

$$E_0^{(0)} = \sum_a \varepsilon_a^{(0)} \tag{6.43a}$$

$$E_0^{(1)} = \sum_a v_{aa} \tag{6.43b}$$

$$E_0^{(2)} = \sum_{ar} \frac{v_{ar}v_{ra}}{\varepsilon_a^{(0)} - \varepsilon_r^{(0)}} \tag{6.43c}$$

Note that the energy of $|\Psi_0\rangle$ (see Eq. (6.34)) is equal to the sum of the zeroth- and first-order energies (i.e., $E_0 = E_0^{(0)} + E_0^{(1)}$).

The summations in Eq. (6.43) are over occupied (hole) and unoccupied (particle) spin orbitals. Since a matrix element $v_{ij} = \langle i|v|j\rangle$ is nonzero only if both spin orbitals i and j have the same spin, for a closed-shell system we can write these expressions as summations over spatial orbitals simply by multiplying by a factor of 2,

$$E_0^{(0)} = 2 \sum_a^{N/2} \varepsilon_a^{(0)} \tag{6.44a}$$

$$E_0^{(1)} = 2 \sum_a^{N/2} v_{aa} \tag{6.44b}$$

$$E_0^{(2)} = 2 \sum_{ar}^{N/2} \frac{v_{ar}v_{ra}}{\varepsilon_a^{(0)} - \varepsilon_r^{(0)}} \tag{6.44c}$$

where the upper limit of summation (i.e., $N/2$) is our shorthand notation indicating that all the summations are over spatial rather than spin orbitals.

Exercise 6.3 Derive

$$E_0^{(2)} = \sum_{ar} \frac{v_{ar}v_{ra}}{\varepsilon_a^{(0)} - \varepsilon_r^{(0)}}$$

starting with the general expression for the second-order energy (Eq. (6.12)) applied to an N-electron system,

$$E_0^{(2)} = \sum_n{}' \frac{|\langle\Psi_0| \sum_i v(i)|n\rangle|^2}{E_0^{(0)} - E_n^{(0)}}$$

where the sum runs over all states of the system except the ground state.
Hint: The states $|n\rangle$ must be single excitations of the type

$$|\Psi_a^r\rangle = |\chi_1^{(0)} \cdots \chi_{a-1}^{(0)}\chi_r^{(0)}\chi_{a+1}^{(0)} \cdots \chi_N^{(0)}\rangle$$

Exercise 6.4 Calculate the third-order energy $E_0^{(3)}$ using the general expression given in Eq. (6.15).

a. Show that

$$B_0^{(3)} = -E_0^{(1)} \sum_n{}' \frac{|\langle \Psi_0|\mathscr{V}|n\rangle|^2}{(E_0^{(0)} - E_n^{(0)})^2} = -\sum_{abr} \frac{v_{aa}v_{rb}v_{br}}{(\varepsilon_b^{(0)} - \varepsilon_r^{(0)})^2}.$$

b. Show that

$$A_0^{(3)} = \sum_{nm}{}' \frac{\langle \Psi_0|\mathscr{V}|n\rangle\langle n|\mathscr{V}|m\rangle\langle m|\mathscr{V}|\Psi_0\rangle}{(E_0^{(0)} - E_n^{(0)})(E_0^{(0)} - E_m^{(0)})} = \sum_{abrs} \frac{v_{ar}v_{sb}\langle \Psi_a^r|\mathscr{V}|\Psi_b^s\rangle}{(\varepsilon_a^{(0)} - \varepsilon_r^{(0)})(\varepsilon_b^{(0)} - \varepsilon_s^{(0)})}.$$

c. Show that

$$
\begin{aligned}
\langle \Psi_a^r|\mathscr{V}|\Psi_b^s\rangle &= v_{rs} & \text{if } a = b \quad r \neq s \\
&= -v_{ba} & \text{if } a \neq b \quad r = s \\
&= \sum_c v_{cc} - v_{aa} + v_{rr} & \text{if } a = b \quad r = s.
\end{aligned}
$$

and zero otherwise.

d. Finally, combine the two terms to obtain

$$E_0^{(3)} = A_0^{(3)} + B_0^{(3)} = \sum_{ars} \frac{v_{ar}v_{rs}v_{sa}}{(\varepsilon_a^{(0)} - \varepsilon_r^{(0)})(\varepsilon_a^{(0)} - \varepsilon_s^{(0)})} - \sum_{abr} \frac{v_{ra}v_{ab}v_{br}}{(\varepsilon_a^{(0)} - \varepsilon_r^{(0)})(\varepsilon_b^{(0)} - \varepsilon_r^{(0)})}.$$

e. Show that for a closed-shell system

$$E_0^{(3)} = 2 \sum_{ars}^{N/2} \frac{v_{ar}v_{rs}v_{sa}}{(\varepsilon_a^{(0)} - \varepsilon_r^{(0)})(\varepsilon_a^{(0)} - \varepsilon_s^{(0)})} - 2 \sum_{abr}^{N/2} \frac{v_{ra}v_{ab}v_{br}}{(\varepsilon_a^{(0)} - \varepsilon_r^{(0)})(\varepsilon_b^{(0)} - \varepsilon_r^{(0)})}.$$

In Subsection 5.3.2, we have used a variety of many-electron approaches to calculate the resonance energy of a cyclic polyene with N carbon atoms ($N = 2n = 4v + 2$, $v = 1, 2, \ldots$) in the framework of Hückel theory. As an illustration of the above formalism we now approach this problem using orbital perturbation theory. Recall that the resonance energy is defined as the difference between the exact total energy (as obtained by diagonalizing the Hückel matrix and adding up the occupied orbital energies) and the energy of $N/2 = n$ localized ethylenic units. The occupied (hole) and unoccupied (particle) orbitals of the ith unit are denoted by $|i\rangle$ and $|i^*\rangle$, respectively (see Eq. (5.137)). The total Hamiltonian is partitioned as

$$\mathscr{H} = \mathscr{H}_0 + \mathscr{V} = \sum_i h_0(i) + \sum_i v(i) \tag{6.45}$$

so that the ethylenic orbitals are eigenfunctions of h_0,

$$h_0|i\rangle = (\alpha + \beta)|i\rangle = \varepsilon_i^{(0)}|i\rangle \tag{6.46a}$$

$$h_0|i^*\rangle = (\alpha - \beta)|i^*\rangle = \varepsilon_{i^*}^{(0)}|i^*\rangle \qquad i = 1, 2, \ldots, n \tag{6.46b}$$

For future reference note that $\varepsilon_i^{(0)} - \varepsilon_{j*}^{(0)} = 2\beta$ independent of i and j. The nonzero matrix elements of the perturbation are

$$\langle i|v|(i \pm 1)^*\rangle = \pm\beta/2 \tag{6.47a}$$

$$\langle i|v|(i \pm 1)\rangle = \beta/2 \tag{6.47b}$$

$$\langle i^*|v|(i \pm 1)^*\rangle = -\beta/2 \tag{6.47c}$$

Since the polyene is cyclic, the zeroth ethylenic unit is the same as the nth while the $(n + 1)$th is just the first. In this model, the exact resonance energy of benzene is 2β, while asymptotically, the exact resonance energy is

$$\lim_{N \to \infty} E_R = (4/\pi - 1)N\beta = 0.2732N\beta \tag{6.48}$$

The matrix elements of the perturbation v, given in Eq. (6.47), and the zeroth-order orbital energies in Eq. (6.46) are all we need to calculate the perturbation energies. We begin with the second-order energy. Since we are dealing with a closed-shell system, the required expression is given by Eq. (6.44c),

$$E_0^{(2)} = 2 \sum_{ar}^{N/2} \frac{v_{ar}v_{ra}}{\varepsilon_a^{(0)} - \varepsilon_r^{(0)}} = 2 \sum_{ar}^{N/2} \frac{\langle a|v|r\rangle\langle r|v|a\rangle}{\varepsilon_a^{(0)} - \varepsilon_r^{(0)}} \tag{6.49}$$

Since the index a runs over all n occupied ethylenic orbitals $|i\rangle$ while the index r runs over all n unoccupied orbitals $|j^*\rangle$, we have

$$E_0^{(2)} = 2 \sum_{i=1}^{n} \sum_{j=1}^{n} \frac{\langle i|v|j^*\rangle\langle j^*|v|i\rangle}{\varepsilon_i^{(0)} - \varepsilon_{j*}^{(0)}} = \frac{1}{\beta} \sum_{i=1}^{n} \sum_{j=1}^{n} \langle i|v|j^*\rangle\langle j^*|v|i\rangle \tag{6.50}$$

where we have used the fact that the difference between orbital energies is always 2β. For fixed i, the summation over j is easily evaluated, since the matrix elements are nonzero only when $j = i \pm 1$ (see Eq. (6.47a)). Thus

$$E_0^{(2)} = \frac{1}{\beta} \sum_{i=1}^{n} \langle i|v|(i + 1)^*\rangle\langle (i + 1)^*|v|i\rangle + \langle i|v|(i - 1)^*\rangle\langle (i - 1)^*|v|i\rangle$$

$$= \frac{1}{\beta} \sum_{i=1}^{n} [(\beta/2)^2 + (-\beta/2)^2] = \frac{n\beta}{2} = \frac{N\beta}{4} = 0.25\, N\beta \tag{6.51}$$

For benzene the second-order resonance energy is 1.5β as compared to the exact value of 2β (i.e., 75%). For larger systems the agreement is even better; by comparing Eqs. (6.48) and (6.51), it can be seen that, for large N, the second-order energy approaches 91.5% of the exact result.

We now consider the above derivation from a slightly different point of view, which is extremely helpful in organizing the calculations of the higher order energies. To evaluate the summation over j, for fixed i, in Eq. (6.50) we note that orbital i can only interact with the orbitals $(i \pm 1)^*$. This can be represented pictorially as

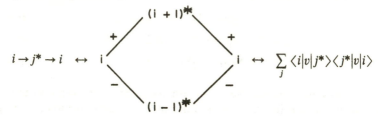

$$i \to j^* \to i \quad \leftrightarrow \quad i \quad \leftrightarrow \quad \sum_j \langle i|v|j^* \rangle \langle j^*|v|i \rangle$$

where the plus and minus signs indicate whether the matrix element between the two orbitals is $\pm \beta/2$. The sum over j can be evaluated as follows. Start at i on the left and go to i on the right by all possible paths. Assign to each path, a number corresponding to the product of the matrix elements encountered, and then add all these contributions. To illustrate, the value of the path $i \to (i + 1)^* \to i$ is $(+\beta/2)(+\beta/2) = \beta^2/4$, while the value of the path $i \to (i - 1)^* \to i$ is $(-\beta/2)(-\beta/2) = \beta^2/4$, so that the summation over j is just $\beta^2/2$ for each value of i in agreement with our previous results.

We now consider the calculation of the third-order energy. The appropriate expression for a closed-shell system is given in Exercise 6.4(e),

$$E_0^{(3)} = 2 \sum_{ars}^{N/2} \frac{v_{ar} v_{rs} v_{sa}}{(\varepsilon_a^{(0)} - \varepsilon_r^{(0)})(\varepsilon_a^{(0)} - \varepsilon_s^{(0)})} - 2 \sum_{abr}^{N/2} \frac{v_{ra} v_{ab} v_{br}}{(\varepsilon_a^{(0)} - \varepsilon_r^{(0)})(\varepsilon_b^{(0)} - \varepsilon_r^{(0)})} \quad (6.52)$$

Proceeding in the same way as we did for $E_0^{(2)}$, the first term of this expression is

$$\frac{2}{(2\beta)^2} \sum_{i=1}^{n} \sum_{j=1}^{n} \sum_{k=1}^{n} \langle i|v|j^* \rangle \langle j^*|v|k^* \rangle \langle k^*|v|i \rangle \quad (6.53)$$

To evaluate the summations over j and k, we use the pictorial representation discussed above,

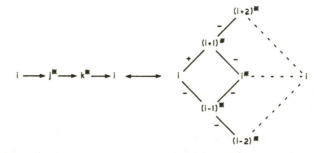

Note that, since i cannot interact with i^* or $(i + 2)^*$ or $(i - 2)^*$ the value of this sum must be zero. A similar argument holds for the second term in Eq. (6.52), so we conclude that the third-order resonance energy is zero.

Actually, we have been somewhat hasty. The above result is true in all cases except benzene. Because of the cyclic nature of the problem, in this case the orbitals $(i \pm 2)^*$ are the same as $(i \mp 1)^*$. Thus the pictorial representation of the sum over j and k with i, say, equal to 1 is

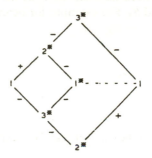

So there are two paths: 1) $1 \to 2^* \to 3^* \to 1$ with value $(\beta/2)(-\beta/2)(-\beta/2) = \beta^3/8$ and 2) $1 \to 3^* \to 2^* \to 1$ with value $(-\beta/2)(-\beta/2)(\beta/2) = \beta^3/8$. Using these results in (6.53), we have for the first term in $E_0^{(3)}$

$$\frac{2}{(2\beta)^2} \sum_{i=1}^{3} [(\beta^3/8) + (\beta^3/8)] = \frac{3\beta}{8}$$

In an entirely analogous way one can show that the second term in Eq. (6.52) is also equal to $3\beta/8$, so that the total third-order energy for benzene is

$$E_0^{(3)}(\text{benzene}) = \frac{3\beta}{4} \tag{6.54}$$

Thus the resonance energy of benzene calculated up to third-order is 2.25β (i.e., 113% of the exact value). In Exercise 6.7, the fourth-order energy

$$E_0^{(4)} = \frac{N\beta}{64} \tag{6.55}$$

is obtained using diagrammatic techniques. Thus the resonance energy of benzene calculated up to fourth order is 2.34β (i.e., 117% of the exact value). Note that the perturbation expansion does not appear to be converging. In fact, it does converge but the convergence is slow and oscillatory. Exercise 6.6 explores this point further. It is interesting to note that the perturbation expansion works better for larger systems. The energy up to fourth order is $0.2656N\beta$ (i.e., $(\frac{1}{4} + \frac{1}{64})N\beta)$) as compared with the asymptotic exact value of $0.2732N\beta$ (i.e., 97% of the exact value).

Exercise 6.5 Show that the second term in Eq. (6.52) is equal to $\frac{3}{8}\beta$ for benzene.

Exercise 6.6 Consider a cyclic polyene with $N = 4v + 2$, $v = 1, 2, \ldots$ carbons. Instead of assuming that all the bonds are identical, suppose they alternate in length. In the context of Hückel theory this means that the resonance integrals between adjacent carbons are not all equal to β but alternate between β_1 and β_2. For example, for benzene we have

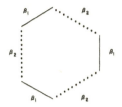

Now it can be shown that the exact energy for a cyclic polyene of this type is

$$\mathscr{E}_0 = N\alpha - 2 \sum_{j=-v}^{v} \left(\beta_1^2 + \beta_2^2 + 2\beta_1\beta_2 \cos \frac{2j\pi}{2v+1} \right)^{1/2}$$

(see for example, L. Salem, *Molecular Orbital Theory of Conjugated Systems*, Benjamin, New York, 1966, pp. 498–500). Note that when $\beta_1 = \beta_2 = \beta$, since $2\cos^2\theta = (1 + \cos 2\theta)$ and β is negative, we recover

$$\mathscr{E}_0 = N\alpha + 4\beta \sum_{j=-v}^{v} \cos \frac{j\pi}{2v+1}$$

which is the result quoted in Subsection 5.3.2. Also note that when $\beta_1 = \beta$ but $\beta_2 = 0$, we have

$$E_0 = N\alpha + N\beta$$

which is just the total energy of the polyene using the localized ethylenic description. The purpose of this exercise is to obtain the perturbation expansion for the resonance energy by expanding the exact energy in powers of β_2/β_1.

a. Show that for benzene ($v = 1$) the exact ground state energy in the alternating short and long bond model is

$$\mathscr{E}_0 = 6\alpha + 2(\beta_1 + \beta_2) - 4(\beta_1^2 + \beta_2^2 - \beta_1\beta_2)^{1/2}$$

Do this first by using the general expression and then by setting up the Hückel matrix, diagonalizing it and adding up the occupied orbital energies. Note that when $\beta_1 = \beta_2 = \beta$ we recover our old result, $6\alpha + 8\beta$.

b. Setting $\beta_1 = \beta$ and $\beta_2/\beta_1 = x$ show that the resonance energy of benzene can be written as

$$E_R = 4\beta(\tfrac{1}{2}x - 1 + (1 + x^2 - x)^{1/2})$$

Note that when $x = 0$, $E_R = 0$ and when $x = 1$, $E_R = 2\beta$ which is exact.

c. Using the relation

$$(1 + y)^{1/2} = 1 + \tfrac{1}{2}y - \tfrac{1}{8}y^2 + \tfrac{1}{16}y^3 - \tfrac{5}{128}y^4 + \cdots \qquad |y| < 1$$

expand E_R to fourth order in x and thus show that

$$E_R = \beta(\tfrac{3}{2}x^2 + \tfrac{3}{4}x^3 + \tfrac{3}{32}x^4 + \cdots)$$

Identifying the coefficient of x^n with the nth-order perturbation result (i.e., $E_0^{(n)}$), we have

$$E_0^{(2)} = \tfrac{3}{2}\beta$$

$$E_0^{(3)} = \tfrac{3}{4}\beta$$

$$E_0^{(4)} = \tfrac{3}{32}\beta$$

Note that $E_0^{(2)}$ and $E_0^{(3)}$ agree with our previously calculated values. This derivation provides some insight into the poor convergence of the perturbation expansion of the resonance energy of benzene. Basically, the perturbation expansion converges rapidly when x is small. However, for our problem x is equal to unity.

The resonance energy calculated up to Mth-order as a function of M is shown below. Note the oscillatory convergence towards the exact value of 2β. The method used above to obtain $E_0^{(n)}$ for $n = 2, 3, 4$ becomes extremely laborious for larger n. The results below were calculated by first showing that $E_0^{(n)} = 4\beta C_n^{-1/2}(\tfrac{1}{2})$, where $C_n^{-1/2}(x)$ is a Gegenbauer polynomial of degree n and order $-\tfrac{1}{2}$, and then using the recursive properties of these polynomials, to show that

$$(n + 1)E_0^{(n+1)} = (n - \tfrac{1}{2})E_0^{(n)} - (n - 2)E_0^{(n-1)}$$

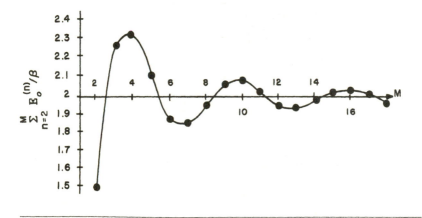

*6.4 DIAGRAMMATIC REPRESENTATION OF ORBITAL PERTURBATION THEORY

In Section 6.2, we introduced a completely general diagrammatic representation of RS perturbation theory. To adapt this to handle orbital perturbations we take the downward and upward lines to represent hole and
particle spin orbitals, respectively, and the dots to correspond to the one-
particle perturbation v. Then we draw the same set of diagrams as before,
labeling the hole lines by indices $a, b, \ldots$ and the particle lines by indices
$r, s, \ldots$. Thus we have

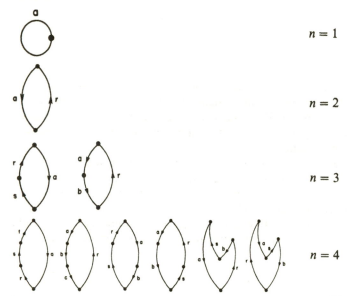

$$n = 1$$

$$n = 2$$

$$n = 3$$

$$n = 4$$

We evaluate these diagrams using the same rules as before (with $E_i^{(0)} \to \varepsilon_i^{(0)}$),
except that we generalize rule 4 to read:

4'. *Sum the expression over all particle and hole indices.*

Using these rules, we have

$$E_0^{(1)} = \begin{matrix} a \\ \bigcirc \end{matrix} = \sum_a v_{aa} \tag{6.56a}$$

$$E_0^{(2)} = {}_a\!\bigcirc\!{}_r = \sum_{ar} \frac{v_{ar}v_{ra}}{\varepsilon_a^{(0)} - \varepsilon_r^{(0)}} \tag{6.56b}$$

Note that these are the same as the results in Eqs. (6.43b, c). For $E_0^{(3)}$ we find

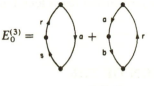

$$= \sum_{ars} \frac{v_{as} v_{sr} v_{ra}}{(\varepsilon_a^{(0)} - \varepsilon_s^{(0)})(\varepsilon_a^{(0)} - \varepsilon_r^{(0)})} - \sum_{abr} \frac{v_{ra} v_{ab} v_{br}}{(\varepsilon_a^{(0)} - \varepsilon_r^{(0)})(\varepsilon_b^{(0)} - \varepsilon_r^{(0)})} \quad (6.56c)$$

which is identical to the result obtained in Exercise 6.4. Finally, for a closed-shell system, since the matrix element v_{ij} is nonzero only if the spin orbitals i and j have the same spin, we can convert the sums over spin orbitals to sums over spatial orbitals by multiplying the value of each diagram by a factor of 2, i.e.,

$$\sum^N = 2 \sum^{N/2} \quad (6.57)$$

We will see in Section 6.7 that, in general, we must multiply the value of the diagram by 2^l, where l is the number of closed loops, to make this conversion.

Exercise 6.7 Find the fourth-order energy for a closed-shell cyclic polyene.

a. Show that

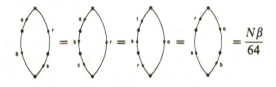

and

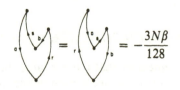

so that

$$E_0^{(4)} = \frac{N\beta}{64}$$

Thus the resonance energy calculated for a cyclic polyene with $N > 6$ up to fourth order is $(1/4 + 1/64)N\beta = 0.2656N\beta$, which compares very favorably with the asymptotically exact value of $0.2732N\beta$ (i.e., 97%).

b. For benzene, show that the diagrammatic result for the fourth-order energy agrees with the independently calculated result found in Exercise 6.6.

6.5 PERTURBATION EXPANSION OF THE CORRELATION ENERGY

In this section, we consider the problem of improving the Hartree-Fock energy of an N-electron system by means of perturbation theory. In other words, we wish to obtain a perturbation expansion for the correlation energy. We partition the Hamiltonian as

$$\mathscr{H} = \mathscr{H}_0 + \mathscr{V} \tag{6.58}$$

where $\mathscr{H}_0$ is the Hartree-Fock Hamiltonian,

$$\mathscr{H}_0 = \sum_i f(i) = \sum_i \left[h(i) + v^{\mathrm{HF}}(i) \right] \tag{6.59}$$

and

$$\mathscr{V} = \sum_{i<j} r_{ij}^{-1} - \mathscr{V}^{\mathrm{HF}} = \sum_{i<j} r_{ij}^{-1} - \sum_i v^{\mathrm{HF}}(i) \tag{6.60}$$

The use of the above partitioning of the Hamiltonian, along with the general expressions of RS perturbation theory, is sometimes called Møller-Plesset perturbation theory.

In this section, we will use the physicists' notation for two-electron integrals rather than the chemists' notation, which we used extensively in Chapter 3. We do this not out of perversity or even laziness but because almost all the literature in this area uses this notation, and we believe that one should develop equal facility with both notations. Recall that in the physicists' notation

$$\int d\mathbf{x}_1 \, d\mathbf{x}_2 \, \chi_i^*(\mathbf{x}_1)\chi_j^*(\mathbf{x}_2) r_{12}^{-1}\chi_k(\mathbf{x}_1)\chi_l(\mathbf{x}_2) = \langle ij|kl \rangle \tag{6.61}$$

It is important to remember that i and k label the spin orbitals which are functions of the coordinates of electron-one while j and l refer to spin orbitals which depend on the coordinates of electron-two, i.e.,

$$\langle i\overset{\frown}{j}|k\overset{\frown}{l} \rangle$$

Recall that the antisymmetrized two-electron integral is defined as

$$\langle ij||kl \rangle = \langle ij|kl \rangle - \langle ij|lk \rangle \tag{6.62}$$

Using this notation, we have

$$\left\langle \Psi_0 \left| \sum_{i<j} r_{ij}^{-1} \right| \Psi_{ab}^{rs} \right\rangle = \langle ab||rs\rangle \qquad (6.63)$$

and

$$v^{HF}(1)\chi_j(\mathbf{x}_1) = \sum_b \langle b|r_{12}^{-1}|b\rangle\chi_j(\mathbf{x}_1) - \sum_b \langle b|r_{12}^{-1}|j\rangle\chi_b(\mathbf{x}_1) \qquad (6.64)$$

Thus

$$\langle i|v^{HF}|j\rangle = v_{ij}^{HF} = \sum_b \langle ib|jb\rangle - \langle ib|bj\rangle = \sum_b \langle ib||jb\rangle \qquad (6.65)$$

The Hartree-Fock wave function $|\Psi_0\rangle$ is an eigenfunction of $\mathcal{H}_0$,

$$\mathcal{H}_0|\Psi_0\rangle = E_0^{(0)}|\Psi_0\rangle \qquad (6.66)$$

with the eigenvalue

$$E_0^{(0)} = \sum_a \varepsilon_a \qquad (6.67)$$

which is just the zeroth-order perturbation energy. The first-order energy is

$$\begin{aligned}
E_0^{(1)} &= \langle\Psi_0|\mathcal{V}|\Psi_0\rangle \\
&= \langle\Psi_0|\sum_{i<j} r_{ij}^{-1}|\Psi_0\rangle - \langle\Psi_0|\sum_i v^{HF}(i)|\Psi_0\rangle \\
&= \frac{1}{2}\sum_{ab}\langle ab||ab\rangle - \sum_a \langle a|v^{HF}|a\rangle \\
&= -\frac{1}{2}\sum_{ab}\langle ab||ab\rangle \qquad (6.68)
\end{aligned}$$

The Hartree-Fock energy is the sum of the zeroth and first-order energies,

$$E_0 = E_0^{(0)} + E_0^{(1)} = \sum_a \varepsilon_a - \frac{1}{2}\sum_{ab}\langle ab||ab\rangle \qquad (6.69)$$

Thus the first correction to the Hartree-Fock energy occurs in the second order of perturbation theory.

The general result for the second-order energy, derived in Section 6.1, is

$$E_0^{(2)} = \sum_n{}' \frac{|\langle 0|\mathcal{V}|n\rangle|^2}{E_0^{(0)} - E_n^{(0)}} \qquad (6.70)$$

where the summation runs over all but the ground state of the system. Clearly, we take $|0\rangle = |\Psi_0\rangle$ but what about $|n\rangle$? These states cannot be single excitations since

$$\begin{aligned}
\langle\Psi_0|\mathcal{V}|\Psi_a^r\rangle &= \langle\Psi_0|\mathcal{H} - \mathcal{H}_0|\Psi_a^r\rangle \\
&= \langle\Psi_0|\mathcal{H}|\Psi_a^r\rangle - f_{ar} = 0
\end{aligned}$$

The first term vanishes because of Brillouin's theorem and the second because the spin orbitals are eigenfunctions of the Fock operator. In addition, triply excited states do not mix with $|\Psi_0\rangle$ because of the two-particle nature of the perturbation. Therefore, we are left with double excitations of the form $|\Psi_{ab}^{rs}\rangle$. Since

$$\mathscr{H}_0|\Psi_{ab}^{rs}\rangle = (E_0^{(0)} - (\varepsilon_a + \varepsilon_b - \varepsilon_r - \varepsilon_s))|\Psi_{ab}^{rs}\rangle$$

and because we can sum over all possible double excitations by summing over all a and all b greater than a and over all r and all s greater than r, the second-order energy is

$$E_0^{(2)} = \sum_{\substack{a<b \\ r<s}} \frac{|\langle\Psi_0|\sum_{i<j} r_{ij}^{-1}|\Psi_{ab}^{rs}\rangle|^2}{\varepsilon_a + \varepsilon_b - \varepsilon_r - \varepsilon_s} = \sum_{\substack{a<b \\ r<s}} \frac{|\langle ab||rs\rangle|^2}{\varepsilon_a + \varepsilon_b - \varepsilon_r - \varepsilon_s} \quad (6.71)$$

Note that the second-order energy can be expressed as a sum of contributions from each pair of electrons in occupied orbitals,

$$E_0^{(2)} = \sum_{a<b} e_{ab}^{FO}$$

where

$$e_{ab}^{FO} = \sum_{r<s} \frac{|\langle ab||rs\rangle|^2}{\varepsilon_a + \varepsilon_b - \varepsilon_r - \varepsilon_s}$$

We have seen in Chapter 5 that e_{ab}^{FO} is the first-order pair energy. Thus at the level of first-order pairs, pair theory gives the same correlation energy as second-order perturbation theory.

The expression for the second-order energy can be transformed into a number of other useful forms. Since the quantity being summed is symmetric in a and b and r and s, and vanishes when $a = b$ or $r = s$, we can write

$$E_0^{(2)} = \frac{1}{4} \sum_{abrs} \frac{|\langle ab||rs\rangle|^2}{\varepsilon_a + \varepsilon_b - \varepsilon_r - \varepsilon_s} \quad (6.72)$$

Furthermore, in terms of regular two-electron integrals, the second-order energy is

$$E_0^{(2)} = \frac{1}{2} \sum_{abrs} \frac{\langle ab|rs\rangle\langle rs|ab\rangle}{\varepsilon_a + \varepsilon_b - \varepsilon_r - \varepsilon_s} - \frac{1}{2} \sum_{abrs} \frac{\langle ab|rs\rangle\langle rs|ba\rangle}{\varepsilon_a + \varepsilon_b - \varepsilon_r - \varepsilon_s} \quad (6.73)$$

Finally, for a closed-shell system, the second-order energy can be written in terms of sums over spatial orbitals as

$$E_0^{(2)} = 2 \sum_{abrs}^{N/2} \frac{\langle ab|rs\rangle\langle rs|ab\rangle}{\varepsilon_a + \varepsilon_b - \varepsilon_r - \varepsilon_s} - \sum_{abrs}^{N/2} \frac{\langle ab|rs\rangle\langle rs|ba\rangle}{\varepsilon_a + \varepsilon_b - \varepsilon_r - \varepsilon_s} \quad (6.74)$$

Exercise 6.8 Derive Eqs. (6.73) and (6.74) starting with Eq. (6.72).

In a similar, but much more laborious way, starting with Eq. (6.15) it can be shown that the third-order energy is

$$E_0^{(3)} = \frac{1}{8} \sum_{abcdrs} \frac{\langle ab||rs\rangle \langle cd||ab\rangle \langle rs||cd\rangle}{(\varepsilon_a + \varepsilon_b - \varepsilon_r - \varepsilon_s)(\varepsilon_c + \varepsilon_d - \varepsilon_r - \varepsilon_s)}$$

$$+ \frac{1}{8} \sum_{abrstu} \frac{\langle ab||rs\rangle \langle rs||tu\rangle \langle tu||ab\rangle}{(\varepsilon_a + \varepsilon_b - \varepsilon_r - \varepsilon_s)(\varepsilon_a + \varepsilon_b - \varepsilon_t - \varepsilon_u)}$$

$$+ \sum_{abcrst} \frac{\langle ab||rs\rangle \langle cs||tb\rangle \langle rt||ac\rangle}{(\varepsilon_a + \varepsilon_b - \varepsilon_r - \varepsilon_s)(\varepsilon_a + \varepsilon_c - \varepsilon_r - \varepsilon_t)} \qquad (6.75)$$

As an illustration of the above formalism, we now calculate the second and third-order energies of minimal basis H_2. In Chapter 4, we showed that the exact correlation energy of H_2 in the minimal basis set is

$$E_{corr} = \Delta - (\Delta^2 + K_{12}^2)^{1/2} \qquad (6.76)$$

where

$$2\Delta = 2(\varepsilon_2 - \varepsilon_1) + J_{11} + J_{22} - 4J_{12} + 2K_{12}$$
$$= 2(\varepsilon_2 - \varepsilon_1) + \langle 11|11\rangle + \langle 22|22\rangle - 4\langle 12|12\rangle + 2\langle 11|22\rangle$$

If we expand the expression for the correlation energy in a Taylor series in the two-electron integrals up to third order, we find

$$E_{corr} = E_0^{(2)} + E_0^{(3)} + \cdots$$

where

$$E_0^{(2)} = \frac{K_{12}^2}{2(\varepsilon_1 - \varepsilon_2)} \qquad (6.77)$$

and

$$E_0^{(3)} = \frac{K_{12}^2(J_{11} + J_{22} - 4J_{12} + 2K_{12})}{4(\varepsilon_1 - \varepsilon_2)^2} \qquad (6.78)$$

We will now show that the second-order energy in Eq. (6.77) is a special case of the general expression given by Eq. (6.74). Since we have but a single hole orbital, $a = b = 1$. Similarly, $r = s = 2$ so that Eq. (6.74) becomes

$$E_0^{(2)} = 2\frac{\langle 11|22\rangle \langle 22|11\rangle}{2(\varepsilon_1 - \varepsilon_2)} - \frac{\langle 11|22\rangle \langle 22|11\rangle}{2(\varepsilon_1 - \varepsilon_2)}$$

$$= \frac{|\langle 11|22\rangle|^2}{2(\varepsilon_1 - \varepsilon_2)} = \frac{K_{12}^2}{2(\varepsilon_1 - \varepsilon_2)}$$

In Section 6.7.2, we will show that the general expression for the third-order energy, given in Eq. (6.75), can be used to obtain the result given in Eqs. (6.78).

Exercise 6.9 Derive Eqs. (6.77) and (6.78) from (6.76).

6.6 THE N-DEPENDENCE OF THE RS PERTURBATION EXPANSION

In the introduction to this chapter, we mentioned that Brueckner was the first to investigate the applicability of the RS perturbation expansion of infinite (macroscopic) systems. He was able to show, by a careful examination of the algebraic expressions that appear in various orders, that $E_0^{(n)}$ for $n = 0, 1, \ldots, 6$ was indeed proportional to the number of particles. He was, however, unable to prove this in general (i.e., for $n = 7, 8, \ldots, \infty$). Here we present a simple illustration of Brueckner's analysis. In Subsection 6.7.2 we shall discuss Goldstone's linked cluster theorem which, using a diagrammatic representation of RS perturbation theory, is proof of Brueckner's conjecture that RS perturbation theory is satisfactory in all orders. We consider a supermolecule consisting of N noninteracting minimal basis H_2 molecules. We will show, using the general expressions derived in Section 6.1, that the first-, second-, and third-order energies of the supermolecule are simply N times the corresponding results for a single molecule. This is precisely the model we used to show, in Chapter 4, that the DCI result for the correlation energy was proportional to $N^{1/2}$ in the limit of large N. Recall that we label the orbitals of the supermolecule as:

$$\underline{2_1} \quad \underline{2_2} \quad \underline{2_3} \quad \cdots \quad \underline{2_N} \quad \varepsilon_2$$

$$\underline{1_1} \quad \underline{1_2} \quad \underline{1_3} \quad \quad \underline{1_N} \quad \varepsilon_1$$

and that all two-electron integrals involving orbitals from different units are zero. The Hartree-Fock wave function for this system is

$$|\Psi_0\rangle = |1_1 \bar{1}_1 1_2 \bar{1}_2 \cdots 1_N \bar{1}_N\rangle \tag{6.79}$$

If $\mathscr{H}_0$ is the Hartree-Fock Hamiltonian as in the previous section, the zeroth- and first-order energies are

$$E_0^{(0)} = \langle \Psi_0 | \mathscr{H}_0 | \Psi_0 \rangle = 2 \sum_{i=1}^{N} \langle 1_i | f | 1_i \rangle = 2N\varepsilon_1 \tag{6.80a}$$

and

$$E_0^{(1)} = \langle \Psi_0 | \mathscr{V} | \Psi_0 \rangle = -\sum_{i=1}^{N} \langle 1_i 1_i | 1_i 1_i \rangle = -N J_{11} \tag{6.80b}$$

The Hartree-Fock energy of the supermolecule,

$$E_0 = \langle \Psi_0 | \mathscr{H}_0 + \mathscr{V} | \Psi_0 \rangle = E_0^{(0)} + E_0^{(1)} = N(2\varepsilon_1 - J_{11}) \tag{6.81}$$

is indeed simply N times the Hartree-Fock energy of one subunit. The general expression for the second-order energy (Eq. (6.12)) is

$$E_0^{(2)} = \sum_n{}' \frac{|\langle 0|\mathscr{V}|n\rangle|^2}{E_0^{(0)} - E_n^{(0)}} \tag{6.82}$$

Clearly, $|0\rangle = |\Psi_0\rangle$ and the state $|n\rangle$ must be a double excitation of the type $|\Psi_{1_i\bar{1}_i}^{2_i\bar{2}_i}\rangle$. For these excitations

$$E_0^{(0)} - E_n^{(0)} = 2(\varepsilon_1 - \varepsilon_2) \tag{6.83a}$$

$$\langle \Psi_0|\mathscr{V}|\Psi_{1_i\bar{1}_i}^{2_i\bar{2}_i}\rangle = \langle 1_i\bar{1}_i|2_i\bar{2}_i\rangle - \langle 1_i\bar{1}_i|\bar{2}_i2_i\rangle = \langle 11|22\rangle = K_{12} \tag{6.83b}$$

and the summation over n can be replaced by a summation over i, so that

$$E_0^{(2)} = \sum_{i=1}^{N} \frac{|\langle\Psi_0|\mathscr{V}|\Psi_{1_i\bar{1}_i}^{2_i\bar{2}_i}\rangle|^2}{2(\varepsilon_1 - \varepsilon_2)} = \frac{NK_{12}^2}{2(\varepsilon_1 - \varepsilon_2)} \tag{6.84}$$

which again is just N times the second-order energy of one unit.

The general expression for the third-order energy (Eq. (6.15)) is

$$E_0^{(3)} = A_0^{(3)} + B_0^{(3)} \tag{6.85}$$

where

$$A_0^{(3)} = \sum_n{}' \sum_m{}' \frac{\langle 0|\mathscr{V}|n\rangle\langle n|\mathscr{V}|m\rangle\langle m|\mathscr{V}|0\rangle}{(E_0^{(0)} - E_n^{(0)})(E_0^{(0)} - E_m^{(0)})} \tag{6.86}$$

and

$$B_0^{(3)} = -E_0^{(1)} \sum_n{}' \frac{|\langle 0|\mathscr{V}|n\rangle|^2}{(E_0^{(0)} - E_n^{(0)})^2} \tag{6.87}$$

At first glance, the third-order energy does not appear to have the correct N dependence, since $B_0^{(3)}$ is proportional to N^2:

$$B_0^{(3)} = -(-NJ_{11}) \sum_{i=1}^{N} \frac{K_{12}^2}{(2\varepsilon_1 - 2\varepsilon_2)^2} = \frac{N^2 J_{11} K_{12}^2}{4(\varepsilon_1 - \varepsilon_2)^2} \tag{6.88}$$

where we have used Eqs. (6.80b) and (6.83a,b). If the third-order energy is to be proportional to N, this term must be cancelled by a part of $A_0^{(3)}$. This is just the type of cancellation that Brueckner found, as discussed in the introduction to this chapter. Let us now examine $A_0^{(3)}$ more closely. It is clear that both $|n\rangle$ and $|m\rangle$ must be states of the type $|\Psi_{1_i\bar{1}_i}^{2_i\bar{2}_i}\rangle$ so that

$$A_0^{(3)} = \sum_{i=1}^{N} \frac{\langle\Psi_{1_i\bar{1}_i}^{2_i\bar{2}_i}|\mathscr{V}|\Psi_{1_i\bar{1}_i}^{2_i\bar{2}_i}\rangle K_{12}^2}{4(\varepsilon_1 - \varepsilon_2)^2} \tag{6.89}$$

where only the diagonal element remains since two-electron integrals involving different units are zero. It can be shown that

$$\langle \Psi^{2\bar{2}}_{1\bar{1}}|\mathscr{V}|\Psi^{2\bar{2}}_{1\bar{1}}\rangle = \langle \Psi^{2\bar{2}}_{1\bar{1}}|\mathscr{H} - \mathscr{H}_0|\Psi^{2\bar{2}}_{1\bar{1}}\rangle$$
$$= -NJ_{11} + J_{11} + J_{22} - 4J_{12} + 2K_{12} \qquad (6.90)$$

so we have

$$A^{(3)}_0 = -\frac{N^2 J_{11} K^2_{12}}{4(\varepsilon_1 - \varepsilon_2)^2} + \frac{NK^2_{12}(J_{11} + J_{22} - 4J_{12} + 2K_{12})}{4(\varepsilon_1 - \varepsilon_2)^2} \qquad (6.91)$$

Thus the N^2 terms do indeed cancel, leaving us with

$$E^{(3)}_0 = A^{(3)}_0 + B^{(3)}_0 = \frac{NK^2_{12}(J_{11} + J_{22} - 4J_{12} + 2K_{12})}{4(\varepsilon_1 - \varepsilon_2)^2} \qquad (6.92)$$

As with $E^{(1)}_0$ and $E^{(2)}_0$, this is just N times the third-order energy of a single H_2 molecule (Eq. (6.78)). Recall that Eq. (6.78) was obtained by expanding the exact correlation energy within the basis in a Taylor series, so that the equivalence of the expressions derived in different ways provides a consistency check. Although this example is by no means a proof, we hope it will inspire some confidence in the statement that RS perturbation theory—in contrast to DCI—yields an approximation to the correlation energy which is size consistent (i.e., has the correct N-dependence).

Exercise 6.10 Derive Eqs. (6.80b) and (6.90).

---

*6.7 DIAGRAMMATIC REPRESENTATION OF THE PERTURBATION EXPANSION OF THE CORRELATION ENERGY

We now introduce a diagrammatic representation of the nth-order energy. Our zeroth-order wave function is taken to be the Hartree-Fock function. Although it is possible to derive a perturbation expansion starting with any single determinantal wave function, the Hartree-Fock description is the most convenient starting point because of Brillouin's theorem. The fact that single excitations do not mix with the Hartree-Fock ground state considerably simplifies the structure of the perturbation expansion. The rules we give for constructing the diagrams will be correct only for Hartree-Fock perturbation theory. Although we will not "derive" the rules for constructing and evaluating the diagrams, we hope that the formalism will appear as a "natural" generalization of our previous results.

6.7.1 Hugenholtz Diagrams

As before, we represent an interaction by a dot and a hole or a particle state by a line with an arrow pointing down or up, respectively. Looking back to

Eq. (6.72), we note that the matrix elements, which appear in the numerator of the second-order energy, now have four indices instead of two. This is an immediate result of the two-particle nature of the perturbation. Therefore, it is reasonable to expect that the diagrams which represent Hartree-Fock perturbation theory have dots with *two* lines going in and *two* lines coming out. This is in fact the case. To obtain the diagrams that contribute to the nth-order energy we connect n vertically ordered dots in all possible ways, such that

1. Each dot has four lines emanating from it.
2. Each diagram is linked.
3. Each diagram is distinct.
4. Diagrams containing more than one dot do not have lines which start and end at the same point. That is, they contain no elements of the type

Requirements (2) and (3) are precisely the same as in Section 6.2, and requirement (4) is a "nonobvious" consequence of Brillouin's theorem. Just as in Section 6.2, the diagrams can be distorted in any way we please as long as the vertical arrangement of the dots is not altered. Using these rules, we find

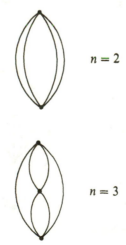

$n = 2$

$n = 3$

There are, however, twelve unique diagrams with four dots as shown in Fig. 6.1. Just as before, we label all the lines with arrows pointing up or down, in all possible ways such that each dot has two lines going into it and two lines coming out of it. In this way, we find only a single second-order

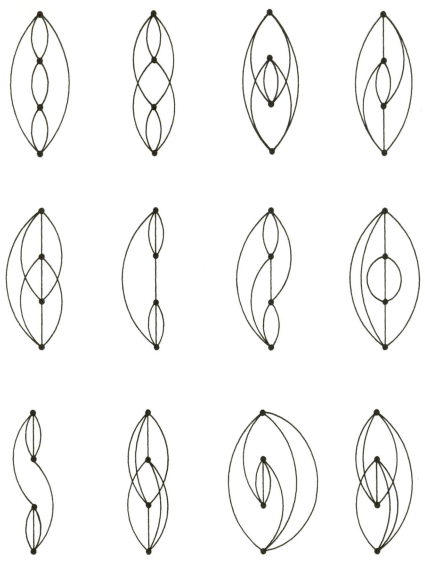

Figure 6.1 Unlabeled fourth-order diagrams.

diagram but three third-order diagrams,

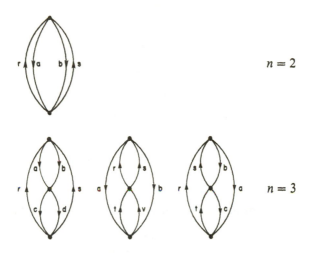

$$n = 2$$

$$n = 3$$

These diagrams are called *Hugenholtz* diagrams.

Fortunately, the rules for translating them into algebraic expressions are virtually identical to our previous rules:

H1. *Each dot contributes an antisymmetrized matrix element* $\langle$*label*-1 *in,* label-2 in$|$$|$label-1 out, label-2 out$\rangle$ *to the numerator. The particular labeling 1 and 2 is arbitrary.*

H2. *Each pair of adjacent dots contributes the denominator factor*

$$\sum \varepsilon_{\text{holes}} - \sum \varepsilon_{\text{particles}}$$

where the sums run over the labels of all hole and particle lines crossing an imaginary horizontal line separating the two adjacent dots.

H3. *The overall sign of the expression is* $(-)^{h+l}$*, where h and l are the number of hole lines and closed loops, respectively. The number of closed loops cannot be determined by looking at the diagram alone. We will give a prescription below as how to find it from the string of matrix elements written down using rule (H1).*

H4. *Sum the expression over all particle and hole indices.*

H5. *Multiply the expression by a weight factor* 2^{-k}*, where k is the number of pairs of equivalent lines in the diagram. Two lines are equivalent when both start and end at the same dot and both go in the same direction.*

You will note that only rule (H5) is substantially different from our previous rules. The only tricky part in applying these rules is the determination of the number of closed loops. How this is done is best illustrated by the following example. Consider the second-order diagram:

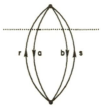

It clearly has two hole lines and two pairs of equivalent lines (r, s) and (a, b). So we can immediately write

Case A:
$$\left(\frac{1}{2}\right)^2 (-)^{2+l_A} \sum_{abrs} \frac{\langle ab||rs\rangle\langle rs||ab\rangle}{\varepsilon_a + \varepsilon_b - \varepsilon_r - \varepsilon_s}$$

Since we said that the order of the two in or two out labels in the antisymmetrized matrix element does not matter, we could equally have written

Case B:
$$\left(\frac{1}{2}\right)^2 (-)^{2+l_B} \sum_{abrs} \frac{\langle ab||rs\rangle\langle sr||ab\rangle}{\varepsilon_a + \varepsilon_b - \varepsilon_r - \varepsilon_s}$$

It is clear that if our rules are any good, these two expressions must be equal. This responsibility belongs to the l factors. The number of closed loops can be found only after we have written down a string of matrix elements. To determine l, we select a particular label and follow it through the matrix elements using the rule

$$\langle i\overrightarrow{j}||\overrightarrow{k}l\rangle \quad \text{and} \quad \langle i\overrightarrow{j}||k\overrightarrow{l}\rangle$$

until we reach the same label. The number of times we have to do this, before we exhaust all the labels, is l. Consider Case A,

$$\langle a\overset{1}{b}||\overset{2}{r}s\rangle\langle rs||ab\rangle \quad : \quad a \overset{1}{\to} r \overset{2}{\to} a$$

$$\langle ab||\overset{1}{r}\overset{2}{s}\rangle\langle rs||ab\rangle \quad : \quad b \overset{1}{\to} s \overset{2}{\to} b$$

Therefore, $l_A = 2$. For Case B, on the other hand, we have

$$\langle a\overset{1}{b}||\overset{2}{r}s\rangle\langle sr||ab\rangle \quad : \quad a \overset{1}{\to} r \overset{2}{\to} b \overset{3}{\to} s \overset{4}{\to} a$$

Since a single "path" involves all the labels, $l_B = 1$. Because

$$\langle rs||ab\rangle = -\langle sr||ab\rangle$$

it is clear that both Cases A and B give the same result namely,

$$E_0^{(2)} = \frac{1}{4} \sum_{abrs} \frac{\langle ab||rs\rangle\langle rs||ab\rangle}{\varepsilon_a + \varepsilon_b - \varepsilon_r - \varepsilon_s}$$

which is identical to the result of Eq. (6.72), which was obtained algebraically. We now evaluate the three third-order diagrams, paying particular attention to the determination of l:

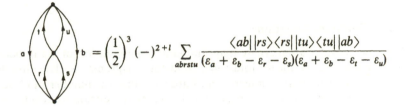

$$= \left(\frac{1}{2}\right)^3 (-)^{4+l} \sum_{abcdrs} \frac{\langle ab||rs\rangle\langle cd||ab\rangle\langle rs||cd\rangle}{(\varepsilon_a + \varepsilon_b - \varepsilon_r - \varepsilon_s)(\varepsilon_c + \varepsilon_d - \varepsilon_r - \varepsilon_s)}$$

The value of l is two for this case since we have the two paths, $a \to r \to c \to a$ and $b \to s \to d \to b$.

$$= \left(\frac{1}{2}\right)^3 (-)^{2+l} \sum_{abrstu} \frac{\langle ab||rs\rangle\langle rs||tu\rangle\langle tu||ab\rangle}{(\varepsilon_a + \varepsilon_b - \varepsilon_r - \varepsilon_s)(\varepsilon_a + \varepsilon_b - \varepsilon_t - \varepsilon_u)}$$

Here $l = 2$ since we have the paths $a \to r \to t \to a$ and $b \to s \to u \to b$. Finally,

$$= (-)^{3+l} \sum_{abcrst} \frac{\langle ab||rs\rangle\langle cs||tb\rangle\langle rt||ac\rangle}{(\varepsilon_a + \varepsilon_b - \varepsilon_s - \varepsilon_r)(\varepsilon_a + \varepsilon_c - \varepsilon_r - \varepsilon_t)}$$

Here $l = 3$ since we have the paths $a \to r \to a$, $b \to s \to b$ and $c \to t \to c$. Thus the third-order energy is

$$E_0^{(3)} = \frac{1}{8} \sum_{abcdrs} \frac{\langle ab||rs\rangle\langle cd||ab\rangle\langle rs||cd\rangle}{(\varepsilon_a + \varepsilon_b - \varepsilon_r - \varepsilon_s)(\varepsilon_c + \varepsilon_d - \varepsilon_r - \varepsilon_s)}$$
$$+ \frac{1}{8} \sum_{abrstu} \frac{\langle ab||rs\rangle\langle rs||tu\rangle\langle tu||ab\rangle}{(\varepsilon_a + \varepsilon_b - \varepsilon_r - \varepsilon_s)(\varepsilon_a + \varepsilon_b - \varepsilon_t - \varepsilon_u)}$$
$$+ \sum_{abcrst} \frac{\langle ab||rs\rangle\langle cs||tb\rangle\langle rt||ac\rangle}{(\varepsilon_a + \varepsilon_b - \varepsilon_s - \varepsilon_r)(\varepsilon_a + \varepsilon_c - \varepsilon_r - \varepsilon_t)}$$

which is the same as the result quoted in Eq. (6.75).

Exercise 6.11 Show that the fourth-order diagram

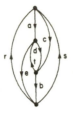

is equal to

$$-\frac{1}{2}\sum_{abcderst}\frac{\langle rs||ac\rangle\langle at||de\rangle\langle dc||tb\rangle\langle eb||rs\rangle}{(\varepsilon_a + \varepsilon_c - \varepsilon_r - \varepsilon_s)(\varepsilon_c + \varepsilon_d + \varepsilon_e - \varepsilon_r - \varepsilon_t - \varepsilon_s)(\varepsilon_b + \varepsilon_e - \varepsilon_r - \varepsilon_s)}$$

6.7.2 Goldstone Diagrams

We now consider an alternate diagrammatic representation of many-body perturbation theory due to Goldstone. Goldstone diagrams translate directly into regular two-electron integrals and are such that the number of closed loops (l) can be determined by inspection. However, there are more Goldstone diagrams than Hugenholtz diagrams in a given order of perturbation theory. In Goldstone diagrams we represent a two-particle interaction not by a dot but by a dashed line. The particle and hole states are still represented by lines with arrows, but now there is one line going into and one line coming out of each end of an interaction (dashed) line. For example, the element

in a Hugenholtz diagram is replaced in a Goldstone diagram by the two elements

This eliminates the antisymmetric matrix elements. One can think of the above correspondence as "pulling apart" the dot and labeling the arrows in two distinct ways. The Goldstone diagrams, which contribute to a given order, can be obtained by pulling apart all Hugenholtz diagrams in that

order. For example,

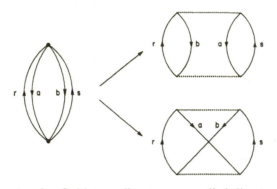

The two second-order Goldstone diagrams are called direct and exchange diagrams, respectively.

Table 6.2 contains all second- and third-order Goldstone diagrams in Hartree-Fock perturbation theory. They have been evaluated using the following rules:

G1. *Each interaction line*

   contributes a matrix element factor $\langle label\text{-}left\ in,\ label\text{-}right\ in\ |\ label\text{-}left\ out,\ label\text{-}right\ out\rangle$ *to the numerator.*

G2. *Each pair of adjacent interaction lines contributes the denominator factor*

$$\sum \varepsilon_{\text{hole}} - \sum \varepsilon_{\text{particle}}$$

   where the sums run over the labels of all hole and particle lines crossing an imaginary horizontal line separating the two adjacent interaction lines.

G3. *The overall sign of the expression is* $(-)^{h+l}$, *where h and l are the number of hole lines and closed loops, respectively.*

G4. *Sum the expression over all particle and hole indices.*

G5. *Diagrams which have a mirror plane perpendicular to the plane of the paper are multiplied by a factor of* $1/2$.

G6. *For closed-shell systems, a summation over spin orbitals is equal to* 2^l *times a summation over spatial orbitals, i.e.,*

$$\sum_{}^{N} = (2)^l \sum_{}^{N/2}.$$

Table 6.2 Second- and third-order Goldstone diagrams

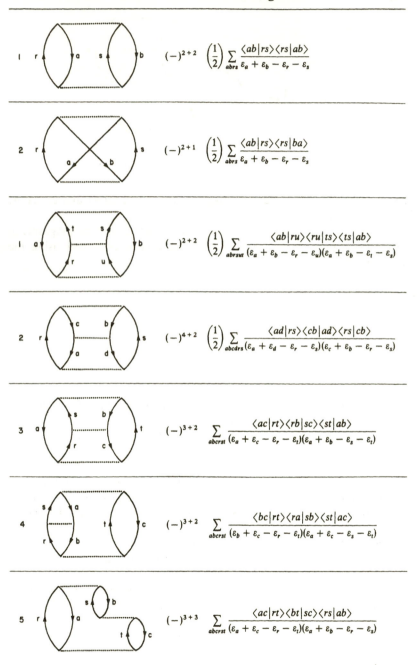

$$(-)^{2+2} \quad \left(\frac{1}{2}\right) \sum_{abrs} \frac{\langle ab|rs\rangle\langle rs|ab\rangle}{\varepsilon_a + \varepsilon_b - \varepsilon_r - \varepsilon_s}$$

$$(-)^{2+1} \quad \left(\frac{1}{2}\right) \sum_{abrs} \frac{\langle ab|rs\rangle\langle rs|ba\rangle}{\varepsilon_a + \varepsilon_b - \varepsilon_r - \varepsilon_s}$$

$$(-)^{2+2} \quad \left(\frac{1}{2}\right) \sum_{abrsut} \frac{\langle ab|ru\rangle\langle ru|ts\rangle\langle ts|ab\rangle}{(\varepsilon_a + \varepsilon_b - \varepsilon_r - \varepsilon_u)(\varepsilon_a + \varepsilon_b - \varepsilon_t - \varepsilon_s)}$$

$$(-)^{4+2} \quad \left(\frac{1}{2}\right) \sum_{abcdrs} \frac{\langle ad|rs\rangle\langle cb|ad\rangle\langle rs|cb\rangle}{(\varepsilon_a + \varepsilon_d - \varepsilon_r - \varepsilon_s)(\varepsilon_c + \varepsilon_b - \varepsilon_r - \varepsilon_s)}$$

$$(-)^{3+2} \quad \sum_{abcrst} \frac{\langle ac|rt\rangle\langle rb|sc\rangle\langle st|ab\rangle}{(\varepsilon_a + \varepsilon_c - \varepsilon_r - \varepsilon_t)(\varepsilon_a + \varepsilon_b - \varepsilon_s - \varepsilon_t)}$$

$$(-)^{3+2} \quad \sum_{abcrst} \frac{\langle bc|rt\rangle\langle ra|sb\rangle\langle st|ac\rangle}{(\varepsilon_b + \varepsilon_c - \varepsilon_r - \varepsilon_t)(\varepsilon_a + \varepsilon_c - \varepsilon_s - \varepsilon_t)}$$

$$(-)^{3+3} \quad \sum_{abcrst} \frac{\langle ac|rt\rangle\langle bt|sc\rangle\langle rs|ab\rangle}{(\varepsilon_a + \varepsilon_c - \varepsilon_r - \varepsilon_t)(\varepsilon_a + \varepsilon_b - \varepsilon_r - \varepsilon_s)}$$

Table 6-2 (*continued*)

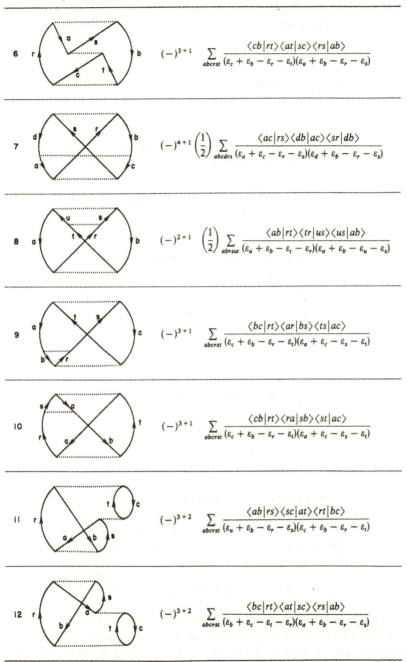

$$6 \qquad (-)^{3+1} \sum_{abcrst} \frac{\langle cb|rt\rangle\langle at|sc\rangle\langle rs|ab\rangle}{(\varepsilon_c + \varepsilon_b - \varepsilon_r - \varepsilon_t)(\varepsilon_a + \varepsilon_b - \varepsilon_r - \varepsilon_s)}$$

$$7 \qquad (-)^{4+1}\left(\frac{1}{2}\right) \sum_{abcdrs} \frac{\langle ac|rs\rangle\langle db|ac\rangle\langle sr|db\rangle}{(\varepsilon_a + \varepsilon_c - \varepsilon_r - \varepsilon_s)(\varepsilon_d + \varepsilon_b - \varepsilon_r - \varepsilon_s)}$$

$$8 \qquad (-)^{2+1}\left(\frac{1}{2}\right) \sum_{abrsut} \frac{\langle ab|rt\rangle\langle tr|us\rangle\langle us|ab\rangle}{(\varepsilon_a + \varepsilon_b - \varepsilon_t - \varepsilon_r)(\varepsilon_a + \varepsilon_b - \varepsilon_u - \varepsilon_s)}$$

$$9 \qquad (-)^{3+1} \sum_{abcrst} \frac{\langle bc|rt\rangle\langle ar|bs\rangle\langle ts|ac\rangle}{(\varepsilon_c + \varepsilon_b - \varepsilon_r - \varepsilon_t)(\varepsilon_a + \varepsilon_c - \varepsilon_s - \varepsilon_t)}$$

$$10 \qquad (-)^{3+1} \sum_{abcrst} \frac{\langle cb|rt\rangle\langle ra|sb\rangle\langle st|ac\rangle}{(\varepsilon_c + \varepsilon_b - \varepsilon_r - \varepsilon_t)(\varepsilon_a + \varepsilon_c - \varepsilon_s - \varepsilon_t)}$$

$$11 \qquad (-)^{3+2} \sum_{abcrst} \frac{\langle ab|rs\rangle\langle sc|at\rangle\langle rt|bc\rangle}{(\varepsilon_a + \varepsilon_b - \varepsilon_r - \varepsilon_s)(\varepsilon_c + \varepsilon_b - \varepsilon_r - \varepsilon_t)}$$

$$12 \qquad (-)^{3+2} \sum_{abcrst} \frac{\langle bc|rt\rangle\langle at|sc\rangle\langle rs|ab\rangle}{(\varepsilon_b + \varepsilon_c - \varepsilon_t - \varepsilon_r)(\varepsilon_a + \varepsilon_b - \varepsilon_r - \varepsilon_s)}$$

As an application of these rules, we evaluate the second-order energy. Consider reflecting the *direct* diagram about the mirror plane perpendicular to the plane of the paper,

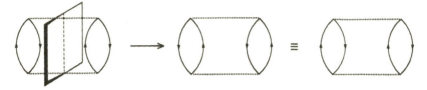

The diagram remains invariant upon reflection, so the weight factor is 1/2. Since we have two hole lines and two closed loops, the value of this diagram is

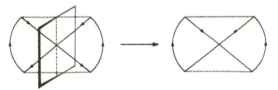

$$= \frac{1}{2}(-)^4 \sum_{abrs} \frac{\langle ab|rs\rangle\langle rs|ab\rangle}{\varepsilon_a + \varepsilon_b - \varepsilon_r - \varepsilon_s}$$

Similarly, the *exchange* diagram has a mirror plane, since

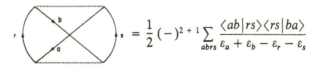

so the weight factor is again 1/2. Since we have two hole lines but only one closed loop $(r \to b \to s \to a \to r)$, the value of this diagram is

$$= \frac{1}{2}(-)^{2+1} \sum_{abrs} \frac{\langle ab|rs\rangle\langle rs|ba\rangle}{\varepsilon_a + \varepsilon_b - \varepsilon_r - \varepsilon_s}$$

Note that the sum of the values of these two diagrams is just the result given in Eq. (6.73). Moreover, for a closed-shell system using rule (G6), we can convert the summation over spin orbitals to spatial orbitals simply by multiplying the value of the direct diagram by 2^2 and the exchange diagram by 2^1 and, thus, recover Eq. (6.74). Table 6.2 contains the algebraic expressions corresponding to all 12 third-order Goldstone diagrams. You should check your understanding of the rules by verifying these results.

Exercise 6.12 We stated that the Goldstone diagrams in Table 6.2 can be obtained by "pulling apart" the second- and third-order Hugenholtz diagrams. This is quite tricky to see but the converse is much easier. For

example, if we push

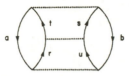

together, it is clear that we obtain

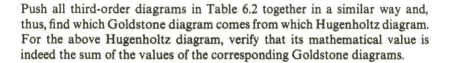

Push all third-order diagrams in Table 6.2 together in a similar way and, thus, find which Goldstone diagram comes from which Hugenholtz diagram. For the above Hugenholtz diagram, verify that its mathematical value is indeed the sum of the values of the corresponding Goldstone diagrams.

As an illustration, we now evaluate all the second- and third-order diagrams in Table 6.2 for minimal basis H_2 and show that the corresponding energies are identical to our previous results (Eqs. (6.77) and (6.78)). To do this we convert to a spatial orbital summation by multiplying by 2^l, where l is the number of closed loops, and then set all hole indices equal to 1 and all particle indices equal to 2. For the two second-order diagrams, we find

$$1 \leftrightarrow (2)^2(1/2)\frac{\langle 11|22\rangle^2}{(2\varepsilon_1 - 2\varepsilon_2)} = \frac{K_{12}^2}{(\varepsilon_1 - \varepsilon_2)}$$

$$2 \leftrightarrow -(2)(1/2)\frac{\langle 11|22\rangle^2}{(2\varepsilon_1 - 2\varepsilon_2)} = -\frac{K_{12}^2}{2(\varepsilon_1 - \varepsilon_2)}$$

so that their sum is indeed identical to the second-order energy of Eq. (6.77). For the third-order diagrams we have

$$1 \leftrightarrow (2)^2\left(\frac{1}{2}\right)\frac{\langle 11|22\rangle\langle 22|22\rangle\langle 22|11\rangle}{4(\varepsilon_1 - \varepsilon_2)^2} = \frac{K_{12}^2 J_{22}}{2(\varepsilon_1 - \varepsilon_2)^2}$$

$$2 \leftrightarrow (2)^2\left(\frac{1}{2}\right)\frac{\langle 11|22\rangle\langle 11|11\rangle\langle 22|11\rangle}{4(\varepsilon_1 - \varepsilon_2)^2} = \frac{K_{12}^2 J_{11}}{2(\varepsilon_1 - \varepsilon_2)^2}$$

$$3 \leftrightarrow -(2)^2\frac{\langle 11|22\rangle\langle 21|21\rangle\langle 22|11\rangle}{4(\varepsilon_1 - \varepsilon_2)^2} = -\frac{K_{12}^2 J_{12}}{(\varepsilon_1 - \varepsilon_2)^2}$$

$$4 \leftrightarrow -(2)^2\frac{\langle 11|22\rangle\langle 21|21\rangle\langle 22|11\rangle}{4(\varepsilon_1 - \varepsilon_2)^2} = -\frac{K_{12}^2 J_{12}}{(\varepsilon_1 - \varepsilon_2)^2}$$

$$5 \leftrightarrow (2)^3 \frac{\langle 11|22\rangle\langle 12|21\rangle\langle 22|11\rangle}{4(\varepsilon_1 - \varepsilon_2)^2} = 2\frac{K_{12}^2 K_{12}}{(\varepsilon_1 - \varepsilon_2)^2}$$

$$6 \leftrightarrow (2) \frac{\langle 11|22\rangle\langle 12|21\rangle\langle 22|11\rangle}{4(\varepsilon_1 - \varepsilon_2)^2} = \frac{K_{12}^2 K_{12}}{2(\varepsilon_1 - \varepsilon_2)^2}$$

$$7 \leftrightarrow -(2)\left(\frac{1}{2}\right) \frac{\langle 11|22\rangle\langle 11|11\rangle\langle 11|22\rangle}{4(\varepsilon_1 - \varepsilon_2)^2} = -\frac{K_{12}^2 J_{11}}{4(\varepsilon_1 - \varepsilon_2)^2}$$

$$8 \leftrightarrow -(2)\left(\frac{1}{2}\right) \frac{\langle 11|22\rangle\langle 22|22\rangle\langle 11|22\rangle}{4(\varepsilon_1 - \varepsilon_2)^2} = -\frac{K_{12}^2 J_{22}}{4(\varepsilon_1 - \varepsilon_2)^2}$$

$$9 \leftrightarrow (2) \frac{\langle 11|22\rangle\langle 12|12\rangle\langle 11|22\rangle}{4(\varepsilon_1 - \varepsilon_2)^2} = \frac{K_{12}^2 J_{12}}{2(\varepsilon_1 - \varepsilon_2)^2}$$

$$10 \leftrightarrow (2) \frac{\langle 11|22\rangle\langle 21|21\rangle\langle 11|22\rangle}{4(\varepsilon_1 - \varepsilon_2)^2} = \frac{K_{12}^2 J_{12}}{2(\varepsilon_1 - \varepsilon_2)^2}$$

$$11 \leftrightarrow -(2)^2 \frac{\langle 11|22\rangle\langle 21|12\rangle\langle 11|22\rangle}{4(\varepsilon_1 - \varepsilon_2)^2} = -\frac{K_{12}^2 K_{12}}{(\varepsilon_1 - \varepsilon_2)^2}$$

$$12 \leftrightarrow -(2)^2 \frac{\langle 11|22\rangle\langle 12|21\rangle\langle 22|11\rangle}{4(\varepsilon_1 - \varepsilon_2)^2} = -\frac{K_{12}^2 K_{12}}{(\varepsilon_1 - \varepsilon_2)^2}$$

One can now verify that the sum of these twelve expressions agrees with the third-order energy of Eq. (6.78).

6.7.3 Summation of Diagrams

In Subsection 6.2.3, we found that certain classes of diagrams could be summed to infinite order, because the values of the diagrams formed a geometric series. The analogous summation can be performed with Goldstone diagrams. Recall that in the N-state problem we summed only certain diagrams that were diagonal; that is, they were formed by adding dots to our second-order diagram in such a way that the labels on the lines entering and leaving the new dot were the same. With Goldstone diagrams we can do the same. We start with the second-order direct and exchange diagrams and add interaction lines in such a way that the labels on a hole or particle line intersected by the added interaction line are the same above and below the interaction line. For example, consider the following set of diagrams

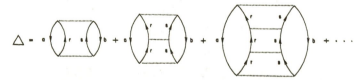

All of these have a weight factor equal to 1/2. Because they all have the same number of hole lines and closed loops, they have the same sign. If we evaluate

them, we find

$$
\Delta = \frac{1}{2} \frac{|\langle rs|ab \rangle|^2}{\varepsilon_a + \varepsilon_b - \varepsilon_r - \varepsilon_s} \left[1 + \frac{\langle rs|rs \rangle}{\varepsilon_a + \varepsilon_b - \varepsilon_r - \varepsilon_s} + \left(\frac{\langle rs|rs \rangle}{\varepsilon_a + \varepsilon_b - \varepsilon_r - \varepsilon_s} \right)^2 + \cdots \right]
$$

$$
= \frac{1}{2} \frac{|\langle rs|ab \rangle|^2}{\varepsilon_a + \varepsilon_b - \varepsilon_r - \varepsilon_s - \langle rs|rs \rangle}
$$

Thus the summation of a geometric series, just as before, results in a shifted energy denominator. Now if one were to sum *all* possible diagrams generated by the prescription given above, one would find

$$
E_{\text{corr}}(\text{EN}) = -\frac{1}{4} \sum_{abrs} \frac{|\langle ab||rs \rangle|^2}{\langle \Psi_{ab}^{rs} | \mathcal{H} - E_0 | \Psi_{ab}^{rs} \rangle} \tag{6.93}
$$

This result is identical to the correlation energy obtained using Epstein-Nesbet pair energies, as discussed in the previous chapter (see Eqs. (5.15) and (5.16)).

Finally, we mention that if one were to sum all *double excitation* diagrams (an imaginary horizontal line crosses only two hole and two particle lines for such diagrams) one would obtain the doubly excited MBPT (D-MBPT(∞)) of Bartlett and coworkers.[2] This approximation is equivalent to the linear CCA discussed in Subsection 5.2.3. Thus, the appropriate correlation energy is (see Eq. (5.65))

$$
E_{\text{corr}}(\text{D-MBPT}(\infty)) = -\mathbf{B}^\dagger (\mathbf{D})^{-1} \mathbf{B} \tag{6.94}
$$

where

$$
(\mathbf{B}^\dagger)_{rasb} = \langle \Psi_0 | \mathcal{H} | \Psi_{ab}^{rs} \rangle \tag{6.95a}
$$

$$
(\mathbf{D})_{rasb,tcud} = \langle \Psi_{ab}^{rs} | \mathcal{H} - E_0 | \Psi_{cd}^{tu} \rangle \tag{6.95b}
$$

Note that if matrix $\mathbf{D}$ is approximated by only its diagonal part, we recover the Epstein-Nesbet second-order energy. Unlike $E_{\text{corr}}(\text{EN})$, $E_{\text{corr}}(\text{D-MBPT}(\infty))$ is exact to third order in perturbation theory. In fourth order, it is missing contributions due to single, triple, and quadruple excitations.

6.7.4 What Is the Linked-Cluster Theorem?

Goldstone's linked-cluster theorem is simply a general proof, using diagrams, of Brueckner's conjecture that RS perturbation theory, order by order, is size consistent. The algebraic terms that are proportional to N^2 can be represented by diagrams consisting of separate pieces (i.e., by *unlinked* diagrams). Goldstone proved, using time-dependent techniques, that such diagrams never appear in the final result for the nth-order energy. Thus he was able to write the nth-order energy in terms of only *linked* diagrams. In this chapter, we used only linked diagrams right from the beginning. We hope that the numerous consistency checks, which compared results obtained using diagrammatic and algebraic methods, have made it convincing (or at least plausible) that all forms of RS perturbation theory (i.e., N-state, orbital, etc.) can be represented by only linked diagrams. It should be emphasized that the term "size consistent" is used in this book to describe theories in which the energy increases linearly with the size of a many-body system. Only if a system is a supermolecule composed of N independent closed-shell monomers and only if a method is invariant to unitary transformations of degenerate orbitals, does size consistency imply that the energy of the supermolecule is N times the monomer energy.

Exercise 6.13 Calculate $E_0^{(3)}$ for a supermolecule consisting of N noninteracting minimal basis H_2 molecules by evaluating the Goldstone diagrams in Table 6.2. Compare your result with that of Eq. (6.92), which was obtained algebraically by explicitly cancelling terms proportional to N^2. *Hint*: Simply show that the value of each Goldstone diagram for the supermolecule is N times the result for a single molecule.

6.8 SOME ILLUSTRATIVE CALCULATIONS

In this section, we use perturbation theory to calculate the correlation energies, potential energy surfaces, equilibrium geometries, and dipole moments of a variety of small molecules.

Table 6.3 compares the correlation energy of H_2 calculated using second- and third-order perturbation theory with the full CI results for a series of basis sets of increasing quality. It is interesting to note that the fraction of the basis set correlation energy obtained at second-order increases significantly with the size of the basis. This appears to be a general phenomenon and is not restricted to two-electron systems. Thus it is dangerous to draw conclusions about the utility of perturbation theory from small basis set calculations. It can be seen that the perturbation expansion converges

Table 6.3 Correlation energy (a.u.) of H_2 at $R = 1.4$

Basis	$E_0^{(2)}$	Percent of full CI	$E_0^{(2)} + E_0^{(3)}$	Percent of full CI	Full CI
STO-3G	−0.0132	64	−0.0180	87	−0.0206
4-31G	−0.0174	70	−0.0226	91	−0.0249
6-31G**	−0.0263	78	−0.0319	94	−0.0339
$(10s, 5p, 1d)^a$	−0.0321	81	−0.0376	95	−0.0397
Exact[b]			−0.0409		

[a] J. M. Schulman and D. N. Kaufman, *J. Chem. Phys.* **53**: 477 (1970), for $E_0^{(2)}$; U. Kaldor, *J. Chem. Phys.* **62**: 4634 (1975) for, $E_0^{(3)}$; C. E. Dykstra, unpublished result, for full CI.

[b] W. Kolos and L. Wolniewicz, *J. Chem. Phys.* **49**: 404 (1968).

Table 6.4 Equilibrium bond length (a.u.) for H_2

Basis set	SCF	$E_0^{(2)}$	Full CI
STO-3G	1.346	1.368	1.389
4-31G	1.380	1.394	1.410
6-31G**	1.385	1.387	1.396
Exact[a]		1.401	

[a] W. Kolos and L. Wolniewicz, *J. Chem. Phys.* **49**:404 (1968).

quickly for the larger basis sets and that the correlation energy obtained through third order is in good agreement with the exact value for the basis.

In Table 6.4 we compare the equilibrium bond lengths of H_2 obtained using SCF, $E_0^{(2)}$ and full CI for three basis sets, with the exact result. For the 6-31G** basis, the SCF result differs from the exact basis set result by only 0.011 a.u.; the second-order energy corrects the SCF bond length in the right direction, but by a rather small amount.

In Table 6.5 we compare the SCF and second-order perturbation theory results for the equilibrium bond lengths of the ten-electron series CH_4, NH_3, H_2O, and FH with experiment. At the 6-31G** level, the largest difference between the SCF and experimental results is 0.03 a.u. In all cases, the second-order energy reduces this error to less than 0.01 a.u. Note that in all the systems discussed so far, large basis set SCF calculations underestimate the equilibrium bond length. Correlation effects tend to lengthen

Table 6.5 Equilibrium bond lengths (a.u.) of the ten-electron series

	STO-3G	4-31G	6-31G*	6-31G**	Experiment
CH_4					
SCF	2.047	2.043	2.048	2.048	2.050
$E_0^{(2)}$	2.077	2.065	2.060	2.048	
NH_3					
SCF	1.952	1.873	1.897	1.897	1.913
$E_0^{(2)}$	1.997	1.907	1.922	1.912	
H_2O					
SCF	1.871	1.797	1.791	1.782	1.809
$E_0^{(2)}$	1.916	1.842	1.831	1.816	
FH					
SCF	1.812	1.742	1.722	1.703	1.733
$E_0^{(2)}$	1.842	1.790	1.765	1.740	

Table 6.6 Equilibrium bond angles for NH_3 and H_2O

	STO-3G	4-31G	6-31G*	6-31G**	Experiment
NH_3					
SCF	104.2	115.8	107.5	107.6	106.7
$E_0^{(2)}$	100.9	113.9	106.3	106.1	
H_2O					
SCF	100.0	111.2	105.5	106.0	104.5
$E_0^{(2)}$	97.2	108.8	104.0	103.9	

those bonds that lead to open-shell fragments when stretched. In Table 6.6 the equilibrium bond angles for NH_3 and H_2O, calculated at the SCF and $E_0^{(2)}$ levels, are compared with experiment. For the larger basis sets, the second-order energy reduces the SCF error by about a half.

Although perturbation theory at the level of the second-order energy predicts excellent bond lengths for bonds involving hydrogen, this is not the case for multiple bonds. Table 6.7 compares the SCF and perturbation theory results for the bond lengths of N_2 and CO with experiment. For the largest basis set, the SCF underestimates the experimental result by about 0.03 a.u.; second-order perturbation theory increases the predicted bond length but it significantly overestimates the correction. The third-order correction partially counteracts the effect of the second-order correction, but the agreement with experiment is still not completely satisfactory.

Before leaving the topic of equilibrium geometries, we consider the calculation of the potential energy of H_2O in the vicinity of its equilibrium geometry using a near-HF-limit one-electron basis set (the 39-STO basis described in Chapter 4). We have used this example both in Chapter 4, in the context of CI, and in Chapter 5, to illustrate pair and coupled-pair theories. In Table 6.8 we compare the correlation energy obtained through fourth order of perturbation theory with those found using a variety of many-electron approaches, discussed in the previous chapters. We note only that the second-order energy gives a respectable fraction of the basis-set correlation energy and is superior to SDCI. A more interesting comparison of these

Table 6.7 Equilibrium bond lengths (a.u.) of N_2 and CO

Basis set	N_2			CO		
	SCF	$E_0^{(2)}$	$E_0^{(2)} + E_0^{(3)}$	SCF	$E_0^{(2)}$	$E_0^{(2)} + E_0^{(3)}$
STO-3G	2.143	2.322	2.222	2.166	2.264	2.216
4-31G	2.050	2.171	2.098	2.132	2.216	2.169
6-31G*	2.039	2.133	2.109	2.105	2.175	2.145
Experiment		2.074			2.132	

Table 6.8 Correlation energies (a.u.) of H_2O at the experimental geometry calculated with the 39-STO basis described in Chapter 4[a]

	E_{corr}
$E_0^{(2)}$	-0.2818
$E_0^{(2)} + E_0^{(3)}$	-0.2850
$E_0^{(2)} + E_0^{(3)} + E_0^{(4)}$	-0.2960
SDCI	-0.2756
IEPA	-0.3274
L-CCA	-0.2908
CCA	-0.2862
Estimated full CI	-0.296 ± 0.001
Exact	-0.37

[a] R. J. Bartlett, S. J. Cole, G. D. Purvis, W. C. Ermler, H. C. Hsieh, and I. Shavitt, *J. Chem. Phys.* **87**: 6579 (1987), and references therein.

many-electron theories is given in Table 6.9, which presents the equilibrium geometry and two force constants for H_2O calculated using the different approaches. Before considering these results, it should be pointed out that the predictions of the various theories should be compared with the exact basis set results (i.e., the results obtained from full CI calculations) and not

Table 6.9 Equilibrium geometry and some force constants of H_2O calculated with the 39-STO basis described in Chapter 4[a]

	R_e (a.u.)	θ_e	$f_{RR}^{[b]}$	$f_{\theta\theta}^{[b]}$
SCF	1.776	106.1	9.79	0.88
$E_0^{(2)}$	1.811	104.4	8.55	0.78
$E_0^{(2)} + E_0^{(3)}$	1.803	104.8	8.80	0.81
$E_0^{(2)} + E_0^{(3)} + E_0^{(4)}$	1.813	104.4	8.42	0.79
SDCI	1.800	104.9	8.88	0.81
L-CCA	1.810	104.6	8.51	0.80
CCA	1.806	104.7	8.67	0.80
Experiment	1.809	104.5	8.45	0.76

[a] R. J. Bartlett, S. J. Cole, G. D. Purvis, W. C. Ermler, H. C. Hsieh, and I. Shavitt, *J. Chem. Phys.* **87**: 6579 (1987), and references therein.

[b] The force constants are in 10^5 dyn/cm with angles measured in radians.

experiment. Unfortunately the exact basis set results are unavailable and hence the best one can do is use the experimental results as a benchmark. The most striking aspect of this Table is that, while SDCI considerably improves the SCF results, all the many-body methods are superior to SDCI for this molecule. The second-order perturbation results are remarkably close to experiment. However, the convergence of the perturbation expansion appears to be oscillatory and the third-order results are relatively poor.

This is in contrast to the correlation energy (see Table 6.8), where the convergence seems to be monotonic. Thus it appears that to obtain highly accurate potential energy surfaces even near equilibrium one needs to go to higher orders in perturbation theory. This is consistent with the improved agreement obtained with the two coupled-cluster methods (CCA and L-CCA). The CCA and the L-CCA are exact to third-order in perturbation theory and also incorporate some important parts of the fourth, fifth, . . . , infinite order energies.

A central problem in quantum chemistry is the calculation of potential energy surfaces far from the equilibrium configuration, for use in describing chemical reactions. A simple example is the potential energy curve of H_2. We have seen in Chapter 3 that restricted Hartree-Fock is not a satisfactory method since it does not predict that H_2 dissociates into two hydrogen atoms. Recall that for the minimal basis set description, in the limit that $R \to \infty$, h_{11} and h_{22} both become equal to the energy of the hydrogen atom in the basis ($E(H)$) while all two-electron integrals (J_{11}, J_{12}, J_{22}, K_{12}) become equal to $\frac{1}{2}(\phi\phi|\phi\phi)$, where ϕ is the hydrogen atom orbital. Thus the HF energy E_0 ($E_0 = 2h_{11} + J_{11}$) behaves as

$$\lim_{R \to \infty} E_0(R) = 2E(H) + \tfrac{1}{2}(\phi\phi|\phi\phi) \qquad (6.96)$$

We have seen that CI rectifies this situation; unfortunately, perturbation theory using restricted HF orbitals makes things even worse. In fact, the second-order energy in the limit $R \to \infty$ diverges. Recall that for minimal basis set H_2 the second-order energy (see Eq. (6.77)) is

$$E_0^{(2)} = \frac{K_{12}^2}{2(\varepsilon_1 - \varepsilon_2)}$$

In the limit $R \to \infty$, K_{12} becomes $\frac{1}{2}(\phi\phi|\phi\phi)$, which is finite, but both ε_1 ($\varepsilon_1 = h_{11} + J_{11}$) and ε_2 ($\varepsilon_2 = h_{22} + 2J_{12} - K_{12}$) go to the same limit (i.e., $E(H) + \frac{1}{2}(\phi\phi|\phi\phi)$). Thus $\varepsilon_1 - \varepsilon_2$ tends to zero and $E_0^{(2)}$ becomes infinite. The simplest way out of this difficulty is to improve the *unrestricted* HF description by means of perturbation theory. Computationally, this involves simply using the unrestricted spin orbitals and orbital energies in the formulas for the second and higher order energies written as sums over spin rather than spatial orbitals (e.g., Eqs. (6.72) and (6.75)). The potential energy curves for minimal basis H_2, obtained in this way, are shown in Fig. 6.2.

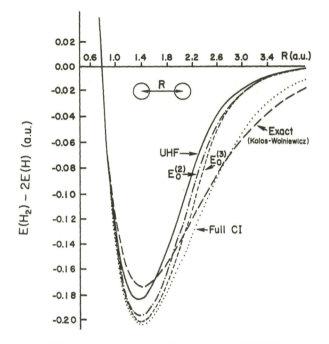

Figure 6.2 Potential energy curves for H_2 using the STO-3G basis set.

The exact result of Kolos and Wolniewicz is shown for comparison. Recall that the UHF and full CI calculations predict a larger dissociation energy than the exact result because the STO-3G basis is so poor for the hydrogen atom. Note that the convergence of the perturbation expansion is excellent for the dissociation energy. At third-order, the perturbation theory result differs only by about 1% from the exact basis set value. The disappointing aspect is that, while the unrestricted second and third-order energies are well behaved at large R (i.e., dissociation is correctly described), perturbation theory starting with the UHF description gives a negligible amount of the correlation energy at large, but finite, R. Thus the potential curve calculated using perturbation theory is not satisfactory at long range. Perturbation theory converges quickly if one has a good starting point. Although the UHF wave function gives the correct energy at $R \to \infty$, it has the incorrect form, i.e.,

$$\lim_{R \to \infty} |\Psi_0^{UHF}\rangle = |\phi_1 \bar{\phi}_2\rangle \tag{6.97}$$

where ϕ_1 and ϕ_2 are the hydrogen atomic orbitals on centers 1 and 2, while the exact singlet wave function at infinite separation is

$$\lim_{R \to \infty} |\Phi_0\rangle = 2^{-1/2}(|\phi_1 \bar{\phi}_2\rangle + |\phi_2 \bar{\phi}_1\rangle) \tag{6.98}$$

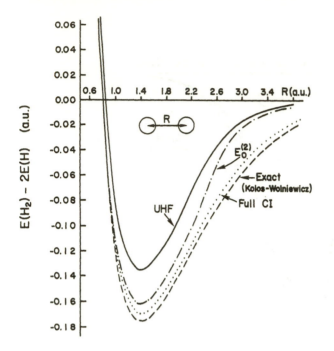

Figure 6.3 Potential energy curves for H_2 using the 6-31G** basis set.

Note that the exact wave function at large R cannot be described by a single determinant. The perturbation theory described in this chapter was obtained assuming that the zeroth-order wave function was a single determinant. It thus appears, that to describe the long range part of potential curves using perturbation theory, one must start with a multiconfigurational zeroth-order wave function. In Fig. 6.3 we present the corresponding results at the 6-31G** level, to show that the above observations are independent of the size of the basis set.

As a final example we consider the calculation of dipole moments in the framework of perturbation theory, using the method of finite perturbations. The idea behind this approach is to calculate the correlated energy of a molecule in the presence of a small but finite electric field and then numerically differentiate the energy with respect to the field to obtain the dipole moment. It should be noted that this method is general and can be used in conjunction with any approximation for the correlation energy. A major advantage of the finite perturbation approach is that it can be computationally implemented with relatively minor modifications of existing programs, which calculate the correlation energy. This is a consequence of the fact that the additional term in the Hamiltonian due to the presence of the electric field contains only one-particle interactions and thus can be incorporated into the HF procedure by a simple modification of the core-Hamiltonian.

To make the above ideas more explicit, consider a molecule in the presence of an electric field $\vec{F}$. The total Hamiltonian in the presence of the field is

$$\mathcal{H}(\vec{F}) = \mathcal{H}(0) + \vec{F} \cdot \sum_i \mathbf{r}(i) \qquad (6.99)$$

where $\mathbf{r}(i)$ is the coordinate of electron i. If we expand the total energy in the presence of the field $\vec{F}$ in a Taylor series, we have

$$E(\vec{F}) = E(0) + \sum_i \left(\frac{\partial E(\vec{F})}{\partial F_i}\right)_0 F_i + \frac{1}{2} \sum_{ij} \left(\frac{\partial^2 E(\vec{F})}{\partial F_i \partial F_j}\right)_0 F_i F_j + \cdots \qquad (6.100)$$

where the sums run over the Cartesian (x, y, and z) components of the field. By definition, the component of the dipole moment vector, $\vec{\mu}$, in the ith direction (i.e. μ_i) is

$$\mu_i = -\left(\frac{\partial E(\vec{F})}{\partial F_i}\right)_0 \qquad (6.101)$$

The minus sign is chosen so that the dipole moment vector points from the negative to the positive part of the molecule. Similarly, the second derivatives of $E(\vec{F})$ evaluated at zero field define the polarizability tensor of the molecule. Thus finite perturbation theory can be used to obtain the polarizability as well as the dipole moment of a molecule.

The dipole moments presented below were obtained by first performing a HF calculation in the presence of the field, then calculating the second-order energy using the field-dependent HF orbitals and orbital energies, and finally, numerically differentiating the total energy calculated up to second-order in perturbation theory. We first consider the dipole moments of the ten-electron series NH_3, H_2O, and FH. Recall that we have seen in Chapter 3 that even at the 6-31G** level the HF dipole moments were poor in comparison with larger SCF calculations near the HF limit. The dipole moments, calculated using the second-order energy for the standard basis sets, are shown in Table 6.10. Note that for the larger basis sets, correlation changes the dipole moment in the right direction but, except for the 6-31G** result for FH, the agreement with experiment is still poor. This illustrates the fact that inadequacies in the basis set cannot be corrected by including correlation effects.

A particularly challenging molecule for any theory of dipole moments is CO. Recall that in Chapter 3 we have seen that the SCF procedure in the limit of a very large basis set gives the incorrect sign for the dipole moment. Does perturbation theory at the level of the second-order energy rectify this? Table 6.11 shows that the larger basis sets are of sufficient quality (i.e., the

Table 6.10 Dipole moments (a.u.) of the ten-electron series

Basis set	NH_3		H_2O		FH	
	SCF	$E_0^{(2)}$	SCF	$E_0^{(2)}$	SCF	$E_0^{(2)}$
STO-3G	0.703	0.692	0.679	0.652	0.507	0.477
4-31G	0.905	0.891	1.026	0.994	0.897	0.861
6-31G*	0.768	0.773	0.876	0.859	0.780	0.747
6-31G**	0.744	0.733	0.860	0.825	0.776	0.728
Experiment	0.579		0.723		0.719	

Table 6.11 Dipole moment (a.u.) of CO^a

Basic set	SCF	$E_0^{(2)}$
STO-3G	0.066	0.510
4-31G	− 0.237	0.165
6-31G*	− 0.131	0.167
Experiment	0.044	

^{a} Positive dipole moment corresponds to C^-O^+.

predicted SCF dipole moment has the wrong sign in agreement with the near-HF results) to be able to answer this question qualitatively. The second-order energy does indeed predict the correct sign of the dipole moment. The basis sets are, however, not sufficiently large to draw any quantitative conclusions. At first glance this success of second-order perturbation theory might appear to be surprising. In Chapter 4 we have seen that, using CI, the correct sign is obtained only if we include *single* excitations. However, because of Brillouin's theorem, only double excitations contribute to the second-order energy. The resolution of this apparent paradox lies in the nature of the numerical differentiation, which was used to obtain the dipole moment. While it is by no means obvious, the answer is that while the second-order energy does indeed contain only double excitations, the HF orbitals used to calculate $E_0^{(2)}$ are field-dependent and the numerical differentiation effectively introduces single excitations involving the HF orbitals of the molecule in the absence of the field.

NOTES

1. R. P. Feynman, The development of the space-time view of quantum electrodynamics, *Science* **153**:699 (1966). This is an amusing as well as inspiring article.

2. R. J. Bartlett, I. Shavitt, and G. D. Purvis, The quartic force field of H_2O determined by many-body methods that include quadruple excitation effects, *J. Chem. Phys.* **71**:281 (1979), and references therein.

FURTHER READING

Bartlett, R. J., Many-body perturbation theory and coupled cluster theory for electron correlation in molecules, *Ann. Rev. Phys. Chem.* **32**: 359 (1981). An excellent review of the theory and applications with two hundred literature references.

Brandow, B. H., Linked-cluster expansion for the nuclear many-body problem, *Rev. Mod. Phys.* **39**: 771 (1967). A classic review which discusses, among many other things, how to use Goldstone-like diagrams with antisymmetric matrix elements associated with Hugenholtz diagrams. This is an alternate way of getting around the problem of not being able to determine the number of closed loops in a Hugenholtz diagram by inspection.

Freed, K. F., Many-body theories of the electronic structure of atoms and molecules, *Ann. Rev. Phys. Chem.* **22**: 313 (1971). An excellent review with almost three hundred references. It is written for the nonspecialist with a historical perspective.

Lindgren, I., and Morrison, J., *Atomic Many-Body Theory*, Springer-Verlag, Berlin, 1982. Chapters 9–12 of this advanced textbook develop diagrammatic perturbation theory in a rigorous yet accessible way. The labels used for occupied and virtual orbitals are the same as ours.

Manne, R., The linked-diagram expansion of the ground state of a many-electron system: A time-independent derivation, *Int. J. Quant. Chem.* **S11**: 175 (1977). A pedagogical article which presents a relatively simple proof of the linked cluster theorem.

Mattuck, R. D., *A Guide to Feynman Diagrams in the Many-Body Problem*, 2nd ed., McGraw-Hill, New York, 1976. A wonderful book which presents diagrammatic perturbation theory from the time-dependent point of view. The author has made the subject as entertaining and as simple as possible. A disadvantage for chemists is that the emphasis is placed on treating infinite systems where the basis functions are plane waves.

Paldus J. and Čížek, J., Time-independent diagrammatic approach to perturbation theory of fermion systems, *Adv. Quantum Chem.* **9**: 105 (1975). An authoritative discussion of the subject.

Robb, M. A., Pair functions and diagrammatic perturbation theory, in *Computational Techniques in Quantum Chemistry*, G. H. F. Diercksen, R. T. Sutcliffe, A. Veillard (eds.), Reidel, Boston, 1975, p. 435. An accessible review which discusses IEPA, CEPA, and CCA in diagrammatic terms.

Wilson, S., *Electron Correlation in Molecules*, Clarendon Press, Oxford, 1984. This book discusses many of the modern approaches to the correlation problem in a way complimentary to our own. It is particularly strong on diagrammatic perturbation theory and also contains an excellent chapter on the group theoretical aspects of electron correlation.

CHAPTER SEVEN

THE ONE-PARTICLE MANY-BODY
GREEN'S FUNCTION

Green's function techniques are one of the most powerful tools of many-body theory. In this chapter, we will consider the one-particle many-body Green's function (MBGF), which contains information about the electron affinities (EA's) and ionization potentials (IP's) of an N-electron system. We will see that MBGF theory provides a systematic framework for improving the IP's and EA's obtained within the Hartree-Fock approximation using Koopmans' theorem. In contrast to perturbation theory, which can be presented without ever mentioning diagrams or using second quantization, Green's function techniques are really quite different from the approaches we have thus far discussed. Therefore, to provide a simple introduction, we have to restrict our scope even more substantially than in previous chapters. We do not discuss the time-dependent aspects of Green's function theory and, consequently, lose the physically appealing interpretation of these functions as propagators. Moreover, we do not consider two-particle MBGF's, which describe the excitation spectrum of an N-electron system. Hence, the one-particle prefix will be implicit when we talk about MBGF's. Finally, we will not discuss the fact that MBGF's, in addition to describing ionization and electron capture, contain information about the one-particle density matrix and ground state energy of an N-electron system. Nevertheless, we hope that our treatment will give you some feeling for Green's functions and serve as a first step towards more advanced discussions.

We begin this chapter by showing how Green's functions naturally arise in the solution of certain inhomogeneous differential equations. We then discuss the properties and the use of these functions in the quantum mechanics of single-particle systems. In Section 7.2, we consider a many-particle system within the framework of the Hartree-Fock approximation and introduce the Hartree-Fock Green's function (HFGF). In Subsection 7.2.1, we consider how to go beyond the HF approximation while retaining a single-particle picture. We introduce an energy-dependent potential, called the self-energy, and set up the machinery for improving on Koopmans' theorem IP's and EA's. In Section 7.3, we apply the formalism to minimal basis H_2 and HeH^+. In Section 7.4, we generalize some of the insights gained in the previous section, and consider the relationship between perturbation theory and Green's function approaches to the calculation of the IP's and EA's of a many-electron system. Finally, in Section 7.5 we present some illustrative calculations.

7.1 GREEN'S FUNCTIONS IN SINGLE-PARTICLE SYSTEMS

Suppose we wish to solve the following matrix equation for **a**

$$(E\mathbf{1} - \mathbf{H}_0)\mathbf{a} = \mathbf{b} \tag{7.1}$$

where E is a parameter, $\mathbf{H}_0$ is an $N \times N$ Hermitian matrix and **a** and **b** are column matrices. The most straightforward way to proceed is to find the inverse of the matrix $E\mathbf{1} - \mathbf{H}_0$, which we denote by $\mathbf{G}_0(E)$,

$$\mathbf{G}_0(E) = (E\mathbf{1} - \mathbf{H}_0)^{-1} \tag{7.2}$$

so that we immediately have

$$\mathbf{a} = (E\mathbf{1} - \mathbf{H}_0)^{-1}\mathbf{b} = \mathbf{G}_0(E)\mathbf{b} \tag{7.3a}$$

or

$$a_i = \sum_j (\mathbf{G}_0(E))_{ij}b_j \tag{7.3b}$$

Note that once we have found $\mathbf{G}_0(E)$, we can solve Eq. (7.1), given any **b**, just by carrying out the matrix multiplication indicated in Eq. (7.3). Recall that in Exercise 1.12c, we showed that $\mathbf{G}_0(E)$ can be expressed in terms of the eigenvalues and eigenvectors of $\mathbf{H}_0$,

$$\mathbf{H}_0\mathbf{c}^\alpha = E_\alpha^{(0)}\mathbf{c}^\alpha \qquad \alpha = 1, 2, \ldots, N \tag{7.4}$$

as

$$(\mathbf{G}_0(E))_{ij} = \sum_\alpha \frac{c_i^\alpha c_j^{\alpha*}}{E - E_\alpha^{(0)}} \tag{7.5}$$

Note that each element of $\mathbf{G}_0(E)$ "blows up" or, more technically, has a *pole* at values of the parameter E equal to the eigenvalues of $\mathbf{H}_0$. We call the matrix $G_0(E)$ the Green's "matrix" associated with the matrix H_0. As we will see, the Green's function for a differential equation is simply a continuous generalization of this matrix.

Exercise 7.1 Suppose we have a matrix H which is

$$H = H_0 + V$$

Show that the associated Green's matrix

$$G(E) = (E\mathbf{1} - H_0 - V)^{-1}$$

obeys the equation

$$G(E) = G_0(E) + G_0(E)VG(E)$$

Suppose we wish to solve the following inhomogeneous differential equation for $a(x)$

$$(E - \mathscr{H}_0)a(x) = b(x) \tag{7.6}$$

where E is a parameter and $\mathscr{H}_0$ is a differential operator which is Hermitian. Note that the equation which determines the first-order wave function in Rayleigh-Schrödinger perturbation theory is of this form (see Eq. (6.9)). If we know the eigenfunctions and eigenvalues of $\mathscr{H}_0$

$$\mathscr{H}_0\psi_\alpha(x) = E_\alpha^{(0)}\psi_\alpha(x) \tag{7.7}$$

we can expand both $a(x)$ and $b(x)$ in terms of these as

$$a(x) = \sum_\alpha a_\alpha\psi_\alpha(x) \tag{7.8}$$

$$b(x) = \sum_\alpha b_\alpha\psi_\alpha(x) \tag{7.9}$$

The coefficients $\{a_\alpha\}$ are to be determined. Since $b(x)$ is given and the set $\{\psi_\alpha(x)\}$ is orthonormal, we can find b_α by multiplying Eq. (7.9) by $\psi_\alpha^*(x)$ and integrating over all x

$$b_\alpha = \int dx' \; \psi_\alpha^*(x')b(x') \tag{7.10}$$

Substituting expansions (7.8) and (7.9) into Eq. (7.6) and using Eq. (7.7), we find

$$\sum_\alpha a_\alpha(E - \mathscr{H}_0)\psi_\alpha(x) = \sum_\alpha a_\alpha(E - E_\alpha^{(0)})\psi_\alpha(x) = \sum_\alpha b_\alpha\psi_\alpha(x)$$

Multiplying both sides of this equation by $\psi_\alpha^*(x)$ and integrating, we have

$$a_\alpha(E - E_\alpha^{(0)}) = b_\alpha$$

Finally, substituting this into Eq. (7.8), we obtain

$$a(x) = \sum_\alpha \frac{b_\alpha}{E - E_\alpha^{(0)}} \psi_\alpha(x) \tag{7.11}$$

and thus we have solved the problem. Let us rewrite this solution by substituting Eq. (7.10) for b_α and interchanging orders of summation and integration,

$$a(x) = \int dx' \left[\sum_\alpha \frac{\psi_\alpha(x)\psi_\alpha^*(x')}{E - E_\alpha^{(0)}} \right] b(x') \qquad (7.12)$$

If we define the quantity in brackets as the Green's function,

$$G_0(x, x', E) = \sum_\alpha \frac{\psi_\alpha(x)\psi_\alpha^*(x')}{E - E_\alpha^{(0)}} \qquad (7.13)$$

then Eq. (7.12) becomes

$$a(x) = \int dx' \, G_0(x, x', E)b(x') \qquad (7.14)$$

Thus if we can find $G_0(x, x', E)$, then we have reduced the problem of solving the inhomogeneous differential equation (Eq. (7.6)), containing *any* function $b(x)$, to an integration (Eq. (7.14)). Note the similarity with the matrix problem discussed in the beginning of this section (compare Eq. (7.14) with Eq. (7.3b) and compare Eq. (7.13) with Eq. (7.5)). We see that, as before, $G_0(x, x', E)$ has poles at values of E equal to the eigenvalues of $\mathcal{H}_0$.

Let us now find the differential equation obeyed by $G_0(x, x', E)$. Notice that if we let $b(x)$ be the Dirac δ function,

$$b(x) = \delta(x - x')$$

then Eq. (7.14) gives

$$a(x) = \int dx'' \, G_0(x, x'', E)\delta(x' - x'') = G_0(x, x', E)$$

Thus the required differential equation is

$$(E - \mathcal{H}_0)G_0(x, x', E) = \delta(x - x') \qquad (7.15)$$

The matrix analogue of this is simply

$$(E\mathbf{1} - \mathbf{H}_0)(E\mathbf{1} - \mathbf{H}_0)^{-1} = \mathbf{1}$$

Exercise 7.2 Consider the differential equation

$$\frac{d^2}{dx^2} a(x) = b(x) \qquad \alpha \le x \le \beta$$

Find $a(x)$ by using the appropriate Green's function.

a. Convince yourself by a graphical argument that $d^2/dx^2 |x|$, where $|x|$ is the absolute value of x, is proportional to the Dirac δ function, i.e., show that it is zero for all x except $x = 0$ where it is infinity.

b. Evaluate the constant of proportionality by integrating both sides over x and thus show that

$$\frac{d^2}{dx^2}|x| = 2\delta(x).$$

c. From this it follows that

$$\frac{d^2}{dx^2}\left(\tfrac{1}{2}|x - x'|\right) = \delta(x - x')$$

and thus the Green's function corresponding to the operator d^2/dx^2 is $\tfrac{1}{2}|x - x'|$. So from Eq. (7.14) we have

$$a(x) = \frac{1}{2}\int_\alpha^\beta dx' \, |x - x'|b(x')$$

Verify, by simple differentiation, *without* using the result in (b), that this is a solution of the above inhomogeneous differential equation.

Suppose that we have an operator

$$\mathscr{H} = \mathscr{H}_0 + \mathscr{V}(x)$$

and we know the Green's function associated with $\mathscr{H}_0$. To find the Green's function corresponding to $\mathscr{H}$,

$$(E - \mathscr{H}_0 - \mathscr{V}(x))G(x, x', E) = \delta(x - x') \tag{7.16}$$

we rewrite Eq. (7.16) in a form identical to Eq. (7.6)

$$(E - \mathscr{H}_0)G(x, x', E) = \delta(x - x') + \mathscr{V}(x)G(x, x', E)$$

Hence, the solution is immediately given by Eq. (7.14),

$$G(x, x', E) = \int dx'' \, G_0(x, x'', E)[\delta(x'' - x') + \mathscr{V}(x'')G(x'', x', E)]$$

or after carrying out the integration involving the δ function

$$G(x, x', E) = G_0(x, x', E) + \int dx'' \, G_0(x, x'', E)\mathscr{V}(x'')G(x'', x', E) \tag{7.17a}$$

Because the unknown function $G(x, x', E)$ appears under the integral sign, this is called an *integral* equation for G. Note that it is the continuous analogue of the result obtained in Exercise 7.1 for matrices,

$$\mathbf{G}(E) = \mathbf{G}_0(E) + \mathbf{G}_0(E)\mathbf{V}\mathbf{G}(E) \tag{7.17b}$$

In fact, Eq. (7.17b) is just the matrix representation of Eq. (7.17a) in some complete orthonormal basis. If we can solve the integral equation (7.17a) for G, we can obtain the eigenvalues of $\mathscr{H}_0 + \mathscr{V}$ by finding the values of the parameter E for which G blows up.

Finally, we rederive Eqs. (7.17a) and (7.17b) using abstract operator notation. We let

$$\mathscr{G}_0(E) = (E - \mathscr{H}_0)^{-1}$$

$$\mathscr{G}(E) = (E - \mathscr{H}_0 - \mathscr{V})^{-1}$$

so that

$$(\mathscr{G}(E))^{-1} = E - \mathscr{H}_0 - \mathscr{V} = (\mathscr{G}_0(E))^{-1} - \mathscr{V} \tag{7.18}$$

Multiplying this on the left by $\mathscr{G}_0(E)$ and on the right by $\mathscr{G}(E)$ and rearranging we have

$$\mathscr{G}(E) = \mathscr{G}_0(E) + \mathscr{G}_0(E)\mathscr{V}\mathscr{G}(E) \tag{7.19}$$

We now give a simple example of the formalism developed so far. Consider a particle moving in one dimension under the influence of an attractive Dirac δ function potential located at the origin,

$$\mathscr{V}(x) = -\delta(x) \tag{7.20}$$

We wish to find the bound state eigenvalues by first calculating the free-particle Green's function $G_0(x, x', E)$, then solving the integral equation (7.17a) for $G(x, x', E)$ and then, finally, finding the values of E for which $G(x, x', E)$ blows up. Since $\mathscr{H}_0 = -\frac{1}{2}d^2/dx^2$, Eq. (7.15) becomes

$$\left(E + \frac{1}{2}\frac{d^2}{dx^2}\right)G_0(x, x', E) = \delta(x - x') \tag{7.21}$$

This equation is most easily solved by first Fourier transforming it and then evaluating the inverse transform by contour integration.[1] The result is

$$G_0(x, x', E) = \frac{1}{i(2E)^{1/2}} \exp(i(2E)^{1/2}|x - x'|) \tag{7.22}$$

where $|x|$ is the absolute value of x.

Exercise 7.3 Show that the result in Eq. (7.22) does indeed satisfy Eq. (7.21). To do so you will need the relation obtained in Exercise 7.2b and should convince yourself that $(d/dx\,|x|)^2 = 1$ for all x.

Substituting Eq. (7.20) into Eq. (7.17a) we have

$$G(x, x', E) = G_0(x, x', E) - \int dx'' \, G_0(x, x'', E)\delta(x'')G(x'', x', E)$$

$$= G_0(x, x', E) - G_0(x, 0, E)G(0, x', E) \tag{7.23}$$

Note that the integral equation has reduced to an algebraic equation. To solve it, we need to know $G(0, x', E)$ on the right hand side. So setting $x = 0$, we have

$$G(0, x', E) = G_0(0, x', E) - G_0(0, 0, E)G(0, x', E) \tag{7.24}$$

Solving this equation for $G(0, x', E)$ and substituting the result back into Eq. (7.23), we find

$$G(x, x', E) = G_0(x, x', E) - \frac{G_0(x, 0, E)G_0(0, x', E)}{1 + G_0(0, 0, E)} \qquad (7.25)$$

Using the result for G_0 in Eq. (7.22), we have

$$G(x, x', E) = \frac{1}{i(2E)^{1/2}} \left(\exp(i(2E)^{1/2}|x - x'|) - \frac{\exp(i(2E)^{1/2}(|x| + |x'|))}{1 + i(2E)^{1/2}} \right)$$

Finally, note that the only nonzero value of E for which G blows up occurs when $i(2E)^{1/2} = -1$, so that $E = -1/2$ is the energy of the only bound state. This result is an example of a general theorem that in one dimension any attractive potential has at least one bound state.

Exercise 7.4 From Eq. (7.13) as applied to $G(x, x', E)$ it follows that

$$\phi_n(x)\phi_n^*(x') = \lim_{E \to E_n} (E - E_n)G(x, x', E)$$

where $\phi_n(x)$ is the normalized eigenfunction of $\mathscr{H}$ with energy E_n. For the attractive δ function potential, show that $\phi(x) = \exp(-|x|)$.

Exercise 7.5 Verify that $\phi(x) = \exp(-|x|)$ is indeed an eigenfunction of

$$\mathscr{H} = -\frac{1}{2}\frac{d^2}{dx^2} - \delta(x)$$

with eigenvalue $-\frac{1}{2}$ by calculating $\mathscr{H}\phi$.

Exercise 7.6 Consider the use of Green's functions in time-dependent quantum mechanics. The time-dependent Schrödinger equation is

$$i\frac{\partial\phi(x, t)}{\partial t} = \mathscr{H}\phi(x, t)$$

a. Given that $G(x, x', t)$ is a solution of the Schrödinger equation

$$i\frac{\partial G(x, x', t)}{\partial t} = \mathscr{H}G(x, x', t)$$

where $\mathscr{H}$ operates on the coordinate x, subject to the initial condition that at $t = 0$

$$G(x, x', 0) = \delta(x - x')$$

show for any $\psi(x)$, that

$$\phi(x, t) = \int dx' \, G(x, x', t)\psi(x')$$

satisfies the time-dependent Schrödinger equation and $\phi(x, 0) = \psi(x)$. Thus given an arbitrary initial wave function $\psi(x)$, we can find the wave function at a later time by using this equation. In this sense, $G(x, x', t)$ "propagates" the initial wave function through time and hence is called a propagator.

b. To make connection with the theory discussed in this section, show that

$$G(x, x', E) = \lim_{\varepsilon \to 0} (-i) \int_0^\infty dt \, e^{iEt} e^{-\varepsilon t} G(x, x', t)$$

$$= -i \int_0^\infty dt \, e^{iEt} G(x, x', t)$$

is a solution of

$$(E - \mathcal{H})G(x, x', E) = \delta(x - x')$$

The parameter $\varepsilon(\varepsilon > 0)$ is a convergence factor which is set equal to zero at the end of the calculation. *Hint:* First multiply both sides of the differential equation obeyed by $G(x, x', t)$ by $-ie^{(iE - \varepsilon)t}$ and integrate over t, and thus obtain

$$\lim_{\varepsilon \to 0} \int_0^\infty dt \, e^{(iE - \varepsilon)t} \frac{\partial G(x, x', t)}{\partial t} = \mathcal{H} \, G(x, x', E)$$

Then integrate the left-hand side by parts to show

$$\lim_{\varepsilon \to 0} \left[e^{(iE - \varepsilon)t} G(x, x', t) \Big|_0^\infty - (iE - \varepsilon) \int_0^\infty dt \, e^{(iE - \varepsilon)t} G(x, x', t) \right] = \mathcal{H} G(x, x', E)$$

c. Redo this problem using abstract operator notation. Manipulate operators as if they were numbers. First, show that $\mathcal{G}(t) = e^{-i\mathcal{H}t}$ satisfies the differential equation

$$i \frac{\partial}{\partial t} \mathcal{G}(t) = \mathcal{H} \mathcal{G}(t)$$

Then show that

$$\mathcal{G}(E) = -i \lim_{\varepsilon \to 0} \int_0^\infty dt \, e^{iEt} e^{-\varepsilon t} \mathcal{G}(t)$$

$$= \frac{1}{E - \mathcal{H}} \equiv (E - \mathcal{H})^{-1}$$

7.2 THE ONE-PARTICLE MANY-BODY GREEN'S FUNCTION

We have seen that in single-particle quantum mechanics the Green's function has poles at values of E equal to the eigenvalues of the Hamiltonian. To generalize Green's function theory to many-particle systems, we first consider an independent particle description, such as the Hartree-Fock (HF),

where

$$\mathscr{H}_0 = \sum_i f(i) \tag{7.26}$$

In the HF approximation for an N-particle system, we obtain a set of spin orbitals and orbital energies by solving the eigenvalue problem

$$f\chi_i(\mathbf{x}) = \varepsilon_i\chi_i(\mathbf{x}) \tag{7.27}$$

Because the χ's and ε's are single-particle quantities, in analogy with Eq. (7.13), it is natural to define the Hartree-Fock Green's function (HFGF) as

$$G_0(\mathbf{x}, \mathbf{x}', E) = \sum_i \frac{\chi_i(\mathbf{x})\chi_i^*(\mathbf{x}')}{E - \varepsilon_i} \tag{7.28}$$

where the summation runs over both occupied and unoccupied spin orbitals,

$$G_0(\mathbf{x}, \mathbf{x}', E) = \sum_a \frac{\chi_a(\mathbf{x})\chi_a^*(\mathbf{x}')}{E - \varepsilon_a} + \sum_r \frac{\chi_r(\mathbf{x})\chi_r^*(\mathbf{x}')}{E - \varepsilon_r} \tag{7.29}$$

The matrix representation of the HFGF in the basis of HF spin orbitals is

$$(\mathbf{G}_0(E))_{ij} = \iint d\mathbf{x}\,d\mathbf{x}'\,\chi_i^*(\mathbf{x})G_0(\mathbf{x}, \mathbf{x}', E)\chi_j(\mathbf{x}') = \frac{\delta_{ij}}{E - \varepsilon_i} \tag{7.30}$$

Thus

$$\mathbf{G}_0(E) = (E\mathbf{1} - \boldsymbol{\varepsilon})^{-1} \tag{7.31}$$

where $\boldsymbol{\varepsilon}$ is a diagonal matrix containing the HF orbital energies. From now on when we talk about MBGF's we will mean the matrix representations of these functions in the basis of HF orbitals of the N-particle system. $\mathbf{G}_0(E)$ has poles at values of E for which the inverse $(E\mathbf{1} - \boldsymbol{\varepsilon})^{-1}$ does not exist (i.e., it is infinite). The inverse of a matrix does not exist when its determinant is zero. Thus $\mathbf{G}_0(E)$ has poles for values of E for which

$$\det(E\mathbf{1} - \boldsymbol{\varepsilon}) = 0$$

Since $\boldsymbol{\varepsilon}$ is a diagonal matrix, we have

$$\det(E\mathbf{1} - \boldsymbol{\varepsilon}) = \prod_i (E - \varepsilon_i) = 0$$

So that the HFGF has poles at the Hartree-Fock orbital energies. This result is also evident from Eq. (7.30).

Recall that Koopmans' theorem says that these orbital energies are related to the ionization potentials and electron affinities of an N-particle system. In particular, if $|^N\Psi_0\rangle$ is the Hartree Fock wave function for the N-particle system and $|^{N-1}\Psi_c\rangle$ is an approximate wave function for the $(N-1)$-electron system obtained by removing an electron from spin orbital

c, then

$$-\text{IP} = \varepsilon_c = \langle {}^N\Psi_0 | \mathcal{H} | {}^N\Psi_0 \rangle - \langle {}^{N-1}\Psi_c | \mathcal{H} | {}^{N-1}\Psi_c \rangle \qquad (7.32a)$$

Similarly

$$-\text{EA} = \varepsilon_r = \langle {}^{N+1}\Psi^r | \mathcal{H} | {}^{N+1}\Psi^r \rangle - \langle {}^N\Psi_0 | \mathcal{H} | {}^N\Psi_0 \rangle \qquad (7.32b)$$

where $|{}^{N+1}\Psi^r\rangle$ is an approximate wave function for the $(N + 1)$-particle system obtained by putting an electron into spin orbital r. As we discussed in Chapter 3, the orbital energies do not give the exact IP's and EA's because of two distinct effects. First, $|{}^{N-1}\Psi_c\rangle$ and $|{}^{N+1}\Psi^r\rangle$ are not the HF wave functions of the $(N - 1)$- and $(N + 1)$-particle systems since they contain the spin orbitals of the N-particle system. In general, the HF orbitals of the $N - 1$, N, and $N + 1$ systems differ. Thus the HF energy of, say, the $(N - 1)$-particle system, which was obtained by removing an electron from the spin orbital c, is

$$^{N-1}E_0(c) = \langle {}^{N-1}\Psi_c | \mathcal{H} | {}^{N-1}\Psi_c \rangle + {}^{N-1}E_R(c) \qquad (7.33)$$

where $^{N-1}E_R(c)$ is called the *relaxation* energy. Second, the correlation energies of the N and $N \pm 1$ systems must be included,

$$^N\mathscr{E}_0 = {}^NE_0 + {}^NE_{\text{corr}} \qquad (7.34a)$$

and for the $(N - 1)$-particle system with an electron in spin orbital c missing

$$^{N-1}\mathscr{E}_0(c) = {}^{N-1}E_0(c) + {}^{N-1}E_{\text{corr}}(c) \qquad (7.34b)$$

Hence, the exact ionization potential can be written as

$$
\begin{aligned}
-\text{IP} &= {}^N\mathscr{E}_0 - {}^{N-1}\mathscr{E}_0(c) \\
&= \varepsilon_c - {}^{N-1}E_R(c) + ({}^NE_{\text{corr}} - {}^{N-1}E_{\text{corr}}(c)) \qquad (7.35)
\end{aligned}
$$

where we have used Eqs. (7.32a, 7.33, 7.34a,b). Thus to get the exact IP, the Koopmans' theorem result must be corrected by the relaxation energy of the $(N - 1)$-particle system and the difference in the correlation energies of the N- and $(N - 1)$-particle system. The situation for electron affinities is analogous.

7.2.1 The Self-Energy

We have seen that the HFGF $(G_0(E))$ has poles at values of E that approximate the energy differences between the N and the $(N \pm 1)$-particle systems (i.e., at the orbital energies). It is not difficult to believe that the HFGF is an approximation to an *exact* MBGF, $G(E)$, which has poles at the exact energy differences between the N and the $(N \pm 1)$-particle systems. While the "one-body" GF of Section 7.1 has poles at the eigenvalues of a Hamiltonian, the "many-body" GF has poles at differences *between* eigenvalues.

If we could obtain $G(E)$, or at least a better approximation to it than $G_0(E)$, then we would be able to improve upon Koopmans' theorem IP's and EA's while retaining the one-particle picture associated with HF theory. At first glance it does not appear possible to construct an *exact* one-particle theory since the many-electron Hamiltonian contains two-particle interactions. F. Dyson surmounted this apparent difficulty by introducing an effective potential which was *energy dependent*, called the *self-energy*. Moreover, he showed that the exact $G(E)$ obeys the integral equation (now called the *Dyson equation*):

$$G(E) = G_0(E) + G_0(E)\Sigma(E)G(E) \qquad (7.36)$$

where $\Sigma(E)$ is the matrix representation of the exact self-energy in the basis of HF spin orbitals. Finally, various terms in the perturbation expansion of $\Sigma(E)$ can be generated,

$$\Sigma(E) = \Sigma^{(2)}(E) + \Sigma^{(3)}(E) + \cdots \qquad (7.37)$$

In particular, the matrix elements of the second-order self-energy ($\Sigma^{(2)}(E)$) are

$$\Sigma_{ij}^{(2)}(E) = \frac{1}{2} \sum_{ars} \frac{\langle rs||ia\rangle\langle ja||rs\rangle}{E + \varepsilon_a - \varepsilon_r - \varepsilon_s} + \frac{1}{2} \sum_{abr} \frac{\langle ab||ir\rangle\langle jr||ab\rangle}{E + \varepsilon_r - \varepsilon_a - \varepsilon_b} \qquad (7.38)$$

We shall spend the rest of this chapter trying to get a feeling for this rather unusual formalism. Comparing the Dyson equation (Eq. (7.36)) with Eqs. (7.17) and (7.19) we see why $\Sigma(E)$ is called an energy-dependent potential. It should be stressed that the above formalism is exact; that is, if we use the exact $\Sigma(E)$ and solve the Dyson equation for $G(E)$, then $G(E)$ will have poles at the *exact* IP's and EA's of the N-particle system. Of course, one has to make approximations and in GF theory one approximates the self-energy. If $\Sigma(E) = 0$, then $G(E) = G_0(E)$. As can be seen from Eq. (7.37), the lowest-order correction to $\Sigma(E)$ occurs in second order of perturbation theory. This is similar to many-body perturbation theory, where the first correction to the HF energy also occurs in second order. In fact, by comparing the expressions for $E_0^{(2)}$ (see Eq. (6.72)), and for $\Sigma_{ij}^{(2)}(E)$ (see Eq. (7.38)) we can see certain common features, i.e., there are two antisymmetrized matrix elements in the numerators and both have single term denominators containing some orbital energies. In Section 7.5, we shall investigate the relationship between GF theory, using $\Sigma^{(2)}(E)$, and second-order perturbation theory for the calculation of ionization potentials. One of the most attractive features of the MBGF formalism is that the dimensionality of all the required matrices is equal to the number of HF spin orbitals. Thus this formalism generalizes HF theory while retaining a single-particle picture. This is accomplished at the expense of introducing a potential which is energy dependent. Before considering the MBGF formalism in more detail, we present the expression for $\Sigma_{ij}^{(2)}(E)$ involving summations over *spatial* orbitals:

$$\Sigma_{ij}^{(2)}(E) = \sum_{ars}^{N/2} \frac{\langle rs|ia\rangle(2\langle ja|rs\rangle - \langle aj|rs\rangle)}{E + \varepsilon_a - \varepsilon_r - \varepsilon_s}$$

$$+ \sum_{abr}^{N/2} \frac{\langle ab|ir\rangle(2\langle jr|ab\rangle - \langle rj|ab\rangle)}{E + \varepsilon_r - \varepsilon_a - \varepsilon_b} \tag{7.39}$$

This form is convenient for the calculation of ionization potentials of closed-shell N-particle systems.

Exercise 7.7 Obtain Eq. (7.39) from Eq. (7.38).

Exercise 7.8 The exact MBGF of an N-particle system can be defined as

$$(G(E))_{ij} = \sum_m \frac{\langle {}^N\Phi_0|a_i^\dagger|{}^{N-1}\Phi_m\rangle\langle {}^{N-1}\Phi_m|a_j|{}^N\Phi_0\rangle}{E - ({}^N\mathscr{E}_0 - {}^{N-1}\mathscr{E}_m)}$$

$$+ \sum_p \frac{\langle {}^N\Phi_0|a_j|{}^{N+1}\Phi_p\rangle\langle {}^{N+1}\Phi_p|a_i^\dagger|{}^N\Phi_0\rangle}{E - ({}^{N+1}\mathscr{E}_p - {}^N\mathscr{E}_0)}$$

where the sums over m and p run over *all* the states of the $N - 1$- and $N + 1$-particle systems, respectively. $|{}^M\Phi_k\rangle$, $M = N$, $N \pm 1$, is the exact full CI wave function containing the HF spin orbitals of the N-particle system, of the kth state of the M-particle system with energy ${}^M\mathscr{E}_k = \langle {}^M\Phi_k|\mathscr{H}|{}^M\Phi_k\rangle$. The creation $(a_i^\dagger)$ and annihilation (a_i) operators add and remove HF spin orbitals, respectively.

Show that this GF reduces to the HFGF given in Eq. (7.30) when

1. $|{}^N\Phi_0\rangle$ is replaced by the HF wave function $|{}^N\Psi_0\rangle$.
2. The states of the $(N - 1)$-particle system are approximated by $a_a|{}^N\Psi_0\rangle$.
3. The states of the $(N + 1)$-particle system are approximated by $a_r^\dagger|{}^N\Psi_0\rangle$.

7.2.2 The Solution of the Dyson Equation

To obtain the IP's and EA's of an N-particle system, we must solve the Dyson equation (Eq. (7.36)) for $G(E)$ and then find the values of E for which $G(E)$ is infinite. Multiplying the Dyson equation by $(G_0(E))^{-1}$ on the left and $(G(E))^{-1}$ on the right we have

$$(G_0(E))^{-1} = (G(E))^{-1} + \Sigma(E) \tag{7.40}$$

Solving this for $G(E)$, we find

$$G(E) = ((G_0(E))^{-1} - \Sigma(E))^{-1} = (E\mathbf{1} - \varepsilon - \Sigma(E))^{-1} \tag{7.41}$$

where we have used the definition of $G_0(E)$ given in Eq. (7.31). Since the inverse of a matrix does not exist when its determinant is zero, we must determine the roots of the equation

$$\det(E\mathbf{1} - \boldsymbol{\varepsilon} - \boldsymbol{\Sigma}(E)) = 0 \tag{7.42}$$

When $\boldsymbol{\Sigma}(E) = \mathbf{0}$, the roots occur at the ε_i's. To find the lowest-order correction to these Koopmans' theorem results, let us ignore the off-diagonal elements of $\boldsymbol{\Sigma}(E)$. Then Eq. (7.42) simplifies to

$$\prod_i (E - \varepsilon_i - \Sigma_{ii}(E)) = 0 \tag{7.43}$$

To find the correction to ε_i, we must solve

$$E = \varepsilon_i + \Sigma_{ii}(E) \tag{7.44}$$

for E. This can be done iteratively by first evaluating $\Sigma_{ii}(E)$ at $E = \varepsilon_i$ to obtain

$$\varepsilon_i' = \varepsilon_i + \Sigma_{ii}(\varepsilon_i) \tag{7.45}$$

and then finding ε_i'' by evaluating $\Sigma_{ii}(E)$ at $E = \varepsilon_i'$ and so on until convergence is found. The lowest-order correction to ε_i is given by Eq. (7.45) with the second-order self-energy,

$$\varepsilon_i' = \varepsilon_i + \Sigma_{ii}^{(2)}(\varepsilon_i) \tag{7.46}$$

We will analyze the results obtained using this equation in some detail.

7.3 APPLICATION OF THE FORMALISM TO H_2 AND HeH^+

A good way of developing a feeling for the way GF theory, using $\Sigma^{(2)}(E)$, improves upon the Koopmans' theorem IP's and EA's is to apply the formalism to some simple cases where the exact results are known. Therefore, we now apply the theory to minimal basis H_2 and HeH^+. In the next section, we shall generalize some of the insights gained here. We focus only on the IP's and the analysis of the EA is left for the exercises.

Recall that for minimal basis H_2 we have one occupied (labeled 1) and one unoccupied (labeled 2) molecular orbital with orbital energies

$$\varepsilon_1 = h_{11} + J_{11} \tag{7.47a}$$

$$\varepsilon_2 = h_{22} + 2J_{12} - K_{12} \tag{7.47b}$$

The HF energy for this model is

$$^N E_0 = 2h_{11} + J_{11} = 2\varepsilon_1 - J_{11} \tag{7.48}$$

while the correlation energy is

$$^N E_{\text{corr}} = \Delta - (\Delta^2 + K_{12}^2)^{1/2} \tag{7.49}$$

where

$$\Delta = (\varepsilon_2 - \varepsilon_1) + \frac{1}{2}(J_{11} + J_{22}) - 2J_{12} + K_{12} \qquad (7.50)$$

Thus the exact energy of this system is $^N\mathscr{E}_0 = {}^NE_0 + {}^NE_{\text{corr}}$. The exact eigenstates of the $(N - 1)$-particle system, i.e., H_2^+, are easily found not only because there is no correlation in this molecule, but also because in the minimal basis the HF orbitals of H_2 are also the optimum orbitals of H_2^+ (i.e., in this model the HF orbitals of the N and $(N - 1)$ systems are the same and the relaxation energy is zero). This follows from the fact that the two HF orbitals of H_2 have different symmetry and hence $h_{12} = 0$. Therefore, the ground state energy of H_2^+ is

$$^{N-1}\mathscr{E}_0 = h_{11} \qquad (7.51a)$$

while the first excited state has energy

$$^{N-1}\mathscr{E}_1 = h_{22} \qquad (7.51b)$$

Thus the exact ionization potential of minimal basis H_2 is

$$-\text{IP} = {}^N\mathscr{E}_0 - {}^{N-1}\mathscr{E}_0 = 2h_{11} + J_{11} + {}^NE_{\text{corr}} - h_{11} = \varepsilon_1 + {}^NE_{\text{corr}} \qquad (7.52a)$$

Note that the ionization potential differs from its Koopmans' theorem value by the correlation energy of H_2. This result is not true in general and is a consequence of the simplicity of the model. It should be pointed out that we have calculated a *vertical* IP, because we have assumed that H_2 and H_2^+ have the same internuclear separation. Finally, it is possible to remove an electron from H_2 and end up with an excited state of H_2^+. The ionization potential for this process is

$$-\text{IP}' = {}^N\mathscr{E}_0 - {}^{N-1}\mathscr{E}_1 = 2h_{11} + J_{11} + {}^NE_{\text{corr}} - h_{22}$$
$$= 2\varepsilon_1 - \varepsilon_2 + (2J_{12} - J_{11} - K_{12}) + {}^NE_{\text{corr}} \qquad (7.52b)$$

We will now apply GF theory to this system and compare the results obtained with the exact answers. The second-order self-energy for this model is easily found from Eq. (7.39) by setting all the hole and particle indices equal to one and two, respectively (i.e., $a = b = 1, r = s = 2$),

$$\Sigma_{11}^{(2)}(E) = \frac{K_{12}^2}{E + \varepsilon_1 - 2\varepsilon_2} = \frac{K_{12}^2}{E - \varepsilon_1 + 2(\varepsilon_1 - \varepsilon_2)} \qquad (7.53a)$$

$$\Sigma_{22}^{(2)}(E) = \frac{K_{12}^2}{E + \varepsilon_2 - 2\varepsilon_1} = \frac{K_{12}^2}{E - \varepsilon_2 - 2(\varepsilon_1 - \varepsilon_2)} \qquad (7.53b)$$

$$\Sigma_{12}^{(2)}(E) = \Sigma_{21}^{(2)}(E) = 0 \qquad (7.53c)$$

where we have used the fact that all two-electron integrals, which do not contain an even number of 1's or 2's, vanish by symmetry. Since $\mathbf{\Sigma}^{(2)}(E)$ is a diagonal matrix, the matrix elements of $G(E)$ calculated from Eq. (7.41) are just

$$G_{11}(E) = (E - \varepsilon_1 - \Sigma_{11}^{(2)}(E))^{-1} \tag{7.54a}$$

$$G_{22}(E) = (E - \varepsilon_2 - \Sigma_{22}^{(2)}(E))^{-1} \tag{7.54b}$$

$$G_{12}(E) = G_{21}(E) = 0 \tag{7.54c}$$

To find the IP's and EA's, we must find the poles of $G_{11}(E)$ and $G_{22}(E)$. The poles of $G_{11}(E)$ occur when

$$E - \varepsilon_1 - \Sigma_{11}^{(2)}(E) = 0 \tag{7.55a}$$

Substituting the expression for $\Sigma_{11}^{(2)}(E)$, given in (7.53a), we have

$$E - \varepsilon_1 - \frac{K_{12}^2}{E - \varepsilon_1 + 2(\varepsilon_1 - \varepsilon_2)} = 0 \tag{7.55b}$$

Before solving this quadratic equation for the two roots, let us find the lowest-order correction to ε_1. Using Eq. (7.46), which is equivalent to setting E equal to ε_1 in $\Sigma_{11}^{(2)}(E)$, we have

$$\varepsilon_1' = \varepsilon_1 + \Sigma_{11}^{(2)}(\varepsilon_1) = \varepsilon_1 + \frac{K_{12}^2}{2(\varepsilon_1 - \varepsilon_2)} = \varepsilon_1 + {}^N E_0^{(2)} \tag{7.56}$$

where we have noticed that $K_{12}^2/2(\varepsilon_1 - \varepsilon_2)$ is just the second-order perturbation result for the correlation energy of H_2 found in Chapter 6 (see Eq. (6.77)). This is a very pleasing result. By comparing with the exact result for the model given in Eq. (7.52a) we see that the GF formalism, using $\Sigma^{(2)}(E)$, in lowest order is correcting the Koopmans' theorem result (ε_1) by the second-order many-body perturbation energy. What happens if we solve Eq. (7.55) exactly? The two roots of the quadratic equation are

$$\varepsilon_{11}^{\pm} = \varepsilon_1 + ((\varepsilon_2 - \varepsilon_1) \pm ((\varepsilon_2 - \varepsilon_1)^2 + K_{12}^2)^{1/2})$$

$$= \varepsilon_1 + \left((1 \pm 1)(\varepsilon_2 - \varepsilon_1) \mp \frac{K_{12}^2}{2(\varepsilon_1 - \varepsilon_2)} \pm \frac{K_{12}^4}{8(\varepsilon_1 - \varepsilon_2)^3} + \cdots \right)$$

$$= \varepsilon_1 + \left((1 \pm 1)(\varepsilon_2 - \varepsilon_1) \mp {}^N E_0^{(2)} \pm \frac{K_{12}^4}{8(\varepsilon_1 - \varepsilon_2)^3} + \cdots \right) \tag{7.57}$$

where we have expanded the square root. What is the significance of these two roots? Consider ε_{11}^-,

$$\varepsilon_{11}^- = \varepsilon_1 + ((\varepsilon_2 - \varepsilon_1) - ((\varepsilon_2 - \varepsilon_1)^2 + K_{12}^2)^{1/2})$$

$$= \varepsilon_1 + {}^N E_0^{(2)} - \frac{K_{12}^4}{8(\varepsilon_1 - \varepsilon_2)^3} + \cdots \tag{7.58}$$

Thus, ε_{11}^- is the negative of the ionization potential of H_2. Note that the correlation energy implicit in the full GF treatment, using $\Sigma^{(2)}(E)$, has the same form as the exact result in Eq. (7.49). The difference is that in the GF result Δ is simply $(\varepsilon_2 - \varepsilon_1)$ (see Eq. (7.50)). The perturbation expansion of this correlation energy is exact in second order and also contains approximate higher order terms. What is the significance of the other root ε_{11}^+? Exercise 7.9 will ask you to show that it is an approximation to the energy for an electron capture process in which H_2^- ends up in an excited state.

Exercise 7.9 Consider the $(N + 1)$-particle system, H_2^-.

a. Show that in the minimal basis description

$$^{N+1}\mathscr{E}_0 - {}^N\mathscr{E}_0 = \varepsilon_2 - {}^N E_{corr}$$

$$^{N+1}\mathscr{E}_1 - {}^N\mathscr{E}_0 = 2\varepsilon_2 - \varepsilon_1 + (J_{22} + K_{12} - 2J_{12}) - {}^N E_{corr}.$$

b. Show that ε_{11}^+ (see Eq. (7.57)) approximates $^{N+1}\mathscr{E}_1 - {}^N\mathscr{E}_0$.

c. Show that the poles of $G_{22}(E)$ occur at

$$\varepsilon_{22}^{\pm} = \varepsilon_2 - ((\varepsilon_2 - \varepsilon_1) \mp ((\varepsilon_2 - \varepsilon_1)^2 + K_{12}^2)^{1/2}).$$

d. Show that ε_{22}^+ approximates $^{N+1}\mathscr{E}_0 - {}^N\mathscr{E}_0$ while ε_{22}^- approximates $^N\mathscr{E}_0 - {}^{N-1}\mathscr{E}_1$ (see Eq. (7.52b)).

Exercise 7.10 The matrix elements of the third-order self-energy, $\Sigma^{(3)}(E)$, for minimal basis H_2 are

$$\Sigma_{11}^{(3)}(E) = \frac{K_{12}^2(J_{22} - 2J_{12} + K_{12})}{(E - 2\varepsilon_2 + \varepsilon_1)^2} + \frac{K_{12}^2(J_{11} - 2J_{12} + K_{12})}{(E - 2\varepsilon_2 + \varepsilon_1)(\varepsilon_1 - \varepsilon_2)}$$

$$+ \frac{K_{12}^2(2J_{12} - K_{12} - J_{11})}{4(\varepsilon_1 - \varepsilon_2)^2}$$

$$\Sigma_{22}^{(3)}(E) = \frac{K_{12}^2(2J_{12} - K_{12} - J_{11})}{(E - 2\varepsilon_1 + \varepsilon_2)^2} + \frac{K_{12}^2(J_{22} - 2J_{12} + K_{12})}{(E - 2\varepsilon_1 + \varepsilon_2)(\varepsilon_1 - \varepsilon_2)}$$

$$+ \frac{K_{12}^2(J_{22} + K_{12} - 2J_{12})}{4(\varepsilon_1 - \varepsilon_2)^2}$$

$$\Sigma_{12}^{(3)}(E) = \Sigma_{21}^{(3)}(E) = 0$$

Note that the third-order self-energy has an *energy-independent* part. Now consider the lowest-order approximation to the first IP (i.e., Eq. (7.45)) using $\Sigma(E) = \Sigma^{(2)}(E) + \Sigma^{(3)}(E)$,

$$\varepsilon_1' = \varepsilon_1 + \Sigma_{11}^{(2)}(\varepsilon_1) + \Sigma_{11}^{(3)}(\varepsilon_1)$$

Show that the right-hand side of this equation is equal to $\varepsilon_1 + {}^N E_0^{(2)} + {}^N E_0^{(3)}$, where ${}^N E_0^{(3)}$ is the third-order many-body perturbation result for the correlation energy for H_2 (see Eq. (6.78)). Similarly, show that the lowest-order correction to the EA,

$$\varepsilon_2' = \varepsilon_2 + \Sigma_{22}^{(2)}(\varepsilon_2) + \Sigma_{22}^{(3)}(\varepsilon_2)$$

is equal to $\varepsilon_2 - ({}^N E_0^{(2)} + {}^N E_0^{(3)})$. Thus for this model, the correlation energy implicit in the GF formalism, using a self-energy correct up to third order and the simplest approximation for finding the roots, is just the many-body perturbation result up to third order.

We now consider the application of GF theory to minimal basis HeH$^+$. This is a slightly more complicated and hence a more interesting system because the HF orbitals of HeH$^+$ are not the optimum orbitals of HeH^{2+} (i.e., the $(N-1)$-particle system). Since in HeH$^+$ one cannot classify the HF orbitals as gerade or ungerade, h_{12} does not vanish by symmetry. In fact

$$h_{12} = -\langle 11|12 \rangle \tag{7.59}$$

since $f_{12} = h_{12} + \langle 11|12 \rangle$ is equal to zero. Thus the ground state energy of HeH^{2+} is not h_{11}, but can be written as

$$^{N-1}\mathscr{E}_0 = h_{11} + {}^{N-1}E_R \tag{7.60}$$

where ${}^{N-1}E_R$ is called the *orbital relaxation* energy. Since the exact energy of HeH$^+$ is

$$^N\mathscr{E}_0 = {}^N E_0 + {}^N E_{\text{corr}} = 2h_{11} + J_{11} + {}^N E_{\text{corr}} \tag{7.61}$$

and the negative of the ionization potential of HeH$^+$ is

$$^N\mathscr{E}_0 - {}^{N-1}\mathscr{E}_0 = 2h_{11} + J_{11} + {}^N E_{\text{corr}} - h_{11} - {}^{N-1}E_R = \varepsilon_1 + {}^N E_{\text{corr}} - {}^{N-1}E_R \tag{7.62}$$

Thus the exact IP differs from the Koopmans' value because of: 1) the correlation in the two-electron system and 2) orbital relaxation in the one-electron system.

We now derive an approximate expression for ${}^{N-1}E_R$. To find the exact ground state energy for HeH^{2+} in the minimal basis, we must solve the eigenvalue problem

$$\begin{pmatrix} h_{11} & h_{12} \\ h_{12} & h_{22} \end{pmatrix} \begin{pmatrix} 1 \\ c \end{pmatrix} = {}^{N-1}\mathscr{E}_0 \begin{pmatrix} 1 \\ c \end{pmatrix} \tag{7.63}$$

Using Eqs. (7.60), (7.59), and (7.47a, b) and proceeding in the standard way, it can be shown that Eq. (7.63) leads to

$$^{N-1}E_R = \frac{|\langle 11|12 \rangle|^2}{\varepsilon_1 - \varepsilon_2 - (J_{11} - 2J_{12} + K_{12}) + {}^{N-1}E_R} \tag{7.64}$$

Instead of solving this equation exactly for $^{N-1}E_R$, we obtain the following approximate expression

$$^{N-1}\tilde{E}_R^{(2)} = \frac{|\langle 11|12\rangle|^2}{\varepsilon_1 - \varepsilon_2} \tag{7.65}$$

by ignoring $^{N-1}E_R$ and the two-electron integrals in the denominator. It is clear from the form of this expression that it can be obtained from some sort of second-order perturbation theory. In the next section, we clarify just what sort of perturbation theory.

Exercise 7.11 Derive Eq. (7.64) from Eq. (7.63).

We now apply GF theory, using $\Sigma^{(2)}(E)$, to HeH^+. From Eq. (7.39) by setting $a = b = 1$ and $r = s = 2$, we find

$$\Sigma_{11}^{(2)}(E) = \frac{K_{12}^2}{E + \varepsilon_1 - 2\varepsilon_2} + \frac{|\langle 11|12\rangle|^2}{E + \varepsilon_2 - 2\varepsilon_1} \tag{7.66a}$$

$$\Sigma_{22}^{(2)}(E) = \frac{|\langle 22|12\rangle|^2}{E + \varepsilon_1 - 2\varepsilon_2} + \frac{K_{12}^2}{E + \varepsilon_2 - 2\varepsilon_1} \tag{7.66b}$$

$$\Sigma_{12}^{(2)}(E) = \Sigma_{21}^{(2)}(E) = \frac{K_{12}\langle 22|12\rangle}{E + \varepsilon_1 - 2\varepsilon_2} + \frac{\langle 11|12\rangle K_{12}}{E + \varepsilon_2 - 2\varepsilon_1} \tag{7.66c}$$

Note that $\Sigma_{12}^{(2)}(E)$ is not zero as it was for H_2. Instead of calculating $G(E)$ using Eq. (7.41), which involves inverting a two-by-two matrix, we find the lowest-order correction to the Koopmans' IP using Eq. (7.46), i.e.,

$$\varepsilon_1' = \varepsilon_1 + \Sigma_{11}^{(2)}(\varepsilon_1) = \varepsilon_1 + \frac{K_{12}^2}{2(\varepsilon_1 - \varepsilon_2)} - \frac{|\langle 11|12\rangle|^2}{\varepsilon_1 - \varepsilon_2} \tag{7.67}$$

Note that $K_{12}^2/2(\varepsilon_1 - \varepsilon_2)$ is the second-order energy for HeH^+ ($^N E_0^{(2)}$) and $|\langle 11|12\rangle|^2/(\varepsilon_1 - \varepsilon_2)$ is the approximate expression for the relaxation energy of HeH^{2+} ($^{N-1}\tilde{E}_R^{(2)}$) in Eq. (7.65). Thus

$$\varepsilon_1' = \varepsilon_1 + {}^N E_0^{(2)} - {}^{N-1}\tilde{E}_R^{(2)} \tag{7.68}$$

Comparing this with the form of the exact result in Eq. (7.62), we see that GF theory, using $\Sigma^{(2)}(E)$, takes into account both the correlation energy of the N-particle system and the orbital relaxation energy of the $(N-1)$-particle system. When the simplest approximation is used to find the poles (i.e., Eq. (7.46)) these effects are included to second order in perturbation theory. If we were to solve the Dyson equation for $G(E)$ and then determine the poles exactly, these effects would be approximately treated in higher orders. Since we considered only two-electron systems in this section, we were not able to determine whether the GF method also takes into account correlation in the $(N - 1)$-particle system. In the next section we shall see that it does account for this effect.

Exercise 7.12 Consider the EA of HeH$^+$.

a. Show that the exact ground state energy of HeH is

$$^{N+1}\mathscr{E}_0 = \varepsilon_2 + {}^N E_0 + {}^{N+1} E_R$$

where $^{N+1} E_R$ is a solution of

$$\begin{pmatrix} 0 & \langle 12|22 \rangle \\ \langle 12|22 \rangle & D \end{pmatrix} \begin{pmatrix} 1 \\ c \end{pmatrix} = {}^{N+1} E_R \begin{pmatrix} 1 \\ c \end{pmatrix}$$

where

$$D = \varepsilon_2 - \varepsilon_1 - 2J_{12} + K_{12} + J_{22}$$

b. Then show that

$$^{N+1} \tilde{E}_R^{(2)} = \frac{|\langle 12|22 \rangle|^2}{\varepsilon_1 - \varepsilon_2}$$

is an approximation to the relaxation energy.
c. Finally show that

$$\varepsilon_2' = \varepsilon_2 + \Sigma_{22}^{(2)}(\varepsilon_2) = \varepsilon_2 + {}^{N+1} \tilde{E}_R^{(2)} - {}^N E_0^{(2)}$$

7.4 PERTURBATION THEORY AND THE GREEN'S FUNCTION METHOD

In the previous section we studied two very simple two-electron minimal basis problems using GF theory with $\Sigma^{(2)}(E)$. We saw that the simplest correction to ε_1, i.e., $\Sigma_{11}^{(2)}(\varepsilon_1)$ includes both the effect of correlation in the two-particle system and orbital relaxation in the one-particle system. Moreover, we found that the magnitudes of these effects were precisely what one would expect from second-order perturbation theory. We would like to know: 1) if our observations hold for two-electron systems described by larger basis sets and 2) if for larger systems GF theory accounts for correlation in the $(N - 1)$-particle system.

In this section we will establish the relationship between the simplest approximation in GF theory for the energy required to remove an electron from spin orbital c,

$$-\text{IP} = \varepsilon_c' = \varepsilon_c + \Sigma_{cc}^{(2)}(\varepsilon_c)$$

$$= \varepsilon_c + \frac{1}{2} \sum_{ars} \frac{|\langle rs||ca \rangle|^2}{\varepsilon_a + \varepsilon_c - \varepsilon_r - \varepsilon_s} + \frac{1}{2} \sum_{abr} \frac{|\langle ab||cr \rangle|^2}{\varepsilon_c + \varepsilon_r - \varepsilon_a - \varepsilon_b} \tag{7.69}$$

THE ONE-PARTICLE MANY-BODY GREEN'S FUNCTION **399**

and perturbation theory. Since the analysis is somewhat involved, it is appropriate to present the result before the derivation. For the N-particle system, we choose our unperturbed Hamiltonian as the HF Hamiltonian and calculate the energy up to second order

$$
\begin{aligned}
{}^N E_0^{(0+1+2)} &= {}^N E_0^{(0)} + {}^N E_0^{(1)} + {}^N E_0^{(2)} \\
&= {}^N E_0 + {}^N E_0^{(2)}
\end{aligned}
\tag{7.70}
$$

using the formalism presented in Chapter 6. We represent the zeroth-order wave function of the $(N-1)$-particle system by a Slater determinant obtained by removing spin orbital c from the HF wave function of the N-particle system, i.e., $|^{N-1}\Psi_c\rangle$. Note that this is *not*, in general, the HF wave function for the $(N-1)$-particle system since it contains the HF spin orbitals of the N-particle system. Then we partition the total Hamiltonian of the $(N-1)$-particle system into an unperturbed and perturbed part in such a way that $|^{N-1}\Psi_c\rangle$ is an eigenfunction of the unperturbed Hamiltonian. Using this partitioning, we calculate the energy of the $(N-1)$-particle system up to second order,

$$
{}^{N-1}E_0^{(0+1+2)}(c) = {}^{N-1}\tilde{E}_0^{(0)}(c) + {}^{N-1}\tilde{E}_0^{(1)}(c) + {}^{N-1}\tilde{E}_0^{(2)}(c)
\tag{7.71}
$$

The tildes serve as a reminder that we are not doing perturbation theory starting with a HF Hamiltonian. Finally, we shall show that

$$
\varepsilon_c' = \varepsilon_c + \Sigma_{cc}^{(2)}(\varepsilon_c) = {}^N E_0^{(0+1+2)} - {}^{N-1}\tilde{E}_0^{(0+1+2)}(c)
\tag{7.72}
$$

which is the main result of this section. In addition, we shall see that the derivation suggests a simple physical interpretation of the various terms that contribute to $\Sigma_{cc}^{(2)}(\varepsilon_c)$. The above result goes some of the way towards demystifying the GF formalism. It shows that the simplest possible approximation in this formalism yields an answer which is identical to the result given by the simplest form of perturbation theory (i.e., second-order using the HF orbitals of the N-particle system for both the N- and $(N-1)$-particle systems).

To establish our notation, we briefly review HF perturbation theory for the N-particle system (see Chapter 6). We choose the unperturbed Hamiltonian, $\mathcal{H}_0^N$, as the HF Hamiltonian,

$$
\mathcal{H}_0^N = \sum_{i=1}^N (h(i) + v_N^{\mathrm{HF}}(i))
\tag{7.73a}
$$

and hence the perturbation is

$$
\mathcal{V}^N = \sum_{i<j}^N r_{ij}^{-1} - \sum_{i=1}^N v_N^{\mathrm{HF}}(i)
\tag{7.73b}
$$

where the HF potential is

$$
v_N^{\mathrm{HF}}(i) = \sum_b \langle \chi_b(2)|r_{12}^{-1}(1-\mathscr{P}_{12})|\chi_b(2)\rangle
\tag{7.73c}
$$

with the sum running over all occupied HF spin orbitals of the N-particle system. The Hartree-Fock wave function

$$|^N\Psi_0\rangle = |ab \cdots c - 1 \, c \, c + 1 \cdots\rangle \qquad (7.74)$$

is an eigenfunction of $\mathscr{H}_0^N$

$$\mathscr{H}_0^N|^N\Psi_0\rangle = {}^NE_0^{(0)}|^N\Psi_0\rangle = \sum_a \varepsilon_a|^N\Psi_0\rangle \qquad (7.75)$$

The first-order energy is

$$^NE_0^{(1)} = \langle^N\Psi_0|\mathscr{V}^N|^N\Psi_0\rangle = -\frac{1}{2}\sum_{ab}\langle ab||ab\rangle \qquad (7.76)$$

and the second-order energy is

$$^NE_0^{(2)} = {\sum_n}' \frac{|\langle^N\Psi_0|\mathscr{V}^N|n\rangle|^2}{\langle^N\Psi_0|\mathscr{H}_0^N|^N\Psi_0\rangle - \langle n|\mathscr{H}_0^N|n\rangle} = {\sum_n}' \frac{|\langle^N\Psi_0|\mathscr{V}^N|n\rangle|^2}{{}^NE_0^{(0)} - {}^NE_n^{(0)}}$$

$$= \frac{1}{4}{\sum_{abrs}}' \frac{|\langle^N\Psi_0|\mathscr{V}^N|^N\Psi_{ab}^{rs}\rangle|^2}{\varepsilon_a + \varepsilon_b - \varepsilon_r - \varepsilon_s} = \frac{1}{4}\sum_{abrs}\frac{|\langle ab||rs\rangle|^2}{\varepsilon_a + \varepsilon_b - \varepsilon_r - \varepsilon_s} \qquad (7.77)$$

Recall, that in calculating the second-order energy the intermediate states, $|n\rangle$, exclude single excitations of the form $|^N\Psi_a^r\rangle$ because of Brillouin's theorem. In this section, we always use $a, b, \ldots$ to label the orbitals which are occupied in the N-particle system and $r, s, \ldots$ to label orbitals which are unoccupied in this system.

For the $(N - 1)$-particle system, we choose the unperturbed Hamiltonian, $\mathscr{H}_0^{N-1}$, as

$$\mathscr{H}_0^{N-1} = \sum_{i=1}^{N-1} (h(i) + v_N^{HF}(i)) \qquad (7.78a)$$

so that the perturbation is

$$\mathscr{V}^{N-1} = \sum_{i>j}^{N-1} r_{ij}^{-1} - \sum_{i=1}^{N-1} v_N^{HF}(i) \qquad (7.78b)$$

It is important to note that our zeroth-order Hamiltonian for the $(N - 1)$-particle system contains the HF potential of the N (*not* the $N - 1$) particle system. We have chosen this partitioning so that the $(N - 1)$-particle wave function, obtained by removing spin orbital c from the HF wave function of the N-particle system,

$$|^{N-1}\Psi_c\rangle = |ab \cdots c - 1 \, c + 1 \cdots\rangle \qquad (7.79)$$

is an eigenfunction of $\mathscr{H}_0^{N-1}$,

$$\mathscr{H}_0^{N-1}|^{N-1}\Psi_c\rangle = {}^{N-1}\tilde{E}_0^{(0)}(c)|^{N-1}\Psi_c\rangle = \sum_{a\neq c}\varepsilon_a|^{N-1}\Psi_c\rangle \qquad (7.80)$$

The tilde on the zeroth-order energy serves as a reminder that, since $\mathcal{H}_0^{N-1}$ is not the HF Hamiltonian of the $(N-1)$-particle system, we are not doing standard HF perturbation theory. In particular, Brillouin's theorem does not hold for the $(N-1)$-particle system, so single excitations will contribute to the second-order energy, $^{N-1}\tilde{E}_0^{(2)}(c)$, of this system. We have chosen the above zero-order Hamiltonian so that the terms in perturbation expansion for the energy of the $(N-1)$-particle system contain two-electron integrals and orbital energies of the N-particle system. Consequently, since we are interested in the difference between energies of the N and $(N-1)$-particle systems, we will be able to explicitly cancel terms in the two perturbation expansions, which result from correlation effects common to the two systems.

Note that since the difference between the zeroth-order energies of the N and $(N-1)$-particle systems is ε_c

$$^NE_0^{(0)} - {}^{N-1}E_0^{(0)}(c) = \sum_a \varepsilon_a - \sum_{a \neq c} \varepsilon_a = \varepsilon_c \tag{7.81}$$

It then follows from Koopmans' theorem

$$\varepsilon_c = \langle {}^N\Psi_0|\mathcal{H}^N|{}^N\Psi_0\rangle - \langle {}^{N-1}\Psi_c|\mathcal{H}^{N-1}|{}^{N-1}\Psi_c\rangle$$
$$= ({}^NE_0^{(0)} + {}^NE_0^{(1)}) - ({}^{N-1}\tilde{E}_0^{(0)}(c) + {}^{N-1}\tilde{E}_0^{(1)}(c))$$

that the first-order energies of the N- and $(N-1)$-particle systems must be equal (i.e., $^NE_0^{(1)} = {}^{N-1}\tilde{E}_0^{(1)}(c)$). Thus to find the lowest order correction to the Koopmans' theorem IP, we must find the difference between the second-order energies of the two systems.

Exercise 7.13 By explicitly evaluating $\langle {}^{N-1}\Psi_c|\mathcal{V}^{N-1}|{}^{N-1}\Psi_c\rangle$ (see Eqs. (7.78b) and (7.79)) show that it is equal to $^NE_0^{(1)}$.

The second-order energy of the $(N-1)$-particle system is

$$^{N-1}E_0^{(2)}(c) = \sum_n{}' \frac{|\langle {}^{N-1}\Psi_c|\mathcal{V}^{N-1}|n\rangle|^2}{\langle {}^{N-1}\Psi_c|\mathcal{H}_0^{N-1}|{}^{N-1}\Psi_c\rangle - \langle n|\mathcal{H}_0^{N-1}|n\rangle}$$
$$= \sum_n{}' \frac{|\langle {}^{N-1}\Psi_c|\mathcal{V}^{N-1}|n\rangle|^2}{\sum_{a \neq c} \varepsilon_a - \langle n|\mathcal{H}_0^{N-1}|n\rangle} \tag{7.82}$$

where the sum runs over all states of the system except $|{}^{N-1}\Psi_c\rangle$. There are three types of excitations which contribute to the sum.

1. Single excitations $|{}^{N-1}\Psi_{ca}^r\rangle$. If $\mathcal{H}_0^{N-1}$ were the HF Hamiltonian of the $(N-1)$-particle system, such excitations would not contribute because of Brillouin's theorem. The effect of these single excitations is to improve upon the occupied orbitals of the $(N-1)$-particle system (which are the HF orbitals of the N-particle system) by making them closer to the optimum orbitals; i.e., they allow the orbitals to "relax." The resulting contribution to the second-order energy, denoted by $^{N-1}\tilde{E}^{(2)}\binom{r}{a}$, can be shown to be

$$
^{N-1}\tilde{E}^{(2)}\binom{r}{a} = \sum_{\substack{a \neq c \\ r}} \frac{|\langle ^{N-1}\Psi_c|\mathcal{V}^{N-1}|^{N-1}\Psi_{ca}^r\rangle|^2}{\varepsilon_a - \varepsilon_r} = \sum_{\substack{a \neq c \\ r}} \frac{|\langle ac||cr\rangle|^2}{\varepsilon_a - \varepsilon_r}
$$

$$
= -\sum_{ar} \frac{|\langle ac||cr\rangle|^2}{\varepsilon_r - \varepsilon_a} \tag{7.83}
$$

where in the last step we used the fact that the antisymmetric matrix element is zero when $a = c$. Note that this quantity is always negative.

Exercise 7.14 Show that for minimal basis HeH$^+$, Eq. (7.83) is equivalent to $^{N-1}\tilde{E}_R^{(2)}$ given in Eq. (7.65).

---

2. Double excitations $|^{N-1}\Psi_{cab}^{rs}\rangle$. These are just the usual types of excitations which occur in HF perturbation theory of the N-particle system, except that a and b cannot equal c. Thus we have

$$
^{N-1}\tilde{E}^{(2)}\binom{rs}{ab} = \frac{1}{4}\sum_{\substack{a \neq c \\ b \neq c \\ rs}} \frac{|\langle ab||rs\rangle|^2}{\varepsilon_a + \varepsilon_b - \varepsilon_r - \varepsilon_s}
$$

This can be rewritten in terms of the second-order energy of the N-particle system (i.e., $^N E_0^{(2)}$) as

$$
^{N-1}\tilde{E}^{(2)}\binom{rs}{ab} = \frac{1}{4}\sum_{abrs} \frac{|\langle ab||rs\rangle|^2}{\varepsilon_a + \varepsilon_b - \varepsilon_r - \varepsilon_s} - \frac{1}{2}\sum_{ars} \frac{|\langle ca||rs\rangle|^2}{\varepsilon_a + \varepsilon_c - \varepsilon_r - \varepsilon_s}
$$

$$
= {}^N E_0^{(2)} + \frac{1}{2}\sum_{ars} \frac{|\langle rs||ca\rangle|^2}{\varepsilon_r + \varepsilon_s - \varepsilon_a - \varepsilon_c} \tag{7.84}
$$

Note that this contribution is always less negative than the second-order energy of the N-particle system.

3. Double excitations $|^{N-1}\Psi_{cab}^{cr}\rangle$. These excitations are possible in the $(N-1)$-particle system since there is no electron in spin orbital c (i.e., c is a virtual or particle orbital in the $N-1$ system). It can be shown that these excitations make the contribution

$$
^{N-1}\tilde{E}^{(2)}\binom{cr}{ab} = \frac{1}{2}\sum_{\substack{a \neq c \\ b \neq c \\ r}} \frac{|\langle ^{N-1}\Psi_c|\mathcal{V}^{N-1}|^{N-1}\Psi_{cab}^{cr}\rangle|^2}{\varepsilon_a + \varepsilon_b - \varepsilon_r - \varepsilon_c}
$$

$$
= -\frac{1}{2}\sum_{\substack{a \neq c \\ b \neq c \\ r}} \frac{|\langle ab||cr\rangle|^2}{\varepsilon_c + \varepsilon_r - \varepsilon_a - \varepsilon_b} \tag{7.85}
$$

to the second-order energy. Note that this term is negative.

Finally, combining all these terms, the difference between the second-order energies of the N- and $(N-1)$-particle system becomes

$$
{}^N E_0^{(2)} - {}^{N-1}\tilde{E}_0^{(2)}(c) = {}^N E_0^{(2)} - \left[{}^{N-1}\tilde{E}^{(2)}\binom{r}{a} + {}^{N-1}\tilde{E}^{(2)}\binom{cr}{ab} + {}^{N-1}\tilde{E}^{(2)}\binom{rs}{ab} \right]
$$

$$
= \sum_{ar} \frac{|\langle ac||cr\rangle|^2}{\varepsilon_r - \varepsilon_a} + \frac{1}{2} \sum_{\substack{a \neq c \\ b \neq c \\ r}} \frac{|\langle ab||cr\rangle|^2}{\varepsilon_c + \varepsilon_r - \varepsilon_a - \varepsilon_b} + \frac{1}{2} \sum_{ars} \frac{|\langle rs||ca\rangle|^2}{\varepsilon_a + \varepsilon_c - \varepsilon_r - \varepsilon_s}
$$

(7.86)

which, upon combining the first two terms and comparing with Eq. (7.69), can be seen to equal $\Sigma_{cc}^{(2)}(\varepsilon_c)$. This completes the proof of the relationship between perturbation theory and the simplest approximation in GF theory that was stated in the beginning of this section.

Exercise 7.15 Fill in the missing steps leading to Eqs. (7.83), (7.84) and (7.85).

Now that we have quickly dispensed with the algebra, we will consider the physical interpretation of the three terms shown in Eq. (7.86), that make up $\Sigma_{cc}^{(2)}(\varepsilon_c)$. For the purposes of the discussion, it is convenient to express our results as a correction to the Koopmans' theorem IP, i.e., IP = $(IP)_{Koopmans} + \Delta IP$, where

$$
\Delta IP = -\sum_{ar} \frac{|\langle ac||cr\rangle|^2}{\varepsilon_r - \varepsilon_a} - \frac{1}{2} \sum_{\substack{b \neq c \\ a \neq c \\ r}} \frac{|\langle ab||cr\rangle|^2}{\varepsilon_r + \varepsilon_c - \varepsilon_a - \varepsilon_b} + \frac{1}{2} \sum_{ars} \frac{|\langle rs||ca\rangle|^2}{\varepsilon_r + \varepsilon_s - \varepsilon_a - \varepsilon_c}
$$

$\quad\quad\quad$ (ORX) $\quad\quad\quad\quad\quad$ (PRX) $\quad\quad\quad\quad\quad$ (PRM) $\quad\quad$ (7.87)

The acronyms representing the algebraic expressions will be explained below. The first term (ORX), which is always negative, arises from the presence of single excitations $a \rightarrow r$ in the second-order energy of the $(N-1)$-particle system. Because of Brillouin's theorem, such excitations would *not* have contributed if we had used the HF orbitals of the $(N-1)$-particle system. Their effect is to optimize or relax the orbitals we did use (i.e., the HF orbitals of the N-particle system) and hence this term is said to arise from orbital relaxation (ORX). Since it lowers the energy of the $(N-1)$-particle system relative to the "frozen" orbital approximation, it decreases the Koopmans' theorem IP. Consider the second term (PRX), which is also always negative. This term arises from double excitations of the type $\begin{matrix} a \rightarrow c \\ b \rightarrow r \end{matrix}$ in the second-order energy of the $(N-1)$-particle system. Such excitations do not occur in the N-particle system because c is occupied. Thus this term accounts for extra correlation in the $(N-1)$-particle system due to the presence of an additional virtual orbital. Using the language of pair theory, we can say that a given pair of electrons in spin orbitals ab in the $(N-1)$-particle system can correlate (in a way not possible in the N-particle system) by being excited into the virtual orbital c. Thus in analogy with orbital relaxation, this effect can be called **pair relaxation (PRX)**. Note that both these effects stabilize the $(N-1)$-particle system and hence decrease the ionization potential. Finally, consider the last term (PRM), which is always positive. This represents the additional correlation in the N-particle system due to the presence of an extra electron. Since we have removed an electron from spin orbital c, pair correlation energies involving this spin orbital are absent in the $(N-1)$-particle system. Hence the N-particle system (which has a larger number of correlating pairs) is stabilized relative to the $(N-1)$-particle system. The corresponding increase in the IP can be said to result from **pair removal (PRM)**. The above interpretation is due to Pickup and Goscinski.[2]

In Table 7.1 we have summarized the results of this analysis. Single excitations of the type $a \rightarrow r$ contribute only to the second-order energy of the $(N-1)$-particle system. Their effect is to stabilize the $(N-1)$-particle system relative to the "frozen" orbital approximation implicit in the Koopmans' theorem IP and thus they tend to decrease the IP. Excitations of the type $\begin{smallmatrix} a \rightarrow r \\ b \rightarrow s \end{smallmatrix}$, where neither a or b are equal to c, are present in both systems and their effect exactly cancels in second order of perturbation theory. The excitations $\begin{smallmatrix} a \rightarrow c \\ b \rightarrow r \end{smallmatrix}$, however, occur only in the $(N-1)$-particle system and change the IP in the same direction as single excitations. Finally, double excitations of the type $\begin{smallmatrix} c \rightarrow r \\ b \rightarrow s \end{smallmatrix}$ contribute exclusively to the second-order energy of the N-particle system. Since they stabilize the N-particle system

Table 7.1 Contributions of various excitations to the second-order energies of the N- and $(N-1)$-particle systems

Excitation	N	$N-1$	Effect on Koopmans' IP	Name
$a \rightarrow r$ $a \neq c$	X	✓	Decrease	Orbital relaxation (ORX)
$a \rightarrow r$ $b \rightarrow s$ $a \neq b \neq c$	✓	✓	None	
$a \rightarrow c$ $b \rightarrow r$	X	✓	Decrease	Pair relaxation (PRX)
$c \rightarrow r$ $b \rightarrow s$	✓	X	Increase	Pair removal (PRM)

relative to the $(N - 1)$-particle system, they tend to increase the Koopmans' theorem IP. Thus the simplest approximation in GF theory incorporates, at least qualitatively, all the physical effects that are missing in the "frozen" orbital approximation. In the next section, we will present the results of calculations on a variety of systems to see how well this simple approximation works quantitatively.

7.5 SOME ILLUSTRATIVE CALCULATIONS

In this section, we present numerical results for the correction to the Koopmans' vertical ionization potentials of H_2, N_2, and the ten-electron series (CH_4, NH_3, H_2O, and FH), calculated using the simplest possible approximation of GF theory, i.e., Eq. (7.46). It is important to note that these calculations are illustrative and do not reflect the full power of the GF approach. We will see that, while GF theory at this level changes the Koopmans' IP in the right direction and, in cases such as N_2, rectifies the incorrect ordering of the two lowest IP's, it tries too hard. That is, if the Koopmans' IP is greater than experiment, the lowest order correction underestimates the experimental value often by a larger amount than Koopmans' theorem overestimated it. This situation is unfortunately not uncommon in quantum chemistry. If the simplest theory works fairly well, the first correction will sometimes make things worse. To obtain satisfactory agreement with experimental ionization potentials within the framework of the GF approach we must use a self-energy, which is correct to at least third order in perturbation theory.

We begin with the application of the formalism to H_2. A two-electron system is special since the ion (e.g., H_2^+) has no correlation. Thus the exact IP differs from the Koopmans' value only because of orbital relaxation in the $(N - 1)$-particle system and correlation in the N-particle system. In the language of the previous section, there is no pair relaxation term (PRX) since the $(N - 1)$-particle system has but a single electron. Moreover, in second order, the pair removal term is just the negative of the second-order perturbation energy of the N-particle system.

Exercise 7.16 Show that for a two-electron system the PRX term in Eq. (7.87) is zero while the PRM term equals $-{}^N E_0^{(2)}$.

In Table 7.2, we compare the results for the vertical IP of H_2 calculated from Koopmans' theorem and GF theory using $\Sigma^{(2)}$, with the exact (full CI) values in the basis. Unless otherwise indicated, all results are obtained using the diagonal part of $\Sigma^{(2)}$ evaluated at the appropriate orbital energy. The exact basis set results were obtained by subtracting the full CI energy of H_2 and the exact energy of H_2^+ in the same basis. Note that, even for the largest basis, there is a significant difference between the exact result in the basis

Table 7.2 Ionization potential (a.u.) of H_2

Basis set	Koopmans'	$\Sigma^{(2)}(\varepsilon)$	Full CI
STO-3G	0.578	0.591	0.599
4-31G	0.596	0.593	0.595
6-31G**	0.595	0.598	0.600
Experiment		0.584	

and experiment. This is a consequence of the fact that the 6-31G** basis for hydrogen is far from complete. Except for the 4-31G basis, the GF results are closer to the full CI values than the Koopmans' theorem results. It is interesting to note that, for the larger bases, the Koopmans' theorem result is close to the full CI result. This arises from the near cancellation of two competing corrections: 1) the orbital relaxation (ORX) in H_2^+, which tends to decrease the IP, and 2) the correlation energy in H_2, which tends to increase the IP. This near cancellation is illustrated in Table 7.3. The exact ORX was obtained by subtracting the energy of H_2^+ calculated using the HF orbitals of H_2 from the exact energy of H_2^+ in the basis. In the minimal basis description, the optimum orbitals of H_2 and H_2^+ are the same by symmetry, so there is no orbital relaxation. The exact pair removal (PRM) contribution is simply the negative of the full CI correlation energy of H_2. Both the ORX and PRM contributions calculated using GF theory (see Eq. (7.87)) are somewhat smaller than the exact results. Nevertheless, for the 6-31G** basis, the GF answer is very close to the exact result in the basis. Note that for the 4-31G basis the Koopmans' result is slightly larger than the exact value. The Green's function formalism tries to remedy this but overshoots slightly.

This behaviour of GF theory at the level of the second-order self-energy is more striking for the lowest ionization potentials of the ten-electron series shown in Table 7.4. At the level of 6-31G** both the Koopmans' and GF results appear to have converged. For all four molecules, Koopmans' theorem overestimates the IP's, and in all cases the GF treatment lowers the IP too far, so that the Koopmans' result actually agrees better with experiment.

Table 7.3 Orbital relaxation (ORX) and pair removal (PRM) contributions (in a.u.) to the correction to Koopmans' theorem IP for H_2

Basis set	ORX	ORX (exact)	PRM	PRM (exact)
STO-3G			0.013	0.021
4-31G	−0.020	−0.026	0.017	0.025
6-31G**	−0.023	−0.029	0.026	0.034

Table 7.4 Lowest ionization potentials (a.u.) of the ten-electron series

		STO-3G	4-31G	6-31G*	6-31G**	Experiment
CH_4	Koopmans'	0.518	0.543	0.545	0.543	0.529
$(1t_2)$	$\Sigma^{(2)}(\varepsilon)$	0.493	0.502	0.507	0.510	
NH_3	Koopmans'	0.353	0.414	0.421	0.421	0.400
$(3a_1)$	$\Sigma^{(2)}(\varepsilon)$	0.275	0.331	0.352	0.353	
H_2O	Koopmans'	0.391	0.500	0.498	0.497	0.463
$(1b_1)$	$\Sigma^{(2)}(\varepsilon)$	0.299	0.388	0.394	0.395	
FH	Koopmans'	0.464	0.628	0.628	0.627	0.581
(1π)	$\Sigma^{(2)}(\varepsilon)$	0.396	0.507	0.509	0.509	

Although the GF formalism, using $\Sigma^{(2)}$, changes the Koopmans' result in the right direction, it is clearly not a quantitative theory. One must use more accurate self-energies if satisfactory results are to be obtained.

As a final example we consider the ionization potentials of N_2. As we have seen in Chapter 3, this is a particularly interesting case, because using large basis sets, Koopmans' theorem is qualitatively wrong. Experimentally, one finds that the lowest ionization potential corresponds to an ejection of an electron from the $3\sigma_g$ orbital (i.e., N_2^+ ends up in a $^2\Sigma_g^+$ state) and that it is more difficult to remove an electron from the $1\pi_u$ orbital. Koopmans' theorem predicts the opposite. How does the simplest approximation of GF theory work? Table 7.5 shows that the GF approach at the level of $\Sigma^{(2)}$ yields ionization potentials that are qualitatively in accord with experiment. For the largest basis set used, the Koopmans' theorem result for ionization out of the $1\pi_u$ orbital is raised slightly so that it is very close to the experimental value. On the other hand, the ionization potential of the $3\sigma_g$ orbital is decreased significantly, in fact far below the experimental result. Thus we see again, the GF approach at the level of the second-order self-energy, while qualitatively correct, is quantitatively deficient. It is interesting to examine

Table 7.5 Ionization potentials (a.u.) of N_2

Basis	$3\sigma_g$ orbital		$1\pi_u$ orbital	
	Koopmans'	$\Sigma^{(2)}(\varepsilon)$	Koopmans'	$\Sigma^{(2)}(\varepsilon)$
STO-3G	0.540	0.463	0.573	0.620
4-31G	0.629	0.517	0.621	0.643
6-31G*	0.630	0.534	0.612	0.627
Experiment	0.573		0.624	

Table 7.6 Orbital relaxation (ORX), pair relaxation (PRX), and pair removal (PRM) contributions (in a.u.) to the correction to Koopmans' theorem IP's for N_2

Basis	$3\sigma_g$ orbital			$1\pi_u$ orbital		
	ORX	PRX	PRM	ORX	PRX	PRM
STO-3G	-0.006	-0.091	$+0.020$	-0.001	-0.008	0.056
4-31G	-0.051	-0.098	0.037	-0.044	-0.008	0.074
6-31G*	-0.056	-0.096	0.055	-0.055	-0.016	0.086

(see Table 7.6) the breakup of the second-order correction into orbital relaxation (ORX), pair relaxation (PRX), and pair removal (PRM) contributions. Consider the 6-31G* basis. For both orbitals, the orbital relaxation term is about the same. For the $1\pi_u$ orbital the PRM term is somewhat larger than it is for the $3\sigma_g$ orbital. This means that double excitations involving the $1\pi_u$ orbital make a larger contribution to the correlation energy of N_2 than excitations involving the $3\sigma_g$ orbital. Although this effect helps, the correct ordering results primarily from the much larger pair relaxation contribution for $3\sigma_g$ orbital. That is, double excitations in which one of the electrons of N_2^+ ends up in the vacant $3\sigma_g$ orbital are much more effective in lowering the energy of the ion than double excitations to the $1\pi_u$ orbital.

As an illustration of the effect of using higher-order self-energies, in Table 7.7 we present some calculations of Cederbaum and co-workers on N_2. These calculations were done using a near-Hartree-Fock-limit basis set of Gaussians. For the second-order self-energy, the results of two types of calculations are shown. The first ($\Sigma^{(2)}(\varepsilon)$) was obtained in the same way as the GF results in the previous tables (i.e., using Eq. (7.46)), while the second ($\Sigma^{(2)}(E)$) was found by solving the Dyson equation, with the second-order self-energy, exactly (i.e., solving Eq. (7.42)). The results shown for the second-plus third-order self-energy were also obtained by solving the Dyson equation exactly. For the $3\sigma_g$ orbital, the third-order calculation raises the IP so that the final result is reasonably close to experiment. Note that while the second-order result undershoots, the third-order result slightly overshoots.

Table 7.7 Ionization potentials (a.u.) of N_2 using second- and third-order self-energies[a]

Orbital	Koopmans'	$\Sigma^{(2)}(\varepsilon)$	$\Sigma^{(2)}(E)$	$\Sigma^{(2)}(E) + \Sigma^{(3)}(E)$	Experiment
$3\sigma_g$	0.635	0.538	0.546	0.584	0.573
$1\pi_u$	0.615	0.626	0.626	0.613	0.624

[a] L. S. Cederbaum and W. Domcke, *Adv. Chem. Phys.* **36**: 205 (1977).

Thus the convergence is oscillatory. For the $1\pi_u$ orbital, the second-order result is fortuitously close to experiment and the third order is thus in somewhat worse agreement with experiment. These calculations show that in order to obtain very accurate results, one must use self-energies that go beyond third order.

In this chapter, we have barely scratched the surface of the Green's function approach. This technique has been widely used, and with sufficient effort (i.e., using sophisticated self-energies) it provides an excellent description of both valence and core ionization potentials of a large class of molecules. We hope that our introduction will help the interested reader penetrate the literature.

NOTES

1. See for example, D. Rapp, *Quantum Mechanics*, Holt-Reinhart-Winston, New York, 1971, p. 463.
2. B. T. Pickup and O. Goscinski, Direct calculation of ionization energies I. Closed shells, *Mol. Phys.* **26**: 1013 (1973). This article also contains a relatively simple derivation of the expression for the second-order self-energy using algebraic rather than diagrammatic techniques.

FURTHER READING

Cederbaum, L. S. and Domcke, W., Theoretical aspects of ionization potentials and photo-electron spectroscopy: Green's function approach, *Adv. Chem. Phys.* **36**: 205 (1977). A comprehensive review article with emphasis on the authors' approach but with numerous literature references.

Fetter, A. L. and Walecka, J. D., *Quantum Theory of Many-Particle Systems*, McGraw-Hill, New York, 1971. An advanced textbook for physicists.

Linderberg, J. and Öhrn, Y., *Propagators in Quantum Chemistry*, Academic Press, New York, 1973. A worthwhile but demanding book.

Mattuck, R. D., *A Guide to Feynman Diagrams in the Many-Body Problem*, 2nd ed., McGraw-Hill, New York, 1976. An entertaining book which discusses many-body Green's functions from the time-dependent point of view. Its only disadvantage to chemists is that the emphasis is on treating infinite systems where the basis functions are plane waves.

Oddershede, J., Propagator methods, *Adv. Chem. Phys.* **69**: 201 (1987). A concise review of the principles and applications of Green's functions and related approaches with numerous references to the literature.

Öhrn, Y., Propagator theory of atomic and molecular structure, in *The New World of Quantum Chemistry* B. Pullman and R. Parr (eds.), Reidel, Boston, 1976 p. 57. A review lecture presented as a major international conference.

Paul, R., *Field Theoretical Methods in Chemical Physics*, Elsevier, Amsterdam, 1982. A valiant attempt to treat a difficult subject in a pedagogical way. Chapter VII deals with the application of the Green's function method to problems of atomic and molecular structure.

INTEGRAL EVALUATION WITH
1s PRIMITIVE GAUSSIANS

In most molecular calculations one uses a fixed molecular coordinate system such that the basis functions are centered at position vectors $\mathbf{R}_A$ in this coordinate system, as shown in Fig. A.1. The value at a position vector $\mathbf{r}$, of a function centered at $\mathbf{R}_A$, will depend on $\mathbf{r} - \mathbf{R}_A$ and we hence write a general basis function as $\phi_\mu(\mathbf{r} - \mathbf{R}_A)$ to denote that it is centered at $\mathbf{R}_A$. In a

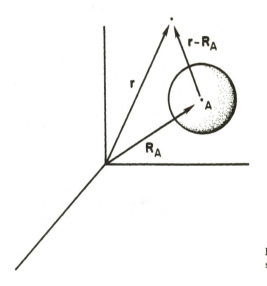

Figure A.1 Molecular coordinate system.

molecular calculation we will need to evaluate a number of one- and two-electron integrals involving functions $\phi_\mu(\mathbf{r} - \mathbf{R}_A)$ at different centers. If we have basis functions at four or more centers, then the two-electron integrals will involve 1-, 2-, 3-, and 4-center integrals. Nuclear attraction integrals can involve at most 3 centers. These multicenter (more than 2) integrations are difficult for Slater-type functions but are relatively simple for Gaussian-type functions. Thus most polyatomic calculations use Gaussian functions.

In a Gaussian calculation the contracted Gaussian basis functions ϕ_μ^{CGF} are normally expanded in a set of primitive Gaussian functions $g \equiv \phi_p^{\text{GF}}$ (see Eq. (3.283) or Eq. (3.212)), and thus we need only consider integral evaluation over the primitive functions. The integrals for contracted functions ϕ_μ^{CGF} are obtained just by summing up integrals over primitives using the appropriate contraction coefficients. One commonly uses only 1s, 2p, 3d, etc. primitive Gaussians, since any function of s, p, d, etc. symmetry can be expanded in just these Gaussians. Here we consider integral evaluations only for 1s primitive Gaussians. Shavitt[1] has given a similar, but complementary, discussion of integral evaluation methods for the Gaussian function. Integrals involving p, d, etc. symmetry functions could be obtained by differentiating our results with respect to the Cartesian coordinates of the centers $\mathbf{R}_A$, $\mathbf{R}_B$, etc. However, integrals for higher angular quantum numbers are more efficiently formulated for computation using newer methods based on Rys polynomials.[2]

The unnormalized 1s primitive Gaussian at $\mathbf{R}_A$ is

$$\tilde{g}_{1s}(\mathbf{r} - \mathbf{R}_A) = e^{-\alpha|\mathbf{r} - \mathbf{R}_A|^2} \tag{A.1}$$

We will use α, β, γ, and δ for orbital exponents of functions centered at $\mathbf{R}_A$, $\mathbf{R}_B$, $\mathbf{R}_C$, $\mathbf{R}_D$, respectively. The reason why Gaussians simplify multicenter integrations is that the product of two 1s Gaussians, each on different centers, is proportional to a 1s Gaussian on a third center. Thus

$$\tilde{g}_{1s}(\mathbf{r} - \mathbf{R}_A)\tilde{g}_{1s}(\mathbf{r} - \mathbf{R}_B) = \tilde{K}\tilde{g}_{1s}(\mathbf{r} - \mathbf{R}_P) \tag{A.2}$$

where the proportionality constant $\tilde{K}$ is

$$\tilde{K} = \exp[-\alpha\beta/(\alpha + \beta)|\mathbf{R}_A - \mathbf{R}_B|^2] \tag{A.3}$$

The third center P is on a line joining the centers A and B,

$$\mathbf{R}_P = (\alpha\mathbf{R}_A + \beta\mathbf{R}_B)/(\alpha + \beta) \tag{A.4}$$

The exponent of the new Gaussian centered at $\mathbf{R}_P$ is

$$p = \alpha + \beta \tag{A.5}$$

Thus for 1s Gaussians any 2-center distribution, i.e., product of functions on 2 different centers, can immediately be converted to a one-center distribution. We now evaluate the basic integrals that we need for any *ab initio* calculation that uses only contractions of primitive $1s$ Gaussians. We will evaluate integrals for *unnormalized* functions and, if we require integrals for normalized functions, we can easily multiply by the appropriate normalization constants.

Consider the 2-center overlap integral first

$$(A|B) = \int d\mathbf{r}_1 \, \tilde{g}_{1s}(\mathbf{r}_1 - \mathbf{R}_A)\tilde{g}_{1s}(\mathbf{r}_1 - \mathbf{R}_B) \tag{A.6}$$

Using (A.2) we immediately have

$$(A|B) = \tilde{K} \int d\mathbf{r}_1 \, \tilde{g}_{1s}(\mathbf{r}_1 - \mathbf{R}_P) = \tilde{K} \int d\mathbf{r}_1 \, \exp[-p|\mathbf{r}_1 - \mathbf{R}_P|^2] \tag{A.7}$$

If we now let $\mathbf{r} = \mathbf{r}_1 - \mathbf{R}_P$ and $d\mathbf{r} = d\mathbf{r}_1$, then

$$(A|B) = \tilde{K} \int d\mathbf{r} \, e^{-pr^2} = 4\pi\tilde{K} \int_0^\infty dr \, r^2 e^{-pr^2} \tag{A.8}$$

This last integral is just $(\pi/p)^{3/2}/4\pi$, so that

$$(A|B) = [\pi/(\alpha + \beta)]^{3/2} \exp[-\alpha\beta/(\alpha + \beta)|\mathbf{R}_A - \mathbf{R}_B|^2] \tag{A.9}$$

The kinetic energy integral

$$(A|-\tfrac{1}{2}\nabla^2|B) = \int d\mathbf{r}_1 \, \tilde{g}_{1s}(\mathbf{r}_1 - \mathbf{R}_A)(-\tfrac{1}{2}\nabla_1^2)\tilde{g}_{1s}(\mathbf{r}_1 - \mathbf{R}_B) \tag{A.10}$$

can be evaluated in a very similar way (after letting ∇_1^2 operate) to give

$$(A|-\tfrac{1}{2}\nabla^2|B) = \alpha\beta/(\alpha + \beta)[3 - 2\alpha\beta/(\alpha + \beta)|\mathbf{R}_A - \mathbf{R}_B|^2][\pi/(\alpha + \beta)]^{3/2}$$
$$\times \exp[-\alpha\beta/(\alpha + \beta)|\mathbf{R}_A - \mathbf{R}_B|^2] \tag{A.11}$$

In order to evaluate the nuclear attraction and two-electron repulsion integrals, we now introduce here a powerful and general technique that is often useful for evaluating integrals, particularly of the type that we will be considering. The technique is to replace each quantity in the integrand by its Fourier transform. If we are given a function $f(\mathbf{r})$ of the vector $\mathbf{r}$, then its three-dimensional Fourier transform $F(\mathbf{k})$ is defined by

$$F(\mathbf{k}) = \int d\mathbf{r} \, f(\mathbf{r})e^{-i\mathbf{k}\cdot\mathbf{r}} \tag{A.12}$$

where the vector $\mathbf{k}$ is the transform variable. The Fourier integral theorem states that

$$f(\mathbf{r}) = (2\pi)^{-3} \int d\mathbf{k} \, F(\mathbf{k})e^{i\mathbf{k}\cdot\mathbf{r}} \tag{A.13}$$

$F(\mathbf{k})$ and $f(\mathbf{r})$ are said to be a Fourier transform pair. All the Fourier transform pairs we shall need here are given in Table A.1. In particular, the Fourier representation of the three-dimensional Dirac delta function is

$$\delta(\mathbf{r}_1 - \mathbf{r}_2) = (2\pi)^{-3} \int d\mathbf{k} \, e^{i\mathbf{k}\cdot(\mathbf{r}_1 - \mathbf{r}_2)} \tag{A.14}$$

Table A.1 Fourier transform pairs

$f(\mathbf{r})$	$F(\mathbf{k})$
$\dfrac{1}{r}$	$\dfrac{4\pi}{k^2}$
$e^{-\alpha r^2}$	$\left(\dfrac{\pi}{\alpha}\right)^{3/2} e^{-k^2/4\alpha}$
$\delta(\mathbf{r})$	1

Recall that the Dirac δ function has the property that

$$\int d\mathbf{r}_1\, \delta(\mathbf{r}_1 - \mathbf{r}_2)h(\mathbf{r}_1) = h(\mathbf{r}_2) \tag{A.15}$$

for any function $h(\mathbf{r})$.

Let us now consider the nuclear attraction integral for our primitive 1s Gaussians.

$$(A|-Z_C/r_{1C}|B) = \int d\mathbf{r}_1\, \tilde{g}_{1s}(\mathbf{r}_1 - \mathbf{R}_A)[-Z_C/|\mathbf{r}_1 - \mathbf{R}_C|]\tilde{g}_{1s}(\mathbf{r}_1 - \mathbf{R}_B)$$

$$= -Z_C \int d\mathbf{r}_1\, e^{-\alpha|\mathbf{r}_1 - \mathbf{R}_A|^2}|\mathbf{r}_1 - \mathbf{R}_C|^{-1}e^{-\beta|\mathbf{r}_1 - \mathbf{R}_B|^2} \tag{A.16}$$

We first combine the two Gaussians to obtain a single Gaussian at $\mathbf{R}_P$,

$$(A|-Z_C/r_{1C}|B) = -Z_C\tilde{K} \int d\mathbf{r}_1\, e^{-p|\mathbf{r}_1 - \mathbf{R}_P|^2}|\mathbf{r}_1 - \mathbf{R}_C|^{-1} \tag{A.17}$$

We then substitute for the two terms in the above integrand using (A.13) and Table A.1,

$$(A|-Z_C/r_{1C}|B) = -Z_C\tilde{K}(2\pi)^{-6}(\pi/p)^{3/2} \int d\mathbf{r}_1\, d\mathbf{k}_1\, d\mathbf{k}_2\, e^{-k_1^2/4p}$$

$$\times\, e^{i\mathbf{k}_1 \cdot (\mathbf{r}_1 - \mathbf{R}_P)}4\pi k_2^{-2}e^{i\mathbf{k}_2 \cdot (\mathbf{r}_1 - \mathbf{R}_C)} \tag{A.18}$$

If we collect all the exponential terms involving our original variable of integration $\mathbf{r}_1$, we obtain

$$(A|-Z_C/r_{1C}|B) = -4\pi Z_C\tilde{K}(2\pi)^{-6}(\pi/p)^{3/2} \int d\mathbf{r}_1\, d\mathbf{k}_1\, d\mathbf{k}_2\, k_2^{-2}e^{-k_1^2/4p}$$

$$\times\, e^{-i\mathbf{k}_1 \cdot \mathbf{R}_P}e^{-i\mathbf{k}_2 \cdot \mathbf{R}_C}e^{i\mathbf{r}_1 \cdot (\mathbf{k}_1 + \mathbf{k}_2)} \tag{A.19}$$

We can now use the definition (A.14) of the delta function to perform the integration over $\mathbf{r}_1$ to obtain

$$(A|-Z_C/r_{1C}|B) = -4\pi Z_C\tilde{K}(2\pi)^{-3}(\pi/p)^{3/2} \int d\mathbf{k}_1\, d\mathbf{k}_2\, e^{-k_1^2/4p}k_2^{-2}$$

$$\times\, e^{-i\mathbf{k}_1 \cdot \mathbf{R}_P}e^{-i\mathbf{k}_2 \cdot \mathbf{R}_C}\delta(\mathbf{k}_1 + \mathbf{k}_2)$$

$$= -Z_C\tilde{K}/(2\pi^2)(\pi/p)^{3/2} \int d\mathbf{k}\, e^{-k^2/4p}k^{-2}e^{-i\mathbf{k} \cdot (\mathbf{R}_P - \mathbf{R}_C)} \tag{A.20}$$

where we have set $\mathbf{k}_2 = -\mathbf{k}_1$ because of the delta function and relabeled the variable as $\mathbf{k}$. If we let $\mathbf{R}_P - \mathbf{R}_C$ lie along the z axis so that $\mathbf{k} \cdot (\mathbf{R}_P - \mathbf{R}_C) = k|\mathbf{R}_P - \mathbf{R}_C|\cos\theta$, then we can easily perform the angular part of the integration over $\mathbf{k}$ to obtain

$$(A|-Z_C/r_{1C}|B) = N \int_0^\infty dk \, e^{-k^2/4p} k^{-1} \sin(k|\mathbf{R}_P - \mathbf{R}_C|) \qquad (A.21)$$

$$N = -2Z_C \tilde{K}(\pi|\mathbf{R}_P - \mathbf{R}_C|)^{-1}(\pi/p)^{3/2} \qquad (A.22)$$

To evaluate the integral in (A.21), we consider the general integral:

$$I(x) = \int_0^\infty dk \, e^{-ak^2} k^{-1} \sin kx \qquad (A.23)$$

and its derivative with respect to x,

$$I'(x) = \int_0^\infty dk \, e^{-ak^2} \cos kx \qquad (A.24)$$

Note that the integrand in Eq. (A.24) is an even function of k. Since $I(0) = 0$, we have

$$\int_0^x dy \, I'(y) = I(x) - I(0) = I(x) \qquad (A.25)$$

The integral in (A.24) can be evaluated by noting that $\cos\theta$ is the real part of $e^{i\theta}$ (i.e., $\cos\theta = \mathscr{R}e[e^{i\theta}]$), so that

$$I'(x) = \frac{1}{2}\int_{-\infty}^\infty dk \, e^{-ak^2} \cos kx = \frac{1}{2}\mathscr{R}e\left[\int_{-\infty}^\infty dk \, e^{-ak^2} e^{ikx}\right] \qquad (A.26)$$

By completing the square, we have

$$I'(x) = \frac{1}{2} e^{-x^2/4a} \mathscr{R}e\left[\int_{-\infty}^\infty dk \, e^{-(a^{1/2}k - 1/2ia^{-1/2}x)^2}\right] \qquad (A.27)$$

Letting $u = a^{1/2}k - \frac{1}{2}ia^{-1/2}x$, we have

$$I'(x) = \frac{1}{2} e^{-x^2/4a} a^{-1/2} \int_{-\infty}^\infty du \, e^{-u^2} = \frac{1}{2}(\pi/a)^{1/2} e^{-x^2/4a} \qquad (A.28)$$

so that

$$I(x) = \frac{1}{2}(\pi/a)^{1/2} \int_0^x dy \, e^{-y^2/4a} \qquad (A.29)$$

Therefore

$$(A|-Z_C/r_{1C}|B) = -2\pi Z_C \tilde{K}(p|\mathbf{R}_P - \mathbf{R}_C|)^{-1} \int_0^{|\mathbf{R}_P - \mathbf{R}_C|} dy \, e^{-py^2}$$

$$= -2\pi Z_C \tilde{K} p^{-1}(p^{1/2}|\mathbf{R}_P - \mathbf{R}_C|)^{-1} \int_0^{p^{1/2}|\mathbf{R}_P - \mathbf{R}_C|} dy \, e^{-y^2} \qquad (A.30)$$

We now introduce the F_0 function, which is defined by

$$F_0(t) = t^{-1/2} \int_0^{t^{1/2}} dy\, e^{-y^2} \tag{A.31}$$

It is related to the error function by

$$F_0(t) = \tfrac{1}{2}(\pi/t)^{1/2} erf(t^{1/2}) \tag{A.32}$$

Therefore, our nuclear attraction integral can be written in terms of the F_0 function as

$$(A|-Z_C/r_{1C}|B) = -2\pi/(\alpha + \beta)Z_C \exp[-\alpha\beta/(\alpha + \beta)|\mathbf{R}_A - \mathbf{R}_B|^2]$$
$$\times F_0[(\alpha + \beta)|\mathbf{R}_P - \mathbf{R}_C|^2] \tag{A.33}$$

The error function (hence $F_0(t)$) is a built-in function on IBM FORTRAN compilers.

Let us now consider the two-electron integral

$$(AB|CD) = \int d\mathbf{r}_1\, d\mathbf{r}_2\, \tilde{g}_{1s}(\mathbf{r}_1 - \mathbf{R}_A)\tilde{g}_{1s}(\mathbf{r}_1 - \mathbf{R}_B)r_{12}^{-1}\tilde{g}_{1s}(\mathbf{r}_2 - \mathbf{R}_C)\tilde{g}_{1s}(\mathbf{r}_2 - \mathbf{R}_D) \tag{A.34}$$

The first step is to combine the Gaussian at $\mathbf{R}_A$ and $\mathbf{R}_B$ into a new Gaussian at $\mathbf{R}_P$ and to combine those at $\mathbf{R}_C$ and $\mathbf{R}_D$ into a new Gaussian at $\mathbf{R}_Q$ as shown in Fig. A.2. Then (A.34) becomes

$$(AB|CD) = \exp[-\alpha\beta/(\alpha + \beta)|\mathbf{R}_A - \mathbf{R}_B|^2 - \gamma\delta/(\gamma + \delta)|\mathbf{R}_C - \mathbf{R}_D|^2]$$
$$\times \int d\mathbf{r}_1\, d\mathbf{r}_2\, e^{-p|\mathbf{r}_1 - \mathbf{R}_P|^2}r_{12}^{-1}e^{-q|\mathbf{r}_2 - \mathbf{R}_Q|^2} \tag{A.35}$$

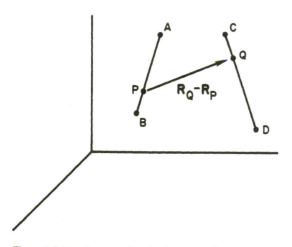

Figure A.2 The six centers involved in a two-electron integral.

Replacing the three terms in the integrand by their Fourier representation gives

$$(AB|CD) = M(2\pi)^{-9} \int d\mathbf{r}_1 \, d\mathbf{r}_2 \, d\mathbf{k}_1 \, d\mathbf{k}_2 \, d\mathbf{k}_3 \, (\pi/p)^{3/2} e^{-k_1^2/4p} e^{i\mathbf{k}_1 \cdot (\mathbf{r}_1 - \mathbf{R}_P)}$$

$$\times \, 4\pi k_2^{-2} e^{i\mathbf{k}_2 \cdot (\mathbf{r}_1 - \mathbf{r}_2)} (\pi/q)^{3/2} e^{-k_3^2/4q} e^{i\mathbf{k}_3 \cdot (\mathbf{r}_2 - \mathbf{R}_Q)} \quad \text{(A.36)}$$

where

$$M = \exp[-\alpha\beta/(\alpha + \beta)|\mathbf{R}_A - \mathbf{R}_B|^2 - \gamma\delta/(\gamma + \delta)|\mathbf{R}_C - \mathbf{R}_D|^2] \quad \text{(A.37)}$$

After collecting the exponential terms in the original variables of integration $\mathbf{r}_1$ and $\mathbf{r}_2$, we have

$$(AB|CD) = 4\pi M(2\pi)^{-9}(\pi^2/pq)^{3/2} \int d\mathbf{r}_1 \, d\mathbf{r}_2 \, d\mathbf{k}_1 \, d\mathbf{k}_2 \, d\mathbf{k}_3 \, e^{-k_1^2/4p} e^{-k_3^2/4q}$$

$$\times \, k_2^{-2} e^{-i\mathbf{k}_1 \cdot \mathbf{R}_P} e^{-i\mathbf{k}_3 \cdot \mathbf{R}_Q} e^{i\mathbf{r}_1 \cdot (\mathbf{k}_1 + \mathbf{k}_2)} e^{i\mathbf{r}_2 \cdot (\mathbf{k}_3 - \mathbf{k}_2)} \quad \text{(A.38)}$$

The integral over $\mathbf{r}_1$ and $\mathbf{r}_2$ now introduces two delta functions (see Eq. (A.14))

$$(AB|CD) = 4\pi M(2\pi)^{-3}(\pi^2/pq)^{3/2} \int d\mathbf{k}_1 \, d\mathbf{k}_2 \, d\mathbf{k}_3 \, e^{-k_1^2/4p} e^{-k_3^2/4q}$$

$$\times \, k_2^{-2} e^{-i\mathbf{k}_1 \cdot \mathbf{R}_P} e^{-i\mathbf{k}_3 \cdot \mathbf{R}_Q} \delta(\mathbf{k}_1 + \mathbf{k}_2) \delta(\mathbf{k}_3 - \mathbf{k}_2) \quad \text{(A.39)}$$

We can now set $\mathbf{k}_1 = -\mathbf{k}_2$ and $\mathbf{k}_3 = \mathbf{k}_2$ and relabel $\mathbf{k}_2$ as $\mathbf{k}$ to obtain

$$(AB|CD) = M/(2\pi^2)(\pi^2/pq)^{3/2} \int d\mathbf{k} \, k^{-2} e^{-(p+q)k^2/4pq} e^{i\mathbf{k} \cdot (\mathbf{R}_P - \mathbf{R}_Q)} \quad \text{(A.40)}$$

This integral is identical in form to that of (A.20) which we encountered in evaluating the nuclear attraction integral. Performing the rest of the algebra we finally obtain

$$(AB|CD) = 2\pi^{5/2}/[(\alpha + \beta)(\gamma + \delta)(\alpha + \beta + \gamma + \delta)^{1/2}]$$

$$\times \exp[-\alpha\beta/(\alpha + \beta)|\mathbf{R}_A - \mathbf{R}_B|^2 - \gamma\delta/(\gamma + \delta)|\mathbf{R}_C - \mathbf{R}_D|^2]$$

$$\times F_0[(\alpha + \beta)(\gamma + \delta)/(\alpha + \beta + \gamma + \delta)|\mathbf{R}_P - \mathbf{R}_Q|^2] \quad \text{(A.41)}$$

These then are the explicit formulas for all the integrals required in a Hartree-Fock calculation provided we are using only $1s$-type primitive Gaussian functions. They are used in the computer program listed in Appendix B.

NOTES

1. I. Shavitt, The Gaussian function in calculations of statistical mechanics and quantum mechanics, in *Methods in Computational Physics*, B. Alder, S. Fernbach, and M. Rotenberg (Eds.), Academic Press, New York, 1963.
2. M. Dupuis, J. Rys, and H. King, Evaluation of molecular integrals over Gaussian basis functions, *J. Chem. Phys.* **65**: 111 (1976).

TWO-ELECTRON
SELF-CONSISTENT-FIELD PROGRAM

This Appendix gives a FORTRAN listing and the sample output (HeH$^+$) of a small program which illustrates *ab initio* Hartree-Fock calculations. Subsection 3.5.3 of the text discusses the HeH$^+$ calculation.

The program will calculate STO-NG (for N = 1, 2, or 3) wave functions for any two-electron diatomic molecule. The input parameters to the main subroutine HFCALC are an option to control the printing, the number of primitive $1s$ Gaussian functions in which a $1s$ Slater function is to be expanded, i.e., the N of STO-NG, the bond length R in atomic units, the two exponents ζ_1 and ζ_2 of the $1s$ Slater functions, and the atomic numbers Z_A and Z_B of the two nuclei. If

$$g_{1s}(\alpha) = (2\alpha/\pi)^{3/4}e^{-\alpha r^2} \tag{B.1}$$

is a normalized primitive $1s$ Gaussian, then the program can use for basis functions any one of the following three least-squares fits to Slater-type functions

$$\phi_{1s}^{CGF}(\zeta = 1.0, \text{STO-1G}) = g_{1s}(0.270950) \tag{B.2}$$

$$\phi_{1s}^{CGF}(\zeta = 1.0, \text{STO-2G})$$
$$= 0.678914g_{1s}(0.151623) + 0.430129g_{1s}(0.851819) \tag{B.3}$$

$$\phi_{1s}^{CGF}(\zeta = 1.0, \text{STO-3G})$$
$$= 0.444635 g_{1s}(0.109818) + 0.535328 g_{1s}(0.405771)$$
$$+ 0.154329 g_{1s}(2.22766) \tag{B.4}$$

The program was most recently run with the FORTRAN IV compiler of a PDP-10 computer. It should, however, translate unaltered to almost any FORTRAN compiler. On IBM machines the function DERF can be replaced by a standard library routine. The program uses double precision arithmetic, but single precision would be sufficient in many cases. The program is specifically written in an inefficient way so that anyone with a knowledge of FORTRAN and the discussions of Chapter 3 should be able to follow completely the details of the *ab initio* Hartree-Fock calculation.

Subroutine INTGRL, after scaling the contraction coefficients and exponents of (B.2) to (B.4) to the input values of ζ_1 and ζ_2, calculates all basic one- and two-electron integrals according to the explicit formulas of Appendix A. The overlap, kinetic energy, nuclear attraction, and two-electron integrals for unnormalized primitives are evaluated by the functions S, T, V, and TWOE, and then summed using modified contraction coefficients that include the normalization constants. The basic integrals are then passed through COMMON storage to COLECT, which forms from the integrals those matrices S, H^{core}, etc. which remain fixed during the SCF iterations. The transformation matrix **X** to orthogonal basis functions is the canonical one described in the text. The SCF iteration procedure uses the core-Hamiltonian for an initial guess at the Fock matrix and convergence is assumed when the standard deviation in a density matrix element is less than 10^{-4} a.u. A maximum of 25 iterations is allowed. No expectation values are calculated other than the energy, but the matrix **PS** is printed for a Mulliken population analysis.

The program is simple, but it contains the basic ingredients of sophisticated programs for large *ab initio* calculations. The integral routines S, T, V, and TWOE are general for any set of $1s$ Gaussian basis functions. The interested and adept student, who fully understands the current program, could extend it to model calculations on polyatomic molecules using a basis set of floating $1s$ Gaussians, such as the "Gaussian lobe" basis sets of Whitten.[1] Calculations with more than two basis functions would require a general matrix diagonalization routine[2] rather than just the 2×2 diagonalization considered here. Most polyatomic programs, in addition to the above, require special techniques for storing and handling the large number of two-electron integrals and an efficient algorithm for forming the matrix **G** from the two-electron integrals and density matrix. Evaluating integrals for p- and d-type Cartesian Gaussians is considerably more difficult than for the $1s$ Gaussians and the writing of an efficient polyatomic program is a major undertaking.[3]

```
C•••••••••••••••••••••••••••••••••••••••••••••••••••••••••••••••••••••••••
C
C     MINIMAL BASIS STO-3G CALCULATION ON HEH+
C
C     THIS IS A LITTLE DUMMY MAIN PROGRAM WHICH CALLS HFCALC
C
C•••••••••••••••••••••••••••••••••••••••••••••••••••••••••••••••••••••••••
      IMPLICIT DOUBLE PRECISION(A-H,O-Z)
      IOP=2
      N=3
      R=1.4632D0
      ZETA1=2.0925D0
      ZETA2=1.24D0
      ZA=2.0D0
      ZB=1.0D0
      CALL HFCALC(IOP,N,R,ZETA1,ZETA2,ZA,ZB)
      END
C•••••••••••••••••••••••••••••••••••••••••••••••••••••••••••••••••••••••••
      SUBROUTINE HFCALC(IOP,N,R,ZETA1,ZETA2,ZA,ZB)
C
C     DOES A HARTREE-FOCK CALCULATION FOR A TWO-ELECTRON DIATOMIC
C     USING THE 1S MINIMAL STO-NG BASIS SET
C     MINIMAL BASIS SET HAS BASIS FUNCTIONS 1 AND 2 ON NUCLEI A AND B
C
C     IOP=0  NO PRINTING WHATSOEVER (TO OPTIMIZE EXPONENTS, SAY)
C     IOP=1  PRINT ONLY CONVERGED RESULTS
C     IOP=2  PRINT EVERY ITERATION
C     N      STO-NG CALCULATION (N=1,2 OR 3)
C     R      BONDLENGTH (AU)
C     ZETA1  SLATER ORBITAL EXPONENT (FUNCTION 1)
C     ZETA2  SLATER ORBITAL EXPONENT (FUNCTION 2)
C     ZA     ATOMIC NUMBER (ATOM A)
C     ZB     ATOMIC NUMBER (ATOM B)
C
C•••••••••••••••••••••••••••••••••••••••••••••••••••••••••••••••••••••••••
      IMPLICIT DOUBLE PRECISION(A-H,O-Z)
      IF (IOP.EQ.0) GO TO 20
      PRINT 10,N,ZA,ZB
   10 FORMAT(1H1,2X,4HSTO-,I1,21HG FOR ATOMIC NUMBERS ,F6.2,6H AND ,
     $F6.2,//)
   20 CONTINUE
C     CALCULATE ALL THE ONE AND TWO-ELECTRON INTEGRALS
      CALL INTGRL(IOP,N,R,ZETA1,ZETA2,ZA,ZB)
C     BE INEFFICIENT AND PUT ALL INTEGRALS IN PRETTY ARRAYS
      CALL COLECT(IOP,N,R,ZETA1,ZETA2,ZA,ZB)
C     PERFORM THE SCF CALCULATION
      CALL SCF(IOP,N,R,ZETA1,ZETA2,ZA,ZB)
      RETURN
      END
C•••••••••••••••••••••••••••••••••••••••••••••••••••••••••••••••••••••••••
      SUBROUTINE INTGRL(IOP,N,R,ZETA1,ZETA2,ZA,ZB)
C
C     CALCULATES ALL THE BASIC INTEGRALS NEEDED FOR SCF CALCULATION
C
C•••••••••••••••••••••••••••••••••••••••••••••••••••••••••••••••••••••••••
```

```
         IMPLICIT DOUBLE PRECISION(A-H,O-Z)
         COMMON/INT/S12,T11,T12,T22,V11A,V12A,V22A,V11B,V12B,V22B,
       $ V1111,V2111,V2121,V2211,V2221,V2222
         DIMENSION COEF(3,3),EXPON(3,3),D1(3),A1(3),D2(3),A2(3)
         DATA PI/3.1415926535898D0/
C        THESE ARE THE CONTRACTION COEFFICIENTS AND EXPONENTS FOR
C        A NORMALIZED 1S SLATER ORBITAL WITH EXPONENT 1.0 IN TERMS OF
C        NORMALIZED 1S PRIMITIVE GAUSSIANS
         DATA COEF,EXPON/1.0D0,2*0.0D0,0.678914D0,0.430129D0,0.0D0,
       $ 0.444635D0,0.535328D0,0.154329D0,0.270950D0,2*0.0D0,0.151623D0,
       $ 0.851819D0,0.0D0,0.109818D0,0.405771D0,2.22766D0/
         R2=R*R
C        SCALE THE EXPONENTS (A) OF PRIMITIVE GAUSSIANS
C        INCLUDE NORMALIZATION IN CONTRACTION COEFFICIENTS (D)
         DO 10 I=1,N
         A1(I)=EXPON(I,N)*(ZETA1**2)
         D1(I)=COEF(I,N)*((2.0D0*A1(I)/PI)**0.75D0)
         A2(I)=EXPON(I,N)*(ZETA2**2)
         D2(I)=COEF(I,N)*((2.0D0*A2(I)/PI)**0.75D0)
      10 CONTINUE
C        D AND A ARE NOW THE CONTRACTION COEFFICIENTS AND EXPONENTS
C        IN TERMS OF UNNORMALIZED PRIMITIVE GAUSSIANS
         S12=0.0D0
         T11=0.0D0
         T12=0.0D0
         T22=0.0D0
         V11A=0.0D0
         V12A=0.0D0
         V22A=0.0D0
         V11B=0.0D0
         V12B=0.0D0
         V22B=0.0D0
         V1111=0.0D0
         V2111=0.0D0
         V2121=0.0D0
         V2211=0.0D0
         V2221=0.0D0
         V2222=0.0D0
C        CALCULATE ONE-ELECTRON INTEGRALS
C        CENTER A IS FIRST ATOM, CENTER B IS SECOND ATOM
C        ORIGIN IS ON CENTER A
C        V12A = OFF-DIAGONAL NUCLEAR ATTRACTION TO CENTER A, ETC.
         DO 20 I=1,N
         DO 20 J=1,N
C        RAP2 = SQUARED DISTANCE BETWEEN CENTER A AND CENTER P, ETC.
         RAP=A2(J)*R/(A1(I)+A2(J))
         RAP2=RAP**2
         RBP2=(R-RAP)**2
         S12=S12+S(A1(I),A2(J),R2)*D1(I)*D2(J)
         T11=T11+T(A1(I),A1(J),0.0D0)*D1(I)*D1(J)
         T12=T12+T(A1(I),A2(J),R2)*D1(I)*D2(J)
         T22=T22+T(A2(I),A2(J),0.0D0)*D2(I)*D2(J)
         V11A=V11A+V(A1(I),A1(J),0.0D0,0.0D0,ZA)*D1(I)*D1(J)
         V12A=V12A+V(A1(I),A2(J),R2,RAP2,ZA)*D1(I)*D2(J)
         V22A=V22A+V(A2(I),A2(J),0.0D0,R2,ZA)*D2(I)*D2(J)
```

```
      V11B=V11B+V(A1(I),A1(J),0.0D0,R2,ZB)*D1(I)*D1(J)
      V12B=V12B+V(A1(I),A2(J),R2,RBP2,ZB)*D1(I)*D2(J)
      V22B=V22B+V(A2(I),A2(J),0.0D0,0.0D0,ZB)*D2(I)*D2(J)
   20 CONTINUE
C     CALCULATE TWO-ELECTRON INTEGRALS
      DO 30 I=1,N
      DO 30 J=1,N
      DO 30 K=1,N
      DO 30 L=1,N
      RAP=A2(I)*R/(A2(I)+A1(J))
      RBP=R-RAP
      RAQ=A2(K)*R/(A2(K)+A1(L))
      RBQ=R-RAQ
      RPQ=RAP-RAQ
      RAP2=RAP*RAP
      RBP2=RBP*RBP
      RAQ2=RAQ*RAQ
      RBQ2=RBQ*RBQ
      RPQ2=RPQ*RPQ
      V1111=V1111+TWOE(A1(I),A1(J),A1(K),A1(L),0.0D0,0.0D0,0.0D0)
     $ *D1(I)*D1(J)*D1(K)*D1(L)
      V2111=V2111+TWOE(A2(I),A1(J),A1(K),A1(L),R2,0.0D0,RAP2)
     $ *D2(I)*D1(J)*D1(K)*D1(L)
      V2121=V2121+TWOE(A2(I),A1(J),A2(K),A1(L),R2,R2,RPQ2)
     $ *D2(I)*D1(J)*D2(K)*D1(L)
      V2211=V2211+TWOE(A2(I),A2(J),A1(K),A1(L),0.0D0,0.0D0,R2)
     $ *D2(I)*D2(J)*D1(K)*D1(L)
      V2221=V2221+TWOE(A2(I),A2(J),A2(K),A1(L),0.0D0,R2,RBQ2)
     $ *D2(I)*D2(J)*D2(K)*D1(L)
      V2222=V2222+TWOE(A2(I),A2(J),A2(K),A2(L),0.0D0,0.0D0,0.0D0)
     $ *D2(I)*D2(J)*D2(K)*D2(L)
   30 CONTINUE
      IF (IOP.EQ.0) GO TO 90
      PRINT 40
   40 FORMAT(3X,1HR,10X,5HZETA1,6X,5HZETA2,6X,3HS12,8X,3HT11/)
      PRINT 50, R,ZETA1,ZETA2,S12,T11
   50 FORMAT(5F11.6//)
      PRINT 60
   60 FORMAT(3X,3HT12,8X,3HT22,8X,4HV11A,7X,4HV12A,7X,4HV22A/)
      PRINT 50, T12,T22,V11A,V12A,V22A
      PRINT 70
   70 FORMAT(3X,4HV11B,7X,4HV12B,7X,4HV22B,7X,5HV1111,6X,5HV2111/)
      PRINT 50, V11B,V12B,V22B,V1111,V2111
      PRINT 80
   80 FORMAT(3X,5HV2121,6X,5HV2211,6X,5HV2221,6X,5HV2222/)
      PRINT 50, V2121,V2211,V2221,V2222
   90 RETURN
      END
C*******************************************************************
      FUNCTION F0(ARG)
C
C     CALCULATES THE F FUNCTION
C     F0 ONLY (S-TYPE ORBITALS)
C
C*******************************************************************
```

```
      IMPLICIT DOUBLE PRECISION(A-H,O-Z)
      DATA PI/3.1415926535898D0/
      IF (ARG.LT.1.0D-6) GO TO 10
C     F0 IN TERMS OF THE ERROR FUNCTION
      F0=DSQRT(PI/ARG)*DERF(DSQRT(ARG))/2.0D0
      GO TO 20
C     ASYMPTOTIC VALUE FOR SMALL ARGUMENTS
   10 F0=1.0D0-ARG/3.0D0
   20 CONTINUE
      RETURN
      END
C****************************************************************
      FUNCTION DERF(ARG)
C
C     CALCULATES THE ERROR FUNCTION ACCORDING TO A RATIONAL
C     APPROXIMATION FROM M. ABRAMOWITZ AND I.A. STEGUN,
C     HANDBOOK OF MATHEMATICAL FUNCTIONS, DOVER.
C     ABSOLUTE ERROR IS LESS THAN 1.5*10**(-7)
C     CAN BE REPLACED BY A BUILT-IN FUNCTION ON SOME MACHINES
C
C****************************************************************
      IMPLICIT DOUBLE PRECISION(A-H,O-Z)
      DIMENSION A(5)
      DATA P/0.3275911D0/
      DATA A/0.254829592D0,-0.284496736D0,1.421413741D0,
     $ -1.453152027D0,1.061405429D0/
      T=1.0D0/(1.0D0+P*ARG)
      TN=T
      POLY=A(1)*TN
      DO 10 I=2,5
      TN=TN*T
      POLY=POLY+A(I)*TN
   10 CONTINUE
      DERF=1.0D0-POLY*DEXP(-ARG*ARG)
      RETURN
      END
C****************************************************************
      FUNCTION S(A,B,RAB2)
C
C     CALCULATES OVERLAPS FOR UN-NORMALIZED PRIMITIVES
C
C****************************************************************
      IMPLICIT DOUBLE PRECISION(A-H,O-Z)
      DATA PI/3.1415926535898D0/
      S=(PI/(A+B))**1.5D0*DEXP(-A*B*RAB2/(A+B))
      RETURN
      END
C****************************************************************
      FUNCTION T(A,B,RAB2)
C
C     CALCULATES KINETIC ENERGY INTEGRALS FOR UN-NORMALIZED PRIMITIVES
C
C****************************************************************
      IMPLICIT DOUBLE PRECISION(A-H,O-Z)
      DATA PI/3.1415926535898D0/
```

```
        T=A*B/(A+B)*(3.0D0-2.0D0*A*B*RAB2/(A+B))*(PI/(A+B))**1.5D0
     $    *DEXP(-A*B*RAB2/(A+B))
        RETURN
        END
C*********************************************************************
        FUNCTION V(A,B,RAB2,RCP2,ZC)
C
C       CALCULATES UN-NORMALIZED NUCLEAR ATTRACTION INTEGRALS
C
C*********************************************************************
        IMPLICIT DOUBLE PRECISION(A-H,O-Z)
        DATA PI/3.1415926535898D0/
        V=2.0D0*PI/(A+B)*F0((A+B)*RCP2)*DEXP(-A*B*RAB2/(A+B))
        V=-V*ZC
        RETURN
        END
C*********************************************************************
        FUNCTION TWOE(A,B,C,D,RAB2,RCD2,RPQ2)
C
C       CALCULATES TWO-ELECTRON INTEGRALS FOR UN-NORMALIZED PRIMITIVES
C       A,B,C,D ARE THE EXPONENTS ALPHA,BETA, ETC.
C       RAB2 EQUALS SQUARED DISTANCE BETWEEN CENTER A AND CENTER B, ETC.
C
C*********************************************************************
        IMPLICIT DOUBLE PRECISION(A-H,O-Z)
        DATA PI/3.1415926535898D0/
        TWOE=2.0D0*(PI**2.5D0)/((A+B)*(C+D)*DSQRT(A+B+C+D))
     $    *F0((A+B)*(C+D)*RPQ2/(A+B+C+D))
     $    *DEXP(-A*B*RAB2/(A+B)-C*D*RCD2/(C+D))
        RETURN
        END
C*********************************************************************
        SUBROUTINE COLECT(IOP,N,R,ZETA1,ZETA2,ZA,ZB)
C
C       THIS TAKES THE BASIC INTEGRALS FROM COMMON AND ASSEMBLES THE
C       RELEVANT MATRICES, THAT IS S,H,X,XT, AND TWO-ELECTRON INTEGRALS
C
C*********************************************************************
        IMPLICIT DOUBLE PRECISION(A-H,O-Z)
        COMMON/MATRIX/S(2,2),X(2,2),XT(2,2),H(2,2),F(2,2),G(2,2),C(2,2),
     $  FPRIME(2,2),CPRIME(2,2),P(2,2),OLDP(2,2),TT(2,2,2,2),E(2,2)
        COMMON/INT/S12,T11,T12,T22,V11A,V12A,V22A,V11B,V12B,V22B,
     $  V1111,V2111,V2121,V2211,V2221,V2222
C       FORM CORE HAMILTONIAN
        H(1,1)=T11+V11A+V11B
        H(1,2)=T12+V12A+V12B
        H(2,1)=H(1,2)
        H(2,2)=T22+V22A+V22B
C       FORM OVERLAP MATRIX
        S(1,1)=1.0D0
        S(1,2)=S12
        S(2,1)=S(1,2)
        S(2,2)=1.0D0
C       USE CANONICAL ORTHOGONALIZATION
        X(1,1)=1.0D0/DSQRT(2.0D0*(1.0D0+S12))
```

```
        X(2,1)=X(1,1)
        X(1,2)=1.0D0/DSQRT(2.0D0*(1.0D0-S12))
        X(2,2)=-X(1,2)
C       TRANSPOSE OF TRANSFORMATION MATRIX
        XT(1,1)=X(1,1)
        XT(1,2)=X(2,1)
        XT(2,1)=X(1,2)
        XT(2,2)=X(2,2)
C       MATRIX OF TWO-ELECTRON INTEGRALS
        TT(1,1,1,1)=V1111
        TT(2,1,1,1)=V2111
        TT(1,2,1,1)=V2111
        TT(1,1,2,1)=V2111
        TT(1,1,1,2)=V2111
        TT(2,1,2,1)=V2121
        TT(1,2,2,1)=V2121
        TT(2,1,1,2)=V2121
        TT(1,2,1,2)=V2121
        TT(2,2,1,1)=V2211
        TT(1,1,2,2)=V2211
        TT(2,2,2,1)=V2221
        TT(2,1,2,2)=V2221
        TT(1,2,2,2)=V2221
        TT(2,2,2,2)=V2222
        IF (IOP.EQ.0) GO TO 40
        CALL MATOUT(S,2,2,2,2,4HS   )
        CALL MATOUT(X,2,2,2,2,4HX   )
        CALL MATOUT(H,2,2,2,2,4HH   )
        PRINT 10
   10 FORMAT(//)
        DO 30 I=1,2
        DO 30 J=1,2
        DO 30 K=1,2
        DO 30 L=1,2
        PRINT 20, I,J,K,L,TT(I,J,K,L)
   20 FORMAT(3X,1H(,4I2,2H ),F10.6)
   30 CONTINUE
   40 RETURN
        END
C***************************************************************
        SUBROUTINE SCF(IOP,N,R,ZETA1,ZETA2,ZA,ZB)
C
C       PERFORMS THE SCF ITERATIONS
C
C***************************************************************
        IMPLICIT DOUBLE PRECISION(A-H,O-Z)
        COMMON/MATRIX/S(2,2),X(2,2),XT(2,2),H(2,2),F(2,2),G(2,2),C(2,2),
     $ FPRIME(2,2),CPRIME(2,2),P(2,2),OLDP(2,2),TT(2,2,2,2),E(2,2)
        DATA PI/3.1415926535898D0/
C       CONVERGENCE CRITERION FOR DENSITY MATRIX
        DATA CRIT/1.0D-4/
C       MAXIMUM NUMBER OF ITERATIONS
        DATA MAXIT/25/
C       ITERATION NUMBER
```

```
         ITER=0
C        USE CORE-HAMILTONIAN FOR INITIAL GUESS AT F, I.E. (P=0)
         DO 10 I=1,2
         DO 10 J=1,2
      10 P(I,J)=0.0D0
         IF (IOP.LT.2) GO TO 20
         CALL MATOUT(P,2,2,2,2,4HP    )
C        START OF ITERATION LOOP
      20 ITER=ITER+1
         IF (IOP.LT.2) GO TO 40
         PRINT 30, ITER
      30 FORMAT(/,4X,28HSTART OF ITERATION NUMBER = ,I2)
      40 CONTINUE
C        FORM TWO-ELECTRON PART OF FOCK MATRIX FROM P
         CALL FORMG
         IF (IOP.LT.2) GO TO 50
         CALL MATOUT(G,2,2,2,2,4HG    )
      50 CONTINUE
C        ADD CORE HAMILTONIAN TO GET FOCK MATRIX
         DO 60 I=1,2
         DO 60 J=1,2
         F(I,J)=H(I,J)+G(I,J)
      60 CONTINUE
C        CALCULATE ELECTRONIC ENERGY
         EN=0.0D0
         DO 70 I=1,2
         DO 70 J=1,2
         EN=EN+0.5D0*P(I,J)*(H(I,J)+F(I,J))
      70 CONTINUE
         IF (IOP.LT.2) GO TO 90
         CALL MATOUT(F,2,2,2,2,4HF    )
         PRINT 80, EN
      80 FORMAT(///,4X,20HELECTRONIC ENERGY = ,D20.12)
      90 CONTINUE
C        TRANSFORM FOCK MATRIX USING G FOR TEMPORARY STORAGE
         CALL MULT(F,X,G,2,2)
         CALL MULT(XT,G,FPRIME,2,2)
C        DIAGONALIZE TRANSFORMED FOCK MATRIX
         CALL DIAG(FPRIME,CPRIME,E)
C        TRANSFORM EIGENVECTORS TO GET MATRIX C
         CALL MULT(X,CPRIME,C,2,2)
C        FORM NEW DENSITY MATRIX
         DO 100 I=1,2
         DO 100 J=1,2
C        SAVE PRESENT DENSITY MATRIX
C        BEFORE CREATING NEW ONE
         OLDP(I,J)=P(I,J)
         P(I,J)=0.0D0
         DO 100 K=1,1
         P(I,J)=P(I,J)+2.0D0*C(I,K)*C(J,K)
     100 CONTINUE
         IF (IOP.LT.2) GO TO 110
         CALL MATOUT(FPRIME,2,2,2,2,4HF'  )
         CALL MATOUT(CPRIME,2,2,2,2,4HC'  )
         CALL MATOUT(E,2,2,2,2,4HE    )
```

```
            CALL MATOUT(C,2,2,2,2,4HC    )
            CALL MATOUT(P,2,2,2,2,4HP    )
        110 CONTINUE
C           CALCULATE DELTA
            DELTA=0.0D0
            DO 120 I=1,2
            DO 120 J=1,2
            DELTA=DELTA+(P(I,J)-OLDP(I,J))**2
        120 CONTINUE
            DELTA=DSQRT(DELTA/4.0D0)
            IF (IOP.EQ.0) GO TO 140
            PRINT 130, DELTA
        130 FORMAT(/,4X,39HDELTA(CONVERGENCE OF DENSITY MATRIX) =
           $ F10.6,/)
        140 CONTINUE
C           CHECK FOR CONVERGENCE
            IF (DELTA.LT.CRIT) GO TO 160
C           NOT YET CONVERGED
C           TEST FOR MAXIMUM NUMBER OF ITERATIONS
C           IF MAXIMUM NUMBER NOT YET REACHED THEN
C           GO BACK FOR ANOTHER ITERATION
            IF (ITER.LT.MAXIT) GO TO 20
C           SOMETHING WRONG HERE
            PRINT 150
        150 FORMAT(4X,21HNO CONVERGENCE IN SCF)
            STOP
        160 CONTINUE
C           CALCULATION CONVERGED IF IT GOT HERE
C           ADD NUCLEAR REPULSION TO GET TOTAL ENERGY
            ENT=EN+ZA*ZB/R
            IF (IOP.EQ.0) GO TO 180
            PRINT 170, EN, ENT
        170 FORMAT(//,4X,21HCALCULATION CONVERGED,//,
           $ 4X,20HELECTRONIC ENERGY = ,D20.12,//,
           $ 4X,20HTOTAL ENERGY =      ,D20.12)
        180 CONTINUE
            IF (IOP.NE.1) GO TO 190
C           PRINT OUT THE FINAL RESULTS IF
C           HAVE NOT DONE SO ALREADY
            CALL MATOUT(G,2,2,2,2,4HG    )
            CALL MATOUT(F,2,2,2,2,4HF    )
            CALL MATOUT(E,2,2,2,2,4HE    )
            CALL MATOUT(C,2,2,2,2,4HC    )
            CALL MATOUT(P,2,2,2,2,4HP    )
        190 CONTINUE
C           PS MATRIX HAS MULLIKEN POPULATIONS
            CALL MULT(P,S,OLDP,2,2)
            IF (IOP.EQ.0) GO TO 200
            CALL MATOUT(OLDP,2,2,2,2,4HPS   )
        200 CONTINUE
            RETURN
            END
C***************************************************************
            SUBROUTINE FORMG
C
```

```
C      CALCULATES THE G MATRIX FROM THE DENSITY MATRIX
C      AND TWO-ELECTRON INTEGRALS
C
C**********************************************************************
       IMPLICIT DOUBLE PRECISION(A-H,O-Z)
       COMMON/MATRIX/S(2,2),X(2,2),XT(2,2),H(2,2),F(2,2),G(2,2),C(2,2),
     $ FPRIME(2,2),CPRIME(2,2),P(2,2),OLDP(2,2),TT(2,2,2,2),E(2,2)
       DO 10 I=1,2
       DO 10 J=1,2
       G(I,J)=0.0D0
       DO 10 K=1,2
       DO 10 L=1,2
       G(I,J)=G(I,J)+P(K,L)*(TT(I,J,K,L)-0.5D0*TT(I,L,K,J))
    10 CONTINUE
       RETURN
       END
C**********************************************************************
       SUBROUTINE DIAG(F,C,E)
C
C      DIAGONALIZES F TO GIVE EIGENVECTORS IN C AND EIGENVALUES IN E
C      THETA IS THE ANGLE DESCRIBING SOLUTION
C
C**********************************************************************
       IMPLICIT DOUBLE PRECISION(A-H,O-Z)
       DIMENSION F(2,2),C(2,2),E(2,2)
       DATA PI/3.1415926535898D0/
       IF (DABS(F(1,1)-F(2,2)).GT.1.0D-20) GO TO 10
C      HERE IS SYMMETRY DETERMINED SOLUTION (HOMONUCLEAR DIATOMIC)
       THETA=PI/4.0D0
       GO TO 20
    10 CONTINUE
C      SOLUTION FOR HETERONUCLEAR DIATOMIC
       THETA=0.5D0*DATAN(2.0D0*F(1,2)/(F(1,1)-F(2,2)))
    20 CONTINUE
       C(1,1)=DCOS(THETA)
       C(2,1)=DSIN(THETA)
       C(1,2)=DSIN(THETA)
       C(2,2)=-DCOS(THETA)
       E(1,1)=F(1,1)*DCOS(THETA)**2+F(2,2)*DSIN(THETA)**2
     $ +F(1,2)*DSIN(2.0D0*THETA)
       E(2,2)=F(2,2)*DCOS(THETA)**2+F(1,1)*DSIN(THETA)**2
     $ -F(1,2)*DSIN(2.0D0*THETA)
       E(2,1)=0.0D0
       E(1,2)=0.0D0
C      ORDER EIGENVALUES AND EIGENVECTORS
       IF (E(2,2).GT.E(1,1)) GO TO 30
       TEMP=E(2,2)
       E(2,2)=E(1,1)
       E(1,1)=TEMP
       TEMP=C(1,2)
       C(1,2)=C(1,1)
       C(1,1)=TEMP
       TEMP=C(2,2)
       C(2,2)=C(2,1)
       C(2,1)=TEMP
```

```
    30 RETURN
       END
C*********************************************************************
       SUBROUTINE MULT(A,B,C,IM,M)
C
C      MULTIPLIES TWO SQUARE MATRICES A AND B TO GET C
C
C*********************************************************************
       IMPLICIT DOUBLE PRECISION(A-H,O-Z)
       DIMENSION A(IM,IM),B(IM,IM),C(IM,IM)
       DO 10 I=1,M
       DO 10 J=1,M
       C(I,J)=0.0D0
       DO 10 K=1,M
    10 C(I,J)=C(I,J)+A(I,K)*B(K,J)
       RETURN
       END
C*********************************************************************
       SUBROUTINE MATOUT(A,IM,IN,M,N,LABEL)
C
C      PRINT MATRICES OF SIZE M BY N
C
C*********************************************************************
       IMPLICIT DOUBLE PRECISION(A-H,O-Z)
       DIMENSION A(IM,IN)
       IHIGH=0
    10 LOW=IHIGH+1
       IHIGH=IHIGH+5
       IHIGH=MINO(IHIGH,N)
       PRINT 20, LABEL,(I,I=LOW,IHIGH)
    20 FORMAT(///,3X,5H THE ,A4,6H ARRAY,/,15X,5(10X,I3,6X)//)
       DO 30 I=1,M
    30 PRINT 40, I,(A(I,J),J=LOW,IHIGH)
    40 FORMAT(I10,5X,5(1X,D18.10))
       IF (N-IHIGH) 50,50,10
    50 RETURN
       END
```

STO-3G FOR ATOMIC NUMBERS 2.00 AND 1.00

R	ZETA1	ZETA2	S12	T11
1.463200	2.092500	1.240000	0.450770	2.164313

T12	T22	V11A	V12A	V22A
0.167013	0.760033	-4.139827	-1.102912	-1.265246

V11B	V12B	V22B	V1111	V2111
-0.677230	-0.411305	-1.226615	1.307152	0.437279

V2121	V2211	V2221	V2222
0.177267	0.605703	0.311795	0.774608

```
    THE S   ARRAY
                        1                   2
         1      0.1000000000D+01    0.4507704116D+00
         2      0.4507704116D+00    0.1000000000D+01


    THE X   ARRAY
                        1                   2
         1      0.5870642812D+00    0.9541310722D+00
         2      0.5870642812D+00   -0.9541310722D+00


    THE H   ARRAY
                        1                   2
         1     -0.2652744703D+01   -0.1347205024D+01
         2     -0.1347205024D+01   -0.1731828436D+01


   ( 1 1 1 1 )   1.307152
   ( 1 1 1 2 )   0.437279
   ( 1 1 2 1 )   0.437279
   ( 1 1 2 2 )   0.605703
   ( 1 2 1 1 )   0.437279
   ( 1 2 1 2 )   0.177267
   ( 1 2 2 1 )   0.177267
   ( 1 2 2 2 )   0.311795
   ( 2 1 1 1 )   0.437279
   ( 2 1 1 2 )   0.177267
```

```
( 2 1 2 1 )   0.177267
( 2 1 2 2 )   0.311795
( 2 2 1 1 )   0.605703
( 2 2 1 2 )   0.311795
( 2 2 2 1 )   0.311795
( 2 2 2 2 )   0.774608
```

```
THE P    ARRAY
                      1                       2
      1      0.0000000000D+00       0.0000000000D+00
      2      0.0000000000D+00       0.0000000000D+00

START OF ITERATION NUMBER = 1


THE G    ARRAY
                      1                       2
      1      0.0000000000D+00       0.0000000000D+00
      2      0.0000000000D+00       0.0000000000D+00


THE F    ARRAY
                      1                       2
      1     -0.2652744703D+01      -0.1347205024D+01
      2     -0.1347205024D+01      -0.1731828436D+01


ELECTRONIC ENERGY =   0.000000000000D+00


THE F'   ARRAY
                      1                       2
      1     -0.2439732411D+01      -0.5158386047D+00
      2     -0.5158386047D+00      -0.1538667186D+01


THE C'   ARRAY
                      1                       2
      1      0.9104462570D+00       0.4136295856D+00
      2      0.4136295856D+00      -0.9104462570D+00


THE E    ARRAY
                      1                       2
      1     -0.2674085994D+01       0.0000000000D+00
      2      0.0000000000D+00      -0.1304313603D+01
```

```
THE C     ARRAY
                            1                     2
        1         0.9291467304D+00   -0.6258569539D+00
        2         0.1398330503D+00    0.1111511265D+01


THE P     ARRAY
                            1                     2
        1         0.1726627293D+01    0.2598508430D+00
        2         0.2598508430D+00    0.3910656393D-01

DELTA(CONVERGENCE OF DENSITY MATRIX) =    0.882867


START OF ITERATION NUMBER = 2


THE G     ARRAY
                            1                     2
        1         0.1262330044D+01    0.3740040563D+00
        2         0.3740040563D+00    0.9889530699D+00


THE F     ARRAY
                            1                     2
        1        -0.1390414659D+01   -0.9732009679D+00
        2        -0.9732009679D+00   -0.7428753661D+00


ELECTRONIC ENERGY =  -0.414186268681D+01


THE F'    ARRAY
                            1                     2
        1        -0.1406043275D+01   -0.3627102456D+00
        2        -0.3627102456D+00   -0.1701365815D+00


THE C'    ARRAY
                            1                     2
        1         0.9649913726D+00    0.2622816249D+00
        2         0.2622816249D+00   -0.9649913726D+00


THE E     ARRAY
                            1                     2
```

```
1        -0.1504626781D+01     0.0000000000D+00
2         0.0000000000D+00    -0.7155307568D-01
```

THE C   ARRAY
```
                     1                  2
1         0.8167630145D+00    -0.7667520795D+00
2         0.3162609186D+00     0.1074704427D+01
```

THE P   ARRAY
```
                     1                  2
1         0.1334203644D+01     0.5166204425D+00
2         0.5166204425D+00     0.2000419373D+00
```

DELTA(CONVERGENCE OF DENSITY MATRIX) =   0.279176

START OF ITERATION NUMBER = 3

THE G   ARRAY
```
                     1                  2
1         0.1201346300D+01     0.3038061741D+00
2         0.3038061741D+00     0.9284329600D+00
```

THE F   ARRAY
```
                     1                  2
1        -0.1451398403D+01    -0.1043398850D+01
2        -0.1043398850D+01    -0.8033954759D+00
```

ELECTRONIC ENERGY =   -0.4226491725562D+01

THE F'   ARRAY
```
                     1                  2
1        -0.1496305530D+01    -0.3629699437D+00
2        -0.3629699437D+00    -0.1529380263D+00
```

THE C'   ARRAY
```
                     1                  2
1         0.9694747516D+00     0.2451911622D+00
2         0.2451911622D+00    -0.9694747516D+00
```

```
THE  E    ARRAY
                        1                    2
      1     -0.1588104746D+01    0.0000000000D+00
      2      0.0000000000D+00   -0.6113881008D-01



THE  C    ARRAY
                        1                    2
      1      0.8030885047D+00   -0.7810630108D+00
      2      0.3351994916D+00    0.1068948958D+01



THE  P    ARRAY
                        1                    2
      1      0.1289902293D+01    0.5383897171D+00
      2      0.5383897171D+00    0.2247173984D+00
```

DELTA(CONVERGENCE OF DENSITY MATRIX) =   0.029662

START OF ITERATION NUMBER = 4


```
THE  G    ARRAY
                        1                    2
      1      0.1194670199D+01    0.2971625826D+00
      2      0.2971625826D+00    0.9218705199D+00



THE  F    ARRAY
                        1                    2
      1     -0.1458074504D+01   -0.1050042442D+01
      2     -0.1050042442D+01   -0.8099579160D+00
```

ELECTRONIC ENERGY = -0.422752275334D+01


```
THE  F'   ARRAY
                        1                    2
      1     -0.1505447474D+01   -0.3630336096D+00
      2     -0.3630336096D+00   -0.1528937446D+00



THE  C'   ARRAY
                        1                    2
      1      0.9698136474D+00    0.2438472663D+00
```

```
2            0.2438472663D+00   -0.9698136474D+00
```

THE E   ARRAY
```
                    1                  2
     1     -0.1596727643D+01    0.0000000000D+00
     2      0.0000000000D+00   -0.6161357601D-01
```

THE C   ARRAY
```
                    1                  2
     1      0.8020052055D+00   -0.7821753152D+00
     2      0.3366806982D+00    0.1068483355D+01
```

THE P   ARRAY
```
                    1                  2
     1      0.1286424699D+01    0.5400393450D+00
     2      0.5400393450D+00    0.2267077850D+00
```

DELTA(CONVERGENCE OF DENSITY MATRIX) =   0.002318

START OF ITERATION NUMBER = 5

THE G   ARRAY
```
                    1                  2
     1      0.1194147845D+01    0.2966515832D+00
     2      0.2966515832D+00    0.9213575914D+00
```

THE F   ARRAY
```
                    1                  2
     1     -0.1458596858D+01   -0.1050553441D+01
     2     -0.1050553441D+01   -0.8104708445D+00
```

ELECTRONIC ENERGY =   -0.4227529096120+01

THE F'   ARRAY
```
                    1                  2
     1     -0.1506156505D+01   -0.3630388891D+00
     2     -0.3630388891D+00   -0.1529058377D+00
```

```
THE C'   ARRAY
                        1                      2
      1        0.9698390734D+00   0.2437461212D+00
      2        0.2437461212D+00  -0.9698390734D+00
```

```
THE E    ARRAY
                        1                      2
      1       -0.1597397746D+01   0.0000000000D+00
      2        0.0000000000D+00  -0.6166459619D-01
```

```
THE C    ARRAY
                        1                      2
      1        0.8019236265D+00  -0.7822589536D+00
      2        0.3367921305D+00   0.1068448236D+01
```

```
THE P    ARRAY
                        1                      2
      1        0.1286163006D+01   0.5401631334D+00
      2        0.5401631334D+00   0.2268578784D+00
```

DELTA(CONVERGENCE OF DENSITY MATRIX) =   0.000174

START OF ITERATION NUMBER = 6

```
THE G    ARRAY
                        1                      2
      1        0.1194108547D+01   0.2966131916D+00
      2        0.2966131916D+00   0.9213190058D+00
```

```
THE F    ARRAY
                        1                      2
      1       -0.1458636156D+01  -0.1050591833D+01
      2       -0.1050591833D+01  -0.8105094301D+00
```

ELECTRONIC ENERGY = -0.422752913203D+01

```
THE F'   ARRAY
                        1                      2
      1       -0.1506209810D+01  -0.3630392881D+00
      2       -0.3630392881D+00  -0.1529068392D+00
```

```
THE C'    ARRAY
                          1                      2
       1       0.9698409800D+00     0.2437385353D+00
       2       0.2437385353D+00    -0.9698409800D+00


THE E     ARRAY
                          1                      2
       1      -0.1597448132D+01     0.0000000000D+00
       2       0.0000000000D+00    -0.6166851652D-01


THE C     ARRAY
                          1                      2
       1       0.8019175078D+00    -0.7822652261D+00
       2       0.3368004878D+00     0.1068445602D+01


THE P     ARRAY
                          1                      2
       1       0.1286143379D+01     0.5401724156D+00
       2       0.5401724156D+00     0.2268691372D+00

DELTA(CONVERGENCE OF DENSITY MATRIX) =    0.000013


CALCULATION CONVERGED

ELECTRONIC ENERGY =    -0.422752913203D+01

TOTAL ENERGY =         -0.286066199152D+01


THE PS    ARRAY
                          1                      2
       1       0.1529637121D+01     0.1119927796D+01
       2       0.6424383099D+00     0.4703628793D+00
```

NOTES

1. J. L. Whitten, Gaussian lobe function expansions of Hartree-Fock solutions for the first-row atoms and ethylene, *J. Chem. Phys.* **44**: 359 (1966).
2. A listing of an efficient diagonalization program is given by J. A. Pople and D. L. Beveridge, *Approximate Molecular Orbital Theory*, McGraw-Hill, New York, 1970, Appendix 2.
3. J. S. Binkley, R. A. Whiteside, R. Krishnan, R. Seeger, H. B. Schlegel, D. J. Defrees, and J. A. Pople, Gaussian 80, program #406, Quantum Chemistry Program Exchange, Indiana University.

ANALYTIC DERIVATIVE METHODS AND GEOMETRY OPTIMIZATION

Michael C. Zerner

Quantum Theory Project
University of Florida

C.1 INTRODUCTION

One of the most successful applications of molecular quantum mechanics has been the reproduction and prediction of molecular conformation. In many cases bond lengths are reproduced to ± 0.02 Å and bond angles to $\pm 5°$ with a variety of simple molecular orbital models, or with minimum basis set *ab initio* calculations. Larger basis sets, especially those of double zeta plus polarization type, and the inclusion of electron correlation are now producing geometries which challenge crystallography for accuracy. Armed with the growing success of conformational calculations, one might even choose the calculated results on isolated molecules over the experimental results obtained in condensed media, as the former may be more appropriate for the gas phase.

In addition to yielding information about global minima of the potential energy surface, quantum mechanical calculations yield information on local minima, which may or may not be observable directly, but which might be involved in reaction pathways. Similarly, information can be obtained about transition states and energy barriers that would be difficult or impossible to obtain in other ways.

The gleaning of all this information from a potential energy surface is difficult. For N atoms, the energy is a function of $3N - 6$ (or $3N - 5$)

degrees of freedom. For detailed statistical calculations, one may have to live with this "$3N$" problem and visit all regions of the surface that are thermally accessible. This appendix, however, is concerned with determining only a small part of the potential energy surface: those points that correspond to minima, representing stable or metastable conformations, and points that correspond to transition states.

C.2 GENERAL CONSIDERATIONS

The energy E of a molecular system obtained under the Born-Oppenheimer approximation is a parametric function of the nuclear coordinates denoted here as $\mathbf{X}^\dagger = (X_1, X_2, \ldots, X_{3N})$. We wish to move from $E(\mathbf{X})$ to $E(\mathbf{X}_1)$, where $\mathbf{q} = (\mathbf{X}_1 - \mathbf{X})$. We expand the energy in a Taylor series about $\mathbf{X}$ as follows:

$$E(\mathbf{X}_1) = E(\mathbf{X}) + \mathbf{q}^\dagger \mathbf{f}(\mathbf{X}) + \tfrac{1}{2}\mathbf{q}^\dagger \mathbf{H}(\mathbf{X})\mathbf{q} + \cdots \qquad (C.1)$$

where the gradient is

$$f_i = \frac{\partial E(\mathbf{X})}{\partial X_i}$$

and the Hessian is

$$H_{ij} = \frac{\partial E(\mathbf{X})}{\partial X_i\, \partial X_j}$$

Note that subscripts on column matrices designate different matrices, $\mathbf{X}_1$, $\mathbf{X}_2$, etc., whereas X_i designates the ith element of $\mathbf{X}$. Although the Taylor series is infinite, close to extrema we expect a quadratic form to be adequate; i.e., for $\mathbf{X} = \mathbf{X}_e$, where $\mathbf{X}_e$ designates a stationary point and by definition is characterized by $\mathbf{f}(\mathbf{X}_e) = \mathbf{0}$,

$$E(\mathbf{X}_1) = E(\mathbf{X}_e) + \tfrac{1}{2}\mathbf{q}^\dagger \mathbf{H}(\mathbf{X}_e)\mathbf{q}$$

In a similar fashion,

$$\mathbf{f}(\mathbf{X}_1) = \mathbf{f}(\mathbf{X}) + \mathbf{H}(\mathbf{X})\mathbf{q} \qquad (C.2)$$

For the point $\mathbf{X}_1 = \mathbf{X}_e$,

$$\mathbf{f}(\mathbf{X}) = -\mathbf{H}(\mathbf{X})\mathbf{q} \qquad (C.3)$$

The solution of Eq. (C.3) is the starting point of the most efficient procedures used to find extrema of functions of several variables where the functional form of $E(\mathbf{X})$ is not explicit in $\mathbf{X}$. If $\mathbf{H}$ is nonsingular, then

$$\mathbf{q} = -\mathbf{H}^{-1}(\mathbf{X})\mathbf{f}(\mathbf{X}) \qquad (C.4)$$

which allows the solution for $\mathbf{X}_e$ from any point $\mathbf{X}$ near enough so that the energy function is nearly quadratic. Similarly, an estimate of $E(\mathbf{H}_e)$ is obtained from

$$E(X_e) = E(X) - \tfrac{1}{2}f(X)^{\dagger}H^{-1}(X)f(X)$$
$$= E(X) - \tfrac{1}{2}q^{\dagger}H(X)q \tag{C.5}$$

For the specific problem of uncovering extrema on the potential energy surface, it should be pointed out that $H^{-1}(X)$ will not exist unless the rotations and translations which represent zero eigenvalues of H have been factored out. This may be accomplished via the B matrix of Wilson and Eliashevich:[1]

$$Y^{\dagger} = X^{\dagger}B \tag{C.6}$$

where X has $3N$ entries and B is $3N \times (3N - 6)$ relating an internal set of coordinates to the original Cartesian coordinate X. The easiest way to uncover the structure of the Taylor series in these new coordinates is to consider the work w, as it is independent of the choice of coordinate system. Then work, which is force times distance, can be represented in either coordinate system by

$$w = f^{\dagger}q = f_y^{\dagger}q_y = f_y^{\dagger}B^{\dagger}q$$
$$f^{\dagger} = f_y^{\dagger}B^{\dagger}$$

or

$$f_y^{\dagger} = f^{\dagger}(B^{\dagger})^{-1} \tag{C.7}$$

where $(B^{\dagger})^{-1}$ satisfies

$$B^{\dagger}(B^{\dagger})^{-1} = 1$$

A general solution of the above equation is given by

$$(B^{\dagger})^{-1} = mB(B^{\dagger}mB)^{-1}$$

where m is an arbitrary $3N \times 3N$ matrix, usually taken as a diagonal matrix containing the reciprocal of each atomic mass three times in the appropriate positions. It may also be chosen as the unit matrix with six (or five) zero entries chosen to prevent translation and rotation. A simple choice of this type is to place atom 1 at the origin, atom 2 on the z axis, and atom 3 in the xz plane. Then the six (or five) coordinates removed are $x_1 = y_1 = z_1 = 0$, $x_2 = y_2 = 0$, and $y_3 = 0$. If $y_3 = 0$ implies $x_3 = 0$ for any choice of third atoms, then the molecule is linear and only five degrees of freedom are chosen. In practice, inverting H will prove to be of little difficulty. The update methods that are discussed below build up H^{-1} directly and are constructed in such a way as to never update the rotations and translations. In the analytic evaluation of H, the rotations and translations can be removed as previously discussed, or, if the inversion proceeds via diagonalization, the six (or five) zero eigenvalues of H can be replaced with an arbitrary large number, essentially uncoupling these modes in the inverse.

C.3 ANALYTIC DERIVATIVES

We now turn to the calculation of **f** and **H** in quantum chemistry. A straightforward approach is to simply numerically differentiate the energy. Alternatively, it is possible to obtain these derivatives by analytic derivative methods. Let us illustrate some of the essential ideas in the framework of Hartree-Fock calculations.

The Hartree-Fock energy depends explicitly on the *occupied* molecular orbital coefficients **C** and on **X**. Its derivative is given by

$$\frac{\partial E}{\partial X_A} = \frac{\partial \bar{E}}{\partial X_A} + \sum_{\mu a} \frac{\partial E}{\partial C_{\mu a}} \frac{\partial C_{\mu a}}{\partial X_A} \tag{C.8}$$

where $\partial \bar{E}/\partial X_A$ represents the derivative of all terms with explicit dependence on the nuclear coordinate X_A, and where the chain rule term arises from the implicit dependence of the molecular orbital coefficients on geometry. Since $\partial E/\partial C_{\mu a} = 0$ is the condition for the Hartree-Fock solutions (Subsection 3.2.1),

$$\frac{\partial E}{\partial X_A} = \frac{\partial \bar{E}}{\partial X_A} \tag{C.9}$$

This realization allows one to ignore to first order the change in molecular orbital coefficients with respect to geometry changes. The total energy of a closed-shell system in the Hartree-Fock approximation can be written as

$$E = \sum_{\mu\nu} P_{\nu\mu} H_{\mu\nu}^{core} + \tfrac{1}{2} \sum_{\mu\nu\lambda\sigma} P_{\nu\mu} P_{\lambda\sigma} (\mu\nu| \,|\sigma\lambda) + V_{NN} \tag{C.10}$$

(see Eqs. (3.184), (3.185), and (3.154)) where we have defined

$$V_{NN} = \sum_{A} \sum_{A>B} \frac{Z_A Z_B}{R_{AB}}$$

and introduced the shorthand notation

$$(\mu\nu| \,|\sigma\lambda) = (\mu\nu|\sigma\lambda) - \tfrac{1}{2}(\mu\lambda|\sigma\nu)$$

For real orbitals, the density matrix is

$$P_{\mu\nu} = 2 \sum_{a}^{N/2} C_{\mu a} C_{\nu a}$$

(see Eq. (3.145)). Differentiating Eq. (C.10) yields

$$\frac{\partial E}{\partial X_A} = \sum_{\mu\nu} P_{\nu\mu} \frac{\partial H_{\mu\nu}^{core}}{\partial X_A} + \frac{1}{2} \sum_{\mu\nu\lambda\sigma} P_{\nu\mu} P_{\lambda\sigma} \frac{\partial(\mu\nu| \,|\sigma\lambda)}{\partial X_A} + \frac{\partial V_{NN}}{\partial X_A}$$

$$+ \sum_{\mu\nu} \frac{\partial P_{\nu\mu}}{\partial X_A} H_{\mu\nu}^{core} + \sum_{\mu\nu\lambda\sigma} \frac{\partial P_{\nu\mu}}{\partial X_A} P_{\lambda\sigma} (\mu\nu| \,|\sigma\lambda) \tag{C.11}$$

Equation (C.11) suggests that derivatives of the coefficients are required, whereas Eq. (C.9) does not! Expanding the last two terms of Eq. (C.11) gives

$$= 4 \sum_{\mu\nu}^{} \sum_{a}^{N/2} \frac{\partial C_{\mu a}}{\partial X_A} H_{\mu\nu}^{\text{core}} C_{\nu a} + 4 \sum_{\mu\nu\lambda\sigma}^{} \sum_{a}^{N/2} \frac{\partial C_{\mu a}}{\partial X_A} P_{\lambda\sigma} (\mu\nu| |\sigma\lambda) C_{\nu a}$$

$$= 4 \sum_{\mu\nu}^{} \sum_{a}^{N/2} \frac{\partial C_{\mu a}}{\partial X_A} \left[H_{\mu\nu}^{\text{core}} + \sum_{\lambda\sigma} P_{\lambda\sigma}(\mu\nu| |\sigma\lambda) \right] C_{\nu a}$$

$$= 4 \sum_{\mu\nu}^{} \sum_{a}^{N/2} \frac{\partial C_{\mu a}}{\partial X_A} F_{\mu\nu} C_{\nu a}$$

$$= 4 \sum_{a}^{N/2} \varepsilon_a \sum_{\mu\nu}^{} \frac{\partial C_{\mu a}}{\partial X_A} S_{\mu\nu} C_{\nu a}$$

To evaluate the derivative of the coefficients, we recall that the ortho-normality condition of the molecular orbitals is

$$\sum_{\mu\nu} C_{\mu a} S_{\mu\nu} C_{\nu b} = \delta_{ab}$$

(see Exercise 3.10). Differentiating this yields

$$2 \sum_{\mu\nu} \frac{\partial C_{\mu a}}{\partial X_A} S_{\mu\nu} C_{\nu a} = -\sum_{\mu\nu} C_{\mu a} C_{\nu a} \frac{\partial S_{\mu\nu}}{\partial X_A}$$

Combining these expressions results in

$$\frac{\partial E}{\partial X_A} = \sum_{\mu\nu} P_{\nu\mu} \frac{\partial H_{\mu\nu}^{\text{core}}}{\partial X_A} + \frac{1}{2} \sum_{\mu\nu\lambda\sigma} P_{\nu\mu} P_{\lambda\sigma} \frac{\partial(\mu\nu| |\sigma\lambda)}{\partial X_A}$$

$$- \sum_{\mu\nu} Q_{\nu\mu} \frac{\partial S_{\mu\nu}}{\partial X_A} + \frac{\partial V_{NN}}{\partial X_A} \tag{C.12}$$

where we have defined

$$Q_{\nu\mu} = 2 \sum_{a}^{N/2} \varepsilon_a C_{\mu a} C_{\nu a}$$

Thus the derivative of the energy can be calculated using the molecular orbital coefficients and the derivatives of the overlap and the one- and two-electron integrals.

Taking derivatives of the integrals that appear in electronic structure theory in an effective manner is a somewhat specialized area, but we can easily understand the general ideas. Most *ab initio* calculations are performed using Cartesian Gaussian functions

$$\phi_{lmn}^{\text{GF}} = N_{lmn} \, x_a^l y_a^m z_a^n e^{-a|\mathbf{r} - \mathbf{R}_A|^2}$$

(see Section 3.6), where N_{lmn} is a normalizing constant given by

$$N_{lmn} = \left[\frac{(8a)^{l+m+n} l! m! n!}{(2l)!(2m)!(2n)!} \right]^{1/2} \left(\frac{2a}{\pi} \right)^{3/4}$$

and where lowercase a refers to electronic coordinates measured from nucleus A. The derivative of ϕ_{lmn}^{GF} with respect to the nuclear coordinate is relatively straightforward, yielding, for example,

$$\frac{\partial \phi_{lmn}^{GF}}{\partial X_A} = [(2l + 1)a]^{1/2} \phi_{l+1,mn}^{GF} - 2l \left(\frac{a}{2l - 1} \right)^{1/2} \phi_{l-1,mn}^{GF}$$

where now X_A refers specifically to the x coordinate of nucleus A. It should be understood that the second term in the above equation is not considered when $l = 0$.

The derivative of all one-center integrals is zero, for it is usually assumed that all orbitals on center A follow the displacement of center A. The kinetic energy operator, and the electron-electron repulsion operator r_{12}^{-1}, are not functions of nuclear coordinates. The derivative of the nuclear-nuclear repulsion energy V_{NN} is again straightforward:

$$\frac{\partial V_{NN}}{\partial X_A} = Z_A \sum_B \frac{Z_B(X_B - X_A)}{R_{AB}^3}$$

For the nuclear-electronic attraction term, one finds

$$\frac{\partial V_{Ne}}{\partial X_A} = -Z_A \sum_i \frac{X_i - X_A}{r_{iA}^3}$$

The above equations are sufficient to evaluate the gradients of all integrals that are met in electronic structure calculations. In actual practice, however, many tricks are used to reduce the computational work. The evaluation of the gradients when Slater-type functions are used (see Sections 3.5 and 3.6) is more difficult, but could proceed along the same lines as for Gaussian functions, at least for the two-center integrals that usually appear in semiempirical models.

The relative simplicity of Eq. (C.12), with no derivative of $\mathbf{P}$ appearing, should not be confused with the Hellmann-Feynman theorem.[2] Given that

$$E = \langle \Phi | \mathcal{H} | \Phi \rangle$$

with $\langle \Phi | \Phi \rangle = 1$, then

$$\frac{\partial E}{\partial X_A} = \left\langle \frac{\partial \Phi}{\partial X_A} \Big| \mathcal{H} \Big| \Phi \right\rangle + \left\langle \Phi \Big| \mathcal{H} \Big| \frac{\partial \Phi}{\partial X_A} \right\rangle + \left\langle \Phi \Big| \frac{\partial \mathcal{H}}{\partial X_A} \Big| \Phi \right\rangle \quad \text{(C.13)}$$

The Hellmann-Feynman condition, then, is that

$$\left\langle \frac{\partial \Phi}{\partial X_A} \,\middle|\, \mathcal{H} \,\middle|\, \Phi \right\rangle + \left\langle \Phi \,\middle|\, \mathcal{H} \,\middle|\, \frac{\partial \Phi}{\partial X_A} \right\rangle = 0 \tag{C.14}$$

which holds only for exact solutions, or certain classes of trial functions. Under the constraints of Eq. (C.14), Eq. (C.13) is simply

$$\frac{\partial E}{\partial X_A} = \left\langle \Phi \,\middle|\, \frac{\partial \mathcal{H}}{\partial X_A} \,\middle|\, \Phi \right\rangle \tag{C.15}$$

Equation (C.15) is the expectation value of a simple one-electron operator plus the derivative of the nuclear repulsion term. Equation (C.12), however, does not depend on Eq. (C.14). The integrals in $\partial \mathbf{H}^{\text{core}}/\partial X_A$ and $\partial(\mu\nu|\sigma\lambda)/\partial X_A$ involve the wave function through "atomic orbital following"—i.e., $\partial\phi_\mu^A/\partial X_A$, where ϕ_μ^A is an atomic orbital on center A—and are far more complicated than those of Eq. (C.15). In practice, the forces evaluated through Eq. (C.15) can be large even when they are found to be zero under Eq. (C.13) and thus represent an extremum of the energy function. Nevertheless, the simplicity of Eq. (C.15) is appealing, and one wonders if the increased inconvenience of meeting the condition of Eq. (C.14) is not repaid in utilizing Eq. (C.15) when the goal is geometry optimization.

For a configuration interaction (CI) wave function involving determinants $|\Psi_I\rangle$,

$$|\Phi_{\text{CI}}\rangle = \sum_I c_I |\Psi_I\rangle$$

one obtains for the energy derivatives

$$\frac{\partial E}{\partial X_A} = \frac{\partial \bar{E}}{\partial X_A} + \sum_{\mu i} \frac{\partial E}{\partial C_{\mu i}} \frac{\partial C_{\mu i}}{\partial X_A} + \sum_I \frac{\partial E}{\partial c_I} \frac{\partial c_I}{\partial X_A} \tag{C.16}$$

where now the first sum is over *all* molecular orbital coefficients. In this case, $\partial E/\partial X_A = \partial \bar{E}/\partial X_A$ only for a multiconfiguration self-consistent field (MCSCF) function. For the general Hartree-Fock plus CI wave function, $\partial E/\partial c_I = 0$ and

$$\frac{\partial E}{\partial X_A} = \frac{\partial \bar{E}}{\partial X_A} + \sum_{\mu i} \frac{\partial E}{\partial C_{\mu i}} \frac{\partial C_{\mu i}}{\partial X_A} \tag{C.17}$$

The evaluation of $\partial C_{\mu i}/\partial X_A$ is complicated, but can be approached through perturbation theory.[3] The contribution to the forces of the second term might be expected to be small for a large CI as the dependence of the energy on $\mathbf{C}$ is downgraded, or for a system without a great deal of bond polarity, or for a system in which the molecular orbitals are determined by symmetry. Under such situations, an initial search can be made of the surface using Eq. (C.9), but for accurate results reliance on this approximation is not satisfactory.

Second derivatives of the Hartree-Fock energy can be obtained directly from Eq. (C.12):

$$\frac{\partial^2 E}{\partial X_A \, \partial X_B} = \sum_{\mu\nu} P_{\nu\mu} \frac{\partial^2 H_{\mu\nu}^{\text{core}}}{\partial X_A \partial X_B} + \frac{1}{2} \sum_{\mu\nu\sigma\lambda} P_{\nu\mu} P_{\lambda\sigma} \frac{\partial^2 (\mu\nu| \, |\sigma\lambda)}{\partial X_A \, \partial X_B}$$

$$- \sum_{\mu\nu} Q_{\nu\mu} \frac{\partial^2 S_{\mu\nu}}{\partial X_A \, \partial X_B} + \frac{\partial^2 V_{NN}}{\partial X_A \, \partial X_B} + \sum_{\mu\nu} \frac{\partial P_{\nu\mu}}{\partial X_B} \frac{\partial H_{\mu\nu}^{\text{core}}}{\partial X_A}$$

$$+ \sum_{\mu\nu\sigma\lambda} P_{\nu\mu} \frac{\partial P_{\lambda\sigma}}{\partial X_B} \frac{\partial (\mu\nu| \, |\sigma\lambda)}{\partial X_A} - \sum_{\mu\nu} \frac{\partial Q_{\nu\mu}}{\partial X_B} \frac{\partial S_{\mu\nu}}{\partial X_A}$$

The last three terms of this expression involve the derivatives of the molecular orbital coefficients and cannot easily be avoided. They are obtained through coupled perturbed Hartree-Fock theory (CPHF).[3]

C.4 OPTIMIZATION TECHNIQUES

There is a rather large literature on numerical methods for finding stationary points of a function of many variables. They may be classified as follows:

(a) methods without gradients
(b) methods with numerical gradients and second derivatives
(c) methods with analytical gradients and numerical second derivatives
(d) methods with analytical gradients and analytical second derivatives

and so forth.

All but the first of these methods are based on the Taylor expansions of the function E and its derivatives $\mathbf{f}$ as given in Section C.2. In practice, they can be applied as "estimate" techniques, or as "iterative" techniques.

Type d methods might be preferred, since they utilize the greatest amount of information, but this is the case only if the analytic first and second derivatives can be obtained at the same time, and with the same ease, as the energy E. It is clear, however, that insofar as our initial estimate of the geometry at an extremum is within the quadratic region of the valence bond force field (the $\mathbf{Y}$ coordinates of Eq. (C.6)), a single application of Eqs. (C.4) and (C.5) gives $\mathbf{X}_e$ and the energy $E(\mathbf{X}_e)$. Such a single application of Eq. (C.4) we shall call an *estimate*. If we are not within the quadratic region of the potential, the estimate may not be very accurate, and it may be desirable to iterate; that is, having determined a new set,

$\mathbf{X}_1$, from the initial guess $\mathbf{X}_0$, we might solve the equations of Section C.2 for $\mathbf{X}_2$. This requires $\mathbf{f}(\mathbf{X}_1)$ and $\mathbf{H}^{-1}(\mathbf{X}_1)$. This procedure might then be repeated until $E_n - E_{n-1}$ is below a given threshold, $\sigma = \mathbf{f}(\mathbf{X}_n)^\dagger \mathbf{f}(\mathbf{X}_n)$ is below a given threshold, $\mathbf{q}_n^\dagger \mathbf{q}_n$ is below a given threshold, or all three.

In practice, type d algorithms are not used routinely in geometry searches, because of the difficulty that arises in analytically obtaining the second derivatives. First derivatives, on the other hand, are generally evaluated in a time comparable to the evaluation of the energy itself. For this reason, type c algorithms are the most popular and are implemented in most modern quantum chemistry programs. Unfortunately, most algorithms that search for transition states require analytic second derivatives, and are thus very expensive in computer resources. Analytic second derivatives are also of use in determining whether an extreme point of the potential energy surface is a minimum (all eigenvalues of the Hessian positive) or a transition state (one and only one eigenvalue negative), and in yielding the vibrational spectrum at a minimum, if this is sought.

C.5 SOME OPTIMIZATION ALGORITHMS

The simplest of the methods are of type a. The simplest of these are the so-called axial iteration or univariant techniques. A set of internal coordinates are chosen, and the potential energy is minimized with respect to each coordinate in turn. After completing the $3N - 6 = m$ independent searches, one returns and repeats the procedure until the change in coordinates is below a given threshold.

One successful procedure of this sort is to step along each coordinate Y_i by a_i. If $E(\mathbf{Y} + a_i\mathbf{e}_i) < E(\mathbf{Y})$, where $\mathbf{e}_i$ is the unit vector along i, repeat the step until $E(\mathbf{Y} + ra_i\mathbf{e}_i) > E(\mathbf{Y} + (r - 1)a_i\mathbf{e}_i)$, r an integer. The new coordinates are $\mathbf{Y} = \mathbf{Y} + (r - 1)a_i\mathbf{e}_i$. If $E(\mathbf{Y} + a_i\mathbf{e}_i) > E(\mathbf{Y})$, step in the other direction until $E(\mathbf{Y} - ra_i\mathbf{e}_i) > E(\mathbf{Y} - (r - 1)a_i\mathbf{e}_i)$. Again, the new coordinates are $\mathbf{Y} = \mathbf{Y} - (r - 1)a_i\mathbf{e}_i$. If $E(\mathbf{Y})$ is of lower energy than both $E(\mathbf{Y} + a_i\mathbf{e}_i)$ and $E(\mathbf{Y} - a_i\mathbf{e}_i)$, then a quadratic is fitted through these three points and the minimum is found. The coordinates $\mathbf{Y}$ are updated and a_i is set to $a_i/4$. This procedure is repeated for all i, and then iterated until all a_i are below a specified threshold.

The most effective of the type a algorithms seems to be of the simplex type, and this is a very useful technique for problems where gradients are not available. The method given below is that of Nelder and Mead.[4] Figure C.1 is a schematic attempt to follow this method for two variables.

Consider m variables. $\mathbf{X}_0, \mathbf{X}_1, \ldots, \mathbf{X}_m$ are the $m + 1$ independent points in this m-dimensional space that defines the "simplex." E_i designates the value of the energy $E(\mathbf{X}_i)$. Let E_h be the highest value of $\{E_i\}$ and E_l

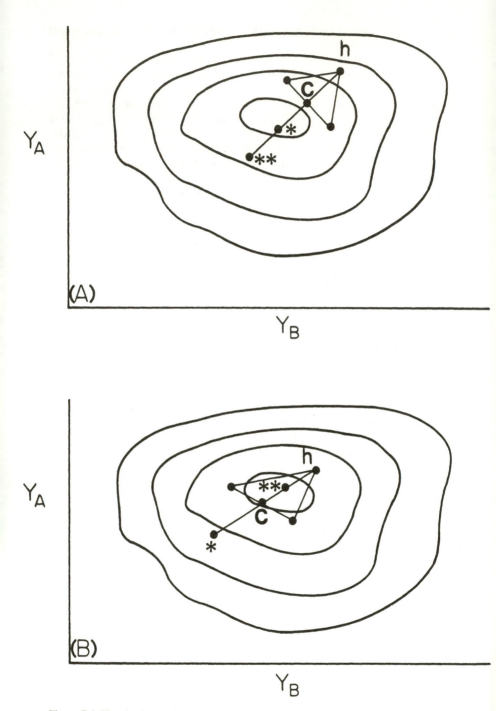

Figure C.1 The simplex method, where h designates E_h, C designates the centroid of points: (A) a successful reflection (*) but failed expansion (**); (B) a failed reflection (*) but successful contraction (**).

446

the lowest. Let $\overline{\mathbf{X}}$ be the centroid of the points $\{\mathbf{X}_i\}_{i \neq h}$ and $[\mathbf{X}_i\mathbf{X}_j]$ the distance between $\mathbf{X}_i$ and $\mathbf{X}_j$.

$$\overline{\mathbf{X}} \equiv \sum_{i \neq h} \frac{\mathbf{X}_i}{m}$$

$$[\mathbf{X}_i\mathbf{X}_j]^2 = \sum_{a=1}^{m} [(\mathbf{X}_i)_a - (\mathbf{X}_j)_a]^2$$

In the above, $(\mathbf{X}_i)_a$ is the ath component of the vector $\mathbf{X}_i$. The reflection of $\mathbf{X}_h$ is denoted $\mathbf{X}^*$ and its coordinates are given by

$$\mathbf{X}^* = (1 + a)\overline{\mathbf{X}} - a\mathbf{X}_h$$

$$a = \frac{[\mathbf{X}^*\overline{\mathbf{X}}]}{[\mathbf{X}_h\overline{\mathbf{X}}]} \tag{C.18}$$

where a is called the *reflection constant* and is positive. $\mathbf{X}^*$ is thus on a line joining $\mathbf{X}_h$ and $\overline{\mathbf{X}}$, but reflected to the far side of $\overline{\mathbf{X}}$ from $\mathbf{X}_h$. Three possibilities ensue. If $E_l < E^* < E_h$, then $\mathbf{X}_h$ is replaced by $\mathbf{X}^*$ and one starts again with a new simplex, reflecting the new $\mathbf{X}_h$, etc.

If $E^* < E_l$, if the reflection has produced a new minimum, then $\mathbf{X}^*$ is "expanded" to $\mathbf{X}^{**}$ by

$$\mathbf{X}^{**} = v\mathbf{X}^* + (1 - v)\overline{\mathbf{X}}$$

$$v = \frac{[\mathbf{X}^{**}\overline{\mathbf{X}}]}{[\mathbf{X}_h\overline{\mathbf{X}}]} > 1 \tag{C.19}$$

where v is the expansion coefficient. If $E^{**} < E_l$, $\mathbf{X}_h$ is replaced by $\mathbf{X}^{**}$ and the procedure is restarted. If $E^{**} > E_l$, then the expansion has failed and $\mathbf{X}_h$ is replaced by $\mathbf{X}^*$ before restarting.

Finally, if $E^* > E_i$ for all i—i.e., if replacing $\mathbf{X}_h$ with $\mathbf{X}^*$ leaves E^* the new maximum—then a new *contraction* is examined (see Fig. C.1B):

$$\mathbf{X}^{**} = \beta\mathbf{X}_h + (1 - \beta)\overline{\mathbf{X}}$$

$$\beta = \frac{[\mathbf{X}^{**}\overline{\mathbf{X}}]}{[\mathbf{X}_h\overline{\mathbf{X}}]} \tag{C.20}$$

The contraction coefficient β lies between 0 and 1. $\mathbf{X}^{**}$ replaces $\mathbf{X}_h$ unless $E^{**} > \min(E(\mathbf{X}_h), E(\mathbf{X}^*))$. In the latter, rather rare case, all $\mathbf{X}_i$ are replaced by $(\mathbf{X}_i + \mathbf{X}_l)/2$ and the process is restarted.

Nelder and Mead[4] suggest the values $a = 1$, $\beta = \frac{1}{2}$, $v = 2$ and discuss the mathematical implications of this strategy.

Exercise C.1

(a) Consider the quadratic function $E = E_0 + \frac{1}{2}K(x - a)^2 + \frac{1}{2}K'(y - b)^2 + K''xy$. Use the Newton-Raphson equation (Eq. (C.4)) to solve for $\mathbf{q} = (x_1 - x, y_1 - y)$; i.e., "evaluate" $\mathbf{H}^{-1}(\mathbf{X})$ and $\mathbf{f}(\mathbf{X})$. You can easily check your result by letting the coupling constant K'' equal zero.

(b) Let $a = 3.00$ a.u., $b = 2.00$ a.u., $K = 1.000$, and $K' = 0.100$. Given the three different coupling cases (1) $K'' = 0$, (2) $K'' = 0.010$, and (3) $K'' = 0.030$, solve the equation of part (a) for (x_e, y_e). Notice that the shorter, "weaker" bond is affected much more by the coupling than is the stronger bond.

Exercise C.2 Consider case 3 of Exercise C.1b. Evaluate the gradient and the Hessian at $x = 3.3$, $y = 1.8$ and solve the Newton-Raphson equation of Exercise C.1a for (x_1, y_1). Comment on this. In practical cases, the quantum mechanical potential energy surface is *not* quadratic, especially for regions as far away from the neighborhood of the minimum as (x, y) in this example.

Exercise C.3 Consider the potential energy function of Exercise C.1. Form the simplex of points $(x, y) = (3.3, 1.8)$, $(3.0, 1.5)$, and $(2.7, 2.1)$. Follow the algorithm of Nelder and Mead, making three successive moves for all three coupling cases. Follow the centroids you calculate as a function of cycle number. In cases where there are few variables, "contractions" often dominate the moves, and this is the case here.

The simplex method is rather easy to program for a computer. If this is done, all three coupling cases yield answers accurate within ± 0.002 for x and y within 15 cycles. For comparison, the univariant method needs 11 energy evaluations for case 1, in which the variables are weakly coupled; and 22 evaluations for case 3, where they are strongly coupled. The advantage of the simplex method over the univariant method becomes even greater when the number of variables increases.

The most efficient methods that use gradients, either numerical or analytic, are based upon quasi-Newton update procedures, such as those described below. They are used to approximate the Hessian matrix $\mathbf{H}$, or its inverse $\mathbf{G}$. Equation (C.4) is then used to determine the step direction $\mathbf{q}$ to the nearest minimum. The inverse Hessian matrix determines how far to move along a given gradient component of $\mathbf{f}$, and how the various coordinates are coupled. The success of methods that use approximate Hessians rests upon the observation that when $\mathbf{f} = \mathbf{0}$, an extreme point is reached regardless of the accuracy of $\mathbf{H}$, or its inverse, provided that they are reasonable.

The most popular quasi-Newton update procedures used at this time are the Murtagh-Sargent[5] variant of the Davidon-Fletcher-Powell method (MS),[6] and the Broyden-Fletcher-Goldfarb-Shanno (BFGS) algorithm.[7]

Consider first the energies E_n and E_{n-1} and the gradients $\mathbf{f}_n$ and $\mathbf{f}_{n-1}$ obtained at two geometries characterized by $\mathbf{X}_n$ and $\mathbf{X}_{n-1}$. If the energy were a quadratic function of the coordinates, then the Hessian ($\mathbf{H}_n = \mathbf{H}_{n-1} = \mathbf{H}$) would relate the coordinates to the gradients by

$$\mathbf{H}_n\mathbf{q}_n = \mathbf{H}\mathbf{q}_n = \mathbf{d}_n$$

$$\mathbf{q}_n \equiv \mathbf{X}_n - \mathbf{X}_{n-1}$$

$$\mathbf{d}_n = \mathbf{f}_n - \mathbf{f}_{n-1}$$

or

$$\mathbf{G}\mathbf{d}_n \equiv \mathbf{H}^{-1}\mathbf{d}_n = \mathbf{q}_n \tag{C.21}$$

Equation (C.21) expresses the quasi-Newton condition, and any procedure that constrains $\mathbf{G}$ to obey Eq. (C.21) is a quasi-Newton procedure.

An "update" procedure is generated by defining a sequence $\mathbf{G}_n$ that approaches $\mathbf{H}^{-1}(\mathbf{X}_e)$ for sufficiently large n. Given $\mathbf{G}_{n-1}$, then

$$\mathbf{G}_n = \mathbf{G}_{n-1} + \mathbf{W}_n \tag{C.22}$$

where $\mathbf{W}_n$ is the correction to $\mathbf{G}_{n-1}$. Inserting Eq. (C.22) into Eq. (C.21) generates a quasi-Newton update procedure:

$$(\mathbf{G}_{n-1} + \mathbf{W}_n)\mathbf{d}_n = \mathbf{q}_n \tag{C.23a}$$

In a "rank 1" procedure, such as MS, $\mathbf{W}$ is chosen as

$$\mathbf{W}_n = a_n \begin{pmatrix} U_1 \\ \cdot \\ \cdot \\ \cdot \\ U_m \end{pmatrix} (U_1, \ldots, U_m) = a_n\mathbf{U}_n\mathbf{U}_n^\dagger \tag{C.23b}$$

Equation (C.23a) becomes

$$\mathbf{G}_{n-1}\mathbf{d}_n + a_n\mathbf{U}_n\mathbf{U}_n^\dagger\mathbf{d}_n = \mathbf{q}_n \tag{C.23c}$$

The number a_n is an arbitrary scale, and can be used to scale $\mathbf{U}_n$ so that

$$a_n^{-1} = \mathbf{U}_n^\dagger\mathbf{d}_n \tag{C.24a}$$

yielding

$$\mathbf{U}_n = \mathbf{q}_n - \mathbf{G}_{n-1}\mathbf{d}_n \tag{C.24b}$$

These equations form the basis of the Murtagh-Sargent procedure.

Equation (C.4) defines the search direction, but the actual step is often scaled:

$$\mathbf{q}_n = \mathbf{X}_n - \mathbf{X}_{n-1} = -\alpha_{n-1}\mathbf{G}_{n-1}\mathbf{f}_{n-1} \tag{C.24c}$$

Exact line searches determine that value of α that minimizes the energy along the search direction defined by $\mathbf{q}_n$. Partial line searches can be designed with less stringent demands, the weakest requiring only that $E_n < E_{n-1}$. The situation is depicted in Fig. C.2. In practice, it is found that exact line searches are seldom worth the extra calculations required to minimize the energy along the line: far better is just to require a reduction in energy.

The popular Murtagh-Sargent algorithm proceeds as follows:

STEP 1: Set $\alpha_0 = \frac{1}{2}$ and $\mathbf{G}_0 = \mathbf{1}$. Use Eq. (C.24c) to obtain a new set of coordinates $\mathbf{X}_1$. If $E_1 > E_0$, repeat this step with $\alpha_0 = \frac{1}{2}\alpha_0$ until $E_1 < E_0$. This is equivalent to the method of steepest descent with "half-steps."

STEP 2: Form

$$\mathbf{U}_k = -\alpha_{k-1}\mathbf{G}_{k-1}\mathbf{f}_{k-1} - \mathbf{G}_{k-1}(\mathbf{f}_k - \mathbf{f}_{k-1})$$

$$= -\mathbf{G}_{k-1}[\mathbf{f}_k + \mathbf{f}_{k-1}(\alpha_{k-1} - 1)]$$

$$a_k^{-1} = \mathbf{U}_k^\dagger\mathbf{d}_k = \mathbf{U}_k^\dagger(\mathbf{f}_k - \mathbf{f}_{k-1})$$

$$T_k = \mathbf{U}_k^\dagger\mathbf{U}_k$$

If $a_k^{-1} < 10^{-5}T_k$ or $a_k\mathbf{U}_k^\dagger\mathbf{f}_{k-1} > 10^{-5}$, $\mathbf{G}_k$ is reset to $\mathbf{1}$ and α_k is reset to $\frac{1}{2}$. These tests ensure the stability of $\mathbf{G}_k$; that is, $\mathbf{G}_k$ remains positive definite after update.[5] With *reasonable* starting geometries, $\mathbf{G}_k$ is seldom reset to $\mathbf{1}$, a fortunate finding, for the reset would mean the loss of all information about the curvature of the surface built up from previous cycles. If these two tests are passed, then $\mathbf{G}_k$ is updated by Eq. (C.22) and α_k is set to unity.

STEP 3: Equation (C.24c) is solved to find a new set of $\mathbf{X}_k$. E_k and $\mathbf{f}_k$ are calculated. If $E_k \le E_{k-1}$, one returns to step 2 until $\sigma/m = \mathbf{f}_k^\dagger\mathbf{f}_k/m < 10^{-3}$ a.u., with m again the number of independent variables, at which point most bond lengths are converged to ± 0.01 a.u. and bond angles to $\pm 1.0°$. If $E_k > E_{k-1}$, $\alpha_{k-1} = \frac{1}{2}\alpha_{k-1}$ (a "backstep") and step 2 repeated without updating the "counter" k.

An important feature of the Murtagh-Sargent procedure is that a stationary value of E is obtained at the latest in $m + 1$ steps even if $\mathbf{H}$ is singular. In practice, far fewer steps than $m + 1$ are required if reasonable guesses on starting geometries are available.

The "rank 2" methods use an update of the form

$$\mathbf{G}_n = \mathbf{G}_{n-1} + a_n\mathbf{U}_n\mathbf{U}_n^\dagger + b_n\mathbf{V}_n\mathbf{V}_n^\dagger \tag{C.25a}$$

Placing Eq. (C.25a) in the quasi-Newton condition of Eq. (C.21) does not uniquely determine $\mathbf{U}$ and $\mathbf{V}$ (one equation, two unknowns!) but leads to

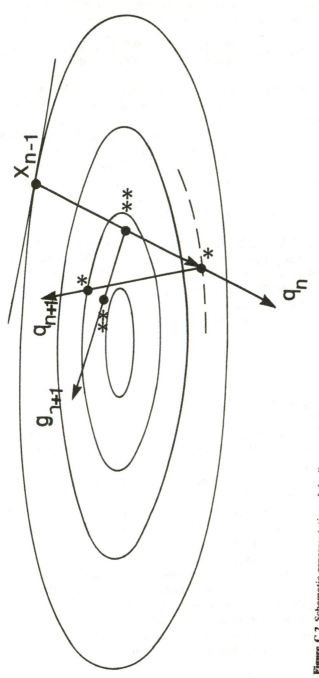

Figure C.2 Schematic representation of the line search. Starting from point X_{n-1}, the direction q_n is determined, perhaps from the MS or BFGS procedure. (**) designates an exact line search; the geometry depicted by (*) is obtained from a partial line search. The directions q_{n+1} and g_{n+1} derive from the geometries obtained through partial and exact line searches, respectively. The picture is constructed to suggest that the extra calculations required by an exact line search may not accelerate convergence to the minimum, an observation made through many test calculations.

a "Broyden family" of solutions. The most successful of these for quantum chemical applications appears to be the BFGS method,[7] defined by

$$G_n = \left(1 - \frac{q_n d_n^\dagger}{q_n^\dagger d_n}\right) G_{n-1} \left(1 - \frac{q_n d_n^\dagger}{q_n^\dagger d_n}\right)^\dagger + \frac{q_n q_n^\dagger}{q_n^\dagger d_n} \qquad \text{(C.25b)}$$

The BFGS method has the important advantage over the MS method that the updated Hessian will remain positive definite (i.e., all eigenvalues of the second derivative matrix will remain positive) and thus a step in the direction given by Eq. (C.4) with G_n obtained from Eq. (C.25b) can always be made to reduce the energy. In practice, the BFGS algorithm is usually more successful than the MS algorithm, especially for larger molecules with many soft vibrational modes. Several line search procedures have been devised for the BFGS method, but again, it appears that partial line searches are the most effective in finding minima.[8]

Exercise C.4 Using the trial energy function of Exercise C.1, and starting with the point $(x, y) = (3.3, 1.8)$, apply the MS method for case 3 of that problem. Assume that the energy E and the derivative f are easy to obtain, but not H. Perform three steps with this algorithm. After three moves, one is remarkably close to the final result. Note also that the energy along the softer mode is the more difficult to minimize.

C.6 TRANSITION STATES

A transition state occurs when $f(X_e) = 0$ (a stationary point) and one and only one eigenvalue of $H(X_e)$ is negative. These two considerations define a simple saddle point. In addition, $E(X_e)$ should be the highest-energy point on a continuous line connecting reactants and products; i.e., X_e should represent the saddle point of highest energy. Such a definition tends to associate clearly one side of the "pass" with reactants, the other with products. In addition X_e must represent the lowest-energy point which satisfies the above three conditions.

Clearly, finding transition states is more difficult than finding simple minima. Unlike the algorithms described for finding minima, algorithms for finding transition states do not always succeed! Part of the problem is that it is difficult to ensure movement along a surface that exactly meets the conditions of a simple saddle point. In addition, a difficulty may reside in the fact that wave functions for a transition state may be considerably more complex than those that describe minima.

Defining, as before

$$\sigma(\mathbf{X}) = \sum_{i=1}^{m} f_i^2(\mathbf{X}) = \mathbf{f}(\mathbf{X})^\dagger \mathbf{f}(\mathbf{X})$$

we seek points for which $\sigma(\mathbf{X}_e) = 0$. Since $\sigma(\mathbf{X}) \geq 0$, least squares minimization procedures are appropriate.[9] Such a procedure, however, will force convergence to any stationary point $\mathbf{X}_e$, and so care must be taken with the (initial) guesses. Chemical intuition is of great use here!

There are many methods of least squares minimization. The general starting point is, again, a Taylor expansion:

$$\sigma(\mathbf{X}_{k+1}) = \sigma(\mathbf{X}_k) + \mathbf{q}_{k+1}^\dagger \mathbf{V}_k + \tfrac{1}{2}\mathbf{q}_{k+1}^\dagger \mathbf{T}_k \mathbf{q}_{k+1} + \cdots$$

$$\mathbf{q}_{k+1} = \mathbf{X}_{k+1} - \mathbf{X}_k \tag{C.26}$$

where $\mathbf{V}_k$ is a column vector, the elements of which are

$$(\mathbf{V}_k)_i = \left(\frac{\partial \sigma(\mathbf{X})}{\partial X_i}\right)_{\mathbf{X}=\mathbf{X}_k} \tag{C.27a}$$

and

$$(\mathbf{T}_k)_{ij} = \left(\frac{\partial^2 \sigma(\mathbf{X})}{\partial X_i \, \partial X_j}\right)_{\mathbf{X}=\mathbf{X}_k} \tag{C.27b}$$

At the minimum value of σ, $\sigma(\mathbf{X}_e) = 0$ and $\mathbf{V}(\mathbf{X}_e) = \mathbf{0}$. Examining the Taylor series expansion for $\mathbf{V}(\mathbf{X}_e)$ in an analogous fashion to Eq. (C.2) suggests the iterative equation

$$\mathbf{q}_{k+1} = -(\mathbf{T}_k)^{-1} \mathbf{V}_k \tag{C.28a}$$

where

$$\mathbf{V}_k = 2\mathbf{H}_k \mathbf{f}_k \tag{C.28b}$$

and

$$\mathbf{T}_k = 2(\mathbf{H}_k^\dagger \mathbf{H}_k + \mathbf{C}_k) \tag{C.28c}$$

with

$$(\mathbf{C}_k)_{ij} = \sum_{a=1}^{m} f_a \frac{\partial^2 f_a}{\partial X_i \, \partial X_j} \tag{C.28d}$$

σ can be minimized in exactly the same fashion in which E itself was minimized, e.g., by using the Murtagh-Sargent or the BFGS procedure already described. A generalized Newton-Raphson method may also be employed, assuming

$$\mathbf{T}_k \simeq 2\mathbf{H}_k^\dagger \mathbf{H}_k$$

Interestingly, T_k by this construction is guaranteed positive semidefinite, and $(T_k)^{-1}$ is a good candidate for the Murtagh-Sargent procedure. If the initial geometry is not sufficiently close to the transition state, minimization of σ will not succeed. Then various geometric algorithms might be used to help get sufficiently close. One such method maximizes the energy along a line connecting reactants with products. Then the energy is minimized in the hyperplane perpendicular to this direction, perhaps by the BFGS or the MS method (see Fig. C.3). This procedure might be repeated until the saddle point is found or, after several moves, σ is minimized as described above.

Another interesting procedure uses the concept of a gradient extre-

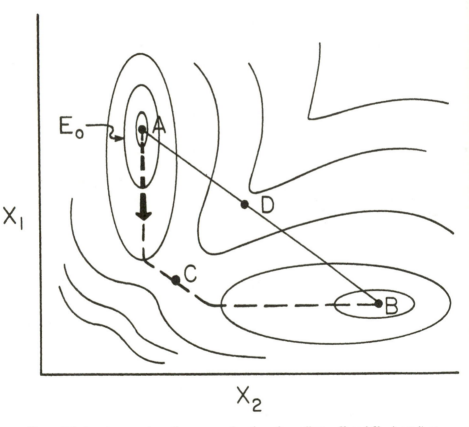

Figure C.3 An energy contour diagram as a function of coordinates X_1 and X_2. A gradient extremum is depicted by the arrow, with its tail on the energy contour with value E_0. Point A represents the reactant, B the product, and C the sought-after transition state. Imagine also on this plot the geometric search described in the text. In this algorithm, the energy along the line AB is maximized, and point D is found. Then a search is made for the minimum energy in the hyperplane perpendicular to line AB and passing through point D.

mal.[10] This is defined as the smallest energy gradient along a constant-energy contour, as shown in Fig. C.4, and can be used to proceed uphill along the softest direction from any point on the potential energy surface including a minimum. The gradient extremal is formed by

$$\frac{\partial}{\partial X_i}\left[\mathbf{f}^\dagger\mathbf{f} - 2\lambda(E - E_0)\right] = 0 \tag{C.29}$$

where 2λ is a Lagrange multiplier. This yields the eigenvalue equation

$$\mathbf{H(X)f(X)} = \lambda\mathbf{(X)f(X)} \tag{C.30}$$

In the quadratic region,

$$\mathbf{H(X)} = \mathbf{H} \tag{C.31a}$$

$$\lambda\mathbf{(X)} = \lambda \tag{C.31b}$$

$$\mathbf{f(X)} = \mathbf{f(X_0)} + \mathbf{Hq} \tag{C.31c}$$

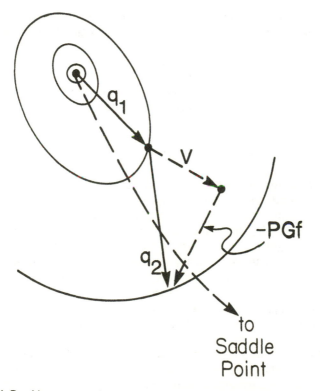

Figure C.4 Graphic representation of the gradient extremal leading from a minimum to a saddle point following Eq. (C.34). The direction $\mathbf{q}_1$ has followed the lowest vibrational frequency at the minimum.

and

$$(H - \lambda 1)Hq = -(H - \lambda 1)f \qquad (C.31d)$$

The equations of group (C.31) are equivalent to the quasi-Newton constraint. Diagonalizing Eq. (C.30), we choose the eigenvector $V = f$, which overlaps with the direction we are interested in, and introduce the projector

$$P = (1 - VV^+) \qquad (C.32)$$

and project the step along the direction

$$PHq = -Pf$$

Since

$$PH = HP = PHP$$

we find

$$Pq = -PGf \qquad (C.33)$$

which is a *projected* Newton-Raphson equation, with direction chosen along the gradient extremal defined by Eq. (C.30). The actual step taken is given by

$$q = -PGf + aV \qquad (C.34)$$

which satisfies Eq. (C.33). The constant a is again an arbitrary scale chosen to correct for the finite size of each step, as indicated in Fig. C.4.[10]

C.7 CONSTRAINED VARIATION

It is clear from the outset that the fewer the degrees of freedom that are varied in the study of the energy surface of a system, the easier it will be to obtain stationary points on that surface. The five or six degrees of freedom representing translation and rotation may always be removed exactly without any real constraints on the optimization procedure. If there exists symmetry in the system, then symmetry-adapted coordinates may be used in the optimization procedure, again simplifying the calculation. Considering formaldehyde, H_2CO, there are 12 coordinates, reduced to 6 by removing translation and rotation. Considering C_{2v} symmetry, only three variables remain, the CO and CH bond lengths and the OCH angle.

Some care must be exercised when using symmetry in optimizing geometry. Except in cases in which the electronic state is degenerate, gradient optimization methods will not reduce symmetry, for the gradients that do so are zero. If this corresponds to a minimum, as, for example, in trigonal NH_3, then considerable effort might be saved; but if a geometry search starts with planar NH_3, the molecule will remain planar. In this case the gradient perpendicular to the plane is zero, and the planar structure is the transition state for the NH_3 inversion. When there is doubt, it is better to let symmetry arise as a result of the optimization, and not as an assumption.

Symmetry constraints do not affect our ability to obtain exact stationary points on the potential energy surface. To ease the calculation, however, we might also consider constraints on the variables suggested by "chemical intuition." In the above example of formaldehyde, we might fix the CH bond at a typical value of 1.1 Å and vary only the CO bond length and the OCH angle. If we are interested in the biphenyl C—C bond between the phenyl moieties, we might fix all the coordinates except this C—C bond length and the dihedral angle between the two phenyl planes. The saving of effort can be substantial, but it is clear that the accuracy of the results obtained will depend on the soundness of our intuition.

Somewhat more dangerous is the use of such intuition for problems that follow pathways on the surface (valleys). In examining internal rotations—for example, of ethane—it is tempting to freeze all bond lengths and angles except the torsional one. This is a reasonably accurate procedure, but if one has started with optimized coordinates for the minimum, the barrier, calculated without relaxing all coordinates, might be too large. In the search for reaction pathways, freezing coordinates will generally lead to an overestimate of barrier energies.

Worse, by freezing coordinates we prejudice the direction of the path, and so can completely miss alternate pathways, the lowest-energy pathways, and perhaps even the most important transition state! New methods that probe the rotational energy surface for stable geometries and for transition states for a wide variety of wave functions are currently becoming available. The literature is now full of interesting applications using these methods, and a dream of theoretical chemistry is being achieved. From a collection of nuclei, and a given number of electrons, quantum chemistry can construct molecules.

NOTES

1. See, for example, E. B. Wilson, J. C. Decius, and P. C. Cross, *Molecular Vibrations*, McGraw-Hill, New York, 1955.
2. H. Hellmann, *Einführung in die Quantenchemie*, Franz Deuticke, Leipzig, 1937; R. P. Feynman *Phys. Rev.* **41:** 721 (1939); A. C. Hurley, *Proc. Roy. Soc.* **A226:** 170, 179 (1954).
3. See, for example, J. A. Pople, H. Krishnan, H. B. Schlegel, and J. S. Binkley, *Int. J. Quantum Chem.* **S13:** 225 (1979), and references therein.
4. J. A. Nelder and R. Mead, *Comput. J.* **7:** 308 (1965).
5. B. A. Murtagh and R. W. H. Sargent, *Comput. J.* **13:** 185 (1970); a principal disadvantage of this method is that for poor starting geometries it is plagued with "resets."
6. W. C. Davidon, AEC Research & Development Report ANL-5990 (1959), and *Comput. J.* **10:** 406 (1968); R. Fletcher and M. J. D. Powell, *Comput. J.* **6:** 163 (1963).

7. R. Fletcher, *Comput. J.* **13:** 317 (1970); D. Goldfarb, *Math. Comput.* **24:** 23, (1970); D. F. Shanno, *Math. Comput.* **24:** 647 (1970); C. G. Broyden, *Math. Comput.* **21:** 368 (1967).
8. J. D. Head and M. C. Zerner, *Chem. Phys. Lett.* **122:** 264 (1985).
9. J. W. McIver, Jr., and A. Komornicki, *J. Am. Chem. Soc.* **94:** 2625 (1972).
10. P. Jørgensen, H. J. A. Jensen, and T. Helgaker, *Theoret. Chim. Acta* **73:** 55 (1988); D. K. Hoffman, R. S. Nord, and K. Ruedenberg, *Theoret. Chim. Acta* **69:** 265 (1986).

FURTHER READING

Head, J. D., Weiner, B., and Zerner, M. C., A survey of optimization procedures for stable structures and transition states, *Int. J. Quantum Chem.* **33:** 177 (1988).

Head, J. D., and Zerner, M. C., Newton-based optimization methods for obtaining molecular conformation, *Adv. Quantum Chem.* **20:** 1 (1988).

Helgaker, T., and Jørgensen, P., Analytical calculation of geometrical derivatives in molecular electronics structure theory, *Adv. Quantum Chem.* **19:** 183 (1988).

Jørgensen, P., and Simons, J., *Geometrical Derivatives of Energy Surfaces and Molecular Properties*, Reidel, Dordrecht, Netherlands, 1986.

Press, W. H., Flannery, B. P., Teukolsky, S. A., and Vetterling, W. T., *Numerical Recipes*, Cambridge University Press, Cambridge, 1986. Chapter 10 discusses the problem of optimizing functions and provides FORTRAN implementations of the Nelder-Mead and BFGS algorithms. Highly recommended!

Pulay, P., Analytical derivative methods in quantum chemistry, *Adv. Chem. Phys.* **69:** 241 (1987).

Schlegel, H. B., Optimization of equilibrium geometries and transition structures, *Adv. Chem. Phys.* **67:** 249 (1987).

MOLECULAR INTEGRALS FOR
H_2 AS A FUNCTION OF
BOND LENGTH

All quantities are in atomic units. The integrals are for a minimal basis STO-3G calculation with Slater exponent $\zeta = 1.24$, as described in the text (Subsection 3.5.2).

R	ε_1	ε_2	J_{11}	J_{12}	J_{22}	K_{12}
0.6	-0.7927	1.3327	0.7496	0.7392	0.7817	0.1614
0.8	-0.7321	1.1233	0.7330	0.7212	0.7607	0.1655
1.0	-0.6758	0.9418	0.7144	0.7019	0.7388	0.1702
1.2	-0.6245	0.7919	0.6947	0.6824	0.7176	0.1755
1.4	-0.5782	0.6703	0.6746	0.6636	0.6975	0.1813
1.6	-0.5368	0.5715	0.6545	0.6457	0.6786	0.1874
1.8	-0.4998	0.4898	0.6349	0.6289	0.6608	0.1938
2.0	-0.4665	0.4209	0.6162	0.6131	0.6439	0.2005
2.5	-0.3954	0.2889	0.5751	0.5789	0.6057	0.2179
3.0	-0.3377	0.1981	0.5432	0.5512	0.5734	0.2351
4.0	-0.2542	0.0916	0.5026	0.5121	0.5259	0.2651
5.0	-0.2028	0.0387	0.4808	0.4873	0.4947	0.2877
7.5	-0.1478	-0.0114	0.4533	0.4540	0.4547	0.3206
10.0	-0.1293	-0.0292	0.4373	0.4373	0.4373	0.3373
20.0	-0.1043	-0.0543	0.4123	0.4123	0.4123	0.3623
100.0	-0.0843	-0.0743	0.3923	0.3923	0.3923	0.3823
∞	-0.0793	-0.0793	0.3873	0.3873	0.3873	0.3873

INDEX

Ab initio calculation, 147
Adiabatic ionization potential, 194
Ammonia
  bond angle
    via MBPT, 372
    via SCF, 202
  bond length
    via MBPT, 371
    via SCF, 202
  dipole moment
    via MBPT, 378
    via SCF, 205
  ionization potential
    via Koopmans', 198
    via MBGF, 407
  orbital energies, 198
  population analysis, 203, 204
  SCF energy, 192
Angström unit, 43
Annihilation operator, 90
Anticommutator, 4, 90, 92
Antisymmetrized two-electron integrals, 67
Antisymmetry principle, 46, 50
Atomic units, 41

Basis, 2
Basis functions, 57, 136
Basis set correlation energy, 234
Basis sets, 180
  STO-NG, 156, 184
  4-31G, 186
  6-31G*, 189
  6-31G**, 189
  double zeta, 186
  minimal, 57, 184
Basis vectors, 2, 9, 10
BeH_2 (*see* Beryllium hydride)
Beryllium atom
  pair energies, 285
Beryllium hydride, 248
Bethe-Goldstone theory, 273
Bohr unit, 42
Born-Oppenheimer approximation, 43
Bra vectors, 10
Brillouin's theorem, 128, 235, 352, 356

Canonical Hartree-Fock equations, 120
Canonical orthogonalization, 144, 173

Carbon monoxide
  bond length
    via MBPT, 372
    via SCF, 201
  dipole moment
    via CI, 251
    via MBPT, 378
    via SCF, 204
  ionization potentials
    via Koopmans', 195
  SCF energy, 192
CCA (*see* Coupled-cluster approximation)
CEPA (*see* Coupled electron pair
    approximation)
CH_3 (*see* Methyl radical)
CH_4 (*see* Methane)
Charge density, 138, 203, 212
Chemists' notation, 67
CI (*see* Configuration interaction)
Closed shell determinant, 83, 101, 132
Cluster form of the wave function, 290
CO (*see* Carbon monoxide)
Commutator, 4
Completeness relation, 3, 11
Configuration interaction (CI), 61, 231
Contracted Gaussian functions, 155, 181
Contractions, 155, 181
Convergence of SCF procedure, 148
Core-Hamiltonian, 64, 114
  matrix, 140
Correlation energy, 61, 231, 238
Coulomb integrals, 85, 209
Coulomb operator, 113, 134, 209
Coupled-cluster approximation (CCA),
    287, 295
Coupled electron pair approximation
    (CEPA), 293, 295
Coupled pair many electron theory
    (CPMET), 288
Creation operator, 89
Cyclic polyenes, 310
  resonance energy, 311

Davidson correction, 267
DCI (*see* Doubly excited CI)
Debye unit, 43
Delta function (*see* Dirac delta function)

Density matrix, 139, 252
  unrestricted, 212
Density operator, 139
Determinant, 7
Diagonal diagrams, 335, 368
Diagrammatic perturbation theory, 327, 356
Dipole moment, 150, 377
Dirac delta function, 25, 385, 413
Dirac notation, 9, 26, 30
Dissociation, 221
D-MBPT (∞) doubly excited infinite order many-body perturbation theory, 293, 369
Double-zeta basis sets, 186
Doublet, 102
Doubly excited CI (DCI), 243
Doubly excited determinants, 59, 236
Doubly excited singlets, 104
Dyadic, 3
Dyson equation, 390

EA (*see* Electron affinity)
Eigenfunction, 28
Eigenvalue, 15, 28
Eigenvalue problem, 22, 145
Eigenvector, 15
Electric field, 37, 339, 377
Electron affinity, 127, 194
Electron volt, 43
Electronic Hamiltonian, 43
Electronic problem, 40
Energy
  electronic, 43
  orbital, 123
  total, 44
Epstein-Nesbet pair energy, 275
Epstein-Nesbet second-order energy, 369
Exchange correlation, 51, 86
Exchange integrals, 86, 210
Exchange operator, 113, 134, 209
Excited determinants, 58, 234
Extended Hückel theory, 148

Fermi hole, 53, 280, 285
FH (*see* Hydrogen fluoride)
Finite perturbation theory, 376
First-order energy, 131, 351
First-order pairs, 276, 352
First-order reduced density matrix, 252
Fock matrix, 137, 140, 214
Fock operator, 54, 114, 208
Force constants, 249

Four center integrals, 154, 415
Frozen orbital approximation, 128, 404
Full CI, 62, 234
  matrix, 235
Functional variation, 115

Gaussian basis functions, 153, 181
  integral evaluation, 154, 410
Gaussian lobes, 155, 181, 418
Generalized valence bond (GVB), 259
Geometric series, 337, 369
Gerade, 57
Goldstone diagrams, 362
  translation rules, 363
Green's functions
  many-body (MBGF), 387, 391
  one-body, 383
GVB (*see* Generalized valence bond)

H (*see* Hydrogen atom)
H$_2$ (*see* Hydrogen molecule)
Hamiltonian, 31, 41
Hartree unit, 42
Hartree-Fock approximation, 53, 108
Hartree-Fock equation, 54, 111
  derivation of, 115
Hartree-Fock Green's function, 388
Hartree-Fock Hamiltonian, 130, 320, 350
Hartree-Fock limit, 55, 62, 146, 191
Hartree-Fock potential, 54, 114
Hartree product, 48
HeH$^+$ (*see* Helium hydride cation)
Helium hydride cation
  electron affinity, 398
  ionization potentials, 396
  potential energy curve, 178
  SCF calculation, 168
HFGF (*see* Hartree-Fock Green's function)
H$_2$O (*see* Water)
Hole orbitals, 54
Hückel theory, 309, 342
Hund's rule, 217, 221
Hugenholtz diagrams, 356
  translation rules, 359
Hydrogen atom, 33, 178
Hydrogen fluoride
  bond length
    via MBPT, 371
    via SCF, 202
  dipole moment
    via MBPT, 378
    via SCF, 205
  ionization potential

via Koopmans', 198
via MBGF, 407
orbital energies, 198
population analysis, 203, 204
SCF energy, 192
Hydrogen molecule
bond length
via CI, 245
via MBPT, 371
via SCF, 201
correlation energy
via CI, 245
via GVB, 260
via MBPT, 370
correlation energy of noninteracting H_2's
via CCA, 289
via CEPA, 294
via DCI, 265, 281, 284
via Epstein-Nesbet pairs, 284
via first-order pairs, 277, 283
via full CI, 265, 266
via IEPA, 277, 281, 282
via L-CCA, 293
via second-order MBPT, 355
via third-order MBPT, 366, 370
GVB wave function, 259
ionization potential
via CI, 406
via Koopmans', 194
via MBGF, 406
MCSCF wave function, 258
minimal basis description, 55
ab initio SCF calculation, 159
correlation energy, 241
electron affinity, 395
Hartree-Fock energy, 82, 135
ionization potentials, 393
MBGF calculation, 393
MO integrals, 164, 459
orbital energies, 135, 136, 165
second-order energy, 353, 367
spin-adapted configurations, 102
spin orbitals, 57
third-order energy, 353, 367
UHF calculation, 222
orbital energies, 195
potential energy curve
via CI, 242, 246
via GVB, 260
via MBPT, 375, 376
via RHF, 166
via UHF, 227
SCF energy, 192

Hydrogenation energies, 193
Hyperfine coupling constant, 218

Idempotent, 139
IEPA (*see* Independent electron pair
approximation)
Independent electron pair approximation
(IEPA), 273
Initial guess, 148, 174, 216
Integral transformation, 164
Intermediate normalization, 237, 323
Ionization potentials, 125, 194
effect of correlation, 128, 389
effect of relaxation, 128, 389
IP (*see* Ionization potential)
Iterative natural orbital method, 257

Ket vectors, 9
Koopmans' theorem, 127
Kronecker delta symbol, 3

Lagrange's method of undetermined
multipliers, 34, 116
L-CCA (*see* Linear coupled-cluster
approximation)
Linear coupled-cluster approximation
(L-CCA), 292, 295
Linear dependence, 144
Linear variational problem, 33
Linked cluster theorem, 321, 369
Löwdin population analysis, 152, 204

Many-body perturbation theory (MBPT),
321, 350
diagrammatic representation, 356
relation to MBGF theory, 398
Matrix
adjoint of, 6
diagonal, 6
eigenvalue of, 16
functions of, 21
Hermitian, 7
inverse of, 7
multiplication, 5
orthogonal, 7
symmetric, 7
trace of, 6, 14
transpose, 6
unit, 6
unitary, 7, 13
Matrix elements
rules for one-electron operators, 70, 72
rules for two-electron operators, 70, 72

Maximum coincidence, 69
MBGF, many-body Green's function (*see* Green's function)
MBPT (*see* Many-body perturbation theory)
MCSCF (*see* Multiconfiguration self-consistent field method)
Methane
  bond angle
    via MBPT, 372
    via SCF, 202
  bond length
    via MBPT, 371
    via SCF, 202
  ionization potential
    via Koopmans', 198
    via MBGF, 407
  orbital energies, 198
  population analysis, 203, 204
  SCF energy, 192
Methyl radical, 216
Minimal basis set, 57, 184
Møller-Plesset perturbation theory (MPPT), 320, 350
MPPT (*see* Møller-Plesset perturbation theory)
Mulliken population analysis, 151, 203
Multiconfiguration self-consistent field (MCSCF), 258

N_2 (*see* Nitrogen)
NH_3 (*see* Ammonia)
Natural orbitals, 252, 254
Nitrogen
  bond length
    via MBPT, 373
    via SCF, 201
  ionization potentials
    via CI, 250
    via Koopmans', 196
    via MBGF, 407, 408
    via UHF, 220
  orbital energies, 197
  SCF energy, 192

O_2 (*see* Oxygen)
Occupied orbitals, 54
One-electron integrals, 65
  notations for, 68
One-matrix, 253
One-particle reduced density matrix, 252

Open-shell, 101, 206
Operators
  adjoint of, 12
  Hermitian, 12, 28
  linear, 3
  matrix representation of, 4, 10, 11
  nonlocal, 27, 113
  one-electron, 68
  second-quantized, 95
  two-electron, 68
Orbital energies, 123
Orbital exponents, 153, 159
Orbital perturbation theory, 339
  diagrammatic representation, 348
Orbital relaxation, 403
Orbitals
  canonical, 122
  Gaussian, 56
  hole, 54
  molecular, 46
  natural, 254
  occupied, 54
  particle, 54
  restricted, 100, 132
  Slater, 56
  spatial, 46
  spin, 47
  unoccupied, 54
  unrestricted, 105, 207
  virtual, 54
Orthogonalization of basis functions, 142
Orthonormal
  functions, 25, 47
  vectors, 3
Overlap integral, 56
Overlap matrix, 137
Oxygen
  UHF orbital energies, 220

Pair energy, 274
Pair function, 273
Pair relaxation, 404
Pair removal, 404
Particle orbitals, 54
Pauli exclusion principle, 46, 50
Permutation, 7
Permutation operator, 67, 114
Perturbation expansion, 130, 322
  of the correlation energy, 350
Physicists' notation, 67
Polarizability, 38, 377
Polarization functions, 189
Pople-Nesbet equations, 210

Population analysis
  Löwdin, 152, 204
  Mulliken, 151, 203
Potential energy surface, 44, 146
Primitive Gaussian functions, 155, 181, 411
Propagator, 387

Rayleigh-Schrödinger perturbation theory
    (RSPT), 322
  diagrammatic representation, 327
Reduced density function, 252
Reference state, 59
Relaxation energy, 299, 307
  via CCA, 305
  via CEPA, 306
  via full CI, 304
  via IEPA, 305
  via L-CCA, 306
  via MBPT, 340
  via SCI, 305
Resonance energy, 309, 311
Resonance energy of benzene
  via CEPA, 317
  via IEPA, 314
  via L-CCA, 317
  via nth-order MBPT, 347
  via SCI, 316
  via second-order MBPT, 343
  via third-order MBPT, 345
Resonance energy of cyclic polyenes
  via fourth-order MBPT, 349
  via IEPA, 314
  via SCI, 316
  via second-order MBPT, 343
Restricted determinants, 83, 100
  energy of 83, 87
Restricted Hartree-Fock, 83, 132
Restricted spin orbitals, 100, 132
RHF (see Restricted Hartree-Fock)
Roothan equations, 138
RSPT (see Rayleigh-Schrödinger perturba-
    tion theory)
Rules for matrix elements, 70, 72
  derivation of, 74

Scalar product, 2, 6, 10, 27
Scaling, 158, 188
SCF (see Self-consistent field)
Schmidt orthogonalization, 16, 173
Schrödinger equation, 31, 40
  time-dependent, 386
SDCI, singly and doubly excited con-
    figuration interaction, 243

Second-order energy, 352, 360, 366
Second-order self-energy, 390, 393
Second quantization, 89
Secular determinant, 17, 18
Self-consistent field method, 54, 140, 145
Self-energy, 389
  second-order, 390, 393
  third-order, 390, 395
Shifted energy denominator, 337, 369
Singlet, 101
Singly excited determinants, 59, 236
  effect on dipole moment, 251
  effect on energy, 248
Size consistency, 261, 271, 369
Slater determinant, 50, 74
  energy of, 71, 73, 88
Spatial orbitals, 46
Spin, 45, 97
Spin-adapted configurations, 100, 234
  matrix elements between, 236
Spin-adapted pairs, 278
Spin angular momentum, 97
  commutation relations, 98
Spin contamination, 107, 228
Spin density, 212, 217
  matrix, 213
Spin functions, 45
Spin operators, 97
Spin orbital pair energies, 278
Spin orbitals, 47
Split valence-shell basis set, 187
Summation of diagrams, 336, 368
Symmetric orthogonalization, 143, 174

Triplet, 102
Two-electron integrals, 66, 141
  notations for, 68

UHF (see Unrestricted Hartree-Fock)
Ungerade, 57
Unitary transformation, 14
Unlinked diagrams, 321, 329, 369
Unoccupied orbitals, 54
Unrestricted determinants, 104
  energy of, 210, 215
Unrestricted Hartree-Fock, 205
Unrestricted spin orbitals, 105, 207

Vacuum state, 93
Valence bond, 167, 259
Variation method, 32
Variation principle, 31

Vertical ionization potential, 194, 393
Virtual orbitals, 54

Water
  bond angle
    via CCA, 297
    via CI, 249
    via L-CCA, 297
    via MBPT, 372, 373
    via SCF, 202
  bond length
    via CCA, 297
    via CI, 249
    via L-CCA, 297
    via MBPT, 371, 373
    via SCF, 202
  correlation energy
    via CCA, 296
    via CI, 248
    via IEPA, 296
    via L-CCA, 296

    via MBPT, 373
  dipole moment
    via CI, 251
    via MBPT, 378
    via SCF, 205
  force constants
    via CCA, 297
    via CI, 249
    via L-CCA, 297
    via MBPT, 373
    via SCF, 249
  ionization potential
    via CI, 250
    via Koopmans', 198
    via MBGF, 407
  orbital energies, 198, 199
  population analysis, 203, 204
  SCF energy, 192

Zero-point energies, 193
Zeroth-order energy, 130, 351